Biological Control Programmes against Insects and Weeds in Canada 1969-1980

EDITED BY J.S. KELLEHER AND M.A. HULME

C·A·B COMMONWEALTH AGRICULTURAL BUREAUX

Commonwealth Agricultural Bureaux
Farnham Royal
Slough SL2 3BN
England
Tel. Farnham Common 2281
Telex 847964

ISBN 0 85198 536 X

Printed by Page Bros (Norwich) Ltd

CONTENTS

PART II – Biological control of weeds in Canada, 1969-1980

PART III – Biological control of forest insect pests in Canada 1969-80

Foreword

The CIBC Technical Communication series was initiated to provide an outlet for regional reviews on biological control. This review, the 8th in the series and the 3rd prepared by Canadian scientists, points to the importance which Canada attaches to biological control and reflects the volume of Canadian activity in this field.

Canada continues to support an active classical biological control programme, relying on the Commonwealth Institute of Biological Control to undertake most of its foreign exploration and research. The collection and redistribution of natural enemies successfully established earlier to hasten control has continued. However, the current volume indicates an acceptance of biological control in its wider context as a component, albeit a very important one, of integrated pest management. Increasing attention and interest are being focused on other biological control approaches, i.e., augmentative and inundative releases as well as amelioration of the environment. The value of food plants or nectar sources for adult parasitoids has been demonstrated.

To date, augmentative and inundative releases in Canada have involved pathogens, both introduced and native, rather than parasitoids or predators. The period under review has seen the implementation, by large field trials, of microbial control, utilizing both bacterial and viral preparations for the control of forestry pests. The use of *Bacillus thuringiensis* has involved the large-scale production of "tailor-made" formulations by the agro-chemical industry and provides a preferred, although more expensive, alternative to the former wide-scale use of chemical pesticides for the control of spruce budworms. *B.thuringiensis* is fully registered as a pesticide in Canada for use against several lepidopteran defoliators. Similarly, research on viruses of neodiprionid sawflies has led to practical field application, with some degree of carry-over from year to year.

The first book reported introductions of control agents against two weeds; the second reviewed attempts against four additional species or species complexes whereas in the present volume, thirty agents have now been released against fourteen weed species in Canada. Increasing efforts have been made to evaluate or quantify the gains which have been or might be derived from biological control of weeds.

The book is timely; by its preparation, scientists and administrators can take stock of achievements to date and thus will be better placed to plan for the future. As M.A. Hulme and G.A. Green point out, the scope for research opportunities is limited not by lack of suitable problems but by the resources available to devote to biological control. The economic climate in Canada, as elsewhere, dictates that emphasis be given to developing management strategies for the pests causing the greatest economic losses and, as pointed out by Dr. P. Harris, proposals for future funding for biological control must, in order to convince administrators to release funds, provide quantitive assessment of the losses inflicted by the target pest or weed and thereby demonstrate the financial benefit which will result if the project is successful. This book attempts to provide an evaluation of each control attempt, and administrators and scientists now should be in a good position to plan for the future.

F.D. Bennett

Director

Commonwealth Institute of

Biological Control

General Introduction

This book is the third review of Canadian biological control work published by the Commonwealth Agricultural Bureaux. It is compiled and edited by Dr. M.A. Hulme and Dr. J.S. Kelleher. Dr. P. Harris coordinated preparation of the section on biological control of weeds.

The first compilation, "A Review of the Biological Control Attempts Against Insects and Weeds in Canada" was published in 1962 and dealt with control attempts against agricultural pests and weeds up to 1959 and against forest insects from 1910–1958. In 1971 a sequel was published entitled "Biological Control Programmes against Insects and Weeds in Canada 1959–1968"; this volume had separate sections on agricultural insects, on weeds, on forest pests, and on the organization of Canadian biological control research. During the late 1970s, a further sequel was proposed and preparation began in 1980. This volume is the end result.

Like its immediate predecessor, this review has two main functions. First, it collates Canadian biological control attempts against insects and weeds from 1969 to 1980, and thus provides a permanent record of otherwise diffuse and relatively inaccessible data. Secondly, it attempts to provide an evaluation of each control attempt in terms of reasons for success or failure, and hence develops recommendations for future work.

Terminology used in biological control research can, at times, be ambiguous. The simple term "parasite" can be particularly troublesome and as an aid to readers its usage is defined more precisely. More effort has been devoted to entomopathogens during the current review period and in some cases both microbial parasites and insect parasites have been examined from the same host insect. To avoid confusion when referring to these entomogenous and entomophagous parasites, the latter have been termed parasitoids or insect parasites rather than just parasites. A parasitoid, following the definition of DeBach, is an insect that develops as a larva on or in a single host insect from eggs laid on, in or near the host; it usually consumes most of the host's tissues and it kills the host. Phytophagous insects that feed on plants regarded as weeds are thus termed parasites, whereas those feeding on crop plants, including trees, are referred to as pests (emphasizing their relationship with man rather than with the host plant).

The term "biological control" also requires some explanation. In the present context the term refers to an applied control strategy that involves the manipulation of living natural enemies for the purpose of regulating the abundance of pest populations. The regulation may be short or long term and the manipulation may include release strategies ranging from spot-introductions to widespread inundation. Introductions of sterile individuals are thus included here; but the use of semiochemicals, notably kairomones and pheromones, is included only where these chemicals influence predator or parasitoid behaviour. Pheromone confusion trials on the pest insect are excluded since they involve direct chemical manipulation of the pest. The use of insect growth regulators is similarly excluded.

The book is divided into three sections which cover agricultural insects, weeds, and forest insects; chapters within each section are generally listed in the alphabetical order of the insect's or weed's generic name; and each of the three sections begins with an overview of the chapters that follow. Addresses of contributors are listed in the Appendix. The book does not include an updated synopsis of the organization and accomplishments of Canadian biological control research in a separate section equivalent to Part IV of the 1971 volume. Accomplishments are summarized in the overview chapter beginning each section and the organizational observations in the 1971 synopsis remain equally valid today without need for further elaboration.

Several changes in the organization of Canadian biological control research have taken place since publication of the 1971 book. The Research Institute of Agriculture Canada at Belleville, Ontario, which had played a pivotal role in importation and applied research of biological control organisms was closed in 1972 and its staff was relocated. The Biocontrol Unit was formed in Ottawa by Agriculture Canada to co-ordinate importations. Material is imported through the Biocontrol Unit, and through Macdonald College in Québec, the Forest Pest Management Institute and the University of Guelph in Ontario, and the Regina Research Station of Agriculture Canada in Saskatchewan. The Department of Fisheries and Forestry became part of a new Department, Environment Canada, in 1971. Forestry research is conducted by the Canadian Forestry Service within this Department; biological control research is still carried out at the same laboratories as those listed in the last review but establishment names have been modified to reflect the new organization.

Many people assisted in the preparation of this review. Special acknowledgement is extended to Ms. R. Boyle and Ms. P. Loshak for helping in correcting the text. Dr. F.D. Bennett, Director, Commonwealth Institute of Biological Control, initiated the preparation, Dr. D.J. Greathead, Assistant Director and Mr. D.J. Girling, Information Officer, assisted in its publication, and Mrs. A.H. Greathead read the proofs and completed the index.

PART I

BIOLOGICAL CONTROL OF AGRICULTURAL INSECTS IN CANADA 1969–80

Current Approaches to Biological Control of Agricultural Insect Pests

J.S. KELLEHER

This part covers the biological control of insect and mite pests of crops, fruit trees, and ornamentals. As in the previous review (Kelleher 1971) it includes inoculation (introduction), augmentation, and inundation, using beneficial insects, entomopathogens, and autocidal methods.

Some organizational changes have occurred since the last review. In 1972, the Research Institute at Belleville, Ontario, was closed and the staff transferred to Research Stations in other parts of Canada. The service function, of obtaining information and collections of biological control organisms from abroad was transferred to Ottawa. The programme on the biological control of weeds was moved to the Regina Research Station. An Integrated Pest Control Section was formed at Winnipeg with many of the scientists from Belleville. Other individuals were stationed at Summerland, Saskatoon, Harrow, and Ottawa.

The adoption of integrated pest management (IPM) systems has placed biological control in a much more important role than before. At one time it was considered as an alternative to chemicals i.e. complete control was necessary. Now, however, our knowledge of insect population dynamics has increased, and it is now considered that even a small additional percentage mortality can have a profound effect on the overall reduction of an insect population.

The initial stages of an IPM programme involve a reduction in pesticide usage, by monitoring pest infestations so that chemicals are applied only when necessary. As a consequence, natural control factors are not so suppressed and are able to exert additional mortality on the pest population. But there will be a point where very little more can be obtained from the systems available and additional factors must be added to attain a greater reduction in pesticide use. This can come in many forms but the introduction of new natural enemies should be an important consideration in many instances.

Most of the studies reported in this part have utilized introduced species of parasitoids, or entomopathogens (*Bacillus thuringiensis* Berliner (*B.t.*), fungi, viruses). Other aspects of biological control such as mass releases of parasitoids e.g. *Trichogramma* species, have been largely neglected. However reports from the USSR, China, and South American countries have attributed considerable successes with this approach and are prompting some consideration of initiating similar work in Canada. Much may depend on the development of techniques to provide a large number of parasitoids during a relatively short period for any particular insect. The main emphasis will be on production and storage but consideration should also be given to a number of target pests whose susceptible stages occur at different times.

Increased awareness of the value of natural enemies has led to research on the differential effects of pesticides on natural enemies. Although pesticides usually affect a wide spectrum, some taxa are less affected by them than others. A working group of the West Palaearctic Regional Section of the International Organization for Biological Control has taken an organized and coordinated approach to this subject. Members of this group have standardized the culturing of various taxa of beneficial insects and are adopting standardized methods for testing the effects of pesticides on these insects in the laboratory and in the field. Eventually these data will be available as recommendations to be placed on the label of pesticide containers and in the bulletins of official agencies.

The following 24 chapters in this part have been arranged in alphabetical order using the scientific name of the pest species. An evaluation of the programme is given in Table 1. Only 10 of the pests were included both in this review and the last one. It can be concluded that no other information is available on the other target pests. This indicates the transient nature of many of the agricultural projects, in contrast to those on forest insects and weeds.

Table 1 Evaluation of biological control measures against agricultural insect and mite pests

Pest	Method	Agent	Degree of Success***
Acyrthosiphon pisum (Harris)	Adventitious	Predator	+
Adelphocoris lineolatus (Goeze)	Inoculation	Parasitoid	–
Agromyza frontella (Rondani)	Inoculation	Parasitoid	+
Artogeia rapae (L.), *Trichoplusia*			
ni (Hübner), & *Plutella*	Inundative	Virus, *B.t.*	++++
xylostella (L.)	Inoculation	Parasitoid	–
Culex pipiens L.	Augmentation	Planaria	++
Culiseta inornata (Williston)	Augmentation	Fungus	+++
Cydia pomonella (L.)	Inundation	Virus	+
Delia antiqua (Meigen)	Autocidal		+
Entomoscelis americana Brown**	–	–	
Euxoa messoria (Harris)	Augmentation	Virus	+
Forficula auricularia (L.)	Inoculation	Parasitoid	+
Hypera postica (Gyll.)	Adventitious	Fungus	+++
Lygus spp.	Inoculation	Parasitoid	–
Mamestra configurata Walker**	–	–	
Manduca quinquemaculata (Haworth)	Inundation	*B.t.*	++++
Melanopus spp., *Camnula pellucida*			
(Scudder) & other grasshoppers	Inundation	Protozoa	+++
Musca domestica L.	Integration	Parasitoid	++++
Oulema melanopus (L.)	Inoculation	Parasitoid	++++
Phyllonorycter blancardella (Fabricius)	Inoculation	Parasitoid	+
Phyllotreta spp.	Inoculation	Parasitoid	–
Tetranychus urticae Koch*	Integration	Predator	+++
Thymelicus lineola (Ochs.)	Inundation	*B.t.*	++++
	Inundation	Virus	+++
	Inoculation	Parasitoid	–
Tipula paludosa (Meigen)	Inoculation	Parasitoid	+
Trialeurodes vaporariorum (Westwood)*	Integration	Parasitoid	++++

Measures started during or extending into period 1969 to 1980
*Greenhouse
**Exploratory studies — no. released
***Degree of Success:

– No control or not yet known if established
+ Slight pest reduction or too early for evaluation of control
++ Local control; distribution restricted or not fully investigated
+++ Control widespread but local damage occurs
++++ Control complete

In the last Review, the use of a nematode, DD−136, was reported (Welch 1971) against several crop pests. This work was not continued although there are still several avenues to explore. Hopefully this will be done in the future. It will also be noted from Table 1 that predators were used in only 2 instances — one adventitiously, the other in a greenhouse situation. Possibly this group is less favored because the members tend to be less specific than parasitoids. However for reasons considered above, even relatively small degrees of mortality can effect population reductions on a much larger scale. But it would also be important to determine if there would be a net increase in pest mortality or if it would be merely compensatory.

Of the 26 agents used, 11 were by inoculation with parasitoids and 7 by inundation with pathogens. Inoculations gave high success in 2 cases; others could not be evaluated as it was too early or recoveries had not yet been made. In contrast, 5 of the 7 pests were controlled well by inundation with pathogens. However this method is primarily for short-term stage control whereas inoculation of new natural enemies is a long-term population control measure.

In many cases, establishment of exotic natural enemies may be hampered by the low numbers available for release e.g. against *Adelphocoris lineolatus* (Goeze), *Phyllotreta* spp., and *Thymelicus lineola* (Ochs.). Propagation has been difficult or unattainable. Sometimes mortality has been high in rearing and/or storage. This has required intense efforts by the CIBC to increase the number of individuals obtained. Some new techniques have been considered e.g. leaving an area undisturbed on the premise that parasitoid populations would increase. Increasing host populations by the addition of their eggs from other areas has also been tried successfully in some cases although the danger of introducing more virulent strains has restricted the use of this technique.

In addition to the work reported in the following chapters there have been some minor efforts which did not warrant separate sections. These are as follows: *Mantis religiosa* L. ootheca were collected near Belleville, Ontario, in 1971 and 1972 and released near St. John's, Newfoundland against various grasshopper species (Acrididae); no recoveries have been recorded (R. Morris, 1981, personal communication).

Coccinella septempunctata L. was found in northern New Jersey, in 1973 apparently as an adventitious introduction (Angalet & Jacques 1975). They were redistributed to various states in subsequent years and to Nova Scotia in 1981. The species had been reported from various parts of Québec (Larochelle 1979) and was found in New Brunswick in 1981 (P.W. Schaefer, 1981, personal communication).

As in previous reviews, this publication has been possible only through the cooperation of those who performed the work. In this part, 28 authors have contributed. Their efforts are gratefully acknowledged, as well as those of editors and typists who are too numerous to name individually.

Literature Cited

Angalet, G.W.; Jacques, R.L. (1975) The establishment of *Coccinella septempunctata* L. in the continental United States. *USDA Cooperative Economic Insect Report* 25, 883−884.

Kelleher, J.S. (1971) Current approaches to biological control of agricultural insect pests. *In*: Biological control programmes against insects and weeds in Canada 1959−1968. *Commonwealth Institute of Biological Control Technical Communication* 4, 1−3.

Larochelle, A. (1979) Les coléoptères Coccinellidae du Québec. *Cordulia,* Supplément 10, 1−111.

Welch, H.E. (1971) Various target species attempts with DD−136. In: Biological control programmes against insects and weeds in Canada 1959−1968. *Commonwealth Institute of Biological Control Technical Communication* 4, 62−66.

Acyrthosiphon pisum (Harris), Pea Aphid (Homoptera: Aphididae)

B.D. FRAZER

Coccinella undecimpunctata L. was first encountered in pea aphid, *Acyrthosiphon pisum* (Harris), infested alfalfa in May 1972. Since then it is routinely found on low-growing, aphid infested plants including cereals, strawberries, and vegetables in numbers about equal to the predominant native species, *C. californica* and *C. trifasciata*. *C. undecimpunctata* is one of the first coccinellids to be found in spring. They appear to be adapted to cooler climates than the lower mainland of British Columbia as there are few aphids available when they are first seen. The effect of their predation in reducing spring aphid numbers can greatly reduce the rate of increase of aphids (Frazer & Gilbert 1976) and thereby extend the time before pesticides are required. Later in the season the effects of *C. undecimpunctata* alone are small because losses from the large number of predators in alfalfa, at least, are compensatory (Charnov *et al.* 1976). If *C. undecimpunctata* were added to a field without them, their aphid consumption would simply result in less for the other competing predators with no net increase in aphid mortality.

Literature Cited

Frazer, B.D.; Gilbert, N. (1976) Coccinellids and aphids: a quantitative study of the impact of adult ladybirds (Coleoptera: Coccinellidae) preying on field populations of pea aphids (Homoptera: Aphididae). *Journal of the Entomological Society of British Columbia* 73, 33–56.

Charnov, E.; Frazer, B.D.; Gilbert, N.; Raworth, D. (1976) Fishing for aphid — the exploitation of a natural population. *Journal of Applied Ecology* 13, 379–389.

Adelphocoris lineolatus (Goeze), Alfalfa Plant Bug (Heteroptera: Miridae)

C.H. CRAIG and C.C. LOAN

Pest Status

The alfalfa plant bug, *Adelphocoris lineolatus* (Goeze), a native of Europe and western Asia, was introduced into North America at Cape Breton, Canada, about 1917. In the succeeding 50 years it spread westward through the agricultural areas of central and western Canada to the foothills of the mountains in western Alberta (Craig 1971). Its principal host plant is alfalfa; it also reproduces on other cultivated legumes such as sweet clover, red clover, sainfoin, and trefoil.

Eggs are laid during mid to late summer in hollow stems of the host plants, overwinter, and hatch about mid-May. Nymphs develop through five instars and the first adults appear in mid-June. Throughout its range of distribution south of about latitude 51°N the species is bivoltine or largely so. North of about latitude 53°N the species is largely univoltine; only about 5% of the eggs hatch the year they are laid, and second generation individuals do not mature before freeze-up (Craig 1963).

A. lineolatus is an important pest of alfalfa grown for seed. Feeding injury to alfalfa vegetative growth, flower buds and flowers, and to immature seed within the pod causes reduced yield and quality of seed. Infestations of 10–12 nymphs per sweep can cause almost total seed crop failure. Its damage to seed production in other cultivated legumes has not been evaluated.

Background

In Europe nymphs of *A. lineolatus* are parasitized by a braconid, *Peristenus adelphocoridis* Loan (Loan 1979). This may be the species of parasitoid referred to by Bilewicz-Pawinska (1975), which attacks only first-generation nymphs of the bivoltine host, and whose incidence of parasitism never exceeded 10%. In Canada in the early 1960s, at Belleville, Ontario, and in northern Saskatchewan, first-generation nymphs of *A. lineolatus* were found to be parasitized by *Peristenus pallipes* (Curtis). The incidence ranged from 40 to 60% in Ontario compared to 3% in Saskatchewan (Craig 1963, Loan 1965). In the years 1976 through 1980 the incidence of euphorine parasitism in *A. lineolatus* nymphal populations in different ecological regions of Saskatchewan and Alberta where the host occurs ranged from 0 to 4% with an average of 0.3%. There were no apparent differences in levels of parasitism in hosts collected from the southern shortgrass prairie, the central mixed prairie, or the northern parkland regions. Parasitoids were found only in first-generation nymphs; second-generation nymphs were entirely parasitoid-free. The identity of the parasitoid has not been confirmed by rearings from these collections, but it is likely that it is *Peristenus pallipes*.

Releases and Recoveries

In the summer of 1979 euphorine parasitoids of *A. lineolatus* were collected and reared in Europe by the Delémont Station of the Commonwealth Institute of Biological Control and shipped to Ottawa as diapausing adults in cocoons. In the spring of 1980 two species, *Peristenus adelphocoridis* and *Peristenus rubricollis* (Thomson), emerged from these cocoons. Emergence was sporadic and not synchronous with *A. lineolatus* development at intended liberation sites in western Canada. However, 47 *P. adelphocoridis* were released near Clavet, Saskatchewan, latitude 52°N, in alfafa containing a second and third instar *A. lineolatus* nymph population of 2 per sweep. A total of 29 *P. adelphocoridis* and 49 *P. rubricollis* were released in a similar situation near Shellbrook, Saskatchewan, latitude 53°N. Liberations in 1981 included 16 *P. adelphocoridis* at the same site at Saskatoon and 12 of the species at Shellbrook. The first attempt to recover these introduced parasitoids at the liberation sites will be made in 1981.

Recommendations

The introduction and release of *P. adelphocoridis* and *P. rubricollis* against *A. lineolatus* in western Canada will continue through 1982 or 1983 when the impact of the programme will be assessed. Because *A. lineolatus* is an excellent candidate host for introduced parasitoids, it is recommended that a search be made for other specific parasitoids of *A. lineolatus* in Europe to increase the probability of successful establishment in western Canada.

Literature Cited

Bilewicz-Pawinska, T. (1975) Distribution of the insect parasites *Peristenus* Foerster and *Mesochorus* Gravenhorst in Poland. *Bulletin de l'Académie Polonaise des Sciences Classe II Série des Sciences Biologiques* 23, 823–827.

Craig, C.H. (1971) Distribution of *Adelphocoris lineolatus* (Heteroptera: Miridae) in western Canada. *Canadian Entomologist* 103, 280–81.

Craig, C.H. (1963) The alfalfa plant bug, *Adelphocoris lineolatus* (Goeze) in northern Saskatchewan. *Canadian Entomologist* 95, 6–13.

Loan, C.C. (1965) Life cycle and development of *Peristenus pallipes* (Curtis) (Hymenoptera: Braconidae, Euphorinae) in five mirid hosts in the Belleville district. *Proceedings of the Entomological Society of Ontario* (1964) 95, 115–121.

Loan, C.C. (1979) Three new species of *Peristenus* Foerster from Canada and western Europe (Hymenoptera: Braconidae, Euphorinae). *Naturaliste Canadian* 106, 387–391.

Agromyza frontella (Rondani), Alfalfa Blotch Leafminer (Diptera: Agromyzidae)

J.C. GUPPY, D.G. HARCOURT, M.O'c. GUIBORD and L.S. THOMPSON

Pest Status

The alfalfa blotch leafminer, *Agromyza frontella* (Rondani), was introduced from Europe to the north eastern United States during the late 1960s (Miller & Jensen 1970) and first appeared in Canada at St-Armand, Québec, in 1972 near the United States border (Harcourt 1973). The insect spread rapidly throughout eastern Canada from Prince Edward Island to Ontario (Thompson 1974, Guppy & Harcourt 1977), and by 1979 it had threatened to become the most important pest of alfalfa in the 10 eastern counties of Ontario (Harcourt & Binns 1980) and in the other eastern provinces. It has continued to spread westward in Ontario and damaging levels were recorded in Hastings County in 1980 (Guppy 1981).

Hosts include: *Medicago, Melilotus,* and *Trifolium,* but alfalfa, *Medicago sativa* L., is apparently preferred (Steyskal 1972, Spencer 1973). The insect overwinters as a partially developed pupa and the adult emerges about mid-May to oviposit in the first-growth alfalfa. In eastern Canada there are three complete generations a year (Guppy 1981), but a few individuals may complete a fourth. The crop is attacked by both the adults and larvae, but most of the damage is done by the larvae which feed on the mesophyll tissues forming blotch mines each of which represents about 27% of the leaflet area.

The crop is attacked throughout the growing season but damage is more severe in the first and second cuttings; during the past 4 years, the percentage of leaflets mined in eastern Ontario averaged 70% and 58%, respectively.

Background

Seven parasitoid species native to eastern Canada have been reared from larvae of *A. frontella,* all of them are hymenopterous species of the family Eulophidae. *Diglyphus websteri* (Crawford) and *D. begini* (Ashmead) were recorded in eastern Ontario and these two plus *D. pulchripes* (Crawford) in Québec. In Prince Edward Island, five species were found, namely, *D. begini, D. intermedius* (Girault), *Notanisomorpha* sp., *Pnigalio maculipes* (Crawford), and *Pnigalio* n. sp. One larval-pupal parasitoid *Cyrtogaster* sp. (Hymentoptera: Pteromalidae) emerged from puparia in Ontario.

Assessment of parasitism has shown that the native species are of little importance in suppressing alfalfa blotch leafminer populations. However, a hemipteron species *Nabis ferus* (Linnaeus), common to eastern Canada, appears to be an important predator of the larvae. At Ottawa, incidence of attack increases with each generation, averaging about 60% in the third.

The Beneficial Insects Research Laboratory, Newark, Delaware, United States Department of Agriculture imported 14 parasitoid species from Europe from 1974 to 1978 for release in the United States (Hendrickson & Barth 1979). By 1978 three hymenopterous species, *Dacnusa dryas* (Nixon) (Braconidae), *Chrysocharis punctifacies* Delucchi (Eulophidae), and *Miscogaster hortensis* Walker (Pteromalidae), were established at the Delaware field plots. These were recolonized in other northeastern states, where they became established (Hendrickson & Barth 1979, and personal communication), and imported for release into eastern Canada. A fourth species, *Diglyphus isaea* (Walker) (Eulophidae), was shipped to Québec for release in 1975; establishment was not successful in either country apparently because it crossed with native species of *Diglyphus* and further attempts of establishment were abandoned (R.M. Hendrickson 1976, personal communication).

Releases and Recoveries *D. dryas, C. punctifacies,* and *M. hortensis* are solitary endoparasitoids. The host is attacked during the larval stage and the parasitoids complete their development during the pupal stage.

All releases in Canada were of adults, laboratory reared in Newark, Delaware, or field collected there or in other northeastern States by cooperators and shipped to Canada directly to release points or for distribution through Agriculture Canada, Research Branch, Biological Control Unit, Research Program Service. Release and recovery data are given in Table 2. *D. dryas,* the only species thus far recovered,

Table 2 Releases and recoveries of parasitoids against the alfalfa blotch leafminer, *Agromyza frontella* (Rondani), in Canada.

	Releases		
Species and Province	Year	Number	Year of recovery
Chrysocharis punctifacies			
Prince Edward Island	1978	4	
Prince Edward Island	1980	142	
Ontario	1980	150	
Québec	1979	249	
Québec	1980	90	
Dacnusa dryas			
Prince Edward Island	1977	68	
Prince Edward Island	1978	70	
Prince Edward Island	1979	546	1980
Ontario	1979	586	1979
Québec	1979	16	
Québec	1980	378	1981
Diglyphus isaea			
Québec	1975	900	
Miscogaster hortensis			
Prince Edward Island	1978	32	

appears to be firmly established in eastern Canada. In eastern Ontario, it was reared from third generation pupae in the year of release, 1979, and from each generation in 1980 after overwintering successfully in diapausing pupae. It was also collected in sweepnet samples throughout the growing season in 1980 at Ottawa.

Recommendations

Since native parasitoids in Canada have proven to be of little significance as mortality factors, it is recommended that:

(1) Releases or recolonization of the three exotic larval-pupal parasitoids, *C. punctifacies, D. dryas,* and *M. hortensis,* that have been established successfully in the northeastern United States be continued.

(2) Recovery attempts and assessment of rates of parasitism be strengthened in those regions where detailed studies are not in progress.

(3) For eastern Ontario, it is recommended that each parasitoid species be colonized separately to assess their individual effectiveness as mortality factors, and in combination, to determine whether competition for the same host stage is compensative or disruptive.

Literature Cited

Guppy, J.C.; Harcourt, D.G. (1977) The alfalfa blotch leafminer in eastern Ontario, 1977. Proceedings of the 14th Annual Northeast Invitational Alfalfa Insects Conference, Burlington, Vermont.

Guppy, J.C. (1981) Bionomics of the alfalfa blotch leafminer, *Agromyza frontella* (Diptera: Agromyzidae) in eastern Ontario. *Canadian Entomologist* 113, 593–600.

Harcourt, D.G. (1973) *Agromyza frontella* (Rondani) (Diptera: Agromyzidae): a pest of alfalfa new to Canada. *Annals of the Entomological Society of Québec* 18, 49–51.

Harcourt, D.G.; Binns, M.R. (1980) A sampling system for estimating larval populations of *Agromyza frontella* (Diptera: Agromyzidae) in alfalfa. *Canadian Entomologist* 112, 375–385.

Hendrickson, R.M., Jr.; Barth, S.E. (1979) Introduced parasites of *Agromyza frontella* (Rondani) in the USA. *New York Entomological Society* 87, 167–174.

Miller, D.E.; Jensen, G.L. (1970) Agromyzid alfalfa leafminers and their parasites in Massachusetts. *Journal of Economic Entomology* 63, 1337–1338.

Spencer, K.A. (1973) Agromyzidae (Diptera) of world importance. The Hague; Dr W Junk, 418 pp.

Steyskal, G.C. (1972) The taxonomy of the alfalfa blotch leafminer, *Agromyza frontella* (Rondani). *Cooperative Economic Insect Report* 22, 134–137.

Thompson, L.S. (1974) Alfalfa blotch leafminer, *Agromyza frontella* (Rondani). *Canadian Agricultural Insect Pest Review* 52, 5.

Artogeia rapae (L.), Imported Cabbageworm (Lepidoptera: Pieridae), *Trichoplusia ni* (Hübner), Cabbage Looper (Lepidoptera: Noctuidae) and *Plutella xylostella* (L.), Diamondback Moth (Lepidoptera: Plutellidae)

R.P. JAQUES and J.E. LAING

Pest Status

Artogeia rapae (L.), the imported cabbageworm, causes damage to cole crops practically wherever these crops are grown in Canada. The pest is quite susceptible to certain chemical insecticides and to insecticides containing *Bacillus thuringiensis* Berliner and, therefore, populations of larvae are usually retained at acceptable densities by recommended control programmes. *Trichoplusia ni* (Hübner), the cabbage looper, is not as widespread but, because it is less susceptible than *A. rapae* to registered insecticides, it is considered to be the more important pest of cole crops, especially in southern Ontario. *Plutella xylostella* (L.) is not considered to overwinter in Canada and thus does not become a pest on cole crops until late in the season.

Use of *Bacillus thuringiensis*

Formulations of *B. thuringiensis (B.t.)* containing *B. thuringiensis* var. *kurstaki,* a high-potency strain of *B.t.,* were registered in Canada in 1972 for use against *A. rapae, T. ni,* and *P. xylostella* on cole crops and other crops. The new formulations (Thuricide® HPC, Sandoz Crop Protection Inc.; and Dipel® HD−1, Abbott Laboratories Ltd.) replacing formulations registered previously, protected cole crops equally as well as or better than did chemical insecticides. These formulations of *B.t.* are now used quite extensively.

The native parasitoid complex of the imported cabbageworm (Michalowicz 1980) and the diamondback moth (Butts 1979, Bolter 1982) have been studied intensively. Two species, *Apanteles rubecula* Marsh. which parasitizes *A. rapae* and *Apanteles plutellae* (Kurdj.) a parasitoid of diamondback moth, have been imported and released in southwestern Ontario.

Field Plot Introductions of Viruses

The granulosis virus of *A. rapae* (ARGV), a naturally occurring virus (Jaques & Harcourt 1971) that often causes high mortality in populations of larvae of *A. rapae* late in the growing season, was evaluated by bioassays and plot tests as an introduced pathogen. In addition, three viruses that kill larvae of *T. ni* were assessed in laboratory and field studies. Two of these viruses, the single-embedded nuclear polyhedrosis virus of *T. ni* (TNNPV-SEV), the predominant naturally occurring virus of *T. ni* (Jaques 1970*a*), and the nuclear polyhedrosis virus of *Autographa californica* (ACNPV), a virus that kills *T. ni* and that closely resembles the multiple-embedded nuclear polyhedrosis virus of *T. ni* (TNNPV-MEV), are considered as candidates for development as microbial insecticides for use against *T. ni.*

Efficacy of the viruses of *A. rapae* and *T. ni* was assessed in 1970–80 using field plots of cabbage (each 24–32 m^2) Jaques 1970*b*, 1972, 1973, 1977; Jaques & Laing 1978).

In these tests:

(1) ARGV (7.5 x 10^{12} granular inclusion bodies (GIB/ha) applied to foliage 4 to 6 times during the growing season reduced populations by 95 to 100% and protected the crop as well as did chemical insecticides applied at the recommended rates.

(2) TNNPV-SEV, TNNPV-MEV and ACNPV were about equally effective against the cabbage looper. Four to 6 applications of 1.5 x 10^{12} polyhedral inclusion bodies (PIB)/ha during the growing season reduced populations of *T. ni* on cabbage by about 75% and protected the crop as well as did methomyl or *B.t.* but not as well as permethrin applied at recommended rates.

(3) Reduced-dosage mixtures of ACNPV, ARGV, and/or *B.t.* with chlordimeform were as effective against *A. rapae* and *T. ni* as were the microbial or chemical insecticides applied alone at the full rate (Jaques & Laing 1978). Mixtures of the viruses or *B.t.* and permethrin at 1/4 or 1/2 full rate were equally as effective as permethrin applied alone in protecting the crop.

(4) The efficacy of ACNPV propagated *in vitro* was similar to that of ACNPV propagated *in vivo*.

(5) Because the viruses persist in soil (Jaques 1974*a*, 1974*b*), application of a high dosage of ARGV and ACNPV when plants were transplanted, suppressed populations of *A. rapae* and *T. ni* on cabbage plants and lessened the frequency of use of insecticides later in the season.

Predation and Parasitism

The native parasitoid complex of the imported cabbageworm and the diamondback moth have been studied intensively (Butts 1979, Michalowicz 1980, Bolter 1982). Major mortality factors for *A. rapae* at Cambridge, Ontario during 1977–78 were *Apanteles glomeratus* (L.), *Phryxe vulgaris* (Fallén), *Compsilura concinnata* (Meigen), and *Pteromalus puparum* L. Major changes in the population occurred during the pupal stage and the key factor responsible for these changes was the pupal parasitoid, *P. puparum. A. glomeratus* which parasitizes the larvae of *A. rapae* also contributed substantially to the total mortality. However, rates of parasitism by both species of parasitoids only reached high levels late in the growing season and did not prevent *A. rapae* from reaching densities which were injurious to the cabbage crop (Michalowicz 1980).

Major mortality factors of the diamondback moth larvae in Ontario are the larval parasitoids, *Microplitis plutellae* (Mues.) and *Diadegma insulare* (Cress.), and the pupal parasitoid, *Diadromus subtilicornis* (Grav.) (Butts 1979, Bolter 1982). In three of four years in which parasitoids were monitored from 1977–81 in Cambridge, Ontario, *D. insulare* was the dominant parasitoid while *D. subtilicornis* was the dominant parasitoid in one of the four years.

In a study of the competition between the larval parasitoids *D. insulare* and *M. plutellae* (Bolter 1982), it was shown that female *D. insulare* produced an average of 814 eggs in a lifespan of 26 days at 23°C, while female *M. plutellae* produced an average of 318 eggs in a lifespan of 20 days at 23°C. *D. insulare* avoided superparasitism and multiple parasitism of hosts already parasitized by *M. plutellae. M. plutellae* generally avoids superparasitism, but cannot detect eggs of *D. insulare* in the host for at least 12 hours after they were oviposited. However, the first-instar larva of *M. plutellae* was shown to be intrinsically superior to the first-instar larva of *D. insulare*.

The native parasitoid complex of the diamondback moth does not provide adequate control of diamondback moth on Brussels sprouts late in the season when conditions favour an increase in the host population.

Parasitoid Introductions

Apanteles rubecula a larval parasitoid of *A. rapae* was introduced from British Columbia to Richmond, Ontario by Harcourt in 1970 and 1971. Establishment of the parasitoid was in doubt. However, four specimens of *A. rubecula* were recovered from a garden plot in Ottawa in 1982. A small number of *A. rubecula,* originally obtained from British Columbia, were released in Guelph, Ontario (40 ♀♀ and ♂♂ in 1978, 85 ♀♀ and 120 ♂♂ in 1979), Harrow, Ontario (180 ♀♀ and 120 ♂♂ in 1978) and Cambridge, Ontario (165 ♀♀ and 175 ♂♂ in 1979). Although pupae of *A. rubecula* were observed in the field in the fall during the year of each release, no *A. rubecula* have been recovered from the release areas during 1980–82.

A second strain of *A. rubecula* obtained from north eastern China in 1981 was released at Guelph during 1982, but establishment of the parasitoids is unknown at the time of this report.

Apanteles plutellae (Kurdj.) a larval parasitoid of the diamondback moth, native to Europe was imported to Guelph in 1981 from the CIBC laboratory in Trinidad. It is presently in culture at the University of Guelph and will be released in Guelph during 1982.

Host-Parasitoid-Pathogen Interactions

The host-parasitoid-pathogen interactions between the parasitoid *Apanteles glomeratus,* a granulosis virus (GV), and their lepidopteran host *Artogeia rapae* larvae depended on the length of time between oviposition by the parasitoid and time of ingestion of the virus by the host. Parasitoids did not survive in hosts which ingested virus on or before the fifth day post-parasitism. Percentage survival of parasitoids gradually increased from 26% when oviposition by parasitoids occurred six days prior to GV infection, to 90% when oviposition by parasitoids occurred nine days prior to GV infection. Survival of parasitoids in non-virus treated hosts remained consistently above 95%. The degree of host infection was strongly correlated with survival of parasitoids: the heavier the host's GV infection, the fewer parasitoids that emerged from the host to pupate and that emerged from the pupae as adults (Levin *et al.* 1981).

After prior oviposition in GV-infected *A. rapae* larvae, *A. glomeratus* transmitted GV to healthy *A. rapae* larvae 73% of the time. Ninety per cent of parasitoids transmitted GV to at least one recipient larva. *A. glomeratus* which developed in and emerged from GV-infected *A. rapae* larvae, transmitted GV to 84% of the larvae in which they initially oviposited. Discrimination between healthy and diseased *A. rapae* larvae by female *A. glomeratus* was not observed. This lack of discrimination is likely to contribute to the dissemination of GV from host to host by *A. glomeratus* in the field and to mortality of the parasitoid's progeny. Thus, the parasitoid is at a disadvantage when competing with GV in the field for hosts (Levin *et al.* 1979, 1983).

Evaluation of Control Attempts

These studies showed that the viruses (ARGV, TNNPV, and ACNPV) and *B.t.* have considerable potential as components in integrated management of *A. rapae* and *T. ni* in cole crops. *A. rubecula* for control of *A. rapae* and *A. plutellae* for control of *P. xylostella* will have to be studied further to evaluate their potential to control these pests of cole crops.

Literature Cited

Bolter, C.J. (1982) Competitive abilities of two parasites attacking larvae of the diamondback moth, *Plutella xylostella* (L.) in Ontario. M.Sc. Thesis, University of Guelph, 79 pp.

Butts, R.A. (1979) Some aspects of the biology and control of *Plutella xylostella* (L.) (Lepidoptera: Plutellidae) in southern Ontario. M.Sc. Thesis, University of Guelph, 97 pp.

Jaques, R.P. (1970*a*) Natural occurrence of viruses of the cabbage looper in field plots. *Canadian Entomologist* 102, 36–41.

Jaques, R.P. (1970*b*) Application of viruses to soil and foliage for control of the cabbage looper and imported cabbageworm. *Journal of Invertebrate Pathology* 15, 328–340.

Jaques, R.P. (1972) Control of the cabbage looper and imported cabbageworm by viruses and bacteria. *Journal of Economic Entomology* 65, 757–760.

Jaques, R.P. (1973) Tests on microbial and chemical insecticides for control of *Trichoplusia ni* (Lepidoptera: Noctuidae) and *Pieris rapae* (Lepidoptera: Pieridae) on cabbage. *Canadian Entomologist* 105, 21–27.

Jaques, R.P. (1974*a*) Occurrence and accumulation of viruses of *Trichoplusia ni* in treated field plots. *Journal of Invertebrate Pathology* 23, 140–152.

Jaques, R.P. (1974*b*) Occurrence and accumulation of the granulosis virus of *Pieris rapae* in treated field plots. *Journal of Invertebrate Pathology* 23, 351–359.

Jaques, R.P. (1977) Field efficacy of viruses infectious to the cabbage looper and imported cabbageworm on late cabbage. *Journal of Economic Entomology* 70, 111–118.

Jaques, R.P.; Harcourt, D.G. (1971) Viruses of *Trichoplusia ni* (Lepidoptera: Noctuidae) and *Pieris rapae* (Lepidoptera: Pieridae) in soil in fields of crucifers in southern Ontario. *Canadian Entomologist* 103, 1285–1290.

Jaques, R.P.; Laing, J.E. (1978) Efficacy of mixtures of *Bacillus thuringiensis,* viruses, and chlordimeform against insects on cabbage. *Canadian Entomologist* 110, 443–448.

Levin, D.B.; Laing, J.E.; Jaques, R.P. (1979) Transmission of granulosis virus by *Apanteles glomeratus* to its host *Pieris rapae. Journal of Invertebrate Pathology* 34, 317–318.

Levin, D.B.; Laing, J.E.; Jaques, R.P. (1981) Interactions between *Apanteles glomeratus* (L.) (Hymenoptera: Braconidae) and granulosis virus in *Pieris rapae* (L.) (Lepidoptera: Pieridae). *Environmental Entomology* 10, 65–68.

Levin, D.B.; Laing, J.E.; Jaques, R.P.; Corrigan, J.E. (1983) The transmission of the granulosis virus of *Pieris rapae* by the parasitoid *Apanteles glomeratus. Environmental Entomology* 12, 116-170.

Michalowicz, W.J. (1980) An ecological investigation of *Pieris rapae* (L.) (Lepidoptera: Pieridae) and its natural enemies on cabbage in southern Ontario. M.Sc. Thesis, University of Guelph, 110 pp.

Culex pipiens L., Northern House Mosquito (Diptera: Culicidae)

J.A. GEORGE

Pest Status

The northern house mosquito, or rain barrel mosquito, *Culex pipiens* L., has long been known to live in close proximity to humans in southern Ontario (Wood *et al.* 1979). Larvae occur most commonly in stagnant water polluted to some degree with varying amounts of decaying organic matter such as vegetation or sewage. *C. pipiens* larvae are usually associated with, or follow the earlier occurring larvae of *Culex restuans* Theobald. Adult populations of *C. pipiens* reach a peak in August, and though breeding continues until October, females begin entering diapause in late August and will not take a blood meal until the following spring (Madder *et al.* 1980, Madder 1981). Mated females spend the winter in any protected shelter available to them. About mid-May females seek their first blood meal and deposit the first egg rafts during the third week of May at Guelph (Madder *et al.* 1980). Though females prefer the blood of birds to that of mammals (Hayes 1961), they will, when abundant, invade homes and feed on humans.

C. pipiens has been implicated in the spread of St. Louis encephalitis (SLE) and it, along with *Culex restuans* (Madder *et al.* 1980), is assumed to be the vector of this virus in eastern North America (Wood *et al.* 1979). The first recorded epidemic of SLE occurred in southern Ontario in 1974–75 and stimulated investigations on surveillance for, and abatement of this insect (Mahdy *et al.* 1979, Helson *et al.* 1980).

Four years (1977–80) of identifying mosquito larvae by the author for the Middlesex-London District Health Unit, London, Ontario survey crews, has helped to establish catch basins, designed to drain away surface water, as major breeding sites for *C. pipiens* in this area. In 1979 catch basins accounted for 96% of all breeding sites found by the survey crew (Pringle *et al.* 1978). Moreover, on randomly selected streets in Brampton, Clarkson, Streetsville, and Port Credit, 70% of the catch basins examined contained *Culex* larvae (Hill, 1978, personal communication).

Catch basins not only provide water high in organic matter throughout the summer, but, in addition, the underground pipes into which they drain provide ideal protected sites for the females to spend the winter months. In addition, the street locations of most catch basins puts them in close proximity to human dwellings. Finally, some catch basins form a part of many farm drainage systems. These also support *Culex* larvae. Thus, the drainage systems, established throughout southern Ontario to remove surface water, provide ideal breeding conditions for *C. pipiens* and *C. restuans,* the suspected vectors of SLE.

Background

The planarian flatworm, *Dugesia tigrina* (Girard) (Tricladida, Turbellaria), has been known to be a predator of mosquitoes since Lischetti (1919) reported it to consume mosquito larvae (Jenkins 1964). A larger species, *Dugesia dorotocephala* (Woodworth), has recently been shown also to act as an effective predator of mosquito eggs, larvae, and pupae in California (Legner & Medved 1972, 1974, Medved & Legner 1974, Yu & Legner 1976). Small *Merostoma* species, under 1 mm in length, of planaria in the order Rhabdocoelida, kill mosquito larvae in the rice fields of California (Case & Washino 1979).

D. tigrina occurs in permanent ponds and streams of southern Ontario but not in catch basins. Hence, it was considered that if local planaria could be transferred to catch basins and survive there, they may provide long lasting reductions of *Culex* larvae, thereby decreasing the probable spread of SLE to humans. Not only are indigenous planaria adapted to the local climatological conditions, extensive tests to establish their safety to vertebrates are unnecessary.

Laboratory tests soon confirmed that local *D. tigrina* readily feed and develop on *Culex* larvae and pupae (George 1978). Furthermore, they proved capable of tolerating high levels of the insecticide temephos, the growth regulator methoprene, and are unaffected by consuming mosquito larvae moribund from ingesting the protein inclusions of *Bacillus thuringiensis* var. *israelensis* (George unpublished).

Releases

D. tigrina were collected from natural sites, reared, and released into catch basins in 1978 and 1979. In 1979, planaria collected from two catch basins inoculated in 1978 were used to initiate a colony for the releases made in 1980. Releases of 150 reared planaria into each of four catch basins were made on 28 July 1978, 300 into each of five catch basins on 5 July 1979, and 80 into each of 31 catch basins on 15 July 1980. Mosquito larvae were monitored in treated and control basins by weekly counts of larvae in 300 ml dips in 1978 and 1979. By 1980 a means of monitoring for planaria by counting the numbers on the bottoms of Styrofoam floats placed in the catch basins was devised and only the planaria were monitored.

In 1978, the *Culex* larvae were reduced from an average of 69±34 S.E. per dip in the controls to 4±2.8 S.E. in the treated catch basins on 28 August 1978 (George 1978). In 1979, the average of all counts from 5 July until 30 September in the controls was 106±32 S.E. compared to 30±7 S.E. with planaria, a reduction of 70%.

In 1980, the presence of planaria was monitored every second week in the 31 basins inoculated from 22 July until 21 October. On 31 October, planaria were still present in 24 of the original 31 catch basins. Home owners had placed waste oil in two of the remaining basins and removed the floats from four. This left only one catch basin that tested negative. The overall average of planaria on the floats for the season was 22 with a high of 138 on one float on 31 October. This is well above the total of 80 that were introduced 15 weeks previously.

No planaria were detected in 1980 in the catch basins inoculated in 1978 and 1979. However, these were all drastically altered in 1980 during the installation of underground gas lines.

Evaluation of Control Attempts

The usefulness of *D. tigrina* in the reduction of *Culex* larvae in catch basins depends on the year round survival of planaria in catch basins with *Culex* larvae. If *D. tigrina* were to survive in all catch basins, seasonal reductions in mosquito larvae of about 70% appears possible. However, planaria could not withstand raw sewage, waste oil, laundry soap, road salt of over 0.5%, and other unknown chemicals found in catch basins during the summer months. Nor can planaria survive in catch basins that become dry for even a short period.

Planaria will survive in catch basins in the absence of mosquito larvae by feeding on insects, such as Collembola, that become stranded on the water's surface.

Continual examination of the 31 catch basins inoculated in 1980 will reveal the ability of planaria to survive in catch basins. If survival proves to be high, new catch basins can readily be inoculated with reared planaria or more simply by transferring floats with planaria to new catch basins.

Recommendations

Research on the abatement of *Culex* spp. in catch basins is required to enable intelligent control measures to be implemented if *Culex*-transmitted SLE virus reappears in Ontario. A study of the ability of predaceous planaria to survive year round in catch basins is under way. A search for *Merostoma* species should be conducted in Ontario.

If planaria fail, the next most promising control in catch basins appears to be a means of slow release of *B.t.i.* or methoprene.

Acknowledgement

This research was assisted by the Ontario Ministry of the Environment through the Ontario Pesticide Advisory Committee.

Literature Cited

Case, T.; Washino, R.K. (1979) Flatworm control of mosquito larvae in rice fields. *Science* 206, 1412–1414.

George, J.A. (1978) The potential of a local planarian, *Dugesia tigrina* (Tricladida, Turbellaria), for the control of mosquitoes in Ontario. *Proceedings of the Entomological Society of Ontario* 109, 65–69.

Hayes, R.O. (1961) Host preferences of *Culiseta melanura* and allied mosquitoes. *Mosquito News* 21, 179–187.

Helson, B.V.; Surgeoner, G.A.; Wright, R.E. (1980) The seasonal distribution and species composition of mosquitoes (Diptera: Culicidae) collected during a St. Louis Encephalitis surveillance program from 1976 to 1978 in southwestern Ontario. *Canadian Entomologist* 112, 865–874.

Jenkins, D.W. (1964) Pathogens, parasites and predators of medically important arthropods — Annotated list and bibliography. *Bulletin of the World Health Organization Supplement* to vol 30.

Legner, E.F.; Medved, R.A. (1972) Predators investigated for the biological control of mosquitoes and midges at the University of California, Riverside. *Proceedings of the California Mosquito Control Association* 40, 109–111.

Legner, E.F.; Medved, R.A. (1974) Laboratory and small-scale field experiments with planaria (Tricladida, Turbellaria) as biological mosquito control agents. *Proceedings of the California Mosquito Control Association* 42, 79–80.

Lischetti, A.B. (1919) Un verme del genero *Planaria*, enemigo natural de las larvas del mosquito. *Physis,* Buenos Aires, 4, 591–595.

Madder, D.J. (1981) Biological studies on *Culex pipiens* L. and *Culex restuans* Theo. (Diptera, Culicidae) in southern Ontario. Ph.D. Thesis, University of Guelph, 125 pp.

Madder, D.J.; MacDonald, R.S.; Surgeoner, G.A.; Helson, B.V. (1980) The use of oviposition activity to monitor populations of *Culex pipiens* and *Culex restuans* (Diptera: Culicidae). *Canadian Entomologist* 112, 1013–1017.

Mahdy, M.S.; Spence, L.; Joshua, J.M. (Eds) (1979) Arboviral encephalitides in Ontario with special reference to St. Louis encephalitis. Ontario Ministry of Health, 364 pp.

Medved, R.A.; Legner, E.F. (1974) Feeding and reproduction of the planarian *Dugesia dorotocephala* (Woodworth), in the presence of *Culex peris* Speiser. *Environmental Entomology* 3, 637–641.

Pringle, T.; Ryan, S.; Smith, L. (1978) Middlesex County mosquito-borne encephalitis control programme (1978). Report of Middlesex-London District Health Unit, London, Ontario, 36 pp.

Wood, D.M.; Dang, P.T.; Ellis, R.A. (1979) The insects and arachnids of Canada. Part 6. The mosquitoes of Canada. Diptera: Culicidae. *Agriculture Canada Publication* 1686, 390 pp.

Yu, Hyo-sok; Legner, E.F. (1976) Regulation of aquatic Diptera by planaria. *Entomophaga* 21, 3–12.

Culiseta inornata (Williston), a Mosquito (Diptera: Culicidae)

J.A. SHEMANCHUK, H.C. WHISLER and S.L. ZEBOLD

Pest Status

Culiseta inornata (Williston) is a very common mosquito species in the Prairie Provinces of Canada. It is found not only in the open country but also in the aspen grove regions. Females of this species pass the winter as fertilized females in rodent burrows, caves, root cellars, rock piles, and abandoned buildings. Generally, the species does not appear in large numbers until late July. There are several generations in a season. *C. inornata* is a common pest of livestock and occasionally of man, and is a vector of western equine encephalomyelitis and California encephalitis (Morgante & Shemanchuk 1967, Shemanchuk & Morgante 1967).

Background

Concern about the environmental contamination with pesticides has led many countries to examine the use of biological agents for the control of mosquitoes. An aquatic fungus, *Coelomomyces psorophorae* Couch (Chytridiomycetes: Blastocladiales), parasitic on larvae of *C. inornata* was discovered in the irrigated areas of southern Alberta in 1956 and has been found there every year since. Larval mortalities up to 80% have occurred in some of the breeding ponds. This fungus was also found infecting *C. inornata* larvae in the irrigated areas of Saskatchewan, which indicates a wide geographic distribution.

The persistence of this fungus in the irrigated areas of southern Alberta and Saskatchewan and the ability to colonize the host species of mosquitoes in the laboratory prompted us to study further this particular host-parasite combination.

Laboratory experiments resulted in the discovery of a previously unknown obligate alternate host the copepod, *Cyclops vernalis*, in the life-cycle of *Coelomomyces psorophorae* (Whisler *et al.* 1974, 1975). In the life-cycle, the zygote seeks out and attaches to the intersegmental membranes of the mosquito larva and penetrates the cuticle leading to the development of hyphal bodies, mycelium, and ultimately thick-walled resistant sporangia in the coelom of the larva, which kill the larva. Under appropriate conditions, these sporangia release zoospores of opposite mating type which infect the alternate host, *Cyclops vernalis*. Each zoospore develops into a thallus and eventually gametangia. Gametes of opposite mating type fuse either in or outside of the copepod to form the mosquito-infecting zygote.

The discovery of the life-cycle led to the development of a successful laboratory procedure for the propagation of the fungus and to a better understanding of its basic biology.

Host Specificity

With the reliable technique which was developed for consistent and significant infections in *C. inornata* in the laboratory, host specificity studies of *Coelomomyces psorophorae* were possible. The possibility of infecting *Aedes vexans* (Meigen) received particular attention because larvae of this mosquito with light infections of *Coelomomyces psorophorae* have occasionally been found in field sites where heavily infected larvae of *C. inornata* were also found. Larvae of *A. vexans, A. dorsalis*

(Meigen), and *A. flavescens* (Muller) were exposed to zygotes of *Coelomomyces psorophorae,* but, despite repeated attempts, infections by this strain were never attained. *C. inornata* larvae in the same experiment were infected (Zebold *et al.* 1979).

These results seemed to support the prevailing opinion that *Coelomomyces* is a highly specific parasite. In further tests, however, infection was achieved in *Culex pipiens* L., *C. quinquefasciatus* Say, *A. aegypti* (L.), *A. sierrensis* (Ludlow), *A. triseriatus* (Say), and *C. inornata* (Louisiana strain).

A. aegypti larvae were more susceptible than the other species, but maximum resistant sporangia production was consistently best in the original host species, *C. inornata.*

Field Trials

In 1979, a preliminary field trial was conducted at 8-Mile Lake near Lethbridge using laboratory-cultured *Coelomomyces psorophorae.* Six artificial pools were established. One pool was inoculated with zoospores and copepods, one received copepods that were infected in the laboratory and were ready to release the zygotes, two pools received zygotes from laboratory-infected copepods, and two pools that received no fungal inoculum served as controls. Into each of the pools, first instar laboratory-reared larvae of *C. inornata* were introduced and exposed for 24 hours after which they were brought back to the laboratory for rearing to final instar and for observation for infection.

The results from this test confirmed our field observation that the fungus is an effective killer of *C. inornata* larvae. The fungus was most effective when introduced into mosquito habitats in the zygote stage and produced a 64% kill.

These results further indicate that this fungus will have practical application in controlling *C. inornata* as a short-term microbial pesticide and as a long-term inoculum in mosquito-breeding habitats where the fungus does not occur naturally.

Recommendations

(1) Expand research on the design and development of techniques for large-scale production and long-term storage of *Coelomomyces psorophorae.*

(2) Evaluate *Coelomomyces psorophorae* as a short-term microbial pesticide and as a long-term inoculum in mosquito-breeding habitats in the irrigated areas and in areas where the fungus does not occur.

(3) Continue searching for *Coelomomyces* infections in other species of mosquitoes and define their life-cycles.

(4) Search for other aquatic fungi that might be domesticated and manipulated for the control of mosquitoes.

Literature Cited

Morgante, O.; Shemanchuk, J.A. (1967) Virus of California Encephalitis Complex: isolation from *Culiseta inornata. Science* 157, 692–693.

Shemanchuk, J.A.; Morgante, O. (1967) Isolation of Western Encephalitis virus from mosquitoes in Alberta. *Canadian Journal of Microbiology* 14, 1–5.

Whisler, H.C.; Zebold; S.L.; Shemanchuk, J.A. (1974) Alternate host for mosquito parasitic *Coelomomyces. Nature* 251, 715–716.

Whisler, H.C.; Zebold, S.L.; Shemanchuk, J.A. (1975) Life history of *Coelomomyces psorophorae. Proceedings of the National Academy of Sciences* 72, 693–696.

Zebold, S.L.; Whisler H.C.; Shemanchuk, J.A.; Travland, L.B. (1979) Host specificity and penetration in mosquito pathogen *Coelomomyces psorophorae. Canadian Journal of Botany* 57, 2766–2770.

Cydia pomonella (L.), Codling Moth (Lepidoptera: Tortricidae)

R.P. JAQUES and J.E. LAING

Pest Status

The codling moth, *Cydia pomonella* (L.), is a major pest of apple in most areas of Canada in which the crop is grown. Several chemical insecticides, notably phosmet, azinphos-methyl, and permethrin, are effective against the pest. Nevertheless, protection of apples against the codling moth is a major concern in development of integrated pest management programmes for apples, not only because of the importance of the pest, but also because the broad-spectrum insecticides used to control it are detrimental to the predaceous and parasitic arthropods in orchards.

Background

A granulosis virus discovered by Tanada (1964) has been shown to be highly infectious for *C. pomonella* larvae in laboratory tests (Keller 1973, Laing & Jaques 1981). The virus was quite effective against the pest in apple orchard tests in Germany (Huber & Dickler 1977), reducing damage to fruit by *C. pomonella* by as much as 89%. Falcon *et al.* (1968) obtained similar crop protection in tests in California but the virus was less effective in orchard tests in Ohio (Sheppard & Stairs 1976).

Orchard Tests

The granulosis virus of *C. pomonella* (CPGV) was applied to apple trees at four locations in Canada between 1974 and 1980 in orchard tests on effectiveness of the virus against the codling moth (Jaques *et al.* 1977, 1981).

Seven-tree plots were sprayed at rates of 1 x 10^9 to 3 x 10^{10} GIB of CPGV or 0.5 g azinphos-methyl per litre of spray in a 4-year test (1975–78) carried out at the Research Station at Kentville, Nova Scotia. A summary of results (Table 3) indicated that CPGV was not as effective as azinphos-methyl in protecting apples against damage by larvae of *C. pomonella*. Total damage to picked and dropped apples harvested from plots treated with CPGV was reduced by an average of 95.5%.

Table 3 Injury to apples by larvae of *Cydia pomonella* (L.) in a 4-year orchard test at Kentville, Nova Scotia, on efficacy of a granulosis virus*

Material	Injury by *C. pomonella*/100 apples			
	Picked fruit		Picked + Dropped fruit	
	SE**	DE**	SE	DE
Virus	1.6	1.0	1.7	1.6
Azinphos-methyl	0.1	0.1	0.2	0.4
Check	1.2	3.0	1.3	12.1

*Tests were carried out in cooperation with C.R. MacLellan and K.H. Sanford, Research Station, Agriculture Canada, Kentville, N.S.

**SE and DE refer to "shallow-entry" and "deep-entry" injury, respectively.

The granulosis virus was applied to plots in apple orchards at the Research Station, Summerland, British Columbia, in 1975 and 1976; at the Research Station, Vineland Station, Ontario, in 1974, 1977, and 1978; and at the University of Guelph, Guelph, Ontario, in 1976–80; Table 4 lists data from some of the tests.

Table 4 Injury to apples by larvae of *Cydia pomonella* (L.) in orchard tests on efficacy of *C. pomonella* granulosis virus*.

Material/litre of spray	Applications/ generation		Injury by *C. pomonella*/100 apples picked at harvest	
	1st	2nd	SE**	DE**
	Summerland, British Columbia			
1975				
CPGV 3.2–7.3 x 10^9 GIB	2	2	55	32
Check	0	0	22	71
	Vineland Station, Ontario			
1974				
CPGV 3.5–6.5 x 10^9 GIB	3	2	21	5
Phosmet 3.1 g	2	1	2	1
Check	0	0	5	22
	Guelph, Ontario			
1979				
CPGV 3.3x10^{10} GIB	3	3	9	2
Check	0	0	11	14
1979				
CPGV 3.6x10^{10} GIB	2	2	11	2
	3	4	8	0
Check	0	0	7	11

*Tests were carried out in cooperation with M.D. Proverbs, Research Station Summerland, British Columbia, and E.A.C. Hagley and R. Trottier, Research Station, Vineland Station, Ontario.

**SE and DE refer to "shallow-entry" and "deep-entry" injury, respectively.

Total damage to apples in plots treated with CPGV was not reduced appreciably in many of the tests. Deep-entry (D.E.) damage was reduced by 55% to 100%, approaching the reductions noted in plots treated with a chemical insecticide. It is noteworthy that shallow entries (stings) (S.E.) were higher in apples from plots treated with the virus than in nontreated apples in some tests, e.g. 1975 Summerland test, presumably because larvae infected by the virus died after they had entered the apple.

The 1979 test at Guelph indicated superior crop protection with increased numbers of applications of the virus. In a plot test in 1980 at Guelph, CPGV was applied 2 or 4 times for each of the two generations of *C. pomonella* larvae; however, crop protection was not increased by the higher frequency of application.

Evaluation of Control Attempts

The granulosis virus of *C. pomonella* did not protect the crop as well as did chemical insecticides in these tests. However, the virus is considered to have substantial potential in integrated pest management because it is specific for the target insect.

Further studies will emphasize timing and dosages of application, persistence of the virus in the orchard habitat, and application of the virus in combination with low dosages of chemical insecticides as means of increasing effectiveness.

Literature Cited

Falcon, L.A.; Kane, W.R.; Bethell, R.S. (1968) Preliminary evaluation of the granulosis virus for control of the codling moth. *Journal of Economic Entomology* 61, 1208–1213.

Huber, J.; Dickler, E. (1977) Codling moth granulosis: Its efficacy in the field in comparison with organophosphorous insecticides. *Journal of Economic Entomology* 70, 557–561.

Jaques, R.P.; MacLellan, C.R.; Sanford, K.H.; Proverbs, M.D.; Hagley, E.A.C. (1977) Preliminary orchard tests on control of codling moth larvae by a granulosis virus. *Canadian Entomologist* 109, 1079–1081.

Jaques, R.P.; Laing, J.E.; MacLellan, C.R.; Proverbs, M.D.; Sanford, K.H.; Trottier, R. (1981) Apple orchard tests on efficacy of the granulosis virus of the codling moth, *Laspeyresia pomonella* (Lepidoptera: Olethreutidae). *Entomophaga* 26, 111–117.

Keller, S. (1973) Mikrobiologische Bekämpfung des Apfelwicklers *Laspeyresia pomonella* (L.) (= *Carpocapsa pomonella)* mit spezifischen Granulosisvirus. *Zeitschrift für angewandte Entomologie* 73, 137–181.

Laing, D.R.; Jaques, R.P. (1981) Codling moth: techniques for rearing larvae and bioassaying granulosis virus. *Journal of Economic Entomology* 73, 831–853.

Sheppard, R.E.; Stairs, G.R. (1976) Effects of dissemination of low dosage levels of a granulosis virus in populations of the codling moth. *Journal of Economic Entomology* 69, 583–586.

Tanada, Y. (1964) A granulosis virus of the codling moth, *Carpocapsa pomonella* (Linnaeus) (Olethreutidae, Lepidoptera). *Journal of Insect Pathology* 6, 378–380.

Delia antiqua (Meigen), Onion Maggot (Diptera: Anthomyiidae)

F.L. McEWEN

The onion maggot, *Delia antiqua* (Meigen), is the most destructive pest of onions throughout the temperate regions of the northern hemisphere and in Canada causes losses of 50–90% (Finlayson 1959). Effective control was not achieved until the late 1940s when the cyclodiene insecticides became available. These, when applied in the furrow with the seed, provided excellent control of the first generation and additional treatments of DDT against the second and third generation adults reduced damage to a minimum. Resistance to these insecticides developed and the cyclodienes were replaced about 1960 by such organophosphorus components as ethion and diazinon. Subsequently chlorfenvinphos, fensulfothion, carbofuran, fonofos, and chlorpyrifos have been used in the seed furrow; and parathion, diazinon, naled, or malathion have been used as sprays against the adults of the second and third generation. Resistance to several of these insecticides has developed and, while the level of resistance has not been as great as with the cyclodienes, commercial control is becoming increasingly difficult (Harris & Svec 1976).

The use of the sterile-male approach for insect control was first successfully applied (Knipling 1960) against a livestock pest and since then its use has been explored for many insect species. In the case of the onion maggot, McClanahan & Simmons (1966) and Noordink (1966) showed that the insect could be sterilized without serious effects on the longevity of the adult. In Holland, Noordink and his associates continued studies on the sterile-male method for control, and by 1980 had completed laboratory and field testing to the point where a commercial venture in marketing the programme to the growers is now in place (Noordink 1971, 1974; Ticheler *et al.* 1974; Loosjes 1976).

In Canada, research on sterile-male control for the onion maggot began as a cooperative venture between Agriculture Canada (C.R. Harris) and the University of Guelph (F.L. McEwen) in 1971. In the initial stages of the study, mass rearing procedures were perfected (Harris & Svec 1976), and laboratory tests determined that when pupae reared under diapausing conditions were irradiated (Cobalt 60) after 6 days post-diapause (24°C) at 2 or 4 krad, the males were sterilized and the females made unreproductive. Tests established that males produced from these pupae were competitive and lived about as long as those emerging from non-irradiated pupae.

Initial field releases of sterilized pupae were begun in 1973 and, using fluorescent dyes as markers on released flies, the field distribution pattern was also determined. The survival of released flies was monitored and eggs collected in the field and hatch determined. The initial results indicated that the flies did not disperse widely (most recaptures were within 75 m of release site) and that the hatchability of eggs in the release area was reduced. In addition, an economic study was undertaken that showed that if the sterile-male programme was effective, it was an economically sound approach (King 1973).

Field tests were expanded with 6 000 000 pupae released in 1974 and 10 000 000 in 1975. While it was clear that egg hatch was reduced, bird predation on released pupae and immigration of wild flies into the release area resulted in inconsistent results. Extensive monitoring indicated, however, that when the ratio of wild to released flies approached 1:1, the egg hatch was reduced near 50%.

Since rearing facilities for continuing field releases at the 1974 and 1975 levels were no longer available, the release portion of the programme was curtailed in subsequent years and efforts concentrated on learning more about the field ecology of wild and released flies, studying mating habits and determining a release procedure in which bird predation would not be such an important negative factor on the programme. These studies have demonstrated that the pupae can be released by air or as adults by air and, hopefully, will not be so subject to predation as ground release of pupae in designated sites. It has been shown also that the female fly mates only once (Martin 1981) and that in the field, much of the survival of third generation larvae to overwintering pupae results from cull and volunteer onions left in the field after harvest. Thus fall sanitation is an essential component in a programme to reduce onion maggot populations.

Future plans are to release approximately 100 000 000 sterilized flies in an isolated onion-growing area containing about 300 ha of onions in each of the years 1982 and 1983. To accommodate this, a new biological control laboratory has been built at the University of Guelph with major facilities and space for mass rearing the onion maggot. These releases will be closely monitored and serve as a demonstration project to determine the feasibility of the sterile-male method of control. If successful on the limited area chosen, the programme will be expanded to the major onion areas in Ontario.

Literature Cited

Finlayson, D.G. (1959) Chemical control of the onion maggot in onions grown from seed in various types of soil in Northwestern America in 1955–56. *Journal of Economic Entomology* 52, 851–856.

Harris, C.R.; Svec, H. (1976) Onion maggot resistance to insecticides. *Journal of Economic Entomology* 69, 617–620.

King, D.L.J. (1973) An economic appraisal of the sterile-male technique of onion maggot control on the Holland Marsh of Ontario. M.Sc. Thesis, University of Guelph, 176 pp.

Knipling, E.F. (1960) The eradication of the screw-worm fly. *Scientific American* 203, 54–61.

Loosjes, M. (1976) Ecology and genetic control of the onion fly *Delia antiqua* Meigen. Wageningen, PUDOC; 94 pp.

Martin, J.S. (1981) Mating behaviour of the onion maggot fly *Hylemya antiqua* (Meigen) in laboratory conditions. M.Sc. Thesis, University of Guelph, 43 pp.

McClanaham, R.J.; Simmons, H.S.(1966) Sterilization of onion maggots by irradiation with cesium-137. *Canadian Entomologist* 98, 931–935.

Noordink, J.Ph.W. (1971) Irradiation, competitiveness, and the use of radioisotopes in sterile-male studies with the onion fly, *Hylemya antiqua* M. In: Sterility principle for insect control or eradication. *International Atomic Energy Agency,* 232–328.

Noordink, J.Ph.W. (1974) Onderzoek met behulp van radiactive isotopen. Jaarversl. 1973. *Mededelingen Instituut voor Plantenziektenkundig Onderzoek, Wageningen*, 119–121.

Ticheler, J.; Loosjes, M.; Noordink, J.Ph.W.; Noorlander, J.; Theunissen, J. (1974) Field experiments with the release of sterilized onion flies *Hylemya antiqua* (Meig.). *In*: The sterile-insect technique and its field of applications International Atomic Energy Agency, 103–107.

Entomoscelis americana Brown, Red Turnip Beetle (Coleoptera: Chrysomelidae)

G.H. GERBER

Pest Status

The red turnip beetle, *Entomoscelis americana* Brown, is an oligophagous insect which feeds on Cruciferae and is an occasional pest of rape, *Brassica campestris* L. and *B. napus* L., mustard, *B. juncea* (L.) Czern. and *B. hirta* Moench, and cruciferous garden crops in western Canada (Stewart 1973, Gerber & Obadofin 1981*a*, 1981*b*). The larvae and adults feed on seedling plants in May and June before the adults aestivate in July, and the adults feed on mature plants in August and September after aestivation. The only damage of economic significance is caused by the adults in June (Gerber 1974, 1976).

The red turnip beetle is native to North America and in Canada occurs south of the Arctic climatic zone and west of longitude 96°W (Stewart 1973). The beetle overwinters in the egg stage on the surface of the soil. The eggs hatch in late April and early May, shortly after the snow has melted, but usually before rape, mustard, and cruciferous garden crops are planted (Gerber & Obadofin 1981*a*). Larval development normally is completed by the end of May. The larvae feed on volunteer rape and mustard seedlings, and on cruciferous weeds in fields that contained these plants the previous year (Turnock *et al.* 1979). The adults emerge during the first 3 weeks of June and feed for 2–3 weeks, usually in the same fields in which the larvae fed. If the supply of food plants is not adequate, the adults will move into nearby new fields containing their host plants. At the end of June, the adults enter the soil to aestivate for about a month. They reappear in late July and August; disperse to new fields of rape, mustard, or garden Cruciferae; and mate and lay eggs until late October.

Background

The only known natural enemies of *E. americana* attack the adult stage: a pathogen (microsporidia (Gerber 1975)) and a predator (adults of the carabid beetle, *Pterostichus adstrictus* (W.J. Turnock, 1975, personal communication)). These two natural enemies have been encountered on only a few occasions in the field and consequently are not considered to be important control agents.

Manolache (1941) reported two tachinid parasitoids in the larvae and pupae of a European species of *Entomoscelis (E. adonidis* Pall.) in Romania: *Meigenia mutabilis* Fall. and *Marchantia chalchonta* Meig. However, the incidence of parasitism by *M. mutabilis* was low (<6.0%) and Manolache considered *E. adonidis* to be only a secondary host for this tachinid (a heteroxenous species). Manolache did not report on the abundance and importance of the other tachinid. Brovdii (1977) found up to 38% of the last instar larvae of *E. adonidis* parasitized by tachinids in the Ukraine, but the tachinids were not identified.

Studies were initiated in 1973 to identify the natural enemies of *E. adonidis* in Europe and to determine whether there are any parasitoids of this beetle which are suitable for introduction into Canada as control agents for *E. americana*. The parasitoids of *E. adonidis* were considered to be potential control agents for *E. americana*, because the life histories, phenology, and host plants of these two species are similar (Manolache 1941; Stewart 1973; Brovdii 1977; Gerber & Obadofin 1981*a*, 1981*b*). These studies have not been successful, because populations of *E. adonidis* have been extremely scarce outside of the Soviet Union since 1973. Only a small number of specimens of a species of *Meigenia* have been found parasitizing the larvae. Collections of *E. adonidis* have not yet been made in the Soviet Union.

Recommendations

It is recommended that every attempt be made to collect additional larvae and adults of *E. adonidis* from Europe, especially from the Soviet Union, to identify the parasitoids of this beetle. If *M. mutabilis* is the only parasitoid which can be collected in sufficient number for release in Canada, research is needed to determine its life history and whether it is truly heteroxenous. If *M. mutabilis* is heteroxenous, it would not be a suitable control agent for *E. americana*.

Efforts also should be made to collect another species of *Entomoscelis, E. suturalis*, which occurs on Cruciferae in Europe. The life history of *E. suturalis* is similar to those for *E. americana* and *E. adonidis* (Brovdii 1977), but there is no information in the literature on its natural enemies.

Literature Cited

Brovdii, V.M. (1977) Fauna of the Ukraine — Chrysomelinae leaf beetles. Ukrainian S.S.R. Academy of Sciences, Institute of Zoology. *Naukova Dumka, Kiev* 19(16), 1–388. (In Ukrainian).

Gerber, G.H. (1974) Red turnip beetle on rape. *Canadex* 149,622.

Gerber, G.H. (1975) Occurrence of a microsporidian infection in the red turnip beetle, *Entomoscelis americana* (Coleoptera: Chrysomelidae). *Canadian Entomologist* 107, 1081–1082.

Gerber, G.H. (1976) Effects of feeding by adults of the red turnip beetle, *Entomoscelis americana* Brown (Coleoptera: Chrysomelidae), during late July and August on the yield of rapeseed (Cruciferae). *Manitoba Entomologist* 10, 31–35.

Gerber, G.H.; Obadofin, A.A. (1981*a*) Growth, development and survival of the larvae of the red turnip beetle, *Entomoscelis americana* (Coleoptera: Chrysomelidae) on *Brassica campestris* and *B. napus* (Cruciferae). *Canadian Entomologist* 113, 395–406.

Gerber, G.H.; Obadofin, A.A. (1981*b*) The suitability of nine species of Cruciferae as hosts for the larvae of the red turnip beetle, *Entomoscelis americana*. (Coleoptera: Chrysomelidae). *Canadian Entomologist* 113, 407–413.

Manolache, F. (1941) Research on the morphology, biology, and control of the insect, *Entomoscelis adonidis* Pall., in Romania. Methods, Guidance and Investigations. *Institute of Agronomic Research Romania (Bucharest)* 71, 1–205. (In Romanian).

Stewart, D.B. (1973) The red turnip beetle, *Entomoscelis americana* Brown (Coleoptera: Chrysomelidae), biology and plant relationships. M.Sc.Thesis, University of Alberta, Edmonton, 86 pp.

Turnock, W.J.; Gerber, G.H.; Bickis, M.; Bennett, R.B. (1979) The applicability of X-ray energy-dispersive spectroscopy to the identification of populations of red turnip beetle, *Entomoscelis americana* (Coleoptera: Chrysomelidae). *Canadian Entomologist* 111, 113–125.

Euxoa messoria (Harris), Darksided Cutworm (Lepidoptera: Noctuidae)

H.H. CHENG

Pest Status

The darksided cutworm, *Euxoa messoria* (Harris), is the most destructive insect pest of tobacco in Ontario and indeed throughout most tobacco-growing areas of Canada. It is a northern species which has also done great damage to a variety of crops. The principal injury is caused by larvae which attack the newly transplanted tobacco seedlings and damage the leaves or growing points and sometimes notch or sever the plants. If left uncontrolled, a low to moderate population level of this pest has resulted in 20–48% damaged plants and 14–25% loss of yield (Cheng 1971*a*, 1971*b*, 1973*c*, 1975, 1980*a*). In the tobacco-growing areas, large populations of this pest are produced in the rye rotation crop, but invasion by migrating larvae is always a problem (Bucher & Cheng 1970).

The bionomics of the darksided cutworm in Ontario has been published (Cheng 1973*a*). It has one generation a year and overwinters as an egg in the soil from 6 to 12 mm below the surface. Depending upon the weather, overwintered eggs begin to hatch from the end of March to early April. Larvae in the early instars feed on the rye rotation crop, later they attack newly transplanted tobacco seedlings. The duration of the larval stage is about 3 months. The mature larvae cease feeding and transform to pupae in the soil from late July to mid-August. The pupal stage lasts from 18 to 28 days. The adults emerge from mid-August to October and lay their eggs in the soil during the same period. These eggs are completely developed before winter and hatch in the following spring.

Research on darksided cutworm control in Canada in the past 10 years has centered in two areas: (1) selection and evaluation of the more effective and safe chemical insecticides (Cheng 1979, 1980*a*, 1980*b*), and (2) search and evaluation of effective biological control agents (Bucher 1970, 1971; Cheng 1973*b*, 1977). These approaches are reviewed in this chapter.

Background

As more information becomes available on the hazards involved in the agricultural use of insecticides, there is a greater demand for alternative methods for controlling insect pests. Consequently, laboratory and field tests with virus pathogens and *Bacillus thuringiensis* preparations were conducted from 1968 to 1972 at the Research Station, Delhi, Ontario (Bucher 1970, 1971, Cheng 1973*b*). The results indicated that these pathogens were significantly less effective than chemical insecticides against *E. messoria* larvae in the field (Cheng 1973*b*).

Studies on the biology and collection of parasitoids, predators, and diseases of *E. messoria* have also been conducted by Agriculture Canada at the Delhi Research Station. The degree of the influence of various biotic factors on *E. messoria* populations and the importance and possible roles of the natural enemies in the tobacco crop and pest relationship have been discussed (Bucher & Cheng 1971, Cheng 1973*a*, 1977). In general, larval mortality in *E. messoria* from natural enemies was low.

Tobacco is the number one cash crop in Ontario. Because of the high value of the crop and the insufficient pressure of the various biotic factors on the darksided cutworm, the use of chemical insecticides for control of this pest is vital to tobacco production. Persistent chemical insecticides, such as DDT and endosulfan (Thiodan®, Niagara Chemicals) cannot be used on tobacco because of the danger of environmental contamination and undesirable residues in tobacco (Cheng & Braun 1977, Frank *et al.* 1977). Therefore, research in the evaluation of insecticidal control of cutworms in tobacco has been directed towards the more effective and safe chemical insecticides (Cheng 1973*c*, 1975, 1979, 1980*a*, 1980*b*). For example, the broad-spectrum pyrethroids, permethrin, and cypermethrin are very toxic to the cutworm larvae and nonhazardous, and have been registered and recommended for both preplanting (on the rye cover crop or on soil surface) and postplanting (on tobacco plants) for control of cutworms.

Parasitoids

From 1973 to 1977, 14 species of insect parasitoids were reared from *E. messoria* larvae or pupae in Ontario; they all were primary and internal parasitoids, namely, *Apanteles laeviceps* Ashmead, *A. militaris* Walsh, *Meteorus communis* (Cresson) and *M. leviventris* (Wesmael) (Hymenoptera: Braconidae); *Arenetra rufipes vernalis* Walley, *Campoletis flavicinctus* (Ashmead), *Campoletis* sp., *Enicospilus* sp., and *Eutanyacra suturalis* (Say) (Hymenoptera: Ichneumonidae); *Copidosoma bakeri* (Howard) (Hymenoptera: Encyrtidae); *Muscina stabulans* (Fall.) (Diptera: Muscidae); *Linnaemya compta* (Fall.), *Winthemia rufopicta* (Bigot), and *W. deilephilae* (O.S.) (Diptera: Tachinidae). Of these 14 species, *Arenetra rufipes vernalis, Copidosoma bakeri*, and *Enicospilus* sp. were the most common species reared from *E. messoria* larvae and were present in considerable numbers every year. The total parasitism by these three species was about 22% (Cheng 1977) and were considered to be the most important parasitoids of *E. messoria*.

From the economic point of view, *A. rufipes vernalis* is considered to be the most effective parasitoid of the darksided cutworm in the current generation in Ontario. It kills the sixth-instar larvae and essentially prevents the cutworm from causing the maximum damage to the tobacco crop, because the seventh instar of this pest does more damage to the tobacco crop than all the other instars combined. *Enicospilus* sp. kills its host in the seventh instar or prepupal stage, and it reduces some losses caused by the cutworm. Conversely, *C. bakeri* increases the crop losses by the current generation, because the cutworms attacked by *C. bakeri* consumed 27.5% more food than did normal cutworms (McMillan 1930); however, it will reduce the population levels of the following generation of this pest. The rest of the parasitoids were present in small numbers or reared only once during the period of this study and it is likely that they are not important.

Predators

Several birds, insects, and spiders were observed to attack the larvae and adults of the darksided cutworm (Cheng 1973*a*). Birds periodically landed in the fields in search of larvae in the early morning; most frequently recorded were starlings and grackles. Two species of ground beetles, *Calosoma calidum* (F.) and *Harpalus caliginosus* F. (Coleoptera: Carabidae), were the most important insect predators. Both the adults and larvae of the ground beetles destroyed a large number of cutworms. Moths of the darksided cutworm were preyed upon by spiders, which were identified as *Leiobunum vittatum* (Say), *Phalangium opilio* (L.), *Hadrobunus maculosus* (Wood), and *Odiellus pictus* (Wood), as the moths rested in their hiding places. However, definitive studies on the effectiveness of these predator complexes in the regulation of *E. messoria* populations are lacking. Nevertheless, they are assumed to be valuable biological control factors.

Diseases

Several insect pathogens have been isolated from the diseased larvae of *E. messoria* collected in the fields at Delhi, Ontario (Bucher & Cheng 1971). The bacterial diseases are common and the bacteria most commonly isolated from infested larvae are *Bacillus cereus* Frankland & Frankland, *Enterobacter cloacae* (Jordan), *Enterobacter aerogenes* (Kruse), *Klebsiella pneumoniae* (Schroeter), *Streptococcus faecalis* (Andrewes & Horder), *Pseudomonas fluorescens* Migula, *Pseudomonas* spp., *Bacillus sphaericus* Neide, and *Achromobacter* spp. The infected haemocoels of cutworm larvae usually contained one or a mixture of two or more bacterial species. *B. cereus* is a well-known pathogen of many insects (Heimpel & Angus 1963); *E. aerogenes* and *P. fluorescens* have been considered as potential pathogens of a number of insects (Bucher 1963); the pathogenicity of the other species is very questionable and they probably occurred by chance in the haemocoel of sick larvae of *E. messoria*. The microsporidial disease, *Nosema* sp., is less common; the fungus disease, *Sorosporella uvella* (Krass.), is rare; and the virus diseases are virtually absent.

Although larvae of *E. messoria* collected in the fields in Ontario are free from virus diseases, they are susceptible to two virus diseases, nuclear polyhedrosis and granulosis virus, originally isolated from a related cutworm, *E. ochrogaster* (Guenée) (Bucher 1970). In cooperation with Dr. G.E. Bucher of Agriculture Canada Research Institute, Belleville, Ontario, field tests using preparations of nuclear polyhedrosis virus and granulosis virus were conducted in 1969 and 1970 at the Research Station, Delhi, Ontario. These tests showed that both viruses would protect tobacco seedlings from severe cutworm damage when applied on the green rye in high concentrations, but they were less effective than chlorpyrifos (Dursban® or Lorsban®, Dow Chemical of Canada Ltd.) treatment and better than the untreated control.

In 1971 and 1972, attempts were made to introduce both virus diseases into healthy populations of *E. messoria* by spraying the virus suspensions on tobacco trap-plants growing in a rye field. Consequently, a considerable number of *E. messoria* larvae collected with tobacco trap-plants from areas where viruses were introduced in 1971 and 1972 were infested with virus. This indicated that a large amount of virus survival occurred and that virus introduction is feasible. Unfortunately, this programme was discontinued after the Research Institute, Belleville, Ontario, closed in the fall of 1972.

Microbial Insecticide

In 1969, laboratory experiments were conducted to determine the susceptibility of the various larval stages of *E. messoria* in Ontario to four commercial preparations of *Bacillus thuringiensis* Berliner: Thuricide® 90TS, Thuricide® HP, Biotrol® BTB 183, and Dipel® (Sandoz Canada). In these experiments, first-instar to third-instar larvae that fed on rye leaves treated with all four *B. thuringiensis* preparations were found to be susceptible at all rates applied (Cheng 1973*b*). Mortality of fourth-instar to seventh-instar larvae fed treated tobacco leaves was low.

From the results of laboratory experiments, it was apparent that preparations of *B. thuringiensis* showed potential as a possible alternative to chemical insecticides for the control of the early instar larvae of *E. messoria*. Therefore, a field test was conducted in 1970 at the Agriculture Canada Research Station, Delhi, Ontario, to determine whether these preparations could be as effective as chlorpyrifos for control of this pest when applied on the rye cover crop in spring. Data showed that *B. thuringiensis* preparations as applied in the field for control of *E. messoria* larvae were relatively ineffective as compared with chlorpyrifos (Cheng 1973*b*). The failure of these preparations in the field trial is probably attributable to the different environmental conditions and the much more complex ecosystem in the field than in the laboratory.

Evaluation of Control Attempts

At present, the available natural enemies, such as parasitoids, predators, and diseases, are not capable of reducing the *E. messoria* population below economic levels (Bucher & Cheng 1971; Cheng 1973*a*, 1977). Attempts at control with virus pathogens or *B. thuringiensis* preparations were unsuccessful since these biological control agents were significantly less effective against cutworms than chemical insecticides (Cheng 1973*b*). Therefore, tobacco growers have to apply chemical insecticides for crop protection. In general, a single application of chlorpyrifos, cypermethrin (Ripcord®, Shell Canada Ltd.) or permethrin (Ambush®, Chipman Inc.) on the green cover crop in mid-April, or on soil surface before planting in mid-May gives excellent control of cutworms for the whole growing season; similarly a single application of cypermethrin, permethrin, or acephate (Orthene®, Chevron Chemical Canada Ltd.) on tobacco plants after transplanting also provides very good control of cutworms without incurring the danger of harmful residues. At present, chemical control for this pest costs about $25 to $35 per hectare and the average monetary return is about $8000 per hectare.

At the Delhi Research Station, monitoring and economic threshold data have been developed for *E. messoria*. The hatching rate of overwintered eggs is determined in the spring at intervals from late March to mid-April. When 85% of the eggs have hatched, the rotation cover crop is treated with a recommended insecticide to minimize the infestation of the pest in tobacco fields. After transplanting (late May to mid-June), the crop is regularly checked for feeding injury and cutworm larvae in the soil near the tobacco plants. Insecticides are applied if there are 2 or more cutworms per 100 plants, or 5% of the tobacco plants are severely damaged.

Recommendations

Research on integrated pest management of tobacco in Ontario should be continued. The programme may include: (1) reducing insecticide applications on tobacco and the cost of insecticidal control measures; (2) search for more effective biological control agents; (3) development of tank mix combinations of insecticides and herbicides for control of both cutworms and weeds; (4) continuing to evaluate the efficacy of candidate insecticides and biological control agents for control of tobacco insects and to keep insecticide residues at a minimum level. Since no biotic agents have yet provided efficient control of cutworms on tobacco, chemical insecticides have been and will continue to be an essential part of tobacco crop protection, and they will also play an important role in future integrated pest management programmes.

Literature Cited

Bucher, G.E. (1963) Nonsporulating bacterial pathogens. In: Steinhaus, E.A. *(Ed.),* Insect pathology, vol 2, New York Academic Press, pp.117–147.

Bucher, G.E. (1970) Control of cutworms in tobacco by virus diseases. *The Lighter* 40(2), 14–19.

Bucher, G.E. (1971) *Euxoa messoria* (Harris), darksided cutworm (Lepidoptera: Noctuidae). In: Biological control programmes against insects and weeds in Canada 1959–1968. *Commonwealth Institute of Biological Control Technical Communication* 4, 16–18.

Bucher, G.E.; Cheng, H.H. (1970) Use of trap plants for attracting cutworm larvae. *Canadian Entomologist* 102, 797–798.

Bucher, G.E.; Cheng, H.H. (1971) Mortality in larvae of *Euxoa messoria* (Lepidoptera: Noctuidae) collected from the tobacco area of Ontario. *Canadian Entomologist* 103, 888–892.

Cheng, H.H. (1971*a*) Assessment of tobacco losses caused by the darksided cutworm, *Euxoa messoria* (Lepidoptera: Noctuidae), Delhi, Ontario. *Canadian Entomologist* 103, 534–541.

Cheng, H.H. (1971*b*) Field studies on the chemical control of the darksided cutworm (Lepidoptera: Noctuidae) on tobacco in Ontario, with particular reference to Dursban®. *Canadian Entomologist* 103, 649–653.

Cheng, H.H. (1973*a*) Observations on the bionomics of the darksided cutworm, *Euxoa messoria* (Lepidoptera: Noctuidae), in Ontario. *Canadian Entomologist* 105, 311–322.

Cheng, H.H. (1973*b*) Laboratory and field tests with *Bacillus thuringiensis* against the darksided cutworm, *Euxoa messoria* (Lepidoptera: Noctuidae), on tobacco. *Canadian Entomologist* 105, 941–954.

Cheng, H.H. (1973*c*) Further field evaluation of insecticides for control of the darksided cutworm (Lepidoptera: Noctuidae) on tobacco in Ontario. *Canadian Entomologist* 105, 1351–1357.

Cheng, H.H. (1975) Control of darksided cutworm on flue-cured tobacco with insecticides applied after transplanting. *Tobacco Science* 19, 157–159.

Cheng, H.H. (1977) Insect parasites of the darksided cutworm, *Euxoa messoria* (Lepidoptera: Noctuidae), in Ontario. *Canadian Entomologist* 109, 137–142.

Cheng, H.H. (1979) Toxicity and biological insecticidal persistence of pyrethroid insecticides for control of cutworms on tobacco seedlings in the greenhouse. *Tobacco Science* 23, 120–122.

Cheng, H.H. (1980*a*) Darksided cutworm (Lepidoptera: Noctuidae): field evaluation of pyrethroid insecticides for protection of tobacco in Ontario. *Tobacco Science* 24, 61–63.

Cheng, H.H. (1980*b*) Toxicity and persistence of pyrethroid insecticides as foliar sprays against darksided cutworm (Lepidoptera: Noctuidae) on tobacco in Ontario. *Canadian Entomologist* 112, 451–456

Cheng, H.H.; Braun, H.E. (1977) Chlorpyrifos, carbaryl, endosulfan, leptophos, and trichlorfon residues on cured tobacco leaves from field-treated tobacco in Ontario. *Canadian Journal of Plant Science* 57, 689–695.

Frank, R.; Braun, H.E.; Holdrinet, M.; Stonefield, K.I.; Elliot, J,M.; Zilkey, B.F.; Vickery, L.S.; Cheng, H.H. (1977) Metal contents and insecticide residues in tobacco soils and cured tobacco leaves collected in Southern Ontario. *Tobacco Science* 21, 74–80.

Heimpel, A.M.; Angus, T.A. (1963) Diseases caused by certain sporeforming bacteria. In: Steinhaus, E.A. *(Ed.),* Insect pathology, vol. 2, New York Academic Press, pp. 21–73.

McMillan, E. (1930) A preliminary study of the polyembryonic cutworm parasite *Berecyntus bakeri* var. *gemma* Girault. M.Sc. Thesis, University of Saskatchewan.

Forficula auricularia L., European Earwig (Dermaptera: Forficulidae)

R.F. MORRIS

Pest Status

This report is an update on the European earwig, *Forficula auricularia* L., and its parasitoid, *Bigonicheta setipennis* (Fallén), in Newfoundland as last reported by Morris (1971). This pest continued to spread throughout Newfoundland, and is now found in the following communities: St. John's, Wedgewood Park, Mt. Pearl, Kilbride, Petty Harbour, Logy Bay, Topsail, Long Pond, Torbay, Renews, Bay Roberts, Clarkes Beach, Grand Bank, Fortune, Grand Falls, Stephenville.

Background

The European earwig was first discovered in Newfoundland at St. John's in 1948. The parasitoid, *Bigonicheta setipennis*, was introduced from British Columbia to Newfoundland in 1951, 1952, and 1953. It was successfully established but parasitism was extremely low during the period 1955–59 (Morris 1971). Because of a slow build-up, supposedly hardier strains of the parasitoid were obtained by the Research Institute at Belleville from Sweden and Switzerland and released at St. John's in 1959, 1961, and 1963.

The use of traps and attractants to complement earwig control with parasitoids were investigated and reported on by Morris (1965).

Releases and Recoveries

No additional releases of *Bigonicheta setipennis* have been made since those recorded by Morris (1971). Studies on its establishment and population increase were continued during the period from 1969 to 1978. Results are shown in Table 5.

Evaluation of Control Attempts

The introduced parasitoid, *Bigonicheta setipennis*, has successfully established itself with a maximum parasitism of 16.4% in 1973. Although the initial parasitism was low, 4.4% in 1969, a steady increase was observed until 1973. There has been a considerable reduction in earwig populations in the study area since 1972, making it impossible to collect as large samples (4000 earwigs) as were used in the earlier years, and this reduction is no doubt due to the high levels of parasitism during the 1972–76 period.

Recommendations

There has been a continuing spread of earwigs from the original site in St. John's with populations now established 420 km and 775 km from St. John's. Studies should be initiated to determine if the parasitoid has been introduced with its host to Grand Bank, Fortune, Grand Falls, and Stephenville. If it has not been introduced naturally, steps should be taken to artificially introduce *Bigonicheta setipennis* to these areas.

Table 5 Recoveries of the parasitoid *Bigonicheta setipennis* (Fallén) from *Forficula auricularia* L. at St. John's, 1969–78.

Year	Dates of observation & % parasitism	Number of earwigs studied	Average % parasitism
1969	13 Aug. (6.0) 21 Aug. (5.0) 28 Aug.–2 Sept. (4.3) 4–5 Sept. (2.2)	4000	4.4
1970	6 Aug. (10.1) 13–20 Aug. (6.3) 31 Aug. (5.1) 8 Sept. (5.1)	4000	6.7
1971	29 July (3.7) 12–13 Aug. (11.3) 18 Aug. (10.2) 27–31 Aug. (6.5)	4000	7.9
1972	4 Aug. (11.4) 21 Aug. (12.3) 28 Aug. (15.9) 6–11 Sept. (10.6)	4000	12.6
1973	9 Aug.–5 Sept. (15.5) 11 Sept. (17.4)	1121	16.4
1974	16 Aug.–20 Sept. (13.7)	548	13.7
1975	14 Aug.–12 Sept. (12.0)	1000	12.0
1976	9 Aug.–26 Aug. (12.0)	1000	12.0
1977	21 July (2.6) 8–12 Aug. (10.0) 19 Aug.–7 Sept. (13.6)	3000	8.7
1978	22–30 Aug. (3.5) 30 Aug.–11 Sept. (7.5)	1800	5.5

Literature Cited

Morris, F. (1965) Note on the control of the European earwig, *Forficula auricularia* L., in Newfoundland with traps and attractants. *Canadian Entomologist* 97, 1075–1076.

Morris, F. (1971) *Forficula auricularia* L., European earwig. In: Biological control programmes against insects and weeds in Canada 1959-1968. *Commonwealth Institute of Biological Control Technical Communication* 4, 18-20.

Hypera postica (Gyllenhal), Alfalfa Weevil (Coleoptera: Curculionidae)

D.G. HARCOURT and J.C. GUPPY

Pest Status

Since its invasion of the mid-Atlantic States during the early 1950s, the eastern strain of the alfalfa weevil, *Hypera postica* (Gyll.), has become the most important pest of alfalfa in eastern North America. Of Eurasian origin, this biotype spread rapidly northward and in little more than a decade reached Ontario and Québec where it began to alarm dairy farmers in the lower Great Lakes region. From 1967 to 1974, populations spread throughout southern Ontario, increasing at geometric rates. However, by the mid-seventies, this pattern of unregulated growth had been checked by the combined action of natural enemies, and during the past 5 years numbers of the pest have been constrained at a density level that fluctuates about the economic damage threshold. As a result, outbreaks have been restricted. However, scattered epicentres of infestations still exist and sudden population flare-ups are prone to occur.

Background

In 1957, the Beneficial Insects Research Laboratory, Northeastern region, U.S.D.A., began to make releases of exotic parasitoids against the weevil. A total of 13 species was released, several of which became established (Abu & Ellis 1976). Discovery of the pest in Ontario in 1967 prompted recolonization here, of those species that had become established in the northeastern United States (Williamson 1971, 1972).

Releases and Recoveries

Parasitoids

Six species of parasitoids were released in various localities in Ontario during 1970 and 1971 (Williamson 1971, 1972): *Bathyplectes anurus* (Thomson), *B. curculionis* (Thomson), *B. stenostigma* (Thomson), *Microctonus aethiopoides* Loan, *M. colesi* Drea, and *Tetrastichus incertus* (Ratzeburg). Recoveries in 1971 included all species except *M. colesi* which was not found until 1978. *B. stenostigma* was not found in follow-up years but all other species became firmly established. Although *B. anurus* was recovered the first year after its release, it has been slow to establish and spread, and its incidence of parasitism is of little importance.

The egg parasitoid, *Patasson luna* (Girault) was not released in Ontario but has spread across the province with its host.

Pathogens

In 1973, an infectious disease caused by the fungus *Entomophthora phytonomi* Arthur (Harcourt *et al.* 1974) was discovered in the Bay of Quinte area of southern Ontario where it caused a spectacular epizootic in populations of the weevil. New to the insect in North America, it was soon found throughout Ontario and in adjacent parts of the United States (Muka & Gyrisco 1976).

Evaluation of Control Attempts

A life table for the pest based on mean values for 15 plot years in the Bay of Quinte area of southern Ontario revealed that during the 5-year period, 1972–76, *P. luna* parasitized trace numbers (<1%) of eggs (Harcourt *et al.* 1977). *B. curculionis* killed 4% of the prepupae, and *T. incertus* a further 2%. The most important mortality agent was the fungus *E. phytonomi* which killed 83% of the larvae during the course of feeding and a further 16% of those in freshly-spun cocoons. It also killed 11% of the pupae. Graphical and key factor analysis demonstrated that the disease was mainly responsible for the population decline that began in the Quinte area in 1974 and was manifest thereafter in other parts of Ontario. Examination of the properties of the various mortality factors revealed that the disease is density dependent but not entirely compensating in its action. At the same time, it was shown that the parasitoids *B. curculionis* and *T. incertus* act in a density independent manner and contribute little to population trend. Consequently, populations have continued to oscillate about the economic threshold with numbers that tend to periodically overshoot levels that growers are able to tolerate.

Recent studies have indicated that the braconid *M. aethiopoides* has the capacity to damp these oscillations. Following its release in Ontario in 1970 and 1971, recoveries were made one or two years later at 5 widely separated points across the province (Dysart & Day 1976). Surveys in 1973 and 1974 by Abu & Ellis (1976) suggested that the parasitoid was slow to spread, but in 1975 intensive sampling at two release sites by the same authors revealed high levels of parasitism. It was discovered throughout the Bay of Quinte area in 1977 and in 1978 that populations of summer adult weevils were heavily attacked.

A detailed survey of 41 southern Ontario counties in 1979 showed that incidence of parasitism caused by the spring generation of *M. aethiopoides* ranged from 13% to 92% and averaged 60% (Harcourt *et al.* 1980). Life table studies at 4 locations during 1980 showed that the attack of overwintered adults by the first brood reduced weevil oviposition by 30% and that parasitism of summer adults by the second brood prior to the onset of diapause was 27%.

M. colesi was first recovered in Ontario in 1978. Life tables in 1979 and 1980 indicated that nearly one-third of the summer adults contained diapausing parasitoids.

Recommendations

As foreshadowed by Loan (1971), *M. aethiopoides* and *M. colesi* have now been colonized in Ontario where they have become important natural control agents. However there is some evidence, at least in the Quinte area, that *M. colesi* may supersede *M. aethiopoides*. Moreover, since it deposits its eggs in the host larval stage it may in turn be forced to compete directly with the fungus disease. Hence, there is some concern about a chain reaction leading to a lower level of biological control with resurgence of the weevil to epidemic population levels.

For the above reasons, it is recommended that the biology, habits, and host relationships of the two euphorine parasitoids be carefully documented.

Literature Cited

Abu, J.F.; Ellis, C.R. (1976) Biology of *Microctonus aethiopoides*, a parasite of the alfalfa weevil, *Hypera postica*, in Ontario. *Environmental Entomology* 5, 1040–1042.

Dysart, R.J.; Day, W.H. (1976) Release and recovery of introduced parasites of the alfalfa weevil in eastern North America. *US Production and Research Report* 167, 61pp.

Harcourt, D.G.; Guppy, J.C.; MacLeod, D.M.; Tyrrell, D. (1974) The fungus *Entomophthora phytonomi* pathogenic to the alfalfa weevil, *Hypera postica*. *Canadian Entomologist* 106, 1295–1300.

Harcourt, D.G.; Guppy, J.C.; Binns, M.R. (1977) The analysis of intrageneration change in eastern Ontario populations of the alfalfa weevil, *Hypera postica* (Coleoptera: Curculionidae). *Canadian Entomologist* 109, 1521–1534.

Harcourt, D.G.; Ellis, C.R.; Guppy, J.C. (1980) Distribution of *Microctonus aethiopoides*, a parasitoid of the alfalfa weevil (Coleoptera: Curculionidae) in Ontario. *Proceedings of the Entomological Society of Ontario* (1979) 110, 35–39.

Loan, C.C. (1971) *Sitona cylindricollis* Fahr., sweetclover weevil, and *Hypera postica* (Gyll.), alfalfa weevil (Coleoptera: Curculionidae). *In*: Biological control programmes against insects and weeds in Canada 1959–68. *Commonwealth Institute of Biological Control Technical Communication* 4, 43–46.

Muka, A.A.; Gyrisco, G.G. (1976) Report presented at 13th Northeast Regional Alfalfa Insects Conference, Newark, Delaware.

Williamson, G.D. (1971) Insect liberations in Canada, 1970. Parasites and predators. *Canada Agriculture Liberation Bulletin* 34, 16 pp.

Williamson, G.D. (1972) Insect liberations in Canada, 1971. Parasites and predators. *Canada Agriculture Liberation Bulletin* 35, 16 pp.

Lygus spp., Plant Bugs (Heteroptera: Miridae)

C.H. CRAIG and C.C. LOAN

Pest Status

About 30 species of *Lygus* are known to occur from coast to coast in Canada (Kelton 1975). Most of the species feed on a variety of herbaceous and woody plants, some are host specific or are limited to related plant species. Many species of *Lygus* are economically important as pests of fruits, vegetables, and tobacco crops, and of legume crops, primarily those grown for seed. Of the dozen species found on alfalfa the most abundant and important as pests are *L. borealis* Kelton and *L. lineolaris* (Palisot de Beauvois), with *L. unctuosus* (Kelton), *L. desertinus* Knight, and *L. shulli* Knight of lesser importance.

Lygus overwinter as adults in the shelter of plant litter on the soil surface. In spring they feed on the early plants and move to alfalfa when it becomes suitable for feeding and oviposition.

In Canada the number of generations of *Lygus* per year is not well documented because nymphal taxonomy is not known. It is generally conceded, however, that the species are univoltine in northern latitudes and bivoltine, or largely so, south of about latitude 51°N.

Lygus feeding injures the vegetative and reproductive parts of the plant causing stunting, bud-blast and premature flower drop, and reduced yield and quality of seed. In most years and in most alfalfa seed fields, *Lygus* infestations exceed the economic threshold and control measures are essential.

Background

In Poland, *Lygus rugulipennis* Poppius, the dominant species on alfalfa, is parasitized by the braconids *Peristenus digoneutis* Loan, *P. stygicus* Loan, and *P. rubricollis* (Thomson). The former two are bivoltine and attack both generations of host; the latter is univoltine and attacks only the first generation host (Bilewicz-Pawinska 1969, 1975).

In Canada, at Belleville, Ontario, *L. lineolaris*, in the first generation on forage legumes, is parasitized by *Peristenus pallipes* (Curtis) and in the second generation on several weed species, by *P. pseudopallipes* (Loan) (Loan 1965, 1970).

In 1975 in a survey of first-generation *Lygus* populations in alfalfa seed fields in Saskatchewan and Alberta, *P. pallipes* was the only euphorine parasitoid recovered. Incidence ranged from 3% to 49%, with an average of 21% in the southern shortgrass prairie region, 28% in the central mixed prairie region, and 11% in the northern parkland-mixed forest region (Loan & Craig 1976). Surveys since that time have noted the occurrence of euphorine parasitism probably by *P. pallipes*, in second-generation *Lygus* in the southern prairie region (Craig, 1980, unpublished data).

Releases and Recoveries

A programme to supplement the natural parasitism in *Lygus* populations in western Canada was started in 1977 with the first collection and shipment to Ottawa of parasitoids of European *Lygus*, and in 1978 with the first liberations of these in western Canada. Table 6 shows the progression of liberations in the period 1978–81 inclusive.

Table 6 Liberations of *Peristenus* spp. in alfalfa against *Lygus* spp. in western Canada.

Date of liberation	Location of liberation site	Species	Number
7 June 1978	Moose Jaw, Saskatchewan; 50°N; shortgrass prairie; dry	*P. digoneutis*	197
9 June 1978 13 June 1978 21 June 1978	Saskatoon, Saskatchewan; 52°N; mixed prairie; dry	*P. digoneutis*	177
14 June 1979 21 June 1979	Tilley, Alberta; 50°30′N; shortgrass prairie; irrigated	*P. digoneutis* *P. stygicus*	541 41
29 June 1979	Shellbrook, Saskatchewan; 53°N; parkland-mixed forest; dry	*P. digoneutis* *P. stygicus*	159 23
29 May 1980	Ardath, Saskatchewan; 51°30′N; mixed prairie; dry	*P. digoneutis*	167
3 June 1980	Shellbrook, Saskatchewan (supplement to 1979 release)	*P. digoneutis*	261
10 June 1980	Saskatoon, Saskatchewan mixed prairie; irrigated	*P. digoneutis*	80
10 June 1980	Saskatoon, Saskatchewan (supplement to 1978 release)	*P. digoneutis*	138
6 June 1981	Yellow Creek, Saskatchewan; 53°N parkland; dry	*P. digoneutis*	125
6 June 1981	Saskatoon, Saskatchewan (supplement to 1980 release)	*P. digoneutis*	115

The imported parasitoids were released into *Lygus* populations breeding in alfalfa fields used for seed production. The lifespan of the sites for sampling purposes is about 5 years, and during that time they are subject to normal alfalfa seed cropping practices. Intensive sampling for presence of adult *Peristenus* and mass collection of host nymphs for parasitoid rearing was started the year following the initial liberations and continued for at least three ensuing years. In 1979–80, a total of 274 parasitoid adults, either captured at the liberation sites or reared from *Lygus* hosts collected from alfalfa at the liberation sites, were the native species, *Peristenus pallipes*. No introduced parasitoids have been recovered.

Recommendations

The introduction and liberation of *Peristenus* parasitoids will terminate in the spring of 1981. During the next few years attempts will be made to recover the introduced parasitoids and to evaluate the programme.

Literature Cited

Bilewicz-Pawinska, T. (1969) Natural limitation of *Lygus rugulipennis* Popp. by groups of *Leiophron pallipes* Curtis on the rye crop fields. *Ekologia Polska, Series A* 17, 811–815.

Bilewicz-Pawinska, T. (1975) Distribution of insect parasites *Peristenus* Foerster and *Mesochorus* Gravenhorst in Poland. *Bulletin de l'Académie Polonaise des Sciences* 23, 823–827.

Kelton, L.A. (1975) The *Lygus* bugs (genus *Lygus* Hahn) of North America (Heteroptera: Miridae). *Memoirs of the Entomological Society of Canada* 95, 101 pp.

Loan, C.C. (1965) Life cycle and development of *Leiophron pallipes* Curtis (Hymenoptera: Braconidae, Euphorinae) in five mirid hosts in the Belleville district. *Proceedings of the Entomological Society of Ontario* (1964) 95, 115–121.

Loan, C.C. (1970) The new parasites of the tarnished plant bug in Ontario: *Leiophron pseudopallipes* and *Euphoriana lygivora* (Hymenoptera: Braconidae, Euphorinae). *Proceedings of the Entomological Society of Ontario* 100, 188–195.

Loan, C.C.; Craig, C.H. (1976) Euphorine parasitism of *Lygus* spp. in alfalfa in western Canada (Hymenoptera: Braconidae; Heteroptera: Miridae). *Naturaliste Canadian* 103, 497–500.

Mamestra configurata Walker, Bertha Armyworm (Lepidoptera: Noctuidae)

W.J. TURNOCK

Introduction

The bertha armyworm, *Mamestra configurata* Walker, was first recognized as a major pest of the rapeseed and canola crops, *Brassica napus* (L.) and *B. campestris* L., on the Canadian prairies in 1971. Little was known of the life history and natural control of this species. In 1972, research programmes were initiated at Agriculture Canada Research Stations in western Canada. Soon after, the Commonwealth Institute of Biological Control station at Delémont, Switzerland, began a study of the closely related European species, *Mamestra brassicae* (L.). To date, no decision has been made to introduce parasitoids of *M. brassicae* to Canada. This report summarizes information on the natural enemies of *M. configurata* and *M. brassicae* and suggests future actions in biological control of the former species in Canada.

Pest Status

M. configurata is native to North America and in Canada occurs from Manitoba to British Columbia (Wylie & Bucher 1977). Its native hosts are unknown, but it has been reported to feed on a wide variety of cultivated plants and introduced weeds (Beirne 1971). On the Canadian prairies sporadic infestations occurred on various crops before 1940. Since then attacks have largely been confined to fields of rapeseed and canola (Turnock & Philip 1979). From 1940 to 1970 the outbreaks were local and of short duration. From 1971 to 1974 an outbreak of unprecedented severity occurred from western Manitoba across the parkland region to Edmonton, and from Lethbridge to the Peace River District in Alberta. Insecticides were applied to about 38 000 ha (17.7% of the seeded area) in 1971, and about 31 000 ha (23.3% of the seeded area) in 1972. The outbreak declined in 1973 and damaging infestations were absent by 1975. Populations increased again in 1978 and 1979, and in 1980 about 49 000 ha (2.4% of the seeded area) were sprayed.

The bertha armyworm is univoltine in Canada and overwinters in the pupal stage. Adults emerge from the first week of June to early August, in most years 50% of the adults have emerged by mid-July. Egg-laying, on the underside of the leaves of host plants, begins soon after emergence and the eggs hatch about 1 week later. There are six larval instars. The larvae feed mainly on the leaves, but the older instars may damage flowers and pods. Fully-fed larvae enter the soil in late August and September and pupate at depths down to 15 cm.

Background

The early reports of bertha armyworm infestations do not include lists of parasitoids or diseases (Wylie & Bucher 1977). Since 1972, collections of larvae have yielded 15 species of parasitoids (Table 7). Wylie (1979) reared an additional 3 species from larvae and 5 species from pupae that had been held in field cages in an area where the bertha armyworm has never been found. Neither egg nor pupal parasitoids has been found in field collections.

Table 7 Parasitoids recovered from field-collected larvae of *Mamestra configurata* Walker in Canada.

Hymenoptera	Diptera
Ichneumonidae	Tachinidae
Banchus flavescens Cress. (M,S,A)*	*Athrycia cinerea* (Coq.) (M,S,A)* *Mericia ampelus* (Wlk.) (M,S,A)
Braconidae	*Exorista mella* (Wlk.) (S,A)
Apanteles xylinus (Say) (M)	*Phryxe pecosensis*(Tnsd.) (M)
Cotesia laevicips (Ashm.) (M)	*P. vulgaris* (Fall.) (M) *Chaetogena* sp. (A)
Eulophidae	*C.* sp. (*claripennis* (Macq.) group) (A)
Eulophus nr. *nebulosus* (Prov.) (M)	*Lespesia* sp. (A)
Euplectrus bicolor Swed. (M)	*Winthemia rufopicta* (Bigot) (M) *W. quadripustulata* (F.) (M)

* Location of collection points in parenthesis. (M = Manitoba, S = Saskatchewan, A = Alberta).

Banchus flavescens Cress. was the most abundant species regardless of host abundance (Table 8). This univoltine parasitoid lays its eggs in the early larval instars and completes its development before the host would normally pupate.

Table 8 Parasitism (%) by *Banchus flavescens* Cress., *Athrycia cinerea* (Coq.), *Mericia ampelus* (Wlk.), and other species on *Mamestra configurata* Walker in Manitoba, Saskatchewan, and Alberta, 1972–80.

				Parasitism (%)			
Year(s)	Province	No. Fields	No. Larvae	*Banchus*	*Athrycia*	*Mericia*	Other
1972–74	Manitoba	119	1915	28.3	20.7	0.4	0.1
	Saskatchewan	1	59	47.5	1.7	3.4	1.7
	Alberta	5	506	42.5	21.7	2.4	0
1975–78	Manitoba	152	100	44.0	25.0	0	2.0
1979	Manitoba	43	410	6.1	9.3	0	1.7
	Saskatchewan*	2	1273	14.0	29.0**		
1980	Manitoba	51	2154	26.3	6.2	0.04	0.3
	Alberta	3	775	25.3	4.4	0	0

* Data provided by A.P. Arthur, Agriculture Canada, Research Station, Saskatoon.
** Mainly *A. cinerea*.

Athrycia cinerea (Coq.) usually was less abundant than *B. flavescens* when host populations were low (Table 8), but it may be as abundant in some fields during outbreaks. *A. cinerea* is univoltine in the Prairie Provinces. The eggs are attached to the integument of larvae in the fourth to sixth instars, the larvae usually develop gregariously, and they mature before the host completes its feeding (Wylie 1977*a*).

Mericia ampelus (Wlk.) was found in a very small percentage of bertha armyworm larvae collected during outbreaks, but has not been collected between outbreaks (Table 8). This species attaches its eggs to the leaves of plants and the larvae hatch 1–2 minutes after oviposition. It appears to be bivoltine in Canada, but only the second generation parasitizes bertha armyworm. Larval survival is low, suggesting that *Mamestra configurata* is not a preferred host (Wylie 1977*b*).

Pathogens have not been a major cause of mortality in the bertha armyworm. Wylie & Bucher (1977) identified fungus pathogens of the genus *Entomophthora* and a nuclear polyhedrosis virus as the main larval diseases in 1972. In addition, a protozoan parasite, possibly of the genus *Nosema*, was recovered from a few larvae.

The cabbage moth, *Mamestra brassicae* (L.), is similar to *M. configurata* in its life history and phenology. *M. brassicae* feeds on a wide range of plants and is a major pest of cabbage in northern Eurasia. In western Europe (Switzerland, Austria, Germany) and in the warmer climatic zones of the Soviet Union (Kiev, Kharkov, Voronetz, and Krasnodar regions), it is bivoltine (CIBC Annual Project Statements 1974–80, Dzhuran 1979, Agarkov 1974, Stenin 1974, Drozda & Garnaga 1978, Tsybul'ko 1973). In cooler parts of the Soviet Union (Moscow, Gorky, Leningrad, Volgoda (Perm), Tomsk, Altai, Irkutsk, and Sakhalin regions), it is univoltine (Hori 1935, Kopvillem 1960, Shchetinin 1974, Prokov'eva 1976, Kostylyova 1979). The univoltine populations lie north and the bivoltine populations lie south of the annual isotherm for 2 200° days over 10°C (Fig. 1).

The parasitoid complex of *M. brassicae* consists of 47 species, of which 20 occur in the USSR (Razumov 1972, Kopvillem 1962). However, only 23 species are reported in the CIBC Annual Project Statements and recent literature from the Soviet Union (Table 9). Only two species, *Ernestia consobrina* (Mg.) and *Exetastes cinctipes* (Retz.), are abundant in both areas.

Ernestia consobrina occupies a similar position in the guild of parasitoids to that occupied by *Mericia ampelus* in North America. *E. consobrina* larviposits on the food plant of the host. Parasitism is most successful in third-instar larvae and *M. brassicae* is the major host (Kopvillem 1960, 1962, CIBC Annual Project Statement 1974); *E. consobrina* is the most important parasitoid of *Mamestra brassicae* in most of the Soviet Union (Kopvillem 1962), but not in western Europe (CIBC Annual Project Statements 1974–80). It does not occur in Primorskiy (far east) (Slabospitskii 1980). In the areas of the Soviet Union where both host and parasitoid are bivoltine, it appears to have problems in synchronization with its host. In the Kiev region, Dzhuran (1979) reports a large decrease in parasitism from the first to the second host generation while Drozda & Garnaga (1978) note an increase in the second generation. In univoltine areas, *E. consobrina* appears to be a more effective parasitoid than in bivoltine areas. High levels of parasitism are reported from the Leningrad, Moscow, and Volgoda (Perm) regions (Kopvillem 1960), the Gorky region (Razumov 1972), and the Irkutsk region of western Siberia (Shchetinin 1974).

Exetastes cinctipes occupies a similar position in the guild of parasitoids to that occupied by *B. flavescens* in North America. *E. cinctipes* is bivoltine in western Europe, where it parasitizes larvae of the second generation (CIBC Annual Project Statement 1974). In the Moscow region, it is univoltine (Kopvillem 1960). This ichneumon mainly parasitizes third-instar larvae and appears to be well synchronized with its host (Kopvillem 1960). *E. cinctipes* is a common parasitoid in both western Europe and the Soviet Union. In western Europe and areas of the Soviet Union with bivoltine hosts, it is more important than *E. consobrina* (Tsybul'ko 1973, Agarkov 1974, CIBC Annual Project Statements 1974–78), but in cooler parts of the Soviet Union it is less abundant (Kopvillem 1962, Razumov 1972, Shchetinin 1974).

Microplitis mediator (Hal.), like *B. flavescens* in North America, attacks host larvae in the early instars but unlike it, *M. mediator* is multivoltine (CIBC Annual Project

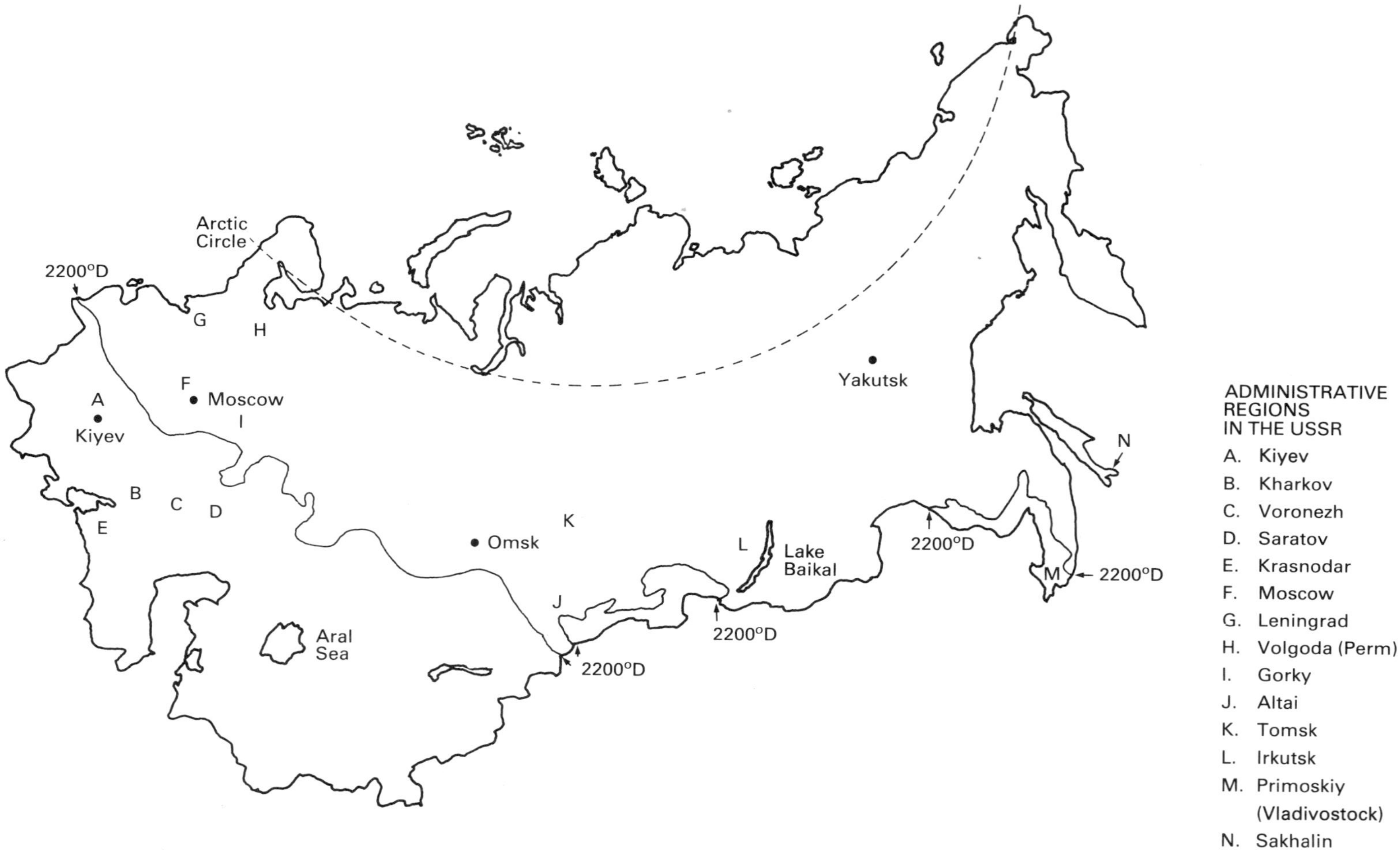

Fig. 1. Location of Administrative Regions in the USSR from which reports of parasitism of *Mamestra brassicae* Wlk. were available and the relation of these locations to the annual isotherm for 2 200° days over 10°C. Adapted from figures in Central Intelligence Agency (1974)

Table 9 Insect parasitoids of *Mamestra brassicae* Walker in western Europe and the USSR (See Fig. 1 for location of regions in the USSR).

	Region of Occurrence	
Family and Species	Host Bivoltine	Host Univoltine
Tachinidae		
Ernestia consobrina (Mg.)	W. Europe, Krasnodar, Kiev, Saratov	Leningrad, Volgoda, Moscow, Gorky, Tomsk
Phryxe vulgaris (Fall.)	W. Europe	Tomsk
Athrycia impressa (Wlp.)		Tomsk
A. erythrocera R.D.	Primorskiy	Tomsk
Siphona cristata (F.)		Tomsk
S. flavifrons Staeger	W. Europe	
Winthemia quadripustulata (F.)		Moscow
Exorista larvarum (L.) (*Tachina larvarum* (L.))	Krasnodar, Veronezh W. Europe	Moscow
Blondelia nigripes (Fall.)	W. Europe	Moscow
Ceromasia sp.		Sakhalin
Ichneumonidae		
Exetastes cinctipes (Retz.)	W. Europe, Kharkov	Moscow, Gorky, Tomsk, Sakhalin
Therion giganteum (Grav.)	Krasnodar	
Campoplegini sp.	W. Europe	
Stenichneumon culpator (Schrank)		Sakhalin
Braconidae		
Microplitis mediator (Hal.)	W. Europe	
M. tuberculifer (Wesm.)	Kharkov, Primorskiy	
Cotesia congesta (Nees)		Gorky
C. glomerata (L.)	Krasnodar, Voronezh	
Meteorus leviventris (Wesm.)		Sakhalin
Eulophidae		
Eulophus larvarum L.	Krasnodar	
Euplectrus sp.		Sakhalin
Euplectrus bicolor Swed.	Krasnodar, Kharkov	
Trichogrammatidae		
Trichogramma evanescens Westw.	W. Europe, Krasnodar, Voronezh, Kharkov, Kiev	Gorky, Irkutsk, Altai

Statement 1978). *M. mediator* is the most abundant parasitoid of *Mamestra brassicae* in Europe (CIBC Annual Project Statements 1974–80) but is not reported from the Soviet Union. It does not enter dispause to overwinter (CIBC Annual Project Statement 1980) and this characteristic may explain its absence from the Soviet Union. *Microplitis tuberculifer* (Wesm.) was the most abundant parasitoid in far eastern regions (Slabospitskii 1980), where *Mamestra brassicae* is bivoltine. *M. brassicae* is univoltine on the island of Sakhaline and is attacked by 5 species of parasitoids (Hori 1935). Only one of these, *E. cinctipes*, also occurs in other parts of the Soviet Union (Table 7).

An egg parasitoid, *Trichogramma evanescens* Westw., occurs in western Europe and the Voronezh, Kharkov, Krasnodar, Gorky, Irkutsk, and Altai regions of the USSR (Razumov 1972, Tsybul'ko 1973, Agarkov 1974, Prokof'eva 1976, CIBC Annual Project

Statement 1978, Kostylyova 1979). In the regions where the host is univoltine, *T. evanescens* generally is very rare (Prokof'eva 1976). It is used as an inundative biological control agent in the USSR.

The major components of the parasitoid guilds of *M. configurata* and *M. brassicae* attacking the early instar larvae include one ichneumon, *B. flavescens*, in Canada; and two ichneumons, *E. cintipes* and *Microplitis mediator,* and one tachinid, *E. consobrina,* in western Europe. In Canada, one tachinid, *A. cinerea*, attacks the late instar larvae of *Mamestra configurata* but no species of comparable effectiveness is reported from Eurasia. No egg parasitoids occur in Canada whereas *T. evanescens* occurs at generally low levels in Eurasia.

Recommendations

The differences in the major components of the parasitoid guilds of *M. configurata* and of *M. brassicae* lead to the following recommendations for parasitoid introductions into Canada:

(1) Parasitoids attacking the early larval instars

(a) Ichneumons: *B. flavescens* in Canada is more effective than either *E. cinctipes* or *Microplitis mediator* in Eurasia. *M. mediator* does not go into diapause and does not occur in those parts of the Soviet Union where *Mamestra brassicae* is univoltine. Neither *E. cinctipes* nor *Microplitis mediator* should be considered for introduction into Canada.

(b) Tachinids: *Mamestra configurata* does not have an effective species attacking the early larval instars as *Mericia ampelus* is poorly adapted. *E. consobrina* is an important parasitoid in Eurasia. It successfully parasitizes *Mamestra configurata* in the laboratory (Wylie 1977*a*) and has potential as a biological control agent. Material for releases in Canada should be obtained from areas where *M. brassicae* is univoltine.

(2) Parasitoids attacking late larval instars

In Canada, *A. cinerea* is common, particularly during outbreaks, but no equivalent species is reported as a major parasitoid of *M. brassicae* in Eurasia. *A. cinerea* could be considered for introduction into the Soviet Union, perhaps as part of an exchange of material involving *E. consobrina*. *M. brassicae* was a suitable host for *A. cinerea* in laboratory tests (Wylie 1977*a*).

(3) Egg parasitoids

The very low level of egg parasitism in *M. brassicae* does not suggest that *T. evanescens* would contribute significantly to the control of *M. configurata* in Canada. However, the local form of *T. evanescens* reported from the Irkutsk region (Kostylyova 1979) should be further investigated. If Canada should develop facilities for producing *Trichogramma* for inundative release, importation of the "Irkutsk" form would be desirable.

Acknowledgements

Copies of the cited literature from the USSR were made available through the efforts of Mr A.I. Sylchenko, Ministry of Agriculture, Moscow, and Dr G. Tsubulskaja, Ukrainian Plant Protection Research Institute, Kiev. Their cooperation is greatly appreciated. Dr A.P. Arthur, Agriculture Canada, Saskatoon, kindly permitted the use of his data on parasitism of *M. configurata* in Saskatchewan.

Literature Cited

Agarkov, V.M. (1974) Effect of entomophages on the abundance of the cabbage moth during a period of low abundance. *Bulletin of the All-Union Scientific Research Institute for the Protection of Plants* 27, 3–6 (Translated from Russian).

Beirne, B.P. (1971) Pest insects of annual crop plants in Canada. *Memoirs of the Entomological Society of Canada* 78, 124 pp.

Central Intelligence Agency (1974) USSR agricultural atlas. U.S. Government Printing Office, 59 pp.

Drozda, V.F.; Garnaga, N.G. (1978) Some biological and ecological characteristics of the tachinid fly, *Ernestia consobrina* (Mg.), a parasite of the cabbage moth caterpillar. *Scientific Paper of the Ukrainian Agricultural Academy* 209, 23–25 (Translated from Russian).

Dzhuran, V.N. (1979) The development of biological means for controlling the cabbage moth. *Scientific Paper of the Ukrainian Agricultural Academy* 230, 25–26 (Translated from Russian).

Hori, M. (1935) The cabbage moth (*Barathra brassicae* L.) in southern Saghalien. *Report of the Saghalien Central Experiment Station Service* 1(3) (Translated from Japanese).

Kopvillem, Kh.G. (1960) Parasites of the cabbage moth (*Barathra brassicae* L.) and the diamond-back moth (*Plutella maculipennis* Curt.) in the Moscow region. *Entomologicheskoe Obozrenie* 39, 584–592.

Kopvillem, Kh.G. (1962) Parasites of the cabbage moth and diamondback moth in the Moscow Region. Biological Methods of Controlling Pests and Diseases of Agricultural Crops 1, 89–113 (Translated from Russian).

Kostylyova, Ye.V. (1979) Experimental application of *Trichogramma* against the caggage moth in the Bratsk district of the Irkutsk region. In: Problems of Plant Protection in Eastern Siberia, Irkutsk, USSR, p. 34–39 (Translated from Russian).

Prokof'eva, N.A. (1976) *Trichogramma* in cabbage fields. *Zaschita Rastenii* 9, 19–20 (Translated from Russian).

Razumov, V.P. (1972) Parasites of the cabbage moth (*Mamestra brassicae* (L.)) in Gorky Oblast. *Proceedings of the Gorky Agricultural Institute* 53, 9–12 (Translated from Russian).

Shchetinin, Yu.V. (1974) Parasites of the cabbage moth and diamond-back moth in the Tomsk region. In: The problems of entomology in Siberia, Novsibirsk, USSR, pp. 131-132 (Translated from Russian).

Slabospitskii, A.I. (1980) (Insect enemies of cabbage pests.) *Zashchita Rastenii* 5, 23 (in Russian).

Stenin, V.I. (1974) The parasites, predators and polyhedrosis of the cabbage moth in the Kuban. *Proceedings of the Kuibyshev Agricultural Institute* 125, 53–55 (Translated from Russian).

Tsybul'ko, V.I. (1973) The main entomophages of the cabbage moth in Kharkov Province. *Proceedings of the Kharkov Agricultural Institute* 182, 15–25 (Translated from Russian).

Turnock, W.J.; Philip, H.G. (1979) The outbreak of bertha armyworm, *Mamestra configurata* (Noctuidae: Lepidoptera) in Alberta 1971 to 1975. *Manitoba Entomologist* 11(1977), 10–21.

Wylie, H.G. (1977*a*) Observations on *Athrycia cinerea* (Diptera: Tachinidae), a parasite of *Mamestra configurata* (Lepidoptera: Noctuidae). *Canadian Entomologist* 109, 747–754.

Wylie, H.G. (1977*b*) Observations on *Mericia ampelus* (Diptera: Tachinidae), an occasional parasite of bertha armyworm, *Mamestra configurata* (Lepidoptera: Noctuidae) in western Canada. *Canadian Entomologist* 109, 1023–1024.

Wylie, H.G. (1979) Insect parasites reared from artificial field populations near Winnipeg, Manitoba. *Manitoba Entomologist* 11(1977), 50–55.

Wylie, H.G., Bucher, G.E. (1977) The bertha armyworm, *Mamestra configurata* (Lepidoptera: Noctuidae). Mortality of immature stages on the rape crop, 1972–1975. *Canadian Entomologist* 109, 823–837.

Manduca quinquemaculata (Haworth), Tomato Hornworm (Lepidoptera: Sphingidae)

H.H. CHENG

Pest Status

The tomato hornworm, *Manduca quinquemaculata* (Haworth), is principally a pest of tomato and tobacco plants in Canada. The distribution of this species is North and South American. On tobacco, the first-instar and second-instar larvae make many small holes in the leaves, later they eat larger areas of the lamina, and by the last instar the larvae may strip the top and mid-leaves of the plants leaving only the midribs and main leaf veins. In Canada, a moderate to high population level of this pest may cause 0.5% to 2.7% loss of the marketable tobacco in untreated fields (Cheng 1977*a*).

The tomato hornworm has one complete generation and part of a second generation each year under southwestern Ontario climatic conditions. Only the first generation of this species causes appreciable damage to tobacco. The insect overwinters as a pupa in the soil. Moth emergence occurs from late June to September. The adults fly at dusk and lay their eggs on the under side of the tobacco leaves during the same period. Eggs hatch within four to six days and the larvae dig into the soil to a depth of 10 to 12 cm and pupate. Most of the pupae remain in the soil until the following year, but some moths may emerge from late August to September to start a second generation. Ten-year blacklight trap studies showed that population levels of *M. quinquemaculata* in southwestern Ontario are generally low to moderate (unpublished data).

Background

The tomato hornworm can be successfully controlled using a number of different chemical insecticides, however, the use of chemical insecticides on tobacco presents a number of problems that may assume greater importance in the future. In particular, applications of certain insecticides on tobacco during late July and August result in insecticide residues (Cheng & Braun 1977), which have undesirable effects on the quality of tobacco and may put the Canadian tobacco export trade in serious jeopardy. Consequently, a search for and development of alternative control methods for hornworms has been carried out during the past 20 years. Among the more promising of the biological agents tested for the control of this pest are strains of the spore forming bacterium *Bacillus thuringiensis* Berliner, which was found to be as effective as chemical insecticides, and has been used successfully for controlling hornworms on tobacco (Guthrie *et al.* 1959, Creighton *et al.* 1961, Begg 1964, Bucher & Cheng 1971, Cheng 1973, 1977*b*, 1978).

Two commercial preparations of *B. thuringiensis*, Dipel® and Thuricide®-HPC, have been registered and recommended for use in controlling hornworms on tobacco since 1973 (Anonymous 1973). A survey conducted in southwestern Ontario in 1977, 1978, and 1979 indicated that about 20% of growers applied *B. thuringiensis* preparations for control of hornworms. When aphids and hornworms occurred in the same period, most of the growers would apply chemical insecticides or a tank mix combination of *B. thuringiensis* preparation for hornworm and a chemical insecticide for aphids because *B. thuringiensis* preparations gave almost no control of aphids.

The most important insect parasitoids of the tomato hornworm are *Apanteles* spp., which parasitize 10–20% of the hornworm larvae in tobacco fields each year. *Apanteles* spp. kill the fourth-instar and fifth-instar larvae and essentially prevent the hornworm from causing maximum damage to the tobacco crop.

Evaluation of Control Attempts

Preparations of *B. thuringiensis* control tomato hornworms as well as chemical insecticides (Bucher & Cheng 1971, Cheng 1973, 1977*b*, 1978) and for approximately the same cost of $20 to $25 per hectare for the material. *B. thuringiensis* has the merits of being nontoxic to man, plants, and non-target animals including beneficial insects such as parasitoids and predators. Thus it should be used by growers when an economic population level of the tomato hornworm larvae occurs in the tobacco fields.

Under the present system of annual cropping of flue-cured tobacco in Ontario, sequential applications of insecticides for hornworm control and of sucker control agents within short periods are common practices in tobacco production. To reduce the cost of control measures, attempts have been made to mix recommended insecticides and sucker control agents in the same tank and apply them to tobacco plants for control of hornworms and suckers in one spray operation. Five year tests indicated that all recommended insecticides used for hornworm control can be mixed with sucker control agents in the same tank before application without losing their efficacy and caused no effects on the yield and quality of flue-cured tobacco in Ontario (Cheng 1977*b*, 1978, 1980).

At the Delhi Research Station, monitoring and economic threshold data have been developed for *M. quinquemaculata* (Cheng 1977*a*). Since this pest overwinters as a pupa in the soil, early detection methods involve the capture of moths in light traps. Larval densities are monitored from mid-July to mid-August and insecticide applications are recommended when 5 or more large larvae (larger than 3 cm) are found per 100 plants. Generally, *M. quinquemaculata* can be controlled for the whole growing season by a single application of *B. thuringiensis* or one of the recommended chemical insecticides.

Recommendations

Detailed studies on the bionomics of the tomato hornworm in Ontario should be conducted. Although preparations of *B. thuringiensis* have provided excellent control of the hornworm larvae, the rapid drop of larvae from plants treated with chemical insecticides is much more dramatic than the cessation of feeding and the slow death of *B.t.*-infected larvae. Growers, therefore, should be better informed of the merits of *B. thuringiensis* to offset their natural preference for a chemical insecticide that has more spectacular effects. A similar approach should be made in other areas of Canada where the tomato hornworm is a problem.

Literature Cited

Anonymous (1973) 1973 Tobacco production recommendations. Ontario Ministry Agriculture and Food, Publ. 298, 32pp.

Begg, J.A. (1964) Microbial and chemical control of hornworms attacking tobacco in Ontario. *Journal of Economic Entomology* 57, 646–649.

Bucher, G.E.; Cheng, H.H. (1971) Comparison of *Bacillus thuringiensis* preparations with carbaryl for hornworm (Lepidoptera: Sphingidae) control on tobacco. *Canadian Entomologist* 103, 142-144.

Cheng, H.H. (1973) Microplot test using microbial and chemical insecticides for control of tomato hornworms on tobacco in Ontario. *The Lighter* 43(3), 10–13.

Cheng, H.H. (1977*a*) Flue-cured tobacco losses caused by the tomato hornworm, *Manduca quinquemaculata* (Lepidoptera: Sphingidae), at various infestation levels in Ontario. *Canadian Entomologist* 109, 1091–1095.

Cheng, H.H. (1977*b*) Insecticides and sucker control chemicals: compatibility and effects on green peach aphid, tomato hornworm and suckers on flue-cured tobacco in Ontario. *Tobacco Science* 21, 108–111.

Cheng, H.H. (1978) Pesticides play important role. *Canadian Tobacco Grower* 26(8), 31–33.

Cheng, H.H. (1980) Effects of tank-mixed combinations of insecticides and sucker control agents on efficacy and on the yield and quality of flue-cured tobacco. *Proceedings of the Entomological Society of Ontario* 111, 7–12.

Cheng, H.H.; Braun, H.E. (1977) Chlorpyrifos, carbaryl, endosulfan, leptophos, and trichlorfon residues on cured tobacco leaves from field-treated tobacco in Ontario. *Canadian Journal of Plant Science* 57, 689–695.

Creighton, C.S.; Kinard, W.S.; Allen, N. (1961) Effectiveness of *Bacillus thuringiensis* and several chemical insecticides for control of budworms and hornworms on tobacco. *Journal of Economic Entomology* 54, 1112–1114.

Guthrie, F.E.; Rabb, R.L.; Bowery, T.G. (1959) Evaluation of candidate insecticides and insect pathogens for tobacco hornworm control, 1956–1958. *Journal of Economic Entomology* 52, 798–804.

Melanoplus spp., *Camnula pellucida* (Scudder), and other Grasshoppers (Orthoptera: Acrididae)

A.B. EWEN and M.K. MUKERJI

Grasshoppers have been important destructive pests since crop production began on the Canadian prairies. Four species: the migratory grasshopper, *Melanoplus sanguinipes* (Fab.); the twostriped grasshopper, *M. bivittatus* (Say); the Packard grasshopper, *M. packardii* Scudder; and the clearwinged grasshopper, *Camnula pellucida* (Scudder); generally are considered to be the most important. Natural enemies and weather conditions cause marked fluctuations in their populations from year to year from one area to another, and several population peaks have been recorded during the more than thirty years that grasshopper surveys have been conducted in Saskatchewan (Gage & Mukerji 1977). High populations and significant damage to crops have been associated with early egg hatch due to an early, warm, and dry spring (Randell & Mukerji 1974).

Grasshoppers host a variety of pathogenic micro-organisms including fungi, bacteria, viruses, and protozoans. Epizootics of fungal pathogens can produce drastic reductions in grasshopper populations. Pickford & Riegert (1964) showed that *Entomophthora grylli* Fres. was a major factor responsible for significant reductions in grasshopper populations, espcially *C. pellucida*, in Saskatchewan in 1963. Although there have been other incidents of fungal epizootics, these pathogens have not given consistent control of grasshopper populations because of their restrictive requirements of specific moisture and temperature conditions for initiation and dissemination of infection in the host populations (Henry 1970). Most bacteria associated with grasshoppers are facultative pathogens which cannot multiply in the host gut but require direct passage into the haemocoele (Bucher 1960). Also, bacteria and many of the fungi pathogenic in grasshoppers are not suitable microbial control agents because they are associated with or pathogenic in other insects and animals (Henry 1970).

Viral and protozoan pathogens have received more attention as potential biological control agents for grasshoppers. An entomopox virus (Henry & Jutila 1966) and a crystalline-array virus of the picornovirus group (Jutila *et al.* 1970) have been isolated from *M. sanguinipes* and *M. bivittatus*, respectively. Although these viruses are uncommon in grasshoppers under normal conditions, observations indicate that they may be capable of lowering the population density of their hosts. Prolonged effects of these pathogens have yet to be assessed. Of the protozoans, *Malameba locustae* (King & Taylor), several species of gregarines, three species of *Nosema (N. locustae* Canning, *N. acridophagus* Henry, and *N. cuneatum* Henry), and a number of other microsporidians have been reported or described in grasshoppers. *Malameba locustae* and the gregarines (the most common protozoans in grasshoppers) cause chronic infections but have little immediate effect on their hosts.

Most research effort in recent years has been directed towards using the *Nosema* spp. as biological control agents for the management of grasshopper populations and, of these, *N. locustae* has received the most attention. In 1980 a quantitative study of the effect of *N. locustae* as a control agent against grasshopper populations was carried out in a short grass pasture in east central Saskatchewan (Ewen & Mukerji 1980). The predominant grasshopper species were *M. sanguinipes, M. packardii,* and *C. pellucida*, and about 50% of these populations were infected between 4 and 5 weeks after application of 2.5–5.0 x 10^9 spores of *N. locustae* per ha. Maxima of 95–100% infection were evident between 9 and 12 weeks after application for all three grasshopper species. This increase in prevalence of infection throughout the season probably was due to the fact that feeding grasshoppers tend to consume grasshopper cadavers that they find, and data (Ewen & Mukerji 1980) indicate that most cadavers would contain mature spores of *N. locustae* later in the season.

M. sanguinipes showed the greatest vulnerability to *N. locustae* infection among the three predominant grasshopper species in the Saskatchewan study (Ewen & Mukerji 1980). At a 50% level of infection, its population was reduced 26%, significantly more than either *M. packardii* (20%), or *C. pellucida* (8%). Overall, the treated populations of *M. sanguinipes* and *M. packardii* were reduced by about 20%, 4 weeks after spore application and by about 60%, 8 weeks later. On the other hand, the population of *C. pellucida*, although showing about the same prevalence of infection as the *Melanoplus* populations, was reduced overall by only 28%, 12 weeks after application of *N. locustae* spores. It seems that the Cyrtacanthacridinae are more vulnerable to *N. locustae* infection than are the Oedipodinae.

Grasshoppers that survived to lay eggs appeared to be infected to the extent that their fecundity was reduced. This was especially true of *M. packardii*, where the females laid about 65% fewer eggs than untreated populations (Ewen & Mukerji 1980). Egg production by the treated *M. sanguinipes* populations also was reduced markedly when compared to the control populations. Not enough data were collected to evaluate the effect of the pathogen on fecundity of *C. pellucida*. Data show that mature spores of *N. locustae* can overwinter (presumably in cadavers,) but it is not known if sufficient numbers survive to adequately infect the next year's populations.

Applications of *N. locustae* spores in Saskatchewan were made to grasshopper populations that were predominantly in the third nymphal instar (Ewen & Mukerji 1980). Applications to younger populations resulted in the insects being killed at much lower levels of infection, and this could have important implications for grasshopper control in crop land. Crop defoliation could be reduced by killing the grasshoppers when they were first-instar or second-instar nymphs, but then fewer spores would be produced for infection of other grasshoppers. Thus, reducing the grasshopper populations earlier would better protect the crop, but a second application of spores a week or two later would be necessary because considerable recruitment would still be possible in these younger populations.

Literature Cited

Bucher, G.E. (1960) Potential bacterial pathogens of insects and their characteristics. *Journal of Insect Pathology* 2, 172–195.

Ewen, Al B.; Mukerji, M.K. (1980) Evaluation of *Nosema locustae* (Microsporidia) as a control agent of grasshopper populations in Saskatchewan. *Journal of Invertebrate Pathology* 35, 295–303.

Gage, S.H.; Mukerji, M.K. (1977) A perspective of grasshopper distribution in Saskatchewan and interrelationship with weather. *Environmental Entomology* 6, 469–479.

Henry, J.E. (1970) Potential microbial control of grasshoppers and locusts. Proceedings of the International Study Conference, Current & Future Problems of Acridology, pp. 465–468.

Henry, J.E.; Jutila, J.W. (1966) The isolation of a polyhedrosis virus from a grasshopper. *Journal of Invertebrate Pathology* 8, 417–418.

Jutila, J.W.; Henry, J.E.; Anacker, R.L.; Brown, W.R. (1970) Some properties of a crystalline–array virus (CAV) isolated from the grasshopper *Melanoplus bivittatus* (Say) (Orthoptera: Acrididae). *Journal of Invertebrate Pathology* 15, 225–231.

Pickford, R.; Riegert, P.W. (1964) The fungous disease caused by *Entomophthora grylli* Fres., and its effects on grasshopper populations in Saskatchewan in 1963. *Canadian Entomologist* 96, 1158–1166.

Randell, R.L.; Mukerji, M.K. (1974) A technique for estimating hatching of natural egg populations of *Melanoplus sanguinipes* (Orthoptera: Acrididae). *Canadian Entomologist* 106, 801–812.

Musca domestica L., House Fly (Diptera: Muscidae)

R.A. COSTELLO

Flies, particularly *Musca domestica* L., associated with egg producing operations have become an increasing problem in British Columbia's Fraser Valley in recent years. Conflicts between farmers and their nonfarming neighbours have intensified with the encroachment of urban development on agricultural areas. Flies can also be a considerable annoyance to farm operators and their families and employees.

Certain parasitoid wasps have been found effective in suppressing nuisance fly populations in egg producing operations in California (Legner & Dietrick 1972, Legner *et al.* 1975). A commercial insectary in that state provided the parasitoids for this study. The pupal parasitoid *Spalangia endius* Walker (Hymenoptera: Pteromalidae) was selected for this trial because of its tendency to burrow deeper into manure than other available parasitoids in search of pupae (Legner 1977). Pre-release sampling indicated that more than 90% of the pupae were further than 3 cm below the surface.

In 1978, a total of 400 000 *Spalangia endius*, in the form of parasitized house fly pupae, were released into a deep pit layer barn in Aldergrove, British Columbia. The barn, measuring 91 m by 12 m, contained 27 000 layers. The pit contained a 9 month accumulation of manure at the start of the trial. Four releases of 100 000 parasitoids each were made at 3 week intervals commencing 20 April. No other control methods were applied in this barn. Fly numbers, as indicated by "spot cards" (Rutz & Axtell 1979), were compared to those in a barn 30 m away where fly control consisted of chemical insecticides applied as space sprays, residual surface sprays, and floor baits. The spot cards consisted of white 7.6 x 12.7 cm unlined file cards. These were pinned to posts in the barns, collected weekly, and the fecal and regurgitation marks counted.

Levels of parasitism were measured by collecting manure samples once a week, floating out fly pupae, and holding them at 30°C until either flies or parasitoids emerged. This also provided an indication of the fly species involved and all pupae collected were *Musca domestica*.

Fly numbers were monitored in both barns from 25 April to 13 June. The spot card counts were similar for the first weeks, averaging 45.3 per card in the barn using chemical control methods and 35.7 in the barn where parasitoids were released. By the end of the third week, mean spot card counts rose to 390.2 in the chemical control barn and fell to 26.9 in the barn containing *Spalangia endius*. Repeated applications of pyrethrum as a thermal fog resulted in a reduction of the mean spot card count to 213.2 by the end of the 6th week when the manure was removed and spot card counts were discontinued. Spot counts in the barn using parasitoids averaged 20 to 25 during the 4th, 5th and 6th weeks of the trial.

Six weeks after the initial introduction of parasitoids 63% of the fly pupae separated from manure samples contained *Spalangia endius* larvae and pupae. This was the highest level of parasitism found during the sampling period.

House fly numbers in the barn containing parasitoids remained at acceptable levels throughout the trial and the remainder of the fly season, while flies in the barn employing chemical control methods were intolerably abundant and eventually necessitated removing the manure to relieve the problem. The commercial value of the parasitoids used in this study was approximately $800, about the same as the operator spent on chemicals the previous year.

Further laboratory and field studies are being carried out to develop an integrated approach to fly control in egg layer barns entailing parasitoid introductions, manure management, and chemical insecticides.

Literature Cited

Legner, E.F. (1977) Temperature, humidity and depth of habitat influencing host destruction and fecundity of muscoid fly parasites. *Entomophaga* 22, 199–206.

Legner, E.F.; Dietrick, E.J. (1972) Inundation with parasitic insects to control filth breeding flies in California. *Proceedings of the California Mosquito Control Association* 40, 129–130.

Legner, E.F.; Bowen, W.R.; Rooney, W.F.; McKeen, W.D.; Johnston, G.W. (1975) Integrated fly control on poultry ranches. *California Agriculture* 29(5), 8–10.

Rutz, D.A.; Axtell, R.C. (1979) Sustained releases of *Muscidifurax raptor* (Hymenoptera: Pteromalidae) for house fly (*Musca domestica*) control in two types of caged-layer poultry houses. *Environmental Entomology* 8, 1105–1110.

Oulema melanopus (L.), Cereal Leaf Beetle (Coleoptera: Chrysomelidae)

D.G. HARCOURT, J.C. GUPPY and C.R. ELLIS

Pest Status

During the early and mid 1970s, the cereal leaf beetle (CLB), *Oulema melanopus* (L.), began to threaten the production of small grains in Ontario. Following its initial discovery in Essex County near the border of Michigan in 1965, it spread rapidly eastward and by 1975 it had occupied most of Ontario south of Hwy 17. Populations reached economic levels in 1973, and in 1974 chemical treatment was required in ca. 75 fields within the triangular area bounded by the counties of Lincoln, Grey, and Durham (Bereza 1974). In 1976, it occurred in damaging numbers on Manitoulin Island and, in 1977, it was found north of Lake Huron in the districts of Algoma, Sudbury, and Nipissing (Ellis *et al.* 1979).

Background

Like many of our introduced pests, the cereal leaf beetle is rarely troublesome in its native Europe where it is attacked by a complex of natural enemies. One of these, the larval endoparasitoid *Tetrastichus julis* (Walker) was introduced against the beetle by research workers in Michigan in 1967. It was recovered in low numbers in 1969 (Stehr 1970) and was extensively recolonized in 1971, when further releases were made in Michigan, Indiana, Ohio, West Virginia, and New York (Dysart *et al.* 1973).

T. julis is mostly bivoltine in Michigan (Gage & Haynes 1975). It emerges from its overwintering sites in late May and flies to nearby grain fields where it deposits its eggs in the first host larvae that appear. The gregarious parasitoid larvae grow slowly within their host while it is feeding. As soon as the host larva enters the soil and forms its pupal cell, the parasitoids quickly complete their development and form naked pupae within the cell. Diapause is facultative, but the incidence increases as the season progresses. Some adults emerge in early summer and give rise to a second generation. The remainder of the first generation and all of the second overwinter as last instar larvae inside the pupal cells of their hosts. In Michigan, roughly 80% of the population enters a second generation; this generation is poorly synchronized with populations of its host since at the time of emergence relatively few beetle larvae are available for the adults to attack (Gage & Haynes 1975).

Studies in Michigan during the early seventies indicated that *T. julis* had poor powers of dispersal (Haynes *et al.* 1974). Natural spread was slow and rates of parasitism in the United States midwest remained well below that needed to maintain populations at subeconomic levels.

Releases and Recoveries in Ontario

Large-scale releases of *T. julis* were made at 4 sites in southcentral Ontario in June 1974 when parasitized host larvae were set out in fields of oats and barley (Harcourt *et al.* 1977). This has been the only parasitoid released in the province.

In 1975, life table studies of the host in the Quinte area of southern Ontario showed that 55% and 43% respectively of the pupal cells in study plots in Northumberland and

Hastings Counties (Harcourt *et al.* 1977) contained larvae of *T. julis*. These findings, which were unexpected in light of the Michigan experience, prompted closer examination of larval collections made during the course of annual CLB population surveys by Ontario Ministry of Agriculture and Food personnel at 8 sites within the province during 1975. The surveys consisted of 250 net sweeps from each of 30 fields of spring grain located within a 75 km^2 area. Eight such areas were sampled in late June following the accumulation of 475°D>9°C.

Table 10 shows that *T. julis* had dispersed widely across Ontario by 1975. Rates of parasitism were highest in southwestern Ontario, lowest in eastern Ontario, and intermediate in other parts of the province. Overall parasitism was 84%. The number of parasitoids per host ranged from 1 to 44 and averaged 5.9. The four host instars were parasitized in roughly the same proportion (75, 87, 84, 84), suggesting that the larvae were attacked when small.

Table 10 Parasitism of the cereal leaf beetle, *Oulema melanopus* (L.), by *Tetrastichus julis* (Walker) in Ontario, 1975.

County	Township	No. host larvae collected	% parasitized
Kent	Merlin	894	95.0
Middlesex	Glencoe	212	91.9
Essex	Malden	526	88.6
Wellington	Fergus	227	83.7
York	Whitchurch	143	83.2
Ontario	Brock	15	73.3
Bruce	Mildmay	156	68.6
Renfrew	Bromley	147	15.0
TOTALS		2320	84.4

Sweepnet samples of spring grain on Manitoulin Island in early July 1976 showed that the parasitism rate was 87% (Ellis *et al.* 1979).

Current Status of *T. julis*

Since 1974, there have been no reports of economic damage by the cereal leaf beetle in Ontario other than on Manitoulin Island. In most fields of spring grain it has been difficult to find larvae. This raised fears that the host population might become too low to maintain the parasitoid which would set the stage for a resurgence of the pest. For this reason, extensive sampling has been carried out in representative sites in Ontario since 1977. Rates of parasitism in the past 4 years have ranged from 31–68%, indicating that *T. julis* has the capacity to maintain itself as an effective biological control agent at very low host densities. There has been no indication of resurgence of CLB populations and no damage from the pest was reported in Ontario during the 1977–80 period.

Because the distribution of *T. julis* in 1975 was more extensive than the area of the 1974 releases and because its population density at most sites was inexplicably high, it is concluded that the parasitoid spread into Ontario as a result of the earlier releases in Michigan and other parts of the United States. It is also clear that the parasitoid has better powers of dispersal than originally reported.

We believe that there are two reasons for the spectacular success of *T. julis* in Ontario relative to that in Michigan. First, the larval period of the cereal leaf beetle occurs after the seasonal accumulation of 235°D>9°C (Fulton & Haynes 1975) throughout its range, and this is therefore later in Ontario than in the United States midwest. Since the developmental threshold of *T. julis* resembles that of its host (Gage & Haynes 1975), it similarly emerges later in its northern range and its larvae complete their development later in the season. Hence, a greater proportion of the first generation enter diapause. Second, in deference to the harsher climate than in the United States midwest, agricultural practice in most parts of Ontario favours spring grains as opposed to winter wheat. These frequently grow in rotation with, and as a nurse crop for, alfalfa and other legumes. Hence, the field is left untilled after harvest and host pupal cells below soil level are not disturbed by plowing and disking, a practice which can destroy more than 95% of the parasitoid population (Haynes *et al.* 1973).

Recommendations

In view of the potential importance of the cereal leaf beetle to culture of spring grains in Canada, rates of attack of *T. julis* relative to numbers of its host should be monitored at regular intervals throughout its range.

Literature Cited

Bereza, K. (1974) Ontario insects. *Canadian Agricultural Insect Pest Review* 52, 16.

Dysart, R.J.; Maltby, H.L.; Brunson, M.H. (1973) Larval parasites of *Oulema melanopus* in Europe and their colonization in the United States. *Entomophaga* 18, 133–167.

Ellis, C.R.; Harcourt, D.G.; Dubois–Martin, D. (1979) The current status in Ontario of *Tetrastichus julis* (Hymenoptera: Eulophidae), a parasitoid of the cereal leaf beetle. *Proceedings of the Entomological Society of Ontario* 109, 23–26.

Fulton, W.C.; Haynes, D.L. (1975) Computer mapping in pest management. *Environmental Entomology* 4, 357–360.

Gage, S.H.; Haynes, D.L. (1975) Emergence under natural and manipulated conditions of *Tetrastichus julis*, an introduced larval parasite of the cereal leaf beetle, with reference to regional population management. *Environmental Entomology* 4, 425–434.

Harcourt, D.G.; Guppy, J.C.; Ellis, C.R.E. (1977) Establishment and spread of *Tetrastichus julis* (Hymenoptera: Eulophidae) a parasitoid of the cereal leaf beetle in Ontario. *Canadian Entomologist* 109, 473–476.

Haynes, D.L.; Brandenburg, R.K.; Fisher, P.D. (1973) Environmental monitoring network for pest management systems. *Environmental Entomology* 2, 889–899.

Haynes, D.L.; Gage, S.H.; Fulton, W. (1974). Management of the cereal leaf beetle pest ecosystem. *Quaestiones Entomologicae* 10, 165–176.

Stehr, F.W. (1970) Establishment in the United States of *Tetrastichus julis*, a larval parasite of the cereal leaf beetle. *Journal of Economic Entomology* 63, 1968–1969.

Phyllonorycter blancardella (F.), Spotted Tentiform Leafminer (Lepidoptera: Gracillariidae)

J.E. LAING

Pest Status and Background

The spotted tentiform leafminer (STLM), *Phyllonorycter blancardella* (F.), is an indirect pest of apple in eastern Canada and the United States. This species probably entered North America from Europe early in the 1900s. Severe infestations (> 20 mines per leaf) may cause early leaf drop, stunting of terminal growth, premature fruit ripening, and fruit drop with a resultant decrease in yield and reduction of fruit set the year following the infestation (Pottinger & LeRoux 1971, Johnson 1975). There are three generations of the leafminer per year throughout its range in North America. The leafminer overwinters as a pupa, emerges in May, and begins to oviposit first-generation eggs. Adults of the first generation are present in orchards from mid-June through July. Second-generation adults appear in August (Johnson 1975, Laing 1977).

Although many species of parasitoids have been recorded from STLM (Pottinger & Le Roux 1971, Johnson *et al.* 1978), this species is able to reach damaging levels in some orchards each year by the third or overwintering generation. Because of the cryptic feeding habits of the larvae, chemical controls must be either systemic or timed to coincide with adult emergence. The latter has been facilitated by the synthesis of a sex attractant for STLM (Roelofs *et al.* 1977) and the development of heat unit requirements above a threshold of approximately 7°C for emergence of the adults (Johnson *et al.* 1979). Although individual growers may obtain good control of STLM at times, reinvasion of commercial orchards likely occurs from hedgerow and abandoned or poorly maintained orchards.

The following parasitoids have been reported from STLM in Ontario and Québec (Pottinger & LeRoux 1971, Johnson *et al.* 1978):

Braconidae	*Apanteles ornigis* Weed
Eulophidae	*Sympiesis marylandensis* Girault
	Sympiesis bimaculatipennis Girault
	Sympiesis sericeicornis (Nees)
	Sympiesis conica (Provancher)
	Pnigalio maculipes (Crawford)
	Pnigalio flavipes (Ashmead)
	Pnigalio uroplatae (Howard)
	Chrysocharis cuspidogaster (Yosh.)
	Chrysocharis tricinctus Ashm.
	Chrysocharis laricinellae (Ratz.)
	Tetrastichus sp.
	Cirrospilus cinctithorax (Girault)
	Zagrammosoma multilineatum (Ashmead)
	Horismenus fraternus (Fitch)
Eupelmidae	*Eupelmella vesicularis* (Retzius)

The most abundant parasitoids of STLM are the braconid *Apanteles ornigis* Weed in the first and third generations, and the chalcids *Sympiesis marylandensis* Gir. and *S.*

sericeicornis (Nees) in the second generation. Numerous parasitoids of STLM have been recorded from Europe (Janjua & Carl 1964). The most important of these is *Apanteles circumscriptus* (Nees) which attacks second-instar and third-instar larvae. Since the report of Janjua & Carl (1964) was completed, the *Apanteles circumscriptus*-group has been revised and now includes such species as *A. pedias* Nixon, *A. arisba* Nixon, *A. laetus* Marshall, and *A. circumscriptus* (Nixon 1973). Some species of parasitoids in the *A. circumscriptus*-group were imported to New Zealand from Europe in the late 1950s, became established, and successfully controlled the oak leafminer, *Phyllonorycter messaniella* (Zell.) (Swan 1973).

Releases and Recoveries

In 1978, parasitoids were imported to Guelph, Ontario, from New Zealand through Agriculture Canada in Ottawa. Only two females of the parthenogenetic species *A. pedias* survived and these were released in a sleeve cage in an orchard at the University of Guelph during May 1978. Recoveries of the parasitoid were made in subsequent generations that year. By the third generation, 2.4% of the overwintering leafmining population in the release orchard had been parasitized by *A. pedias*. One year later, in the fall of 1979, the rate of parasitism of STLM by *A. pedias* had increased to 25.7% in the trees surrounding the original release site (Laing & Heraty 1981). The level of parasitism by *A. pedias* in the release orchard has remained between 25–30% of the third (overwintering) generation of STLM through 1980. Parasitism by the native *A. ornigis* has remained at 8–10% in the third generation as it has since 1975 (Laing & Heraty 1981). *A. pedias* has spread to other orchards in the Guelph area. In the spring of 1980, two years after its release, *A. pedias* was found in orchards as far as 43 km south of Guelph (Laing & Heraty 1981).

In July 1981, 100 adult *A. circumscriptus* from northern Italy were released in an orchard in Guelph, Ontario. No data are yet available on the establishment of this species at this time.

Recommendations

(1) Annual evaluations of the uniparental strain of *Apanteles pedias* should be continued for four or five years.

(2) Further importations of this species should be attempted from Europe where the strain is biparental on poplar.

(3) Population sampling of STLM should be continued to determine if *A. circumscriptus* becomes established and, if so, to monitor its effects on STLM.

(4) *A. pedias* should be released in other apple-growing regions of Canada inhabited by STLM.

Literature Cited

Janjua, A.M.; Carl, K.P. (1964) The natural enemies of *Lithocolletis blancardella* F. and some related species in Europe (a literature review). Commonwealth Institute of Biological Control, European Station, Annual Project Statement.

Johnson, E.F. (1975) Influence of temperature on development of *Lithocolletis blancardella* Fabricius and *Apanteles ornigis* Weed. M.Sc. Thesis, University of Guelph, 91 pp.

Johnson, E.F.; Laing, J.E. Trottier, R. (1978) The seasonal occurrence of *Lithocolletis blancardella* (Gracillariidae), and its major natural enemies in Ontario apple orchards. *Proceedings of the Entomological Society of Ontario* (1976) 107, 31–45.

Johnson, E.F., Trottier, R.; Laing, J.E. (1979) Degree-day relationships to the development of *Lithocolletis blancardella* (Lepidoptera: Gracillariidae) and its parasite *Apanteles ornigis* (Hymenoptera: Braconidae). *Canadian Entomologist* 111, 1177–1184.

Laing, J.E. (1977) Spotted tentiform leafminer. Ontario Ministry of Agriculture and Food, Agdex Factsheet, 3 pp.

Laing, J.E.; Heraty, J.M. (1981) Establishment in Canada of the parasite *Apanteles pedias* Nixon on the spotted tentiform leafminer *Phyllonorycter blancardella* (Fabr.). *Environmental Entomology* 10, 933–935.

Nixon, G.E.J. (1973) A revision of the north-western European species of the *vitripennis, pallipes, octonarius, triangulator, fraternus, foniosus, parasitellae, metacarpalis,* and *circumscriptus*-groups of *Apanteles* Forster (Hymenoptera: Braconidae). *Bulletin of Entomological Research* 63, 169–228.

Pottinger, R.P.; LeRoux, E.J. (1971) The biology and dynamics of *Lithocolletis blancardella* (Lepidoptera: Gracillariidae) on apple in Québec. *Memoirs of the Entomological Society of Canada* 77, 437 pp.

Roelofs, W.L.; Reissig, W.H. Weires, R.W. (1977) Sex attractant for the spotted tentiform leafminer moth *Lithocolletis blancardella*. *Environmental Entomology* 6, 373–374.

Swan, D.I. (1973) Evaluation of biological control of the oak leafminer *Phyllonorycter messaniella* (Zell.) (Lepidoptera: Gracillariidae) in New Zealand. *Bulletin of Entomological Research* 63, 49–55.

Phyllotreta spp., Flea Beetles (Coleoptera: Chrysomelidae)

H.G. WYLIE, W.J. TURNOCK and L. BURGESS

Pest Status

Damage to rape crops in western Canada by flea beetles is caused primarily by adults of two species, *Phyllotreta cruciferae* (Goeze) and *Phyllotreta striolata* (F.). A third flea beetle, *Psylliodes punctulata* Melsh., occurs in virtually all seedling rape crops on the prairies but by itself does not constitute a serious problem. *Phyllotreta robusta* LeC. and *Phyllotreta albionica* (LeC.) occur occasionally on rape crops in Saskatchewan (Burgess 1977*a*) and three other described crucifer-feeding species in the genus *Phyllotreta* are known to occur in that province (Burgess 1981*b*).

P. cruciferae was introduced on the west coast of North America in the early 1920s (Milliron 1953) and was abundant on crucifers in the Prairie Provinces by the late 1930s and early 1940s (Westdal & Romanow 1972). *P. cruciferae* has usually been the dominant flea beetle in rape fields across the prairies, particularly in the more southerly parts of the rape-growing area.

P. striolata apparently was introduced to North America before 1801 (Chittenden 1923) and was present in Canada from the Atlantic provinces to British Columbia by the early 1900s (Burgess 1977*a*). This species infests rape crops across the parkland agricultural area from southern Manitoba to north of the Peace River in Alberta (Burgess 1977*a*). *P. striolata* has recently gained in relative importance as a pest in both Saskatchewan and Alberta, being more abundant than *P. cruciferae* in some Saskatchewan rape fields after 1976 (Burgess 1981*a*) and in some fields in central Alberta since 1979 (H.G. Philip 1981 personal communication).

Flea beetle problems in rape crops have increased in severity in the past decade. In Manitoba and Saskatchewan most fields of rape are threatened annually by these insects. In Alberta, flea beetle problems have tended to be less acute than in Saskatchewan and Manitoba, but in 1979 and 1980 heavy infestations were reported from most rape-growing areas (H.G. Philip 1981 personal communication). The estimated percentages of rape acreage treated with insecticides in 1979 were 75% in Manitoba, 59% in Saskatchewan, and 45% in Alberta (Turnock & Samborski 1980). Although yield loss due to flea beetle damage is difficult to estimate, yield reductions of 15–50% have occurred in untreated plots in insecticide trials (Turnock & Samborski 1980). The combination of flea beetles and poor conditions for germination that occurred in 1975 and 1980 in Manitoba completely destroyed many fields of rape.

Background

This is the first biological control programme set up against flea beetles of the genus *Phyllotreta*. The programme was initiated because of the low level of natural control of *P. cruciferae* and *P. striolata* and because these are introduced species. Studies in Manitoba since 1975 confirmed only three native parasitoids of rape-infesting flea beetles: a braconid, *Microctonus vittatae* Mues.; a mermithid; and an allantonematid, *Howardula* sp. *M. vittatae* parasitizes adults of both *P. cruciferae* and *P. striolata*, but the incidence of parasitism is usually less than 5%. Various factors limit the impact of this braconid in Manitoba, including its short adult life and low oviposition rate, and poor temporal synchronization with the host's population in mid-summer. Neither the mermithid nor the allantonematid is sufficiently abundant to limit populations of the rape-infesting species. Occasional predation of flea beetle adults by adults of *Collops vittatus* Say (Coleoptera: Melyridae) and *Geocoris bullatus* (Say) (Hemiptera: Lygaeidae) and by larvae of *Chrysopa* sp. (Neuroptera: Chrysopidae) have been reported (Gerber & Osgood 1975; Burgess 1977*b*, 1980).

In Europe, a large complex of flea beetle species infest rape and other crucifers. Usually, the most destructive species are *Psylliodes chrysocephala* (L.) and *Phyllotreta undulata* Kutsch. (Carl 1973, Carl & Sommer 1975). Neither species occurs in Canada; *P. chrysocephala* was recorded once in Newfoundland (Anon. 1953) but apparently did not become established. Neither *P. striolata* (sometimes called *P. vittata* F. by European entomologists; see Barber 1947) nor *P. cruciferae* is considered to be an important pest in Europe. *P. striolata* is usually more abundant than *P. cruciferae* (Carl & Lehmann 1980).

Studies in Austria, Germany, and Switzerland identified three species of euphorine braconids, three mermithid species, and an allantonematid, *Howardula phyllotretae* Oldham, as parasitoids of *Phyllotreta* spp. Although *H. phyllotretae* parasitized up to 70% of the flea beetle adults and reduced oviposition by parasitized females, it had little or no effect on beetle survival (Carl & Sommer 1975). The impact of the mermithids was also negligible. One of the braconid species was uncommon and was not identified. A second species, *Microctonus* sp. *Z* (Sommer 1981) which parasitized less than 10% of the flea beetles, was morphologically indistinguishable from *M. vittatae* (C.C. Loan 1981 personal communication). The third braconid species, *Microctonus bicolor* Wesm., parasitized up to 50% of the flea beetle adults in summer. Small numbers of this braconid that were imported in Canada in 1977, and tested in the laboratory at Winnipeg, parasitized adults of *P. cruciferae* and *P. striolata* and developed on both species. Therefore, *M. bicolor* was considered to be a suitable candidate species for importation and release.

Releases and Recoveries

Parasitoid cocoons received from the European Laboratory of the Commonwealth Institute of Biological Control were reared at the Research Program Service, Central Experimental Farm, Ottawa. Those cocoons which produced hyperparasites of the genus *Mesochorus* were destroyed, and adults of *Microctonus bicolor* and *M. vittatae* were shipped by air express to Winnipeg.

Details of releases of *Microctonus bicolor* Wesm. are shown in Table 11. Parasitoids were released in the open in experimental plots of rape, *Brassica napus* L., cv. Tower, at the Glenlea Field Station of the Agriculture Canada Research Station, about 20 km south of Winnipeg. With the exception of three small releases, one in July 1979 and two in July 1980 when host numbers were low, all parasitoids were released in May, June, August, or September when the plots were heavily infested by *P. cruciferae* and lightly infested by *P. striolata*. No insecticides were applied to the plots each year for several weeks before the parasitoids were released, and none were applied after releases began.

Table 11 Open releases and recoveries of parasitoids of *Phyllotreta* spp.

Species and Province	Year Origin	Number	Year of Recovery
Microctonus bicolor Wesm.	1978 Austria	182	
Manitoba	1979 Austria	375	
	1980 Austria	653	

Several attacks by the parasitoid females on adults of *P. striolata* were observed on rape plants in the plots during releases.

Flea beetles from the release sites were collected in 1979 and 1980 and reared in the laboratory for evidence of parasitoid establishment. Numbers collected were 8502 *P. cruciferae*, 261 *P. striolata*, and smaller numbers of other rape-infesting beetles in June 1979 from the 1978 release site; 3082 *P. cruciferae* and 674 *P. striolata* during July–October 1979 from the 1979 release site; and 5039 *P. striolata* and 418 *Psylliodes punctulata* Melsh. in April 1980 from overwintering sites adjacent to the 1979 release site. No *M. bicolor* emerged from these beetles.

Some adults of *M. bicolor* and *M.* sp *Z* were retained for laboratory studies (Table 12). These studies showed that females of *M. bicolor* parasitized more *P. striolata* than *P. cruciferae* when equal numbers of the two hosts were present.

Table 12 Imported parasitoids of *Phyllotreta* spp. used for laboratory studies

Species	Year	Origin	Number
Microctonus bicolor Wesm.	1977	Germany	43
	1978	Austria	131
		Switzerland	20
	1979	Austria	26
	1980	Austria	8
Microctonus sp. *Z*	1977	Germany	9
	1978	Switzerland	2
	1979	Austria	13
	1980	Austria	1

Evaluation of Control Attempts

Several factors have limited the probability of establishment. First, small numbers of parasitoids were released. Second, scarcity of the preferred host, *P. striolata*, at Glenlea may have limited oviposition by the females after release. Third, dispersal of the beetles in autumn to overwintering sites and in spring in search of food, scattered the population of parasitized beetles and thereby reduced the probability of female parasitoids that emerged in the spring being fertilized. If *M. bicolor* has become established, the small numbers released and scattering of the population would minimize our chances of detecting establishment until the incidence of parasitism has increased substantially.

Recommendations

Larger colonies of *M. bicolor* (> 500 females per colony) should be released. Although large numbers of parasitized flea beetles have been collected by the CIBC, only small numbers of parasitoid adults have been reared and female colony size has never exceeded 100. There are two reasons for the small parasitoid colonies. First, many of the immature parasitoids enter diapause and eventually die when the hosts are reared in cages in Europe. Second, parasitoid emergence from cocoons obtained in Europe has usually been less than 50%. The possibility that higher temperature and longer photoperiod prevent diapause should be investigated, as well as the factors responsible for poor survival of the parasitoid larvae and pupae in cocoons.

The possibility of obtaining additional candidate parasitoid species for release in Canada should be investigated in discussions planned in 1981 between staff of the CIBC and scientists from the USSR.

Release sites in western Canada that have higher populations of the preferred host *P. striolata*, than the release site at Glenlea should be used for future releases.

Releases of parasitoid adults should be supplemented with releases of flea beetles parasitized in the laboratory, using flea beetles collected in Manitoba and females of *M. bicolor* imported from Europe.

Recovery studies for evidence of establishment should be continued at Glenlea and initiated at other releases that are set up.

Literature Cited

Anonymous (1953) Outbreaks and new records. *FAO Plant Prot. Bull.* 2, 28–29.

Barber, H.S. (1974) *Diabrotica* and two new genera. *Proceedings of the Entomological Society of Washington* 49, 151–161.

Burgess, L. (1977*a*) Flea beetles (Coleoptera: Chrysomelidae) attacking rape crops in the Canadian prairie provinces. *Canadian Entomologist* 109, 21–32.

Burgess, L. (1977*b*) *Geocoris bullatus*, an occasional predator of flea beetles (Hemiptera: Lygaeidae). *Canadian Entomologist* 109, 1519–1520.

Burgess, L. (1980) Predation on adults of the flea beetle *Phyllotreta cruciferae* by lacewing larvae (Neuroptera: Chrysopidae). *Canadian Entomologist* 112, 745–746.

Burgess, L. (1981*a*) Winter sampling to determine overwintering sites and estimate density of adult flea beetle pests of rape. *Canadian Entomologist* 113, 441–447.

Burgess, L. (1981*b*) Crucifer-feeding flea beetles that occur in the province of Saskatchewan, Canada. *Coleopterists' Bulletin* 35, 307-310.

Carl, K.P. (1973) Rape insects. Commonwealth Institute of Biological Control, European Station, Annual Project Statement.

Carl, K.P.; Lehmann, C. (1980) Rape insects. Commonwealth Institute of Biological Control, European Station, Annual Project Statement.

Carl, K.P.; Sommer, G. (1975) Rape insects. Commonwealth Institute of Biological Control, European Station, Annual Project Statement.

Chittenden, F.H. (1923) Notes on the distribution and habits of North American *Phyllotreta* (Coleop.). *Proceedings of the Entomological Society of Washington* 25, 13–139.

Gerber, G.H.; Osgood, C.E. (1975) *Collops vittatus* (Coleoptera: Melyridae): a predator of flea beetle adults in rapeseed. *Manitoba Entomologist* 9, 61.

Milliron, H.E. (1953) A European flea beetle injuring crucifers in North America. *Journal of Economic Entomology* 46, 179.

Sommer, G. (1981) Biologie und Parasitenkomplex der Halticenen Gattung *Phyllotreta*. Ph.D. Thesis, Albert-Ludwigs University of Freiburg.

Turnock, W.J.; Samborski, D.J. (1980) Management of plant pests. *In*: Prairie Production Symposium. Soil and Land Resources. Advisory Committee to the Canadian Wheat Board, Winnipeg, Manitoba.

Westdal, P.H.; Romanow, W. (1972) Observations on the biology of the flea beetle, *Phyllotreta cruciferae* (Coleoptera: Chrysomelidae). *Manitoba Entomologist* 6, 35–45.

Tetranychus urticae Koch, Twospotted Spider mite (Acarina: Tetranychidae)

R.J. McCLANAHAN

Pest Status

The twospotted spider mite, *Tetranychus urticae* Koch, continues to be a major problem in greenhouse crops, particularly cucumbers, chrysanthemums, and roses. The floral crops are fairly well protected by granular applications of the systemic insecticide aldicarb, but cucumber growers are having a difficult time keeping mites under control. Resistance to dicofol and Pentac® (Zoecon Industries Ltd.) has developed in some greenhouses. The effective material oxythioquinox was withdrawn from use on cucumbers in 1974.

Background

The predaceous mite *Phytoseiulus persimilis* Athias-Henriot has been the only biological control agent seriously considered for control of *T. urticae* in greenhouses, although other mite predators are effective in orchards. The European workers have had encouraging results with *P. persimilis* and mass rearing is carried out at a number of government and commercial establishments. Russian sources report a production of 83 million *P. persimilis* per year from one rearing unit (Askretkov & Tolmacheva 1980). Probably the most extensive use of *P. persimilis* on greenhouse crops is in the Netherlands where some 400 ha (60% of total production) of cucumbers is protected by the predator (Koppert 1980).

Research on the interaction of prey mites and *P.persimilis* has mostly been reported from Europe. Methods of introduction have been worked out for tomatoes (French *et al*. 1976) and cucumbers (Gould 1977), while the optimum conditions of temperature and humidity were defined by Stenseth (1979*b*) in Norway. Many references pertain to the compatibility of various insecticides and fungicides and the use of *P.persimilis* (e.g. Jeppson *et al*. 1975), but these results were superseded by investigations using a strain of *P. persimilis* resistant to phosphate insecticides (Stenseth 1979*a*).

At Harrow, laboratory studies proved that some acaricides were selective in favor of the predator (McClanahan 1971). However, the best materials were experimental or not widely accepted by growers for twospotted mite control. Since then oxythioquinox has been withdrawn from use on greenhouse cucumbers, and the twospotted mites have developed varying degrees of resistance to acaricides (McClanahan 1980).

Releases and Recoveries

It is difficult to detail releases of *P. persimilis* in Canada, since there are several sources, including commercial enterprises, and only part of the production is used in Canada. A colony has been maintained at the Harrow Research Station for many years, and limited numbers are available on request by research establishments or educational projects. Acknowledgements of such shipments usually record successful establishment. *P. persimilis* are reared and sold by Better Yield Insects in Ontario and by Applied Bio-Nomics Ltd. in British Columbia. The number of predators available at the release point is dependent on the number of prey mites in the shipment and on the interval between packaging and releasing. Thus the number actually released can be much different than the number shipped, but under favorable conditions reproduction is very rapid with a generation in 5 days.

Four commercial greenhouses near Leamington were used in a demonstration experiment in 1980. Two *P. persimilis* were placed on each mite-infested cucumber plant in February. Control of *T. urticae* was achieved and maintained until late May, but then a change in greenhouse environment favored the pest mite and the growers had to switch to chemical control. Probably the humidity was lowered when vents were open throughout the day and this prevented *P. persimilis* eggs from hatching.

Recommendations

Techniques of using *P. persimilis* in Canadian greenhouses will have to be refined and adjusted to greenhouse conditions. Introduction of predators onto the fall crop of cucumbers might avoid a resurgence of *T. urticae*.

Research projects could include modification of the greenhouse environment by misting to produce favorable conditions for the predator, or spraying only a portion of the cucumber plants with an acaricide while predators were active on the unsprayed portion.

Rearing under artificial conditions, as carried out in Russia (Storozhkov & Kaznacheeva 1980) should be tried in Canada for increased mass production.

Literature Cited

Askretkov, A.V.; Tolmacheva, N.P. (1980) A sovkhoz biolaboratory. *Zashchita Rastenii* 2, 36 (in Russian).

French, N.; Parr, W.J.; Gould, H.J.; Williams, J.J.; Simmonds, S.P. (1976) Development of biological control methods for the control of *Tetranychus urticae* on tomatoes using *Phytoseiulus persimilis*. *Annals of Applied Biology* 83, 177–189.

Gould, H.J. (1977) Biological control of glasshouse whitefly and red spider mite on tomatoes and cucumbers in England and Wales 1975–76. *Plant Pathology* 26, 57–60.

Jeppson, L.R.; McMurtry, J.A.; Mead, D.W.; Jesser, M.J.; Johnson, H.G. (1975) Toxicity of citrus pesticides to some predaceous phytoseiid mites. *Journal of Economic Entomology* 68, 707–710.

Koppert, P.C. (1980) Biological control in glasshouses in the Netherlands. Proceedings of the International Symposium on Integrated Control in Agriculture and Forestry, Vienna 1979, p. 484.

McClanahan, R.J. (1971) Selective acaricides for integrated control of *Tetranychus urticae*. International Organization for Biological Control, West Palearctic Regional Section, Proceedings of a Conference on Integrated Control in Glasshouses, pp. 13–22.

McClanahan, R.J. (1980) Why has integrated control practice levelled off in Canada? *Bulletin SROP* 1980/III, 141–144.

Stenseth, C. (1979*a*) Effect of fungicides and insecticides on a resistant strain of *Phytoseiulus persimilis* Athias-Henriot (Acarina: Phytoseiidae). *Forskning og Forsok i Landbruket* 30, 77–83.

Stenseth, C. (1979*b*) Effect of temperature and humidity on the development of *Phytoseiulus persimilis* and its ability to regulate populations of *Tetranychus urticae* (Acarina: Phytoseiidae, Tetranychidae). *Entomophaga* 24, 311–318.

Storozhkov, Yu.V.; Kaznacheeva, V.P. (1980) Rearing of *Phytoseiulus* in hydroponic greenhouses. *Zashchita Rastenii* 2, 34 (in Russian).

Thymelicus lineola (Ochsenheimer), European Skipper (Lepidoptera: Hesperiidae)

J.N. McNEIL

Pest Status

In the last review dealing with the European skipper, *Thymelicus lineola* (Ochs.), Arthur (1971) reported that the distribution of this insect in Canada was limited to Ontario, Québec, New Brunswick, and Nova Scotia in the east, with one localized population at Terrace, British Columbia. At that time economic losses caused by *T. lineola* had been recorded in Grey and Hastings Counties, Ontario. The distribution of the skipper has increased considerably and is now found in Prince Edward Island (Thompson 1974), Newfoundland (Jackson 1978), and in Manitoba (Preston & Westwood 1981). Also the population in British Columbia has now spread into the Lake Shuswap area (J. Proctor, 1982, personal communication).

Important hay losses have been recorded in Ontario (K. Bereza & W. Riley, 1978, personal communication, Québec (McNeil *et al.* 1975), and Prince Edward Island (Thompson 1977), establishing it as a major pest species of timothy in the more northerly areas of its distribution. While the skipper is present throughout most northeastern states (Arthur 1971) the only incidence where *T. lineola* has reached pest status occurred in northern Michigan (R.F. Ruppel, 1977, personal communication).

Background

Bacillus thuringiensis Berliner, has proven effective in controlling larval skipper populations (Arthur & Angus 1965, Arthur 1968, McNeil *et al.* 1977) and is now recommended by the Conseil des productions végétales du Québec (1×10^{10} I.U./ha) if control measures are required. It not only offers satisfactory foliage protection when compared with chemical insecticides (Thompson 1977, Letendre & McNeil 1980), but also reduces the possible impact on non-target species, such as bees.

The discovery of a nuclear polyhedrosis virus (NPV) affecting a high percentage of larval populations around Normandin, Québec (Smirnoff 1974) stimulated research to evaluate its potential as a biological control agent for *T. lineola*. Smirnoff *et al.* (1976) and Duchesne (1980) demonstrated the efficacy of this virus, however its action is much slower than *B. thuringiensis*. Following an aerial application of *B. thuringiensis* 1×10^{10} I.U. at 18.71 l/ha, infected larvae were observed within 2 days and 100% mortality was obtained in 6 days (Fig. 2*a*). Defoliation levels at the time of harvest were 20%, the same as pre-treatment levels, while in the controls defoliation averaged 90%. By comparison an aerial application of the NPV, 5×10^6 polyhedra/ml at 18.71 l/ha, gave the same mortality but required considerably more time (Fig. 2*b*). Also defoliation at harvest was 50%, and while being lower than the 100% in the controls, was significantly higher than the pre-treatment level of 20%. Thus the virus has less potential than *B. thuringiensis* as a curative means of control, but could be introduced as a preventative measure if populations are beginning to increase but have not reached very high larval densities. This would be advantageous as the virus persists in the population and has an effect over several years (Duchesne 1980). While few other pathogens have been found infecting skipper larvae in Québec, two entomopathogenic fungi *Zoophthora radicans* and *Entomophthora egressa* have been isolated from larvae collected at Amqui in 1977 and 1978 (McNeil & MacLeod 1982).

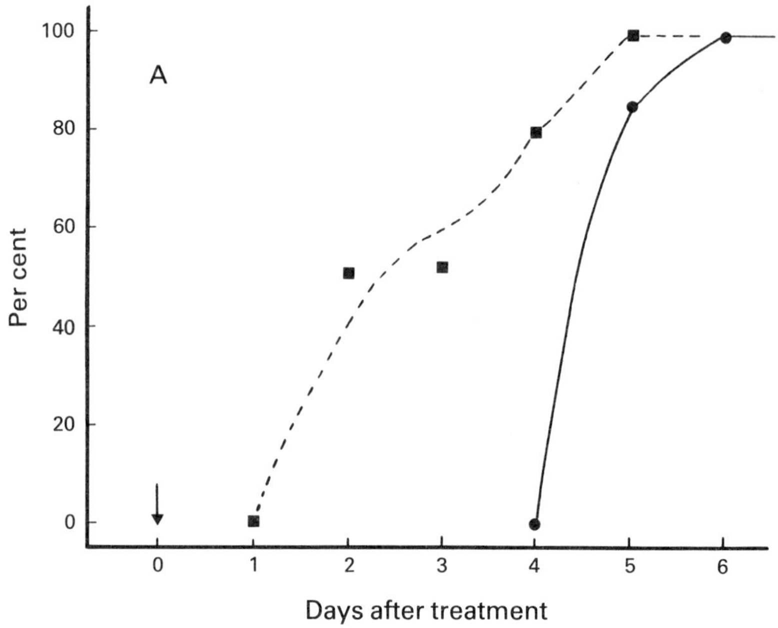

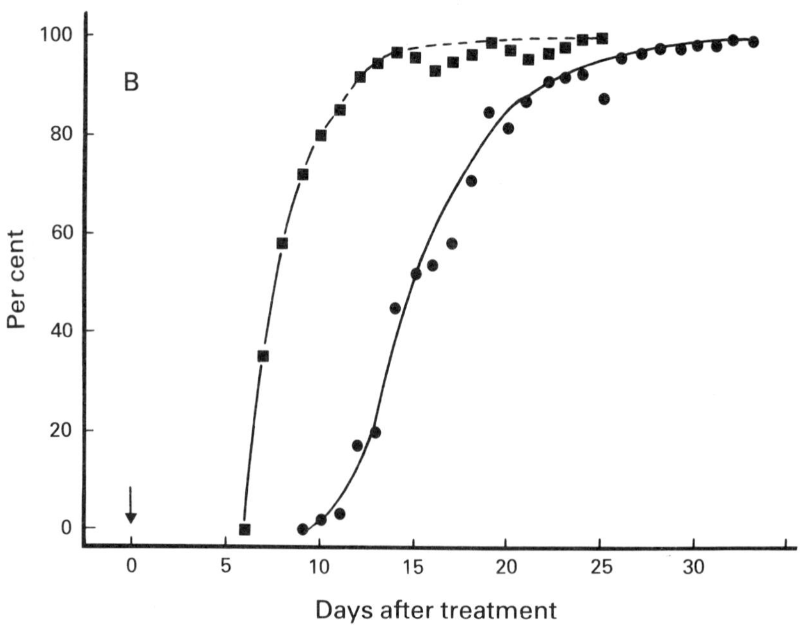

Fig. 2. The incidence of infected (■– – – –■) and dead (● ———— ●) skipper, *Thymelicus lineola* (Ochs.), larvae following an aerial application of (a) *Bacillus thuringiensis* Berl., 1 x 10^{10} I.U. at 18.71 l/ha (b) a nuclear polyhedrosis virus, 5 x 10^{6} polyhedra/ml at 18.71 l/ha.

Imported Parasitoids

Sixteen species of parasitoids have been reared from skipper populations in Québec since 1972 and only 3 species are common to the 22 species found in Ontario (Arthur 1962). As reported for Ontario total parasitism levels are very low and the only parasitoid recovered consistently in any numbers is *Itoplectis conquisitor* (Say). Based on observations over the last 7 years the importance of indigenous parasitoids in the population dynamics of *T. lineola* would have to be considered as negligible. While the diversity of parasitoid species was lower in Europe, Carl (1968) reported high levels of parasitism and felt that the principal larval parasitoids *Phryxe vulgaris* Fall. and *Rogas tristis* Westm., as well as the major pupal parasitoid *Syspasis* (=*Stenichneumon*) *scutellator* (Grav.) would be suitable candidates for introduction into Canada. He believed that *S. scutellator* would be the most promising due to the fact that it was well synchronised with the skipper and, as it is univoltine, would not require alternate hosts as do the larval parasitoids.

Specimens of *P. vulgaris* have been received regularly from Europe since 1972 (Table 13), but unfortunately shipments usually arrived when local skipper populations had already pupated. Thus efforts were made to maintain laboratory cultures using *Artogeia rapae* (L.) larvae as an alternate host. Female parasitoids readily oviposted and parasitoid larvae developed successfully, however difficulties in maintaining healthy host colonies inhibited successful mass rearing of the parasitoid. In 1974, adults arrived soon enough to be released directly in the field at Normandin (Table 14), but no *P. vulgaris* have been recovered in the area. Releases of adults and parasitized *A. rapae* larvae were made at Rawdon in 1974 and 1975 (Table 14), but were too late to coincide with the presence of skipper larvae. However cabbages were quite widely cultivated around the release site and an abundance of imported cabbageworm larvae were available as alternate hosts. In subsequent sampling no skipper larvae parasitized by *P. vulgaris* were recovered. It is of interest to note that *P. vulgaris* is an indigenous species and has been reared from *A. rapae* in Ontario (Harcourt 1966), yet has never been reported attacking *T. lineola* in either Ontario or Québec. Carl (1968) suggested that different geographic races of *P. vulgaris*, with restricted host ranges may exist, which would explain this situation.

Shipments of *S. scutellator* have been received from Delémont since 1974 and as seen in Tables 13 and 14, numbers were rather low in 1974 and 1975 but increased subsequently. In an effort to increase the numbers collected, Dr Carl treated fifth-instar skipper larvae with fluorescent powder and released them at the collection site at Plancher-Bas. It was felt that as the last larval skin remains attached to the pupae individuals could be easily located using a black light. While the recovery of marked individuals was poor, it was observed that skipper pupae that had been parasitized were more readily located with a black light than unparasitized ones, even without marking. Using this technique, night time collections yielded higher numbers than in previous years. Adults held under laboratory conditions throughout the winter (Table 13) were kept at 2 ± 1°C in plastic boxes containing leaf litter and peat moss. Males died rapidly and although females lived much longer only one survived until spring but died before *T. lineola* pupae were available. From 1977 all adults not used in open releases were held on arrival in an insectary and supplied with a 10% sugar solution as well as water. By mid-September all males and some females had died. At this time all surviving females were placed in large field cages containing logs, hay bales, and leaf litter. In 1977–78 there were no survivors while in 1978–79 we recovered 5 of the 200 females in mid-June. These were fed and held in the insectary until skipper pupae were available. One female stung the first pupa that she encountered, remaining 45 minutes with her ovipositor in the host and upon withdrawal, died. The stung pupa was held but no parasitoid emerged. The other 4 females however, showed no interest in skipper pupae. New hosts were provided each day until the parasitoids died. All exposed pupae were held but only *T. lineola* adults emerged. In 1979 the holding cages were destroyed during a storm and most parasitoids

Table 13 Laboratory and field cage studies of introduced parasitoids of *Thymelicus lineola* (Ochs.) in Québec

Species	Origin	Year	Number
Phryxe vulgaris Fall.	France	1972	16*
	France	1973	16*
	France	1974	104*
	France	1975	12*
	France	1977	174*
	France	1978	53*
Syspasis scutellator (Grav.)	France	1974	7 ♂ 5 ♀*
	France	1975	6 ♂ 11 ♀*
	France	1977	87 ♂ 215 ♀**
	France	1978	177 ♂ 217 ♀**
	France	1979	72 ♂ 110 ♀***
	France	1980	68 ♂ 125 ♀**

* Laboratory rearing.
** Overwintering studies in field cages.
*** Unplanned release when cage was destroyed during a storm at Québec, Québec County 46°49′, 71°14′.

Table 14 Open releases of introduced parasitoids against *Thymelicus lineola* (Ochs.) in Québec

Species	Origin	Year	No. released	Location
Phryxe vulgaris	France	1974	98 adults	Rawdon, Montcalm County 46°04′N, 73°44′W
			160 adults	Normandin, Normandin County 48°50′N, 72°32′W
	France	1975	60 adults	Rawdon, Montcalm County 46°04′N, 73°44′W
			200 parasitized *A. rapae* larvae	Rawdon, Montcalm County 46°04′N, 73°44′W
Syspasis scutellator	France	1974	17♂ 32♀ *	Rawdon, Montcalm County 46°04′N, 73°44′W
	France	1975	15♂ 44♀ *	Rawdon, Montcalm County 46°04′N, 73°44′W
	France	1978	19♀ **	Amqui, Matapedia County 49°28′N, 67°26′W
			150 stung *T. lineola* pupae	Amqui, Matapedia County 49°28′N, 67°26′W

* Adults in reproductive diapause.
** Non-diapausing gravid females.

were killed or escaped. Three females were recovered but did not successfully overwinter. In the 1980 shipment, mortality was very high, only 32 females were placed in overwintering cages and none were recovered in the spring of 1981.

Open releases of adults in reproductive diapause or gravid females (together with stung *T. lineola* pupae) were made in 1974, 1975, and 1978 at localities where hayfields were surrounded by woodlots (Table 14). Also as indicated previously there was one unplanned release at Québec in 1979 when the holding cage was destroyed. Intensive pupal sampling in all release areas has not recovered any *S. scutellator* adults.

Recommendations

Beirne (1975) stated that the chances of successful colonization of introduced biological control agents increases when the numbers released are high and felt that chances of success would be small if numbers were below some minimum, about 5000 individuals. If this estimate is reasonable, it is not suprising that *S. scutellator* has not been recovered following open releases. However I believe that we should continue to try and establish *S. scutellator* in Canada. The cage studies showed that this parasitoid is able to overwinter here and, while survival was very low, it is possible that the overwintering sites provided were not the most appropriate for the species. An effort must be made to increase the number of parasitoids available each year. This now presents a serious problem as the skipper population at Plancher-Bas, the only area where densities were high enough to make parasitoid collections worthwhile, have crashed. Therefore, unless another collecting area is located, we will have to wait and see if the population increases and *S. scutellator* may once again be obtained, this time in adequate numbers.

Literature Cited

Arthur, A.P. (1962) A skipper, *Thymelicus lineola* (Ochs.) (Lepidoptera: Hesperiidae) and its parasites in Ontario. *Canadian Entomologist* 94, 1082–1089.

Arthur, A.P. (1968) Further information on the control of the introduced European skipper, *Thymelicus lineola*, with *Bacillus thuringiensis*. *Journal of Invertebrate Pathology* 10, 146–150.

Arthur, A.P. (1971) *Thymelicus lineola* (Ochs.), an introduced skipper (Lepidoptera: Hesperiidae). In: Biological control programmes against insects and weeds in Canada 1956–1968. *Commonwealth Institute of Biological Control Technical Communication* 4, 51–54.

Arthur, A.P.; Angus, T.A. (1965) Control of a field population of the introduced European skipper, *Thymelicus lineola* (Ochsenheimer) (Lepidoptera: Hesperiidae) with *Bacillus thuringiensis* Berliner. *Journal of Invertebrate Pathology* 7, 180–183.

Beirne, B.P. (1975) Biological control attempts by introductions against pest insects in the field in Canada. *Canadian Entomologist* 107, 225–236.

Carl, K.P. (1968) *Thymelicus lineola* (Lepidoptera: Hesperiidae) and its parasites in Europe. *Canadian Entomologist* 100, 785–801.

Duchesne, R.-M. (1980) Possibilités de contrôle à court terme et à long terme de l'hespérie des graminées, *Thymelicus lineola* (Lepidoptera: Hesperiidae), à l'aide d'un virus de la polyhédrie nucléaire endogène. Ph.D. Thesis, Laval University, 142 pp.

Harcourt, D.G. (1966) Major factors in survival of the immature stages *Pieris rapae* (L.). *Canadian Entomologist* 98, 653–662.

Jackson, B.S. (1978) Records of the European skipper in Newfoundland. *Canadian Field-Naturalist* 92,200.

Letendre, M.; McNeil, J.N. (1980) Field evaluation of insecticides for the control of the European skipper, *Thymelicus lineola* (Lepidoptera: Hesperiidae), in Québec. *Canadian Entomologist* 111, 1313–1317.

McNeil, J.N.; Duchesne, R.–M.; Comeau, A. (1975) Known distribution of the European skipper, *Thymelicus lineola* (Lepidoptera: Hesperiidae) in Québec. *Canadian Entomologist* 107, 1221–1225.

McNeil, J.N.; MacLeod, J.M. (1982) Two species of fungus infecting the European skipper, *Thymelicus lineola* (Ochs.) (Lepidoptera: Hesperiidae), in Québec. *Canadian Entomologist* 114, 87–88.

McNeil, J.N.; Smirnoff, W.A.; Letendre, M. (1977) *Bacillus thuringiensis* as a means of control of the European skipper, *Thymelicus lineola* (Lepidoptera: Hesperiidae). *Canadian Entomologist* 109, 37–38.

Preston, W.B.; Westwood, A.R. (1981) The European skipper, *Thymelicus lineola* (Lepidoptera: Hesperiidae) in Manitoba and northwestern Ontario. *Canadian Entomologist* 113, 1123–1124.

Smirnoff, W.A. (1974) Une épizootie virale chez l'hespérie européenne, *Thymelicus lineola* (Ochs.) (Lepidoptères: Hesperiidae), dans la région du Lac St-Jean, Québec. *Phytoprotection* 55,135–138.

Smirnoff, W.A.; McNeil, J.N.; Valero, J.R. (1976) Experiments on the control of the European skipper, *Thymelicus lineola* (Lepidoptera: Hesperiidae) by aerial application of a nuclear polyhedrosis virus. *Canadian Entomologist* 108, 1221–1222.

Thompson, L.S. (1974) Prince Edward Island insects. *Canadian Agricultural Insect Pest Review* 52,5.

Thompson, L.S. (1977) Field tests with chemical and biological insecticides for control of *Thymelicus lineola* on timothy. *Journal of Economic Entomology* 70, 324–326.

Tipula paludosa Meigen, European Cranefly (Diptera: Tipulidae)

A.T.S. WILKINSON

Pest Status

The early history of the European crane fly, *Tipula paludosa* Meigen, in Canada has been recorded already (Fox 1957, Wilkinson & MacCarthy 1967, Wilkinson 1971). In Nova Scotia, this pest peaked about 1965 (Creelman 1971) but had declined to a low level by 1971. High populations of the larvae (leatherjackets) damaged golf courses and lawns around St. John's, Newfoundland until 1970 (Creelman 1969, 1970). In the lower Fraser Valley of British Columbia, it has spread as far east as Hope. On Vancouver Island, it was found near Duncan and Victoria in 1972 and caused damage to golf courses and lawns in Victoria in 1976. It damaged lawns in Nanaimo in 1978 and in Alberni in 1979. In 1980, it damaged lawns and a golf course in Prince Rupert (Fig. 3). Populations were highest when the pest was first found in 1965, but as the infestation was spread by gravid females from the introduction sites near Vancouver into the lower Fraser Valley and south into Washington State, populations declined, damage became less, and chemical controls became unnecessary. When new isolated infestations such as those on Vancouver Island and at Prince Rupert occur, they behave like those near Vancouver did and require control measures, at least in the first few years. The spread south in Washington was recorded by Jackson & Campbell in 1975, when crane flies had been found 180 km south of the British Columbia-Washington border.

No indigenous parasitoids have been found in larvae of *T. paludosa* but many predaceous ground beetles and birds have been observed feeding on them. The starling continues to be the best predator of both the larvae and adults. Many birds feed on the adults.

One disease, or combinations of the many diseases found in the larvae, may have been responsible for reducing populations and keeping them below the economic level, but this has not been proven. The following diseases have been identified in *T. paludosa* larvae from the Vancouver area by Dr. P.L. Sherlock, Rothamsted Experimental Station, Harpenden, Herts., England: *Gregarina longa* Leger, *Hirmocystis ventricosa* (Leger) Labbe, *Actinocephalus tipulae* (Hammerschmidt) Leger, *Diplocystis* sp. and *Nosema binucleatum* Weissenberg. All except *N. binucleatum* were also found in larvae from Newfoundland. In British Columbia, up to 50% of the larvae examined harbour all protozoans mentioned.

Background

The literature indicated that the parasitoid *Siphona geniculata* De Geer was considered the primary control agent of *T. paludosa*. In England, Rennie & Sutherland (1920) recorded 21.3% and 18.9% parastism for 2 consecutive years for the first generation and 27.7% parasitism for the second generation. The latter was based on a low host population. Parasitism by *S. geniculata* in other parts of Europe was not as high as that reported by Rennie & Sutherland. The Commonwealth Institute of Biological Control was retained by Agriculture Canada, Research Branch, in 1967 to study natural controls in Germany where *T. paludosa* occasionally causes damage in pastures. Populations of *T. paludosa* examined for parasitoids in 3 sites in Germany by Carl (1972) showed parasitism of 15.5, 0.8, and 6.0% for the first generation and 4.9, 4.5, and 4.0% for the second generation. *S. geniculata* has been recorded in Europe from *Mamestra brassicae* (L.), *T. oleracea* L., *T.*

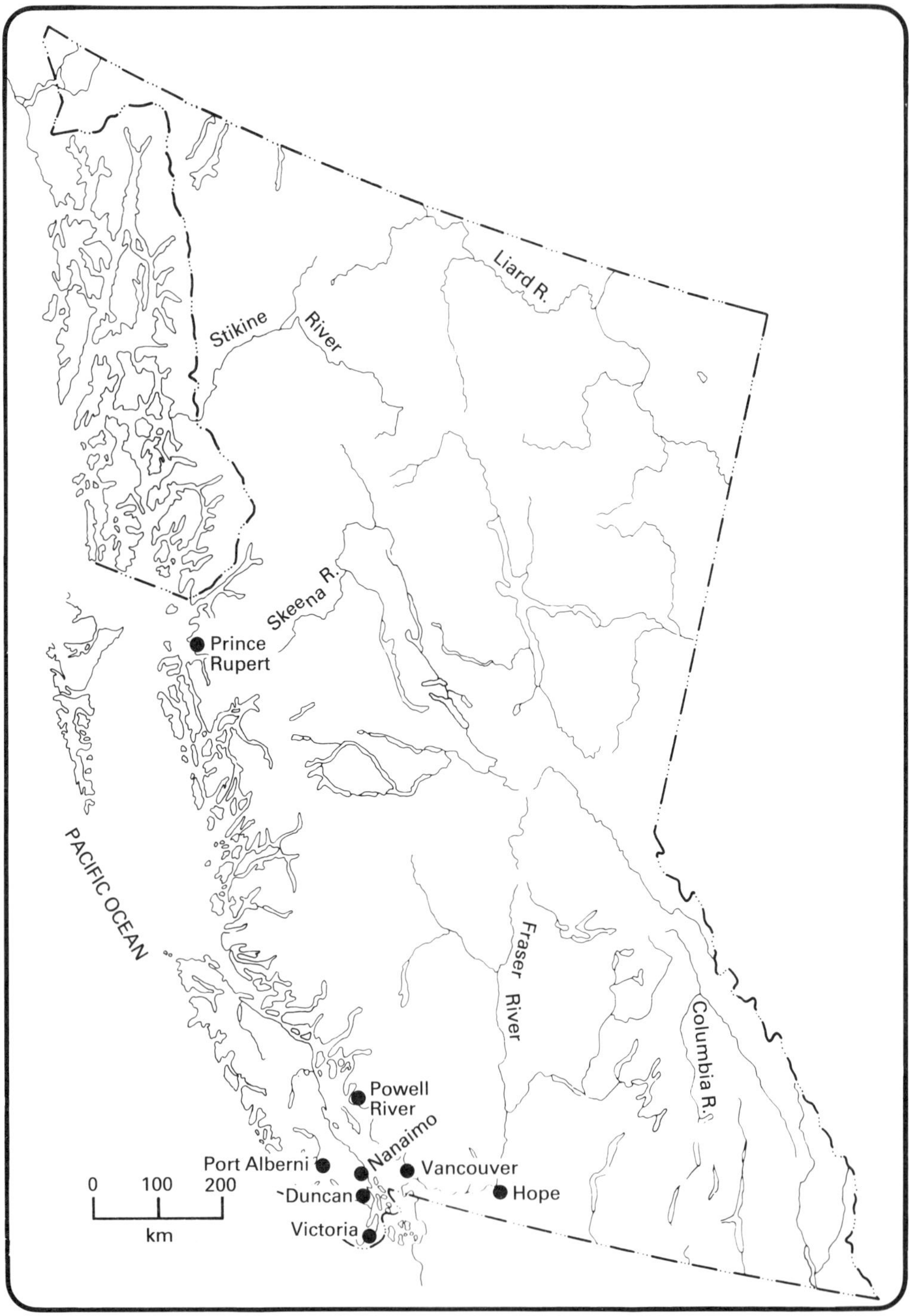

Fig. 3. Map of British Columbia showing localities where *Tipula paludosa* Mg. occurs

fulvipennis De Geer, *T. vittata* Meigen. *T. subnodicornis* Zetterstedt, *T. montium* Egger, and *T.maxima* Poda (Rennie & Sutherland 1920, Chiswell 1956, Coulson 1962, Alma 1975) as well as from *T. paludosa*. It has been reared from only *T. paludosa* in British Columbia.

A phorid, *Megaselia paludosa* (Wood), was found in *T. paludosa* by Coggins (1970) and by Carl (1972). It is rare in Europe and requires damp conditions.

Many insecticides tested for the control of leatherjackets were effective, but only diazinon, chlorfenvinphos, and parathion were registered. Diazinon is used primarily on home and public lawns and the other two are used on pastures. Except in the more recent infestations at Nanaimo, Port Alberni, Prince Rupert, and Victoria, there is very little need for chemical control. If these infestations follow the pattern of those in the lower Fraser Valley, populations will decline and chemical controls will no longer be needed.

The B-exotoxin of *Bacillus thuringiensis* Berliner and the DD-136 nematode of *Neoplectana carpocapsae* Weiser were investigated by Lam & Webster (1972) as possible controls for *T. paludosa*. Both B-exotoxin and DD-136 were successful in killing leatherjackets in the laboratory, but the experimental evidence suggests that the heavy application required to achieve high mortality would not be practical for control of *T. paludosa* in the field.

Carter (1973*a*, 1973*b*) showed that *Tipula* iridescent virus (TIV) can be transmitted when healthy larvae feed on a TIV-infected larva. Field trials by Carter (1978) showed that TIV can be introduced into field populations of *T. paludosa* but with low efficiency.

Releases and Recoveries

In 1968, 331 adult *S. geniculata* were received from the Commonwealth Institute of Biological Control (CIBC), Switzerland, from which 565 adults were reared and released near Cloverdale, British Columbia. Between 1971 and 1975 an additional 7000 *S. geniculata* received from the CIBC or reared at the Vancouver Research Station were released in the lower Fraser Valley. A technique for breeding *S. geniculata* was developed by Carl (1972). Leatherjackets were inoculated with larvae dissected from field-collected or laboratory-reared gravid females of *S. geniculata*. The first recovery was in 1972 when 4 flies were reared from puparia attached to a leatherjacket cadaver. Counts of leatherjackets and parasitism were taken from January to August at Cloverdale from 1974 to 1977 and at Ladner from 1975 to 1979. Parasitism was never very high. At Cloverdale, the highest parasitism recorded for the first generation (December to March) was 4.3% with an average of 1.6%, and for the second generation (June and July) the highest parasitism recorded was 5.1% in 1974 and for the 4 years parasitism averaged 2.4%. At Ladner during 5 years the highest parasitism was 3.3% for the winter generation with an average of 2.3%, and for the second generation the highest parasitism recorded was 6% with an average of 2.8%. As in Europe, the percent parasitism determined for the second generation was based on low numbers of the host because of the marked decrease in the larval population late in the season.

In 1973, 100 *S. geniculata* adults were air expressed to Newfoundland; 79 survived and were released. In 1974, 600 were sent and released. None has yet been recovered.

Megaselia paludosa was found in some of the leatherjackets received at the Vancouver laboratory from CIBC, Switzerland. It was reared for several generations but the colony died and the parasitoid was never released.

Tipula iridescent virus-infected larvae were released near Ladner in 1975 but the virus has not been found in any larvae examined from this area.

Evaluation of Control Attempts

The population of *T. paludosa* has declined to a non-economic level in most areas. This cannot be attributed to parasitism by *S. geniculata* because the decline started before the parasitoid was established and populations have decreased in areas where the parasitoid does not occur. The parasitoid has been found as far east as Abbotsford and as far south as the British Columbia-Washington border. *S. geniculata* has not increased to the population levels reported in Europe, but may do so when it has spread over most of the area. It may aid in keeping the leatherjacket population at its present low level.

The TIV was difficult to establish in a colony in the laboratory under controlled conditions and has not been recovered from a release made in the field.

M. paludosa is rare in Europe and it was difficult to maintain a laboratory colony. Because it is so rarely found, even in high populations, it is doubtful that it is an effective parasitoid.

Recommendations

It is recommended that adult *S. geniculata* be collected in the lower Fraser Valley and released in the more isolated areas on Vancouver Island and at Prince Rupert. Further attempts should be made to establish it in Newfoundland.

Literature Cited

Alma, P.J. (1975) Parasitization of *Tipula* spp. (Dipt., Tipulidae) by *Siphona geniculata* (De Geer) (Dipt., Tachinidae). *Entomologist's Monthly Magazine* 111, 105–107.

Carl, K.P. (1972) Marsh-crane fly (*Tipula paludosa*) work in Europe in 1972. Commonwealth Institute of Biological Control, European Station, Annual Project Statement.

Carter, J.B. (1973*a*) The mode of transmission of *Tipula* iridescent virus. I. Source of infection. *Journal of Invertebrate Pathology* 21, 123–130.

Carter, J.B. (1973*b*) The mode of transmission of *Tipula* iridescent virus. II. Route of infection. *Journal of Invertebrate Pathology* 21, 136–143.

Carter, J.B. (1978) Field trials with *Tipula* iridescent virus against *Tipula* spp. larvae in grassland. *Entomophaga* 23, 169–174.

Chiswell, J.R. (1956) A taxonomic account of the last instar larvae of some British Tipulinae (Diptera: Tipulidae). *Transactions of the Royal Entomological Society of London* 108, 409–484.

Coggins, R.E. (1970) *Megaselia paludosa* (Wood) (Dipt., Phoridae) parasitizing *Tipula paludosa* Mg. (Dipt., Tipulidae). *Entomologist's Monthly Magazine* 106, 108.

Coulson, J.C. (1962) The biology of *Tipula subnodicornis* Zetterstedt, with comparative observations of *Tipula paludosa* Meigen. *Journal of Animal Ecology* 31, 1–21.

Creelman, I.S. (Ed.) (1969) Pests of forage grasses and legumes. *Canadian Agricultural Insect Pest Review* 47, 7.

Creelman, I.S. (Ed.) (1970) Pests of shade trees, ornamentals, and turf. *Canadian Agricultural Insect Pest Review* 48, 31.

Creelman, I.S. (Ed.) (1971) Pests of shade trees, ornamentals, and turf. *Canadian Agricultural Insect Pest Review* 49, 34.

Fox, D.J.S. (1957) Note occurrence in Cape Breton Island of *Tipula paludosa* Meig. (Diptera: Tipulidae). *Canadian Entomologist* 89, 288.

Jackson, D.M.; Campbell, R.L. (1975) Biology of the European crane fly *Tipula paludosa* Meigen, in western Washington (Tipulidae; Diptera). *Washington State University Technical Bulletin* 81.

Lam, A.B.Q.; Webster, J.M. (1972) Effect of DD-136 nematode and of B-Exotoxin preparation of *Bacillus thuringiensis* var. *thuringiensis* on leatherjackets, *Tipula paludosa* larvae. *Journal of Invertebrate Pathology* 20, 141–149.

Rennie, J.; Sutherland, C.H. (1920) On the life history of *Bucentes (Siphona) geniculata* (Diptera: Tachinidae) parasite of *Tipula paludosa* (Diptera) and other species. *Parasitology* 12, 199–211.

Wilkinson, A.T.S.; MacCarthy, H.R. (1967) The marsh crane fly, *Tipula paludosa* Mg., a new pest in British Columbia (Diptera: Tipulidae). *Journal of the Entomological Society of British Columbia* 64, 29–34.

Wilkinson, A.T.S. (1971) *Tipula paludosa* (Meig.), European crane fly (Diptera: Tipulidae). *In*: Biological control programmes against weeds and insects in Canada 1959–1968. *Commonwealth Institute of Biological Control Technical Communication* 4, 54–57.

Trialeurodes vaporariorum (Westwood), Greenhouse Whitefly (Homoptera: Aleyrodidae)

R.J. McCLANAHAN

Pest Status

The greenhouse whitefly, *Trialeurodes vaporariorum* (Westwood), can be a problem wherever greenhouse vegetable and flower crops are grown. In major greenhouse areas of Canada, the damage has been alleviated considerably by extensive use of the parasitoid *Encarsia formosa* Gahan. When whiteflies are not controlled they cause reduction in plant vigour, formation of sooty mould on leaves and fruit, and reduced yield. Insecticide sprays, particularly the synthetic pyrethroids, are effective against the adults but repeated applications are required to keep the population at sub-economic levels.

Much of the increased acreage under plastic and glass in the 1970s has been used for flower crops and bedding plants. These products may be infested with whitefly when they leave the greenhouse, in which case they have a lower value as well as serving to spread whitefly into new areas. Greenhouse whiteflies are also a problem in small hobby greenhouses or solar rooms of houses in which plants are grown. Biological control is the best means of combatting the pest in these situations.

Background

A research project on control of whiteflies was conducted at the Harrow Research Station from 1966 to 1974. An integrated control schedule was devised in which *E. formosa* were released, then sprays of oxythioquinox were applied to keep whiteflies under control until the parasitoids were established (McClanahan 1970, 1972). The withdrawal of oxythioquinox in 1974 shifted the emphasis to biological control alone. By this time mass rearing was well established and sufficient parasitoids were available to provide good whitefly control on greenhouse tomatoes and cucumbers (McClanahan 1973).

In Europe, renewed interest in biological control of whiteflies was generated by a need for control methods compatible with the use of *Phytoseiulus persimilis* Athias-Henriot for twospotted mite control. Extensive research was conducted in England and Holland on the interaction between *E. formosa* and whiteflies. A good review of this European research was given by Vet *et al.* (1980). Although there has been interest expressed in finding a whitefly parasitoid more effective than *E. formosa* at low temperatures, none has been found to date.

The greenhouse whitefly was introduced to Japan about 1974 and within 4 years it became an important pest of greenhouse crops. A number of native parasitoids were found, not including *E. formosa* (Nakazawa & Hayashi 1977). Biological control is being considered in Japan.

Releases and Recoveries

The use of *E. formosa* for whitefly control is so extensive that it is difficult to estimate the numbers released (Table 15). In the early 1970s production and distribution were handled by the Harrow Research Station. Parasitized whiteflies on tobacco leaves were packaged in lots estimated to contain 2, 5, or 10 thousand.

A number of small businesses have been involved in *E. formosa* production and sales. Greenhouse Bio-Controls (1972–77) mainly supplied hobby greenhouse owners, with most sales to points in the United States. In 1973 and 1974 Greenhouse Bio-Controls supplied some of the needs of the Ontario Greenhouse Vegetable Producers' Marketing Board (OGVPMB) members, with 650 000 and 947 000 parasitoids sold to the Board and distributed by them. This accounts for the increased Canadian distribution from this source in those years. In 1974–75, OGVPMB contracted *E. formosa* production to a grower, but that was not as successful as a subsequent arrangement whereby they provided partial salary for a technician supervised by Ontario Ministry of Agriculture and Food, with the rearing done in an isolated greenhouse at the Harrow Research Station.

Table 15 Releases of *Encarsia formosa* Gahan in Canada against *Trialeurodes vaporariorum* (Westwood).

Source	Numbers released* (thousands)											
	1969	1970	1971	1972	1973	1974	1975	1976	1977	1978	1979	1980
Harrow Research Station	10	150	943	3133	2148		1485	1157	6	15	12	19
Greenhouse Bio-Controls				14	886	1028	47	41	81			
OGVPMB**						500	750	2065	2160	1055	1265	550
Better Yield Insects											25	40
Applied Bio-Nomics										450	1750	588

* Parasitized whitefly pupae from which there is 90–98% emergence of adult parasitoids.
** Ontario Greenhouse Vegetable Producers' Marketing Board.

In 1979 a similar arrangement was set up in British Columbia so growers there and in Alberta are supplied with *E. formosa* from Applied Bio-Nomics Ltd. The rearing facilities were supplied by Agriculture Canada and technical support was given by the British Columbia Department of Agriculture.

Better Yield Insects was recently established in Ontario to produce and sell *E. formosa* and other biological control agents. This source serves mainly hobby greenhouse owners in the United States, but plans were made to supply greenhouse vegetable growers in Ontario through OGVPMB, in 1981.

Most releases have resulted in establishment, as judged by the very few failures reported. There have been cases where parasitoids were eliminated by an insecticide application, but with just one spray they have survived in the pupal stage and continued to multiply on the surviving whiteflies.

Evaluation of Control Attempts

The overall incidence of whitefly in Essex County has been reduced to the point where it is no longer a limiting factor in greenhouse tomato and cucumber production, and it is seldom a problem on field vegetables. *E. formosa* are easily found outdoors every summer.

The need for supplies of *E. formosa* has decreased as whitefly populations declined, with a turning point about 1974 in Essex County, Ontario. Other centres of greenhouse production have not realized the full potential of biological control of whiteflies, but they are aware of it and are using increasing numbers of parasitoids, particularly in Québec and British Columbia.

Growers have stated that they have had increased yields of tomatoes since they have stopped using insecticides. There are others who regard biological control as the only solution to problems of allergenic reactions to insecticides.

Commercial large scale production of *E. formosa* is desirable (McClanahan 1980) and this goal is being achieved with some short term government assistance.

Recommendations

The production units for *E. formosa* should not be geared to the needs of a particular area of greenhouse production, since the demand will be at a peak for a month or two then fall off. A wider market is desirable and exportation of parasitoids to other countries, particularly the United States, would provide a steady market. Present regulations require United States purchasers to get an import authorization, because of the unparasitized whitefly pupae in the shipment. Ideally shipments should be exempt from this requirement since whiteflies are univeral and the regulation slows shipments.

Literature Cited

McClanahan, R.J. (1970) Integrated control of the greenhouse whitefly on cucumber. *Journal of Economic Entomology* 63, 599–601.

McClanahan, R.J. (1972) Integrated control of the greenhouse whitefly. *Agriculture Canada Publication* 1469, 6 pp.

McClanahan, R.J. (1973) Integrated control of greenhouse pests. *Canada Agriculture* 18, 34–35.

McClanahan, R.J. (1980) Why has integrated control practice in the greenhouse levelled off in Canada? *Bulletin SROP* 1980/III, 141–144.

Nakazawa, K.; Hayashi, H. (1977) Studies on the biology and control of the greenhouse whitefly. 3. Present condition of the natural enemy complex of greenhouse whitefly in Hiroshima Prefecture. *Bulletin of the Hiroshima Prefecture Agricultural Experiment Station* 39, 35–42.

Vet, L.E.M.; Van Lenteren, J.C.; Woets, J. (1980) The parasite-host relationship between *Encarsia formosa* (Hymenoptera: Aphelinidae) and *Trialeurodes vaporariorum* (Hymenoptera: Aleyrodidae): 9. A review of the biological control of the greenhouse whitefly with suggestions for future research. *Zeitschrift für angewandte Entomologie* 90, 26–51.

PART II

BIOLOGICAL CONTROL OF WEEDS IN CANADA 1969–80

Current Approaches to Biological Control of Weeds

P. HARRIS

Success of the Programme

A total of 30 agents has been released against 14 weed species in Canada. A third of the agents have failed to become established, a third are established but inflict relatively minor damage to the weed, and a third inflict major damage. This gives a crude measure of the success of the programme, but it is not adequate to determine whether investment in biological control of weeds is justified in economic terms. The opportunity is taken in this chapter to discuss this and other current issues.

The success of any pest control programme depends on a combination of its effectiveness and its cost. In biological control both of these vary greatly with the agent selected. It is easiest to predict the cost of an agent; so other things equal, the least expensive agent is the best one. Indeed, van Lenteren (1980) suggested that there are too many variables and unknowns to enable an effective agent to be selected on a scientific basis. Scientific or not, there are rules of thumb for increasing the chances of success: matching the climate of the collection and release area (Wapshere 1980), collecting from the host plant species or strain on which the agent is to be released, and Harris (1973) suggested a number of desirable agent characteristics. These approaches can now be tested and modified against the record in Julien's (1982) world catalogue of releases. The procedures that work in practice can then be followed, even if the reasons for success are not understood.

Biological control of weeds, in common with government sponsored chemical weed control programmes such as those against leafy spurge and knapweed in Canada, has been derelict in not evaluating the benefits of the programme in economic terms. Without economic cost:benefit data it is difficult to determine whether a particular strategy against a weed makes economic sense or it is being pursued for political, philosophical, or historical reasons. Several chapters in this section represent initial evaluation of the prospective cost:benefits of biological control of possible target weeds. Hopefully, this is a step towards substantiating the intuition that biological control is economically superior to any other method of control for certain types of weed problems.

The usual practice in biological control is to rate individual agents as failures, partial or complete successes depending on their impact on the density of the host (see Laing & Hamai 1976). This is neat, but a problem, because it is sometimes the combined cropping pressure of several agents on a weed that achieves its density reduction; also the effects on weed density are often delayed by the seed accumulated in the soil so that the project cannot be rated until several years after the agent has achieved its maximum effect. I suggest that it is more useful to rate biological control agents of weeds in terms of the proportion of the annual production removed or destroyed. The agents are similar in effect to cows in a pasture: if consumption is below a certain level, future yield is not reduced. Thus the good farmer restrains utilization of grasses to 40–60% of annual production depending on the species and area. The good biological control worker, on the other hand, must try to exceed the threshold for the target weed (which might be determined by clipping tests). Most forbs are less tolerant than grasses to utilization (Jameson 1963), but many shrubs will tolerate browsing up to 75% of annual production (Ellison 1960). If the threshold of a target weed is assumed to be 50% of annual production, it will not be controlled by an agent that removes 30% of production. Nevertheless the addition of new defoliators, seed or root feeders, will normally increase the total amount of the weed consumed. Thus the practice is to add agent species until the threshold is exceeded. For this reason I consider any agent that has become numerous enough to remove a substantial proportion of annual production as a success. I would prefer to rate them by the actual amount removed but do not have the data for all the species established in Canada.

The successes of the Canadian programme are listed in Table 16. Several of the agents have in fact achieved density reductions of the target weed that have solved some of the problems from it. For example, the seedhead weevil, *Rhinocyllus conicus* Froel., has reduced the density of *Carduus nutans* L. sufficiently on Saskatchewan rangeland that it does not threaten the cattle stocking rate. On the other hand, the thistle remains numerous on disturbed sites such as gravel pits, and even at low densities can be a nuisance in parks and around beaches. Similarly *Hypericum perforatum* L. presently has little effect on the stocking rate of British Columbia rangeland: the dense stands at Elko, British Columbia, are not of concern for cattle production as stocking is kept low to retain the area as a wildlife overwintering refuge. The weed, however, is still perceived to be a problem by the ranchers, possibly because they fear its spread.

Table 16 Agents removing a substantial proportion of the annual production of the target weed in one or more regions of Canada.

Target Weed	Agent	Country Screening Agent
Carduus nutans L.	*Rhinocyllus conicus* Froel.	Canada
Centaurea diffusa Lam.	*Urophora affinis* Frfld.	Canada
C. maculosa Lam.	*U. quadrifasciata* (Mg.)	Canada
Cirsium vulgare (Savi) Ten.	*Urophora stylata* L.	Canada
Euphorbia cyparissias L.	*Hyles euphorbiae* (L.)*	Canada
Hypericum perforatum L.	*Anaitis plagiata* (L.)	Canada
	Chrysolina hyperici (Forst.)	Australia
	C. quadrigemina (Suffr.)	Australia
Senecio jacobaea L.	*Tyria jacobaeae* L.	New Zealand
	Longitarsus jacobaeae (Wat.)	United States

* The inclusion of this insect is possibly an optimistic estimate of its effect.

The initial work on four of the successful agents (Table 16) was done by other countries. Insects successful in Australia or California, such as *Chrysolina quadrigemina* (Suffr.), have tended to be successful in Canada despite differences in climate and the converse is true of Canadian pioneered species such as *R. conicus*. Indeed, the most effective and widely employed technique for selection of an effective agent is to choose a proven winner. Such agents are also inexpensive as most of the pre-release studies required by Canada have been done. Stock for the *Chrysolina* spp. established in Canada was obtained from California following success there. It would almost certainly have achieved an impact on the weed in Canada sooner if the releases had been made with climatically pre-adapted stock; but there is a trade-off between speed of success and cost. In North America, savings can be achieved on weeds common to both Canada and the United States by demonstrating the effectiveness of the agent in one country and only distributing those that are successful.

There are 11 agent species established in Canada that are not removing a substantial proportion of the host plant production (Table 17). The reasons for their failure to increase to a high population density vary. Some were released so recently that there has not been time. *Sphenoptera jugoslavica* Zell. which was released in 1977 has increased to a high density within 0.25 km of the release point but is still too restricted in distribution to be included in Table 16. Other species may require a period of climatic adaption through selection. Studies on the successful agents *Tyria jacobaeae* L., *Hyles euphorbiae* (L.), and *Chrysolina quadrigemina* indicated that a period of 2–13 generations of natural selection was required to produce a population adapted to the release area. Thus several of the insects in Table 17 are possible successes.

Table 17 Agents established in Canada but not removing a substantial proportion of annual production of target weed.

Target Weed	Agent	Year Established
Carduus nutans L.	*Trichosirocalus horridus* Panz.	1978
Centaurea diffusa Lam.	*Sphenoptera jugoslavica* Zell.	1977
C. maculosa Lam.	*Metzneria paucipunctella* Zell.	1974
Acroptilon repens (L.)	*Paranguina picridis* Kirj. & Ivan.	1977
Cirsium arvense Scop.	*Ceutorhynchus litura* F.	1968
	Urophora cardui (L.)	1976
Euphorbia esula-virgata complex	*Oberea erythrocephala* Schr.	1981
Hypericum perforatum L.	*Aphis chloris* (Koch)	1981
Linaria vulgaris Mill.	*Calophasia lunula* (Hufn.)	1966
Sonchus arvensis L.	*Cystiphora sonchi* (Bremi)	1982
Carduus acanthoides L.	*Rhinocyllus conicus* Froel.	

The failure of *R. conicus* to attack a high proportion of *Carduus acanthoides* L. flower heads seems to be related to a poor adaptation to this thistle. The weevil was a success on *C. nutans* in Canada, the host plant from which it was collected in Europe. The association of both specialized insects and pathogens with a host plant is an extremely stable relationship. The completion of a number of generations on a marginal host normally does not improve survival unless the genetic potential to overcome host resistance already exists in the agent population. There are host races of *R. conicus* on a number of European thistles (Zwölfer & Preiss 1983); but it is uncommon on *C. acanthoides* and was not found in the samples examined by Zwölfer (1965). Thus the *C. nutans* strain of *R. conicus* is likely to remain marginal on *C. acanthoides* as the species does not have the genetic potential to exploit it successfully.

Two of the 12 failures listed in Table 18 have been established in Canada on the plant from which they were collected in Europe. Thus only ten or a third of the agents have failed to become established at all. This compares with a failure rate of about two thirds

Table 18 Agents not established in Canada.

Target Weed	Agent	Reason for Failure	Country Screening Agent
Cirsium arvense (L.) Scop.	*Altica carduorum* Guer.	Predation	Canada
Euphorbia esula-virgata	*Hyles euphorbiae* (L.)	Predation	Canada
complex	*Chamaesphecia empiformis* Esp.	Host unsuitable	Canada
	C. tenthrediniformis (Schiff.)	Host unsuitable	Canada
Hypericum perforatum L.	*Chrysolina varians* (Schall.)	Too dry in summer	Canada
	Agrilus hyperici (G.)	Did not breed	Australia
	Zeuxidiplosis giardi Kieff.	Winter kill	Australia
Salsola pestifer A. Nels.	*Coleophora parthenica* Meyr.	Summer too cool	United States
	C. klimeschiella Toll.	Flooding	United States
Senecio jacobaea L.	*Hylemya seneciella* (Meade)	Did not breed	New Zealand
Sonchus arvensis L.	*Tephritis dilacerata* Loew.	Winter kill	Canada
Linaria dalmatica (L.) Mill.	*Calophasia lunula* (Hufn.)	?	Canada

of the agents released against arthropod pests (Hall & Ehler 1979). The failure of *Calophasia lunula* Hufn. on *Linaria dalmatica* (L.) Mill. and the agents released on *Euphorbia esula-virgata* complex may be related to a poor adaptation to the target weed, as already discussed for *R. conicus*. It seems often that it is better to import from the actual target weed rather than to attempt to establish a race from a related host plant. Unfortunately this is difficult to do with leafy spurge as it is not clear just where the exact counterpart of the North American weed occurs in Europe.

It would be nice to improve the cost: benefit ratio of biological control by reducing the number of failures. However, the host specificity studies on five of the failed agents were done by other countries, so the cost of trying them in Canada was not large. Some of the failures seem to be related to the climate but not necessarily to the Canadian winter. The main saving that might have been achieved is not to have introduced the two insects against *Salsola* until the United States had completed their pilot release studies. To date, the results do not justify spending time in Canada on either species.

Effects Achieved with Classical Biological Control of Weeds

A basic effect of predatory and parasitic organisms is to increase diversity at lower trophic levels. In other words the presence of phytophages tends to increase the diversity of plants in the area. Stanley (1973) suggested that the flora of the Precambrian Era, with its extreme lack of diversity, was typical of an uncropped community, and the rapid diversification that occurred at the end of the era resulted from the appearance of phytophages. This theory in part is derived from the observable tendency of grazing pressure to transform simple plant communities to multi-species mixtures.

In effect, phytophages appear to be a driving force in plant succession. For example, Fox (1958) showed that insecticidal treatment of hayland in Nova Scotia to kill wire-worms delayed the take-over by spruce trees, the next successional stage. The climax community may be dominated by a single plant species if there is little cropping pressure on it, or by a highly diverse flora of many plant species each at a low density where the cropping pressure is high. Plant introductions to new regions, if they are adapted to the habitat and without their specialized phytophages, have a competitive advantage over native plants, which are almost certainly cropped by native insects and pathogens. The result is that an introduced species may displace the native flora to form a new climax or sub-climax. This particularly happens when the vigour of the native vegetation is depressed by overgrazing or disturbance. Canadian examples are *H. perforatum, Centaurea diffusa* Lam., *C. maculosa* Lam. in the grasslands of the interior of British Columbia; *Carduus nutans* in the mid-grass prairie of Saskatchewan; *Euphorbia esula-virgata* complex in Manitoba; and *E. cyparissias* L. on limestone soils in Ontario, North American plants such as *Solidago gigantea* Ait, behave similarly in parts of Europe. These plants do not form extensive monocultures in their native region where they are cropped by phytophages. For example, *E. cyparissias* normally forms only a small portion of the herbaceous community on limestone soils in Europe compared to 25–50% at Braeside, Ontario. The introduction of phytophages against *H. perforatum* in British Columbia and Ontario reduced it to a small percentage of its former abundance at most sites, and *C. nutans* has been reduced to less than 10% on Saskatchewan rangeland. In part, the effectiveness of a control agent depends on the level of competition from other plants (Harris 1981*a*) which is normally greatest in uncultivated habitats and least in disturbed sites.

Specialized phytophages and their hosts normally have a mutual density dependence. Thus cropping pressure is high when the plant is abundant and low when it is scarce and *vice versa* for the mortality of the phytophage. The result is that neither becomes exterminated. A Canadian biological control example is the cinnabar moth, *Tyria jacobaeae*, on *Senecio jacobaea* L. in British Columbia in which the moth has tracked the density of its host (Lakhani & Dempster 1981). Similarly Ralph (1977) found that the highest densities of milkweed *(Asclepias syriaca* L.) supported the highest densities of *Oncopeltus fasciatus* (Dall.), and that below a certain density that plant escaped attack. This has been called 'escape in space' by Feeney (1976).

To summarize, classical biological control of weeds is most appropriate in terms of its impact and its effects against introduced weed species that dominate large areas of range or other uncultivated land.

Number of Phytophages and Weed Control

Cropping pressure on a weed can be increased by establishing several phytophage species on it (Harris 1981*a*). Even two closely competing seed head flies on *C. diffusa* reduced seed production more when together than either did alone (Myers & Harris 1980). Internationally, introductions have tended to continue over a period of years until an average of four species has been established on the weed (Harris 1979). This process of adding agents can only go so far as there is a species area-asymptote: the number of phytophagous insects and pathogens cropping a plant species varies with the log of its abundance (Lawton & Schroeder 1977, Strong & Levins 1979). That is to say the number of phytophages that can be supported by a plant species doubles for every tenfold increase in its abundance. In southern Britain five continental species of Heteroptera have become established on pine following large plantings. The pines displaced juniper and as a result one juniper-feeding Heteroptera, which formerly existed in sizeable colonies, has become extinct (Southwood 1957). Simberloff (1978) achieved similar results experimentally by altering the size of mangrove islands. In some plants an expansion of range is matched by an increase in the species consuming it. The recruits are derived from other plants in the new range or from the host in the native habitat. This occurred within 50 years on cacao (Strong 1974). With other introduced plants such as *E. cyparissias* in North America, there has been negligible recruitment of insect consumers even after a hundred years. Presumably most native insects are unable to overcome the physical and chemical defenses of the plant and the infestations are too remote for specialized insects to reach them from Europe.

The number of niches available to insect species is partly a function of the plant architecture, so a species native to several parts of the world tends to support similar numbers of leaf chewers, stem borers, and insects in other guilds throughout its range, even though the insects may be in different taxonomic orders (Lawton 1978). The niches can be regarded as sites at which the nutrient pool in the plant can be tapped. The pool normally cannot be drained through any one tap but they all lower its level. This implies competition between agents in different niches. For example, Moran & Southwood (1982) found that there was interaction between chewers and sap suckers on the tree species studied.

The best targets for classical biological control are introduced plants on which the consumer pressure is below that found in the native region. It should be possible to raise consumption to at least the native region level although the number of phytophages that can be established will be lower as their density will not be depressed by specialized parasites. Probably each agent established on a weed increases the difficulty of establishing additional species so those with the most potential should be introduced first.

It is unlikely that native weeds can be controlled by classical biological control unless their abundance has increased. This has occurred with ragweed, *Ambrosia artemisiifolia* L. (Harris & Piper 1970). If the phytophage-ragweed area asymptote is below that of ragweeds that have not increased since European settlement, it should be possible to establish additional agents from South America; however the increased consumer pressure on the weed may not reduce allergenic hay-fever since the symptoms in man are relatively dose independent. Usually the best hope for the biological control of a native weed is the periodic application of an agent (probably a native pathogen) to keep it at an artificially high level. This is called augmentative biological control and is discussed under the heading 'The Future'. Hall *et al.* (1980) found no significant difference in the success of biological control against native and exotic insect pests, but whether the native pests had increased in adundance with modern agriculture was not considered.

Demonstration that Candidate Agents are not a Threat to Desirable Plants

The use of a plant species as a host by a phytophage depends on its suitability as a nutrient source and as a place to live, the opportunity of reaching the plant, and the survival advantage of using it (Harris 1981*b*). If one of these requirements is unfavourable, the agent will not reach high levels on it. The strategy is to show that the organism cannot develop on any plant outside a taxonomic group that does not contain desirable species (Harris & Zwölfer 1968, Zwölfer & Harris 1971, Wapshere 1974). Unfortunately in the laboratory, organisms develop on a broader range of plants than they attack in nature. It is then necessary to show that the organism either does not have the opportunity of establishing on the desirable plant in nature or that individuals doing so are at a selective disadvantage. In effect there is a 3 tier system of testing: desirable plants that fail the first test because they will support development, are considered at the next tier of opportunity and if this is positive, at the last, and most difficult, tier of advantage.

The opportunity of an insect to attack a plant within its distributional range normally depends on the presence in the plant of sensory tokens for attraction and acceptance by the ovipositing female. This is the result of many generations of selection to optimize survival and is not easily changed. Indeed, it is often difficult to establish an agent if the strain or species of the target weed is different from the one on which it was collected. This is evident in the programme for the biological control of leafy spurge in Canada. Similarly Goeden (1978) found that the race of *R. conicus* established on *C. nutans* in North America did not attack *Silybum marianum* (L.) Gaertn. in the field, although a strain collected from this plant established readily. This programming of an insect for a particular plant tends to break down in the laboratory and so is best field tested in the native region.

In contrast to the active dispersal of most insects, the passive dispersal of pathogens and to a slightly lesser extent certain insects like aphids, results in their deposition on many plant species which they will frequently attempt to attack. The host range of pathogens is normally genetically stable so most of the pathogen propagules landing on other plant species are lost. However, if the genetic capacity (possibly as a result of hybridization) to attack another plant species exists in a pathogen population, the occasionally successful individual can, by asexual reproduction, increase rapidly. It is not a good omen for biological control if dense stands of a desirable plant on which a pathogen can develop in the laboratory, grow near the target weed.

The advantage of a particular host species to a phytophage depends on its potential for increase on it. The innate reproductive capacity of the phytophage on various plants can be measured in the laboratory, but the rate of increase in the field is influenced by mortality, competition from other consumers, the density of the host, and other factors that cannot be measured in the laboratory. Thus a good laboratory host is not necessarily at risk in nature, and a relatively poor laboratory host might be regularly damaged if there are large amounts growing as a monoculture. This is hard to determine experimentally in the new habitat without danger that the organism will escape. For this reason the development of the organism on an introduced crop plant in the laboratory is justification for rejecting it as a biological control agent; but development on a native wild plant is less cause for concern. For example, the European moth *Phyllonorycter blancardella* (Fabr.) attacks *Crataegus* spp., apple, and several related plants in its native range but, where it has been introduced into Québec, Pottinger & LeRoux (1971) were unable to find it on *Crataegus* spp., an abundant native, although it was common on the introduced apple. Similarly Moran (1982, personal communication) found that insect pests on South African crops derived from native plants were, with rare exceptions, attacked by native species. On the other hand, introduced crops were attacked by both native and introduced insects.

There are examples in which introduced insects and pathogens have become pests of native plants. These phytophages are either polyphagous, like the gypsy moth, so they readily accept and develop on a wide range of plants; or the native lacks resistance, as in the case of the chestnut blight, so the organism has a higher reproductive capacity on it than on the original host. The best indication of the genetic capability of phytophages to utilize plants in the new habitat is their host range in the native region, and if necessary desirable plants of the new habitat can be tested there by planting them at their normal density. Thus a species such as *Lema cyanella* L., which in Europe is only found on *Cirsium arvense* (L.) Scop. (Peschken & Johnson 1979) is unlikely to attack native North American thistles to any extent even though it will accept them in the laboratory.

Selection of Target Weeds for Classical Biological Control

Classical biological control cannot escape some measure of politics because it depends on public funding. Also, as an agent will spread over many properties, it has to be introduced as a matter of public interest.

Originally the priorities for biological control in Canada were largely determined by popular demand: the squeaky wheel approach. This did not necessarily target the best or the most urgent weeds for biological control as is evident from some of the examples in this volume.

It is axiomatic that biological control is only justified if it is likely to produce a satisfactory return on investment which is greater than can be obtained by other means of control. The biological control of a suitable target weed is likely to require 20 scientist years ($2 million at present costs) (Harris 1979). These costs are less if host specificity testing has been done elsewhere. Determination of benefit requires quantification of savings and increased yields minus any detrimental effects of control such as the reduction of a nectar source. Costs and benefits that cannot be quantified in monetary units should be expressed in ecological terms. The study should be published before biological control is started. This gives an opportunity for objections to be voiced and the proposal debated.

The Future

Classical biological control of weeds is likely to continue to be directed primarily against introduced species that form extensive dense stands on uncultivated land. For this type of problem there is a hgh rate of success and the return is excellent, particularly when considered on a national or a continental basis. There are, however, a limited number of weed species that are prime targets in Canada. As work on these is completed or the possibilities for biological control are exhausted, as is happening for *C. arvense*, the programme will be directed increasingly against weeds of more minor importance. One way of maintaining a favorable benefit-cost ratio is to tackle them as joint projects with the United States or other countries so that the costs are shared and the benefits accrue on an international scale. At present international cooperation is informal and this will probably need to be replaced by more formal cooperative agreements against specific target weeds.

There is room to increase the rate at which present projects are completed but the long-term prospect is for a continued small Canadian participation in classicial biological control of weeds. There has been an increase in staffing since the 1968 review (Harris 1971) that reflects the economic seriousness of some of the projects such as knapweed and leafy spurge. Thus Agricultural Canada has increased its participation from 2 to 4 full-time scientists; McGill University (Macdonald College) has an active programme that emphasizes pathogens; the Province of Alberta is recruiting a scientist, and the University of British Columbia has a programme elucidating the effects of some of the agents. The programme will probably stay at this level with more emphasis placed on the selection of both the target weeds and the agents.

The main area for expansion is in augmentative biological control in which a pathogen is applied periodically as a bioherbicide. Two centres, Macdonald College and the Regina Research Station have started investigations in this area. The work has not been reported in this review as it has yet to lead to the licensing of a bioherbicide in Canada.

The few bioherbicides currently in use in the United States are for weeds of crops that are hard to control by other means (Templeton 1982). One of the attractions of bioherbicides is that they can be produced in small amounts at a reasonable cost. Thus they can be used for weed problems of minor crops in which the area involved does not justify the development of a specific herbicide. The method has a profit potential so it is likely to be developed primarily by industry and this could occur rapidly. The main obstacle to date has been the lack of licensing regulations. These have now been drafted in the United States and it is probable that Canada will adopt similar requirements. There is an initial role for government research to get the industry started. Subsequently the government role may be largely regulatory.

Literature Cited

Ellison, L. (1960) Influence of grazing on plant succession of range plants. *Botanical Review* 26, 1–66.

Feeney, P. (1976) Plant apparency and chemical defense. In: Wallace, J.W.; Mansell, R.L. (*Eds*.). Biochemical interactions between plants and insects. *Recent Advances in Phytochemistry* 10, 1–40. New York; Plenum Press.

Fox, C.J.S. (1958) Some effects of insecticide on the wireworms and vegetation of grassland in Nova Scotia. *Proceedings of the 10th International Congress of Entomology* 3, 297–300.

Goeden, R.D. (1978) Initial analysis of *Rhinocyllus conicus* (Froelich) (Coleoptera: Curculionidae) as an introduced natural enemy of milk thistle (*Silybum marianum* (L.) Gaertner) and Italian thistle (*Carduus pycnocephalus* L.) in southern California: 39–50. In: Biological control of the genus *Carduus* in the United States. USDA, Stoneville, Miss., USA, 50 pp.

Goeden, R.D.; Ricker, D.W. (1977) Establishment of *Rhinocyllus conicus* on milk thistle in southern California. *Weed Science* 25, 288–292.

Hall, R.W.; Ehler, L.E. (1979) Rate of establishment of natural enemies in classical biological control. *Bulletin of the Entomological Society of America* 25, 280–282.

Hall, R.W., Ehler, L.E.; Bisabri-Ershadi, B. (1980) Rate of success in classical biological control of arthropods. *Bulletin of the Entomological Society of America* 26, 111–114.

Harris, P. (1971) Current approaches to biological control of weeds. In: Biological control programmes against insects and weeds in Canada 1959–1968. *Commonwealth Institute of Biological Control Technical Communication* 4, 67–76.

Harris, P. (1973) The selection of effective agents for the biological control of weeds. *Canadian Entomologist* 105, 1495–1503.

Harris, P. (1979) The cost of biological control of weeds by insects in Canada. *Weed Science* 27, 242–250.

Harris, P. (1981*a*) Stress as a strategy in the biological control of weeds. Biological control in crop production. In: Papavizas, G.C. (*Ed.*). BARC Symposium 5, pp. 333–340.

Harris, P. (1981*b*) Host preferences of phytophagous insects and their economic implications. *Proceedings of Symposia 9th International Congress of Plant Protection* 1, 105–109.

Harris, P.; Piper, G. (1970) Ragweed, *(Ambrosia* spp.: Compositae), its North American insects and the possibilities for biological control. *Commonwealth Institute of Biological Control Technical Bulletin* 13, 117–140.

Harris, P.; Zwölfer, H. (1968) Screening of phytophagous insects for biological control of weeds. *Canadian Entomologist* 100, 295–303.

Jameson, D.A. (1963) Responses of individual plants to harvesting. *Botanical Review* 29, 532–594.

Julien, M.H. (Ed.) (1982) Biological control of weeds. Commonwealth Institute of Biological Control, 108 pp.

Laing, J.E.; Hamai, J. (1976) Biological control of insect pests and weeds by imported parasites, predators and pathogens. In: Huffaker C.B.; Messenger, P.S. (*Eds.*) Theory and practice of biological control, Academic Press, pp. 685–743.

Lakhani, K.H.; Dempster, J.P. (1981) Cinnabar moth and its food plant, ragwort: further analysis of a simple interacton model. *Journal of Animal Ecology* 50, 231–249.

Lawton, J.H. (1978) Host-plant influences on insect diversity: the effects of space and time. In: Mound, L.A.; Waloff, N. (*Eds.*) Diversity of insect faunas. *Symposium of the Royal Entomological Society of London* 9, 105–125.

Lawton, J.H.; Schroeder, D. (1977) Effects of plant type, size of geographical range and taxonomic isolation on numbers of insect species associated with British plants. *Nature* 265, 137–140.

Lenteren, J.C. van (1980) Evaluation of control capabilities of natural enemies: does art have to become science? *Netherlands Journal of Zoology* 30, 369–381.

Moran, V.C.; Southwood, T.R.E. (1982) The guild composition of arthropod communities in trees. *Journal of Animal Ecology* 51, 289–306.

Myers, J.H.; Harris, P. (1980) Distribution of *Urophora* galls in flower heads of diffuse and spotted knapweed in British Columbia. *Journal of Applied Ecology* 17, 359 367.

Peschken, D.P.; Johnson, G.R. (1979) Host specificity and suitability of *Lema cyanella* (Coleoptera: Chrysomelidae), a candidate for the biological control of Canada thistle (*Cirsium arvense). Canadian Entomologist* 111, 1059–1068.

Pottinger, R.P.; LeRoux, E.J. (1971) The biology and dynamics of *Lithocolletis blancardella* (Lepidoptera: Gracillariidae) on apple in Québec. *Memoirs of the Entomological Society of Canada* 77, 437 pp.

Ralph, C.P. (1977) Effect of host plant density on populations of a specialized seed-sucking bug, *Oncopeltus fasciatus*. *Ecology* 58, 799–807.

Simberloff, D.S. (1978) Colonization of islands by insects: immigration, extinction and diversity. In: Mound, L.A.; Waloff, N. (*Eds.*). Diversity of insect faunas. *Symposium of the Royal Entomological Society of London* 9, 139–153.

Southwood, T.R.E. (1957) The zoogeography of the British Heteroptera. *Proceedings of the South London Entomological and Natural History Society* 1956, 111–136.

Stanley, S.M. (1973) An ecological theory for the sudden origin of multicellular life in the late Precambrian. *Proceedings of the National Academy of Science* 70, 1486–1489.

Strong, D.R. (1974) Rapid asymptotic species accumulation in phytophagous insect communities: the pests of cacao. *Science* 185, 1064–1065.

Strong, D.R.; Levin, D.A. (1979) Species richness of plant parasites and growth form of their hosts. *American Naturalist* 114, 1–22.

Templeton, G.E. (1982) Biological herbicides: discovery, development, deployment. *Weed Science* 30, 430–433.

Wapshere, A.J. (1974) A strategy for evaluating the safety of organisms for biological weed control. *Annals of Applied Biology* 77, 201–211.

Wapshere, A.J. (1980) Recent thoughts on exploration and discovery for biological control of weeds. *Proceedings of the 5th International Symposium on Biological Control of Weeds*. Brisbane, Australia, pp. 75–79.

Zwölfer, H. (1965) Preliminary list of phytophagous insects attacking wild Cynareae (Compositae) species in Europe. *Commonwealth Institute of Biological Control Technical Bulletin* 6, 81-154.

Zwölfer, H.; Harris, P. (1971) Host specificity determination of insects for biological control of weeds. *Annual Review of Entomology* 16, 159–178.

Zwölfer, H.; Preiss, M. (1983) Host selection and oviposition behaviour in west-European ecotypes of *Rhinocyllus conicus* Froel. (Col.: Curculionidae) *Zeitschrift fur angewandte Entomologie* 95, 133-122

Acroptilon repens (L.) DC., Russian Knapweed (Compositae)

A.K. WATSON and P. HARRIS

Pest Status

Acroptilon repens (L.) DC. (*Centaurea repens* L.) is native to Mongolia, Western Turkestan, Iran, Turkish Armenia and Asia Minor (Moore & Frankton 1974), and was first introduced into Canada about 1900 as a contaminant of Turkestan alfalfa seed (Groh 1940). Russian knapweed is now widespread in the southern parts of the four western provinces and is also found in southern Ontario, although most infestations are relatively small (50% of infestations in Saskatchewan are less than one acre in size) (Watson 1980). This noxious weed has a well developed, extensive root system which is the major means of reproduction and spread (Frazier 1944). It forms dense infestations in cultivated fields, in grain and alfalfa fields, in pastures, along roadsides and irrigation ditches and in waste places. Russian knapweed is apparently indifferent to crop association and is able to survive in almost any crop in tillable soil (Rogers 1928). Infested fields commonly have knapweed densities of 100−300 shoots/m^2 which suppress and essentially eliminate other plant growth (Selleck 1964). In addition to its intense competitive ability, Russian knapweed is poisonous to horses (Young *et al.* 1970). Actual crop losses and other losses due to Russian knapweed have not been determined nor estimated in Canada.

Because of its serious threat to Canadian agriculture, Russian knapweed was designated as a "prohibited noxious weed" in the Federal Seeds Act (Agriculture Canada 1967). The benefit of this is debatable as it is difficult to collect viable seed in Canada (Watson 1975) and germination rarely occurs in the field (Selleck 1964). Control of this persistent perennial weed is difficult as its extensive root system is not adversely affected by cultivation and the weed is relatively resistant to commonly used herbicides. Picloram (4-amino-3,5,6-trichloropicolinic acid) and glyphosate [N-(phosphonomethyl) glycine] are two of the more effective herbicides for Russian knapweed control (Alley & Humberg 1979). The biology of Russian knapweed has recently been reviewed (Watson 1980).

Background

In North America, Russian knapweed is relatively free of specialized parasites and is not extensively attacked by polyphagous feeders, but in its native range, Russian knapweed is the host of a number of specialized organisms (Watson 1980). The potential biological control agents of Russian knapweed include seven organisms that attack the seed head: a seed gall mite (*Aceria acroptiloni* V. Shev. & Kov.), three Diptera (*Dasyneura* sp., *Urophora maura* (Frfld.), *Urophora kasochstanica* V. Richter) and three Coleoptera (*Larinus bardus* Gyll., *Larinus jaceae* Fabr., *Rhynchaenus distans* Faust); a stem gall former (*Aulacida acroptilonica* Beliz); a leaf and stem rust (*Puccinia acroptili* Syd.); and a leaf and stem gall nematode (*Paranguina picridis* Kirj. & Ivan.) (Watson 1980).

Releases and Recoveries

***Paranguina picridis* Kirj. & Ivan. (Tylenchida: Tylenchidae)**

(a) Ecology

Populations of Russian knapweed in the southwestern regions of the USSR (primarily Tadzhikistan) are attacked by *P. picridis* (Ivanova 1966, Kirjanova & Ivanova 1969). The nematode produces galls on the stems, leaves, and root-collar of infected plants. Infective second stage larvae free-living in the soil penetrate emerging knapweed shoots in the early spring and a gall is formed at the site of penetration. Larvae feed, mature, and reproduce within the gall structure. Two or more generations are completed during the growing season and second stage larvae become dormant and overwinter in the gall. Larvae are revived by moisture and migrate out of the decaying galls into the soil in search of emerging knapweed shoots. The nematode is reported to be damaging to its host in the USSR with up to 100% of the shoots infected, 20% destroyed, and 30% severely damaged (Ivanova 1966). *P. picridis* is sprayed in a water suspension in an augmentation biological control programme in the USSR (Kovalev *et al.* 1973).

P. picridis was reported to be specific to Russian knapweed (Ivanova 1966, Kirjanova & Ivanova 1969), but laboratory investigations in Canada demonstrated that in addition to Russian knapweed, galls were formed on *Centaurea diffusa* Lam., *Centaurea maculosa* Lam., *Centaurea* x *pratensis* Thuill., *Carduus nutans* L., *Cirsium flodmanii* (Rydb.) Arthur, *Cynara scolymus* L., *Onopordum acanthium* L., *Echinops ritro* L., and *Gerbera jamesonii* Bolus (Watson 1975, Watson 1977). Reproduction of the nematode did not occur in the latter two species, and reduced reproduction and extensive necrosis was observed in galls from plants other than Russian knapweed. It is not uncommon that host ranges of organisms are artificially extended in controlled laboratory studies.

(b) Releases

Prior to unrestricted release of *P. picridis* in North America, additional information was requested on (a) the persistence of the nematode under field conditions in the absence of knapweed, and (b) the effect of the nematode on various crop plants under field conditions. To obtain these data, *P. picridis* was released on small field plots at the Regina Research Station, Saskatchewan, and in confined plots on the Macdonald Campus of McGill Univer-

Table 19 Releases and recoveries against *Acroptilon repens* (L.) DC.

Species	Province	Year	Origin	Number	Year of Recovery
Paranguina picridis	Saskatchewan	1977	Kazakhstan USSR	400 000	1978
	Québec	1977	Kazakhstan USSR	300 000	1978*
	Alberta	1978	Kazakhstan USSR	90 900	1979**

* Released in contained plots
** Knapweed destroyed by herbicide spraying

sity in Ste-Anne-de-Bellevue, Québec, in the spring of 1977. In 1978, an additional small release was made on a natural stand of *A. repens* at Jenner, Alberta, to determine (c) the effects of the nematode on the knapweed. The releases are summarized in Table 19.

Persistence trials

A total of 12 400 nematodes was put into each of 6 clay tiles sunk into the soil at Regina and 10 000 nematodes into each of 8 tiles at Ste-Anne-de-Bellevue in the spring of 1977. The nematode population was sampled at bimonthly intervals through the growing seasons of 1977-79 by taking 3 soil cores (2.5 cm x 10 cm deep) and the nematodes extracted, identified, and counted.

The results (Table 20) show that *P. picridis* was present through the growing season of 1977 in both Saskatchewan and Québec, but did not persist into the second season. Some nematodes were recovered from the soil in 1979, but they were all in other genera.

Table 20 Number of *Paranguina picridis* Kirj. & Ivan. larvae extracted from the soil microplots.

Ste-Anne-de-Bellevue		Regina	
Date sampled	No. of *P. picridis*	Date sampled	No. of *P. picridis*
15.06.77	13	14.06.77	1
04.07.77	16	04.07.77	83
13.07.77	10	18.07.77	4
28.07.77	3	02.08.77	108
15.08.77	5	15.08.77	8
31.08.77	20	29.08.77	30
15.09.77	2		
16.05.77	0		
28.05.79	0	07.79–08.79	0

Effect on crop plants

A total of 10 000 nematodes per m² was inoculated into each of eight m² plots of knapweed in the spring of 1977 at Regina. The plots were then overplanted with the following crops, each replicated three times: flax, lettuce, sunflower, wheat, safflower, carrot, Jerusalem artichoke and globe artichoke. In 1978, kohlrabi, onion, and potato were substituted for the first three plants on the 1977 list. Eight plots sown in the same manner but not inoculated with nematodes served as the control. At the end of the growing season, the crop plants and the knapweed were examined for galls and swellings, and these were extracted to determine the number of nematodes present. Samples of the crop plants from each plot were also extracted to determine if nematodes were present without causing galls.

No galls were found in 1977 on either the knapweed or crop plants. Under most conditions the nematode needs to overwinter in the soil before it is able to attack its host, so the absence of galls on the crop plants was not meaningful. In 1978, there was scattered galling on the knapweed (Table 21), and except for two small swellings on a bolting lettuce stem, nothing suspicious was found on the crop plants. No nematodes were extracted from these swellings or from any other of the crop plants. Swellings of this type in lettuce may be caused by minor insect damage. There was a large increase in the *P. picridis* population in 1979 (Table 21) and no swellings or distortions on the crop plants.

Table 21 Recovery of *P. picridis* Kirj. & Ivan. on Russian knapweed, *A. repens* (L.) DC., at the Regina Research Station.

Year	Galls/m²	*P. picridis*/m²	*A. repens* stems/m²	Precipitation September–May (mm)
1977	0	0		167.5
1978	0.67	2.9	15.0	212.5
1979	3.0	3 599	52.0	213.0
1980	no obvious galls but some swollen stems	16		145.7
1981	0.63		18.6	108.6

The nematode population declined in the springs of 1980 and 1981, which was likely the result of the dry conditions. In the five years from release, no *P. picridis* were found in the control strip of knapweed planted 3 m away from the test plots.

At the Ste-Anne-de-Bellevue site, 6 m^2 plots were established in the early spring of 1977 and 100 000 nematodes/m^2 were inoculated in three of the plots. Russian knapweed was transplanted into all plots and eight crop species were sown in rows between the Russian knapweed plants in 1977. In 1978, globe artichoke and safflower were sown in all plots. A few nematode galls were observed on Russian knapweed in 1977 and heavy galling occurred on Russian knapweed in 1978, 1979, and 1980. In 1981, gall formation on Russian knapweed was reduced. Galls did not develop on any of the test plant species in 1977, 1978, or 1979.

Effect of P. picridis *on the knapweed*

P. picridis was innoculated into 3 m^2 plots of pure Russian knapweed at Jenner, Alberta, at a rate of 30 000/m^2, and three control plots were established with a 50 cm buffer zone from the treated area.

No galls were found on the knapweed in the 1978 growing season even though the nematodes had been kept in soil for six weeks before release. However, in the spring of 1979, most of the knapweed shoots in the treated plots were galled, although those appearing later in the summer were not infected. The plots were sampled at the end of the growing season, the knapweed counted, weighed, and the nematodes extraced (Table 22). The results show that, apart from distorting the knapweed stems, the nematode had a negligible effect. This is in contrast to the results reported from the USSR (Kovalev *et al.* 1973). The other notable feature about the results is that in spite of the close proximity of the treated and control plots, there had been little movement of the nematode into the control plots.

This trial was terminated at the end of 1979, as the knapweed on the surrounding area has been treated with glyphosate and apparently eliminated.

Table 22 Effect of *P. picridis* Kirj. & Ivan. on *A. repens* (L.) DC. at Jenner, Alberta in 1979.

Treatment	No. stems/ 0.25 m²	No. seed heads/ 0.25 m²	Dry weight of foliage/0.25 m²	No. nematodes/ 0.25 m²
P. picridis	25.3 ± 3.9	73 ± 20.7	42 ± 6.2	2259 ± 163
Control	33.2 ± 4.3	55 ± 16.0	44.8 ± 7.0	5.7 ± 3.9

Evaluation of Control Attempts

Fears were expressed following submission of the laboratory screening tests that if *P. picridis* was able to adapt to attack desirable knapweeds or even globe artichoke, it might be extremely difficult to eradicate. The results from the persistence trials and the lack of movement of the nematode from the treated to control plots indicate that these fears are groundless. Removal of the host plant for two growng seasons will result in eradication of the nematode with little likelihood of reinfestation from surrounding populations without human assistance. Furthermore, the host range under field conditions was narrower than in the laboratory. Thus, there were no galls found on *Cynara scolymus* in the plots in either Saskatchewan or Québec, and at Regina stray plants of *Carduus nutans* and *Cirsium arvense* that appeared in the treated plots were also not infected. Lettuce was not a host under laboratory conditions (Watson 1975) and the two small swellings were almost certainly from some other cause, as no nematodes were found in them. From a safety point of view, the nematode should be no problem to any plant except Russian knapweed and it can be easily eradicated if desired.

The effectiveness of *P. picridis* is another matter. The lack of movement in or on the soil suggests that it would have to be spread thoroughly over a knapweed stand in much the same manner as a granular herbicide. The nematode also requires moist spring conditions and there is no purpose in using it on the Canadian prairies unless the area to be treated normally receives a good winter snow cover. The nematode requires a winter before it will infect Russian knapweed, so the fall seems to be the best time for application.

The lack of knapweed suppression in the plots treated in Alberta was disappointing. Almost certainly the effect would have been larger if the knapweed had some grass competition, as this would have helped suppress the later growing shoots. However, the main reason for lack of an effect is that the impact may be delayed for a year. The stem growth in the current year depends on the root reserves of the previous year. The nematode creates a metabolic sink in the stem that will tend to divert reserves from the roots so the plant should be less robust for the following year.

Recommendations

1 The nematode, *P. picridis*, should be released on selected major natural infestations of Russian knapweed in British Columbia, Saskatchewan, and Manitoba.

2 Since Russian knapweed populations in Canada do not reproduce extensively from seed, the importation of biological control agents that attack the seed head is not justified.

3 Since biological weed control programmes usually require more than one biological control agent to be successful (Harris 1979) and since *P. picridis* may not provide sufficient stress to control Russian knapweed in all habitats, studies could be initiated to evaluate the potential of the stem gall former *Aulacida acroptilonica* and the potential of the rust *Puccinia acroptili* as possible additional biological control agents of Russian knapweed.

Literature Cited

Agriculture Canada (1967) Seeds Act and Regulations. Ottawa; Queen's Printer, 50 pp.

Alley, H.P.; Humburg, N,E, (1979) Research in weed science, 1978. *Agricultural Experiment Station of the University of Wyoming Research Journal* 137, 98 pp.

Frazier, J.C. (1944) Nature and rate of development of the root system of *Centaurea picris. Botanical Gazette* (*Chicago*) 105, 345–351.

Groh, H. (1940) Turkestan alfalfa as a medium of weed introduction. *Science in Agriculture* 21, 36–43.

Harris, P. (1979) Costs of biological control of weeds by insects in Canada. *Weed Science* 27, 242–250.

Ivanova, T.S. (1966) Biological control of mountain bluet (*Acroptilon picris* C.A.M.). *Izvestiya Akademii nauk Tadzhikskoi SSR Otdelenie Biologicheskikh Nauk* 2, 51–63 (in Russian).

Kirjanova, E.S.; Ivanova, T.S. (1969) New species of *Paranguina* Kirjanova, 1955 (Nematoda: Tylenchidae) in Tadzhikistan. Ushchel's Kondara, *Akademii nauk Tadzhikskoi SSR* 2,200–217 (in Russian).

Kovalev, O.V.; Danilov, L.G.; Ivanova, T.S. (1973) Method of controlling Russian knapweed. *Opisanie Izorbriteniia Kavtorskomu Svedelet'stuv Byulletin* 38, 2pp (in Russian).

Moore, R.J.; Frankton, C. (1974) The thistles of Canada. *Canadian Department of Agriculture Monograph no.* 10, 111 pp.

Rogers, C.F. (1928) Canada thistle and Russian knapweed and their control. *Colorado Experiment Station Bulletin* 348, 42 pp.

Selleck, G.W. (1964) A competition study of *Cardaria* spp. and *Centaurea repens. Proceedings of the 7th British Weed Control Conference*, 569–576.

Watson, A.K. (1975) The potential of a nematode, *Paranguina picridis* Kirjanova & Ivanova, as a biological control agent of Russian knapweed (*Acroptilon repens*(L.) DC.). Ph.D. Thesis, University of Saskatchewan, Saskatoon, 136 pp.

Watson, A.K. (1977) The biological control of Russian knapweed with a nematode. pp. 221–223 In: Freeman T.E. (*Ed.*) Proceedings of the 4th International Symposium on Biological Control of Weeds, University of Florida, Gainesville (1976), 299 pp.

Watson, A.K. (1980) The biology of Canadian weeds. 43. *Acroptilon* (*Centaurea*) *repens. Canadian Journal of Plant Science* 60, 993–1004.

Young, S.; Brown, W.W.; Klinger, B. (1970) Nigropallidal encephalomalacia in horses fed Russian knapweed (*Centaurea repens* L.). *American Journal of Veterinary Research* 31, 1393–1404.

Ambrosia artemisiifolia L., Common Ragweed (Compositae)

M.G. MAW

Pest Status

There are about 15 species of *Ambrosia* (Gleason 1958) but if *Ambrosia* and the closely related *Franseria* are considered together, there is a world total of about 41 species of which 31 are native to North America and 8 to South America (Harris & Piper 1970). Four species of *Ambrosia* are found in Canada. *A. chamissonis* (Less.) found on the coast of British Columbia and *A. psilostachya* DC., a prairie species, but in recent years has been found in Ontario and Québec (Bassett *et al.* 1976), are perennials and not major concerns. The two annuals, *A. trifida* L. and *A. artemisiifolia* L., cause the greatest problems both as agricultural weeds and as major causes of aeroallergenic hay fever throughout eastern and central United States, and southern Ontario and Québec.

A. trifida, also called great ragweed, kinghead, and buffalo-weed, is typically found on rich alluvial soils, roadsides, in agricultural fields, and in waste places in towns from Québec to British Columbia. It is much less abundant than *A. artemisiifolia,* but it is the most frequently occurring ragweed in southern Manitoba.

A. artemisiifolia, common or short ragweed, is found in all the provinces of Canada, but is most abundant in the densely populated areas of Ontario and Québec. It is a pioneer species of disturbed ground, and rapidly invades sites with a wide variety of soil and moisture conditions. It is a weed of town and country in cultivated fields, vacant lots, roadsides, and fence rows, and is particularly evident in stubble fields after harvest.

The quantity of pollen produced by *A. artemisiifolia* is prodigious. One plant can generate more than a trillion pollen grains and one hectare of ragweed can produce more than 66 kg of pollen during a season (Weber 1977).

Background

Biological control of *Ambrosia* was considered in the 1970s with the publication of a list of insects attacking *Ambrosia* spp. (Harris & Piper 1970). This work also speculated on the prospects of biological control of *Ambrosia* as well as the possible best locations for obtaining control agents. Goeden & Ricker (1974*a*, 1974*b*, 1975, 1976) followed with a series of papers dealing with the phytophagous insect fauna on a number of *Ambrosia* species indigenous to southern California.

From November 1975 to April 1976, the Commonwealth Institute of Biological Control, West Indian Station, surveyed the phytophagous insects on *Ambrosia* in the Tucuman area of Argentina.

Several insects were found on *A. tenuifolia* but lack of manpower and the closing of the Tucuman station precluded their testing in Argentina. However, an unnamed species of *Liothrips*, a curculionid and a cerambycid (both stem borers) were sent to Regina. The beetles entered diapause from which they failed to come out, but the thrips thrived in the laboratory and attacked both *A.artemisiifolia* and *A. psilostachys* with devastating results.

Fifty-two plant species were screened with *Liothrips* sp.. While the thrips fed to a limited extent on many of the test plants, reproduction only occurred on *Ambrosia* spp., sunflower, *Helianthus* sp., and cosmos, *Cosmos bipinnatus* Cav. Feeding and reproduction was inconsistent on sunflower in the tests and seems to parallel the experience with *Zygogramma tortuosa* Rogers, a California ragweed chrysomelid (Goeden & Ricker 1979). It too fed on sunflower in laboratory tests but not in the field. The *Liothrips* is not known to attack sunflower in Argentina (R.E. McFadyen, 1980, personal communication) but the risk in releasing the agent seemed too great and so the *Liothrips* colony was destroyed.

Discussion

In the short time devoted to the survey it was not possible for the workers in Tucuman to investigate fully the insects attacking *Ambrosia* nor was it possible to extend the survey to other potentially promising areas such as the Andean valleys. Several species of *Ambrosia* are native to the Andean region extending into Peru, Bolivia, Chile, and Ecuador and with very diverse climates. These areas should yield potential biological control agents. It is, however, estimated that two to three years would be required to do a thorough study of the biotic agents on ragweed in the Andean region (McFadyen 1976).

Recent advances in the use of endemic plant pathogens as biological control agents of weeds has prompted investigation of endemic pathogens associated with common ragweed. *Albugo tragopogi* (Pers.) S.F. Gray, a white rust, was studied by Hartmann & Watson (1980) and it was found that the rust could severely damage at least a portion of the ragweed population. It reduced pollen production, seed production, and top weight by more than 99, 98, and 79% respectively. It was concluded that further research on *A. tragopogi* was warranted.

Recommendations

(1) Even though *A. artemisiifolia* responds to chemical and cultural methods, these measures are difficult to implement in urban areas. Thus, biological control employing both arthropods and pathogens should be pursued especially in high population density areas.

(2) Studies to determine species, control value, and application of endemic pathogens should be encouraged.

(3) A study should be made to determine the feasibility of obtaining biological control agents from South America.

Literature Cited

Bassett, I.J.; Crompton, C.W.; Frankton, C. (1976) Canadian havens from hay fever. *Canada Agriculture publication* 1570, 23 pp.

Gleason, H.A. (1958) Illustrated flora of northeastern United States and adjacent Canada, vol. 3. Lancaster, Pennsylvania; Lancaster Press, 595 pp.

Goeden, R.D.; Ricker, D.W. (1974*a*) The phytophagous insect fauna of the ragweed, *Ambrosia acanthicarpa* in southern California. *Environmental Entomology* 3, 827–834.

Goeden, R.D.; Ricker, D.W. (1974*b*) The phytophagous insect fauna of the ragweed, *Ambrosia chamissonis*, in southern California. *Environmental Entomology* 3, 835–839.

Goeden, R.D.; Ricker, D.W. (1975) The phytophagous insect fauna on the ragweed, *Ambrosia contertiflora*, in southern California. *Environmental Entomology* 4, 301–306.

Goeden, R.D.; Ricker, D.W. (1976) The phytophagous insect fauna on the ragweed, *Ambrosia dumosa*, in southern California. *Environmental Entomology* 5, 45–50.

Goeden, R.D.; Ricker, D.W. (1979) Life history of *Zygogramma tortuosa* Rogers on the ragweed, *Ambrosia eriocentra* (Gray) Payne, in southern California (Coleoptera: Chrysomelidae). *Pan-Pacific Entomologist* 55, 261–266.

Harris, P.; Piper, G.L. (1970) Ragweed (*Ambrosia* spp.: Compositae); its North American insects and possibilities for its biological control. *Commonwealth Institute of Biological Control Technical Bulletin* 13, 117–140.

Hartmann, H.; Watson, A.K. (1980) Damage to common ragweed (*Ambrosia artemisiifolia*) caused by the white rust fungus (*Albugo tragopogi*). *Weed Science* 28, 632–635.

McFadyen, R.E. (1976) Report on a survey of the insects attacking *Ambrosia tenuifolia* in Tucuman, Argentina. Commonwealth Institute of Biological Control, West Indian Station, Trinidad, Unpublished Report.

Weber, J.B. (1977) Hay fever – a weed problem. *Weeds Today* 8, 20–22, 30.

Artemisia absinthium L., Absinth (Compositae)

M.G. MAW and D. SCHROEDER

Pest Status

Artemisia absinthium L. is indigenous to temperate Europe and Asia and was introduced to North America as a medicinal and flavoring herb some time before 1832. By 1841, absinth had escaped from gardens and was recognized as an established weed in the United States (Mitich 1975). The weed was first recorded in Canada at Fort Garry, Manitoba in 1860 (Scoggan 1957). By 1883, absinth was considered to be naturalized in numerous locations from Newfoundland to western Ontario (Mitich 1975). Although it is abundant in the Prairie Provinces, it was not recognized as a serious weed until 1954 (Frankton & Mulligan 1970).

Absinth can now be found in all provinces of Canada and in many of the northern states but with few exceptions it is confined to farm yards, around grain elevators, and along the edges of fields, railways, and roadsides. It is controlled by cultivation but can be spread into pastures, hayfields, and crop lands. Rapid increase in the spread of the weed on the prairies is associated with relatively moist conditions (Selleck & Coupland 1961).

Although absinth may flavor milk, and grain from infested fields may be rejected because the flour might be tainted (Selleck & Coupland 1961), such occurrences are rare. Cattle will not graze on absinth by choice, but may inadvertently consume it in hay. Absinth is a health hazard as the odor can cause illness in those working in infested fields, and its pollen causes great discomfort in those sensitized by it.

While absinth is controlled by cultivation, the farming techniques of minimal or zero tillage might permit the weed to become a serious problem in some areas.

Background

At Canada's request, surveys of absinth insects in Europe were undertaken by the Commonwealth Institute of Biological Control. A literature review suggested that *Euzophera cinerosella* (Zeller) (Lepidoptera: Pyralidae), a stem and root miner, was specific to *A. absinthium* (Miotk 1973). Field surveys in Austria, Germany, France, and the Swiss Valais showed that *E. cinerosella* occurs in most populations of absinth throughout its European range, and that it causes serious damage to its host (Schroeder 1979).

E. cinerosella emerges about the end of May to the end of July and mating takes place within 24 hours of the female's emergence. Males mate several times; females just once. Oviposition starts the day following mating with 12 eggs laid daily for the first four days. Over 100 eggs may be laid during June and half that number during July.

The eggs are deposited on the lower leaves and hatching occurs in 8–10 days. The young larvae commence to feed on the upper cuticle of the leaf but soon bore into the leaf petiole destroying the bud in the leaf axil, then downwards, feeding on the vascular tissue of the shoots. As the season advances, the larvae reach the roots and destroy the outer parts before mining into the woody tissue. The winter is passed as hibernating larvae and pupation takes place in the spring within a silken cocoon in a pupal cell near the root crown.

Damage to the host plant depends on the development of the larvae. First and second instars cause little damage, but third and fourth instars mine the cambium and vascular tissue, and may completely girdle small shoots, and larger shoots wilt and die prematurely. The greatest damage is caused by the fifth and sixth instar larvae when they destroy the cortical tissue of the root especially when several feed gregariously.

While larvae nibbled on several plants in feeding tests (Schroeder 1979), development beyond the second instar only occurred on plants in the tribe Anthemidae. A single adult from 40 larvae tested developed on *Artemisia vulgaris* L., and limited development was obtained on *Achillea filipendulina* Lam., *A. ptarmica* L., and on *Tanacetum vulgare* L. However, because *E. cinerosella* eggs or larvae were not found on any plant in the field except *A. absinthium*, it was considered to be sufficiently host specific to be recommended for biological control of *A. absinthium* in Canada (Schroeder 1979).

Subsequent testing with *E. cinerosella* on North American species of *Artemisia* and other Compositae showed that the moth was not specific in its choice of oviposition sites, and complete development was not confined to larvae feeding on *A. absinthium* (Maw & Schroeder 1981).

Although larvae develop best and consistently on *A. absinthium*, adults were obtained from *Artemisia cana* Pursh, *Artemisia longifolia* Nutt., and *Artemisia dracunculus* L. Adults were also obtained on *Matricaria matricarioides* (Less.), *Anthemis tinctoria* L., *Tagetes patula* L., *Lactuca sativa* var. *capita* L., and *Cosmos bipinnatus* Cav.

Discussion

A. absinthium is not a problem except in very localized instances but it has the potential to be a serious weed. It is controlled by cultivation, but minimal or zero tillage techniques may soon alter the situation.

The stem and root moth, *E. cinerosella*, is potentially a good biological control agent, but its seemingly wide host range is a concern when species of *Artemisia* other than *absinthium* are liable to attack. In many areas of the western range and desert regions of the United States, *Artemisia* spp. are the dominant vegetation, and furnish forage for cattle, and food and cover for many wild animals and birds. In addition, *A. dracunculus* is an important flavoring and salad herb.

The experimental host range of *E. cinerosella* is undoubtedly an artifact of the testing method, but in the absence of alternative criteria and in view of the relatively low priority of the weed, the risk in releasing the moth is at this time too great.

Recommendations

(1) The range and status of *A. absinthium* should continue to be monitored.
(2) The biology of the plant should be studied.
(3) *E. cinerosella* could be an important control agent and should be re-examined for release in the event *A. absinthium* becomes a serious agricultural problem.

Literature Cited

Frankton, C.; Mulligan, G.A. (1970) Weeds of Canada. *Canadian Department of Agriculture Publication* 948, 217 pp.

Maw, M.G.; Schroeder, D. (1981) *Euzophora cinerosella* (Zeller) (Lep: Pyralidae) not suitable for release to control *Artemisia absinthium* L. (Compositae) in Canada. *Zeitschrift für angewandete Entomologie* 92, 178–184.

Miotk, P. (1973) Phytophagous insects associated with weeds in central Europe. Part I. Weed Project. *Commonwealth Institute of Biologicial Control Progress Report* 31, 19pp.

Mitich, L.W. (1975) Absinth wormwood – a problem weed. Forage and Range Weed Control. *Proceedings of the North Central Weed Control Conference* 30, 41–42.

Schroeder, D. (1979) Investigations on *Euzophora cinerosella* (Zeller) (Lep.: Pyralidae), a possible agent for the biological control of the weed *Artemisia absinthium* L. (Compositae) in Canada. *Mitteilungen der Schweizeriscehn Entomologischen Gesellschaft* 52(1), 91–101.

Scoggan, H.J. (1957) Flora of Manitoba. *National Museum of Canada Bulletin* 140, 619 pp.

Selleck, G.W.; Coupland, T.R. (1961) Studies in the life history of *Artemisia absinthium*. *Weeds* 9(3), 485–490.

Carduus nutans L., Nodding Thistle and *C. acanthoides* L., Plumeless Thistle (Compositae)

P. HARRIS

Pest Status

Carduus nutans L. and *C. acanthoides* L. are thistles of European origin that form dense stands on dry uncultivated grasslands in many parts of North America. Three interbreeding subspecies of *C. nutans* occur in Canada (Moore & Frankton 1974) which are treated as full species by some authors (Kazmi 1964). There is also some hybridization between *C. nutans* and *C. acanthoides* (Mulligan & Frankton 1954), however, as far as the practical aspects of biological control are concerned there are two species: *C. nutans* with large heads (2–7 cm diameter) and a relatively short but intense flowering period that, on Saskatchewan rangeland, is largely finished by the end of July; *C. acanthoides* with small heads (1.2–2.5 cm diameter) and a prolonged flowering period that, in southern Ontario, produces a succession of flower heads from early June until the plant is killed by frost in October or later. The seed of both thistles germinates as conditions permit throughout the growing season and they may be summer annuals, winter annuals, or biennials. The flowering height in *C. nutans* ranges from a few centimeters to over 2 m, so there is considerable plasticity in the species as well as genetic variability as the result of hybridization.

Distribution of the thistle on the Canadian prairies is limited by a need for winter protection of the rosettes and relatively low grass competition for seedling establishment. Thus it is mainly found on light soils in the mid-grass prairie vegetation zone, in places that are covered by snow drifts in winter, such as gullies, fence lines, brush patches, and the lee side of stone piles. The dead stems also trap the snow in winter. Established stands tend to be self-perpetuating as the death of the flowering stems in August creates a seedbed largely devoid of competing vegetation. The occurrence of the thistle on heavier soils is a sign that the site was disturbed within the past few years.

The status of *C. nutans* has changed in Saskatchewan since the establishment of the seed-head weevil, *Rhinocyllus conicus* Froel., for its biological control. In 1970, stands of 150 000/ha of flowering *C. nutans* plants were common in pasture and extended for many kilometers in roadside borrow pits. At this density, there is no grazing within the stand. In 1980, the thistle was largely restricted to breaks in the pasture sward such as ground squirrel diggings and along cattle trails. Stands of over 15 000 plants/ha were uncommon and small roadside stands that used to be a feature of the landscape have been reduced to scattered plants. On the 4 point scale, modified from Vere & Medd (1979), the most densely infested region has declined from category 4 to categories 1–2: 1 = no *C. nutans*, 2 = scattered thistles or < 1 stand/ha, 3 = scattered thistles with 1 to 50 stands/ha, 4 = continuous thistles or >50 stands/ha. There was negligible grazing loss from category 2 infestations in Australia, a loss of 8.3% from category 3, and 16.7% loss from category 4. No comparable data are available for Saskatchewan but the loss is presumably similar. Currently there is little grazing loss from *C. nutans* on permanent pasture although dense patches remain on recently disturbed and abandoned ground which is often on roadsides. At low densities it is still a nuisance in some places such as parks. *C. nutans* continues to spread in the mid-grass prairie zone but on a narrower range of sites.

C. nutans occurs as a problem on some thin gravel soils in Québec. Letendre *et al.* (1976) reported that around 325 km^2 were infested in the Lac St-Jean region. *R. conicus* has been established but whether the status of the thistle has changed is not known. In Alberta, scattered foci of *C. nutans* ssp. *macrolepis* (Petrm.) Kazmi continue to invade from Montana. These are treated chemically. Similarly in Manitoba, there is an invasion of *C. nutans* from North Dakota. The smaller stands have been treated chemically and *R. conicus* has been established on the others.

C. acanthoides forms dense stands, commonly 90 000 plants/ha, in permanent pasture on shallow gravel or limestone soils in southern Ontario and parts of Québec. The Ontario Stockmen's Association periodically request its biological control, but there are no data on losses caused by it or the size of the area infested that can be used to justify biological control in economic terms.

Neither thistle tolerates regular cultivation. On permanent pasture the preferred method of *C. nutans* control is by pasture improvement to maintain a dense and vigorous sward (Hull & Evans 1973) but this alone did not completely solve the problem. The presence of *R. conicus* has changed the situation of *C. nutans* in Saskatchewan but probably not that of *C. acanthoides* in Ontario and Québec. The thistles can be controlled with 1 kg/ha of 2,4-D applied in the spring and fall, but this is normally not economically justified. Vere & Medd (1979) determined that in New South Wales, Australia, the total loss from *C. nutans* was $A0.5 million/year while the cost of chemical control would be $A3.5 million after a programme of pasture improvement.

Background

The programme for the biological control of *Carduus* was reported by Harris & Zwölfer (1971). The Canadian sponsored studies on the seed-head weevil *R. conicus* have led to its release in New Zealand (Jessep 1975) and the parts of the United States with a *Carduus* problem (Kok & Surles 1975, Goeden & Ricker 1977, Rees 1977, Batra 1978, Puttler *et al.* 1978). It has generally resulted in a marked reduction in seed producton by *C. nutans* as well as that of several other thistles.

Another *Carduus* insect, *Trichosirocalus* (*Ceuthorhynchidius*) *horridus* (Panz.) was approved for release in North America after host specificity testing by Ward *et al.* (1974). There is still a need for more agents for the control of *C. acanthoides* as well as for *C. nutans* in certain places in the United States where *R. conicus* has not been effective; but further testing has been discouraged by the concern that the agents might also damage native *Cirsium* spp. This fear has some foundation but seems to be exaggerated. In the laboratory, most *Carduus* insects will develop on the genera *Cirsium*, *Silybum*, and sometimes *Onopordum* although they do not necessarily do so in the field. *R. conicus* breeds on all four genera in Europe; however, Zwölfer & Preiss (1983) found that there were a number of host races involved, each with a predominant or exclusive preference for a single thistle species. In North America, the *C. nutans* strain of the weevil has attacked native and introduced *Cirsium* spp.; but in Saskatchewan, the attack is restricted to the area around a *C. nutans* stand, usually 20 m, and the weevils developed are smaller than those from *C. nutans* (Zwölfer & Harris, in press). In Montana, Rees (1978) found that the attack on *Cirsium* spp, was widespread but they were not preferred for oviposition and the larvae suffered a higher mortality than on *C. nutans*. He concluded that the weevil was not a threat to native *Cirsium* spp. The attack of *Cirsium* seems to be a spill-over from the high weevil densities on a declining *C. nutans* population, and I expect it will decrease as the weevils approach equilibrium with the *C. nutans* population.

Releases and Recoveries

***Rhinocyllus conicus* (Froel.) (Coleoptera: Curculionidae)**

(a) Ecology

The biology and host specificity of *R. conicus* has been treated in detail by Zwölfer & Harris (in press). The weevil is native to Europe, western Asia, and North Africa between latitudes 30°N and 50°N. Its main host is *C. nutans* but there has been race formation on other thistles in the genera *Carduus*, *Cirsium*, *Silybum*, and *Onopordum*.

Teneral *R. conicus* emerge in July to early August in southern Ontario and August in Saskatchewan. The weevil has a second generation if the day length is over 16 hrs. Individuals released in a cage in southern Ontario on the 27 June when the day length was 15 hr 48 min did not breed, but those released in mid-July in Saskatchewan with a day of 16 hr 5 min did so. Laing & Heels (1978) reported a second generation from weevils released in southern Ontario in August, but these were mated in the laboratory, presumably under long day conditions. Normally, in Canada, there is a single generation emerging after mid-summer and then hibernating in the soil litter. They appear again in the spring to feed on the leaves of bolting thistles and oviposit on the involucral bracts of the flower buds. Oviposition starts slightly before the buds are available in the spring with some wastage of eggs on the leaves enclosing the terminal flower bud and the first heads receive an abundance of eggs. Rees (1977) counted over 500 eggs on a single head.

The eggs are covered with a cap of chewed thistle and the larva bores directly into the flower bud from under the cap. It mines the receptacle and sometimes the peduncle as well as feeding on the young ovules. Typically the receptacle mine fills with callus on which the larva feeds (Shorthouse, 1982, personal communication). Thus, like a gall-former, the larva stimulates production of its food supply. The mature larva forms a hard pupal chamber in the thistle head which remains after emergence so the number that developed in a head can be counted.

(b) Releases

Most of the *R. conicus* stock established in Canada was collected from *C. nutans* growing in the French Rhine Valley around Mulhouse. As the initial results with this stock on *C. acanthoides* were disappointing, additional weevils were imported from the small-headed thistle *C. personata* (L.) Jacquin as well as stocks from *Cirsium* spp. Day length at the release date probably prevented many of the colonies from breeding until the year following release (Table 23). With hindsight, it would have been better to have held them in laboratory storage until the following spring or mated them under long day conditions in the laboratory before release.

***Trichosirocalus horridus* (Panz.) (Coleoptera: Curculionidae)**

(a) Ecology

The biology of *T. horridus* was reported by Trumble & Kok (1979) and a bibliography of the weevil compiled by Trumble & Kok (1980). The weevil is found from Italy to Poland, in Turkey and the USSR. The adults have been found feeding on a number of thistle genera and the larvae developed on several of these in the laboratory (Kok 1975), however, in the thistle garden at the Regina Research Station, they have only been found on *Carduus* spp. and not on either European or native *Cirsium* spp. Thus the field host

Table 23 Open releases and recoveries of *R. conicus* Froel. against *Carduus* spp.

Province	Host	Release date	Place	Source	Number	Initial recovery
Saskatchewan	*C. nutans*	26.7.68	Aylesbury	1	400	1969
		16.7.69			1898	
	C. nutans	16.7.69	Findlater	1	1897	1969
Ontario	*C. acanthoides*	25.7.68	Read (site A)	1	320	1969
		23.5.69		1	282	
		10.6.70		2	350	
		5.10.69	Read (Site B)	3	74	
		5.10.69		1	80	
		31.6.70	Read (Site C)	1	3500	1971
		22.5.69	Stoco Lake	1	350	1970
		11.7.69	Roslin	1	4320	1970
		22.5.70		1	1500	
		1.5.70	Moira	4	200	
	C. acanthoides x *C. nutans*	3.6.70	Flesherton	1	250	not checked
	C. nutans	9.6.75	Guelph	5	2185	1976
Manitoba	*C. nutans*	17.6.74	Somerset	5	500	1975
		4.9.74	Darlingford	5	4000	1975
Québec	*C. nutans*	17.6.74	Lac St-Jean	5	1100	1975
	C. acanthoides	23.7.78	Huntingdon	5	7731	1982
		16.6.80		5		10130
British Columbia	*C. nutans*	16.7.79	Williams Lake	5	2600	site destroyed

1. Mulhouse, France ex *C. nutans*
2. St. Hippolyte, France ex *C. personata*
3. Krymsk, USSR ex *Cirsium spinosum*
4. Nantes, France ex *C. vulgare*
5. Findlater, Saskatchewan ex *C. nutans*

range of the strain introduced is more restricted than that of *R. conicus*. Under field conditions in Virginia, United States, it had a strong preference for *C. nutans* over *C. acanthoides* and was more readily established on stands of the former species (Sieburth & Kok 1982). On *C. nutans*, high densities of *T. horridus* and *R. conicus* were found in the same stands (Kok 1980).

In Canada, the adult weevil overwinters in the soil litter. It becomes active in early spring to lay in the mid-vein of large *Carduus* leaves. This differs from the life cycle in Rome, Italy, where the weevil starts ovipositing in October and continues over winter (Bolt & Campobasso 1981). The larva bores down the mid-vein into the crown and develops just below the apical leaves. This causes the plant to produce several lateral shoots, which in Saskatchewan may be attacked again when about 15 cm high; the larvae are also sometimes found in the lateral vegetative buds. The result is to produce a bushy thistle in which flowering is probably delayed. The larvae pupate in the soil and emerge as adults in reproductive diapause in the early summer. They feed on the thistle leaves until the fall and can be frequently found on the bracts of the flower buds.

(b) Releases

The stock released at Aylesbury, Saskatchewan, was collected in eastern Austria, around Morel, Switzerland, and Neuenburg, Germany (Table 24). In addition, 22 weevils from

Neuenburg, released on a 4 m^2 stand of *C. nutans* maintained at the Regina Research Station, were the source of the stock released elsewhere in Canada. At Regina, survival of the weevil was most easily confirmed by removing the apical rosette leaves to expose the larvae during the spring wheat seeding period, usually around mid-May. Later in the year, the weevil was hard to find and even its damage was obscure. Thus, it was hard to confirm establishment at Aylesbury as the road to the site was often impassable until late May. It is suspected that the reason that this cryptic weevil has not been reported established at other sites is that the thistles have been inspected too late in the season.

Table 24 Open releases and recoveries of *T. horridus* (Panz.) against *Carduus* spp.

Province	Site	Host	Year	Source	Number	Recovery
Saskatchewan	Aylesbury	*C. nutans*	1975	Austria-Germany-Switzerland	65	1979
	Aylesbury	*C. nutans*	1979	Saskatchewan	36	
	Regina	*C. nutans*	1975	Germany	22	1976
Québec	Huntingdon	*C. acanthoides*	1977	Saskatchewan	26	
	Huntingdon	*C. acanthoides*	1980	Saskatchewan	6	
Ontario	West Huntingdon	*C. acanthoides*	1978	Saskatchewan	33	
British Columbia	Williams Lake	*C. nutans*	1979	Saskatchewan	22	site destroyed
Manitoba	Snowflake	*C. nutans*	1980	Saskatchewan	16	

Evaluation of Control Attempts

Density of *R. conicus*

The general effects of *R. conicus* on *C. nutans* in Saskatchewan have been described in the Pest Status section. More detailed studies were done at the Findlater and Aylesbury release sites (Table 25 and 26). Findlater was an unused gravel pit and Aylesbury a gully in a permanent pasture. Both sites were sampled in early August: Findlater on a permanent transect along the spoil pile, and Aylesbury in four directions from the release point. The density of flowering thistles was determined with m^2 samples taken at 5 or 10 m intervals, or they were calculated by the distance between nearest neighbour for a population with a uniform distribution (Southwood 1978). The latter estimates are indicated by the absence of an error term in Table 26. In 1978 and 1979, the estimates obtained from the nearest neighbour method were: Aylesbury 1.4, 1.0, and at Findlater 48.2, 8.3/m^2. These are similar to the values obtained from counting on the m^2 plots (Table 26). However, in 1976 at Aylesbury, the distribution was clumped with most of the site free of thistles but some patches containing up to 65/m^2. In this year the nearest neighbour method, assuming a uniform distribution, gave a density of 2.7/m^2 compared to 6.6/m^2 obtained by averaging the m^2 samples. All heads were clipped from the thistle nearest to the right corner of the plot, the receptacle diameter of each head was measured, and the number of *R. conicus* or its pupal chambers were counted. If there were not enough thistles in the transect samples to determine weevil density, the thistle nearest to each sampling interval was used.

At Findlater, the density of *R. conicus* maturing in the flower heads climbed steadily from the second year to reach a plateau of 3.4 to 6.4 weevils/head (Table 25). The weevils released in 1968 at Aylesbury (Table 23) received a day of less than 16 hrs in length and so probably did not breed until the following year. Thus the initial increase at Aylesbury was similar to that at Findlater; but after reaching a peak in 1975, the density fell. This may have been caused by cattle consuming the terminal flower heads, which contained the highest numbers of *R. conicus*. The cattle did this to an increasing extent as the density of the stand declined. The site was not grazed in 1982 and the density of weevils increased.

Table 25 Establishment pattern of *R. conicus* Froel. on *C. acanthoides* L. and *C. nutans* L.

C. acanthoides (Site A) at Read, Ontario. Released 320 *R. conicus* 25 July 1968, 282 on 23 May 1969, 350 on 10 June 1970.

Year	Eggs/Thistle	Calc. *R. conicus* per head	No. thistles sampled
1968	0	0	
1969	0.29	0.017	1453
1971	0.38	0.023	285

C. acanthoides at Roslin, Ontario. Released 4320 *R. conicus* on 11 July 1969.

Year	Eggs/thistle	Calc. *R. conicus* per head	No. thistles sampled
1969	0	0	
1970	0.52	0.03	1531
1970*	1.56	0.09	50
1971	0.18	0.01	1272

* On scattered *C. nutans* in *C. acanthoides* stand.

C. nutans at Aylesbury, Saskatchewan. Released 400 *R. conicus* 26 July 1968, 1898 on 16 July 1969.

Year	*R. conicus*/head	No. heads sampled
1969	0.08	317
1970	0.04	1055
1971	0.06	688
1972	0.26	847
1973	0.79	956
1974	2.72	311
1975	3.68	288
1976	1.43	182
1977	0.43	353
1978	2.70	251
1979	0.86	349
1980	0.40	293
1981	0.06	54
1982	3.18	105

Table 25 continued

C. nutans at Findlater, Saskatchewan. Released 1897 *R. conicus* on 16 July 1969.

Year	*R. conicus*/head	No. heads sampled
1969	0.03	119
1970	0.002	509
1971	0.01	194
1972	0.19	241
1973	0.46	168
1974	1.45	211
1975	4.53	52
1976	4.01	96
1977	1.84	99
1978	3.38	86
1979	5.02	177
1980	3.65	373
1981	3.51	59
1982	6.36	375

Laing & Heels (1978) reported achieving similar population densitites on *C. nutans* in Ontario and the initial results reported by Letendre *et al.* (1976) for Québec followed the same pattern. Thus the Saskatchewan impact of *R. conicus* on *C. nutans* is probably representative of populations elsewhere in Canada except for British Columbia where the release site was destroyed by road building.

Density of *C. nutans*

The density of *C. nutans* declined at Aylesbury from 15.7 flowering plants/m^2 in 1969 to 0.5/m^2 in 1982 (Table 26). The habitat was initially invaded by Russian pigweed, *Axyris amaranthoides* L., and stinkweed, *Thlaspi arvense* L. The stinkweed was cropped by the red turnip beetle, *Entomoscelis americana* Brown, which also developed on *C. nutans* rosette leaves during this transitional period. Finally perennial grasses, *Agropyron repens* (L.) Beauv. and *Stipa comata* Trin. & Rupr., occupied the area and the thistle was confined to breaks in the sward.

At Findlater, the density of the thistle fluctuated from dense, following a moist fall and/or spring, to sparse when it was dry (Table 26). The invasion of the transect on the spoil pile by other plants was slow, so after dry winters the site remained open for recolonization by *C. nutans* when conditions were more favourable for its seedlings. Gradually clumps of crested wheat grass, *Agropyron cristatum* (L.) Gaertn., appeared and by 1982 covered the transect. The *C. nutans* stand was then confined to the south-western slope of the spoil pile, which was still free of grass.

The decline of *C. nutans* stands in Virginia, United States, occurred more rapidly than in Saskatchewan (Kok & Surles 1975). The summer rainfall was greater and its distribution more even so that there would be a greater growth of grass. Thus both studies support the conclusion that control of *C. nutans* has been achieved by a combination of *R. conicus* and competition from other plants.

At neither Saskatchewan site has the lowered density of the thistle reduced the numbers of *R. conicus*/head. Scattered thistles often escape attack and the weevil may not be able to maintain itself on them. For example, 120 *R. conicus* released in 1974 on a stand of *C. nutans* in Regina that consisted of 368 scattered plants at 0.07 plants/m^2 bred in 1975 but not in 1976 when the density was 0.02 plants/m^2. However, the weevil

Table 26 *C. nutans* L. stands at Aylesbury and Findlater following establishment of *R. conicus* Froel. (± S.E.M.).

Year	No. *C. nutans*/m^2	No. heads/plant	Head diameter	No. plants sampled
Aylesbury				
1969	15.9 ± 1.7	5.7 ± 0.8	2.87 ± 0.02	56
1970	11.8 ± 1.1	7.5 ± 0.9	1.68 ± 0.03	125
1971	6.8 ± 0.7	6.2 ± 0.8	2.02 ± 0.03	100
1972	6.4 ± 0.8	8.0 ± 0.9	1.65 ± 0.02	106
1973		7.4 ± 1.1	1.79 ± 0.02	129
1974	4.1	6.6 ± 1.3	1.73 ± 0.02	47
1975	4.9	4.4 ± 0.7	1.94 ± 0.02	65
1976	6.6 ± 1.4	3.4	1.73 ± 0.01	54
1977	1.3 ± 0.3	6.0 ± 0.3	1.55 ± 0.02	60
1978	1.2 ± 0.5	4.3 ± 0.8	1.52 ± 0.03	58
1979	1.4 ± 0.4	3.9 ± 0.7	1.38 ± 0.02	89
1980	1.9 ± 0.7	2.3 ± 0.1	1.16 ± 0.05	124
1981	0.9 ± 0.4	1.5 ± 0.1	1.41 ± 0.02	60
1982	0.5 ± 0.1	2.8 ± 0.5	1.56 ± 0.04	38
Findlater				
1969	13.5 ± 2.6	11.9 ± 1.8	1.80 ± 0.04	10
1970	19.1 ± 2.5	10.0 ± 1.3	1.98 ± 0.02	51
1971	19.8 ± 2.8	4.9 ± 0.5	1.69 ± 0.04	40
1972	8.0 ± 1.2	6.9 ± 0.7	1.70 ± 0.02	35
1973		5.3 ± 1.0	1.87 ± 0.04	32
1974	19.9	5.9 ± 1.2	1.95 ± 0.01	35
1975	52.1	2.9 ± 0.7	1.86 ± 0.06	18
1976	7.2	4.2 ± 0.6	1.81 ± 0.03	24
1977	0.4 ± 0.2	5.2 ± 0.2	2.03 ± 0.03	19
1978	33.9 ± 5.7	2.3 ± 0.3	1.40 ± 0.05	38
1979	6.7 ± 0.1	5.2 ± 0.8	1.19 ± 0.03	33
1980	3.0 ± 0.7	3.6 ± 0.3	1.19 ± 0.03	104
1981	0.6 ± 0.3	4.0 ± 1.6	1.39 ± 0.06	24
1982	0.03*	11.6 ± 1.8	1.83 ± 0.02	32

* 1 thistle in 37 m^2

persisted at another site in Regina on a 10 m^2 stand of *C. nutans* with around 50 flowering thistles. If the weevil requires dense stands, even if they are confined to a few m^2, the roadside and gravel pit stands may be necessary for maintaining a population.

Size of *C. nutans*

The size of flowering *C. nutans* plants as indicated by the number of heads/plant declined (Table 26). The slope of the decline was not significantly different at Aylesbury and Findlater, so the data were combined for the regression in Table 27 (Equation 1). The only major departure from the regression was the 1982 value from Findlater. This value was not used for the regression as it was not from the permanent transect. The equation shows that on the average a plant produced 0.34 fewer heads each year. Thus thistle size has declined more than its density. For example at Aylesbury, the density of the thistle in 1982 was 97% less than in 1969, but the number of heads had declined by nearly 99%.

Table 27 Regression equations of the effect of *R. conicus* Froel. on *C. nutans* L.

		d.f.	r	P
1. Average no. heads/plant at Aylesbury and Findlater	= 994.5 − 0.50 x	26	-0.79	<0.001
2. Average head diameter at Aylesbury	= 79.8 − 0.40 x	12	-0.71	< 0.01
3. Average head diameter at Findlater	= 7.3 − 2.89 x	13	-0.38	N.S.

x = no. years r = correlation coefficient

It is thought that this size decline has been an indirect effect of the weevil. Formerly only large biennial rosettes were able to survive the intense intraspecific competition within the thistle stand. It is reasoned that as this competition lessened with the reduction of thistle seed, greater numbers of small summer annual plants were able to survive. The present restriction of the thistle to disturbed soil and openings in pastures combined with a low seed production results in little competition for the seedlings and presumably a high survival of small plants. The perennation of *C. nutans* was followed on a 20 m^2 permanent plot at Aylesbury over a 4 year period. The results (Table 28) show a considerable variation between years depending on the occurrence of fall rains. The average for the last three years of the period was 45% biennial plants, 27.5% winter annuals, and 27.5% summer annuals.

Table 28 Perennation of *C. nutans* L. at Aylesbury

Year	No. biennials	No. winter annuals	No. summer annuals
1975	140*		0
1976	40	38	28
1977	0	0	8
1978	10	1	0
1979	14	0	3

* Biennials and winter annuals not distinguished

Size of *C. nutans* seed heads

There was a decline in the diameter of the seed heads produced since establishment of the weevil that was significant at Aylesbury and not significant at Findlater (Table 27, Equations 2 and 3). The decline is probably a function of thistle size and it represents a further reduction of seed.

Reasons for the impact of *R. conicus*

Lashley (1969) and Sagar (1972) calculated that over 98% of the seed would have to be destroyed to achieve control of *C. nutans* if 50% of the seed was viable and 10% of the plants survived. At Aylesbury 80.6% ± 3.6 of the plump seed germinated on moist filter paper at room temperature, so the viability factor they used was low. Regression equations of seed production per head versus head size and the number of *R. conicus* gave reductions that varied from 5 to 25 seeds per weevil. The variation may arise from the formation of callus tissue acting as a metabolic sink in the heavily attacked heads; but even the highest estimate would only reduce seed production by around 50%.

Analysis of one year's results of a four year study of thistle survival at Aylesbury indicated that no seedlings survived in a grass sward. Presumably the stand became established initially after a natural or man-made disturbance exposed the ground. The thistle then perpetuated itself and even at the stand margin spread by producing enough plants to smother the site. A constant percentage of plants was lost through the growing season regardless of their density or age and around 3% survived to the flowering stage. Apparently the weevil reduced seed production to the point at which the thistle was not able to completely occupy the site at all times. This allowed the entry of competing vegetation which eventually displaced the thistle. Further reductions of *C. nutans* can almost certainly be achieved by pasture improvement to reduce the number of bare spots.

Effect of *R. conicus* on *C. acanthoides*

C. acanthoides has remained a problem in Ontario since the introduction of *R. conicus*. It is known that the weevil has persisted on the thistle as there is no difficulty in finding flower heads with eggs on them in the early spring; however, it is probable that seed production is reduced by a minor amount. In Virginia, only 12.4% of *C. acanthoides* heads are attacked compared to 81.4% of those on *C. nutans* (Surles & Kok 1977). Much of this difference relates to the synchronization of egg production with flowering. In Saskatchewan, oviposition covers the flowering period and both are largely completed in July. In contrast, on 7 August 1969 at Read, Ontario, after *R. conicus* had ceased oviposition, 22% of the heads of *C. acanthoides* had finished flowering and would have been available to *R. conicus*, 7% were in bloom, and 70.7% were still in bud (unpublished data, T. New). Thus, 78% of the heads produced by *C. acanthoides* were not available to *R. conicus*. Similarly in Virginia, Surles & Kok (1977) found 19.1% of the heads were available to the weevil.

Dowd & Kok (1981) found a density-related weight reduction of *R. conicus* in *C. acanthoides* that did not occur in *C. nutans*. Also, the weevil responded to *C. nutans* more readily than to *C. acanthoides* for oviposition: in Table 25 there was a 3-fold difference in the numbers of eggs laid on the two species while Surles & Kok (1977) found a 4-fold difference. The difference in flower head size does not seem to be the important factor as some races of *R. conicus* achieve high densities on the small-headed thistles, *C. pycnocephalus* and *C. tenuiflorus*.

R. conicus has not adapted to *C. acanthoides* in Europe: Batra *et al.* (1981) found it rarely on the thistle in France and it was not recorded in the surveys on *C. acanthoides* by Zwölfer (1965). In my opinion, it is unlikely to adapt to become an effective control agent of the thistle in North America.

Effect of *T. horridus*

No assessment of the impact of *T. horridus* on *C. nutans* was made at Aylesbury beyond the fact that few plants are attacked. There are no reports on the value of the more dense populations of the insect in Virginia, and this default is an indication that its effect is small. On the cultivated plot of thistles at the Regina Research Station, most of the large *C. nutans* rosettes were attacked in the spring, but they survived to produce seed.

Recommendations

(1) *R. conicus* should be established in all regions of Canada with a *C. nutans* problem. The effect of the weevil can almost certainly be enhanced with pasture improvement.

(2) There seems to be little purpose in distributing *T. horridus* unless there are sites where *R. conicus* is ineffective.

(3) Dense roadside and gravel pit stands of *C. nutans* may provide an essential reservoir of the weevil and so should not be heedlessly eliminated.

(4) For a number of reasons, *R. conicus* is poorly adapted to *C. acanthoides* and *T. horridus* is not a promising alternative. It is recommended that *Urophora solstitialis* (L.) be screened as a control agent for the thistle. This gall fly will tend to displace *R. conicus* from heads that it attacks, but this will not necessarily be detrimental for biological control of *C. nutans*. For example, the destruction of diffuse knapweed seed was increased by establishing two closely competing *Urophora* spp. on it (Myers & Harris 1980).

(5) The magnitude of the *C. acanthoides* problem in Ontario and Québec should be assessed for the purpose of establishing a priority and an economic justification for its biological control.

Acknowledgements

The releases and follow-up studies in Québec were done by A.K. Watson, in British Columbia by R. Cranston, and in Manitoba by D. Webster and R. Christoff. I thank these people for their cooperation.

Literature Cited

Batra, S.W.T. (1978) *Carduus* thistle distribution and biological control in the northeastern states. In: Frick, K.E. (*Ed.*) Biological control of thistles in the genus *Carduus* in the United States. USDA, 50 pp.

Batra, S.W.T.; Coulson, J.R.; Dunn, P.H.; Bolt, P.E. (1981) Insects and fungi associated with *Carduus* thistles (Compositae). *USDA Technical Bulletin* 1616, 100 pp.

Bolt, P.E.; Campobasso, G. (1981) Biology of two weevils, *Ceutorhynchus trimaculatus* and *Trichosirocalus horridus*, on *Carduus* spp. in Europe. *Environmental Entomology* 10, 691–696.

Dowd, P.F.; Kok, L.T. (1981) Impact of larval crowding on survival of *Rhinocyllus conicus* (Coleoptera: Curculionidae). *Environmental Entomology* 10, 472–473.

Goeden, R.D.; Ricker, D.W. (1977) Establishment of *Rhinocyllus conicus* on milk thistle in southern California. *Weed Science* 25, 288–292.

Harris, P.; Zwölfer, H. (1971) *Carduus acanthoides* L., welted thistle and *C. nutans* L., nodding thistle (Compositae). *In*: Biological control programmes against insects and weeds in Canada 1959–1968. *Commonwealth Institute of Biological Control Technical Communication* 4, 76–79.

Hull, A.C.; Evans, J.O. (1973) Musk thistle (*Carduus nutans*): an undesirable range plant. *Journal of Range Management* 26, 383–385.

Jessep, C.T. (1975) Introduction of a weevil *Rhinocyllus conicus* for biological control of nodding thistle *Carduus nutans*. *Proceedings of the 28th New Zealand Weed Pest Control Conference*, 205–206.

Kazmi, S.M.A. (1964) Revision der Gattung *Carduus* (Compositae). Teil 11. *Mitteillungen aus der Botanischen Staatssammlung, Muenchen*, 5, 279–550.

Kok, L.T. (1975) Host specificity studies on *Ceuthorhynchidius horridus* (Panzer) (Coleoptera: Curculionidae) for the biocontrol of musk and plumeless thistle. *Weed Research* 15, 21–25.

Kok, L.T. (1980) Compatibility of *Rhinocyllus conicus*, *Trichosirocalus horridus* and 2,4-D for *Carduus* thistle control. *Proceedings of the 5th International Symposium on Biological Control of Weeds*, 441–445.

Kok, L.T.; Surles, W.W. (1975) Successful biocontrol of musk thistle by an introduced weevil, *Rhinocyllus conicus*. *Environmental Entomology* 4, 1025–1027.

Laing, J.E.; Heels, P.R. (1978) Establishment of an introduced weevil, *Rhinocyllus conicus* (Coleoptera: Curculionidae) for the biological control of nodding thistle, *Carduus nutans* (Compositae) in southern Ontario. *Proceedings of the Entomological Society of Ontario* 109, 3–8.

Lashley, R. (1969) Musk thistle control. *Proceedings of the North Central Weed Control Conference* 24, 99–100.

Letendre, M.C.; Ritchot, M.; Guibord, O'C.; Leduc, C. (1976) Essai d'éradication du chardon penché, *Carduus nutans* L., á l'aide du charançon *Rhinocyllus conicus* Froel. *Phytoprotection* 57, 47–54.

Moore, R.J.; Frankton, C. (1974) The thistles of Canada. *Canadian Department of Agriculture Monograph* 10, Ottawa, 111 pp.

Mulligan, G.A.; Frankton, C. (1954) The plumeless thistles (*Carduus* spp.) in Canada. *Canadian Field-Naturalist* 68, 31–36.

Myers, J.H.; Harris, P. (1980) Distribution of *Urophora* galls in flower heads of diffuse and spotted knapweed in British Columbia. *Journal of Applied Ecology* 17, 359–367.

Puttler, B.; Long, A.H.; Peters, E.J. (1978) Establishment in Missouri of *Rhinocyllus conicus* for the biological control of musk thistle (*Carduus nutans*). *Weed Science* 26, 188–190.

Rees, N.E. (1977) Impact of *Rhinocyllus conicus* on thistles in southwestern Montana. *Environmental Entomology* 6, 839–842.

Rees, N.E. (1978) Interactions of *Rhinocyllus conicus* and thistles in the Gallatin valley, 31-38. In: Frick, K.E. (*Ed.*) Biological control of thistles in the genus *Carduus* in the United States. Stoneville, Miss.; USDA, 50 pp.

Sagar, G.R. (1972) On the ecology of weed control. In: Jones D.P.; Solomon M.E. (*Eds.*) Biology in pest and disease control. British Ecological Society, 42–50.

Sieburth, P.J.; Kok, L.T. (1982) Oviposition preference of *Trichosirocalus horridus* (Coleoptera: Curculionidae). *Canadian Entomologist* 114, 1201–1202.

Southwood, T.R.E. (1978) Ecological methods. Chapman & Hall, 524 pp.

Surles, W.W.; Kok, L.T. (1977) Ovipositional preferences and synchronization of *Rhinocyllus conicus* with *Carduus nutans* and *C. acanthoides*. *Environmental Entomology* 6, 222–224.

Trumble, J.T.; Kok, L.T. (1979) *Ceutorhynchidius horridus* (Coleoptera: Curculionidae): life cycle and development on *Carduus* thistles in Virginia. *Annals of the Entomological Society of America* 72, 563–564.

Trumble, J.T.; Kok, L.T. (1980) A bibliography of *Ceuthorhynchidius horridus* (Panzer) [= *Trichosirocalus horridus* (Panzer)], an introduced weevil for the biological control of *Carduus* thistles. *Bulletin of the Entomological Society of America* 62, 464–467.

Vere, D.T.; Medd, R.W. (1979) Estimating the economic loss caused by *Carduus nutans* in New South Wales. *Proceedings of the 7th Asian-Pacific Weed Science Conference*, 421–423.

Ward, R.H.; Pienkowski R.L.; Kok, L.T. (1974) Host specificity of the first-instar of *Ceuthorhynchidius horridus*, a weevil for the biological control of thistles. *Journal of Economic Entomology* 67, 735–737.

Zwölfer, H. (1965) Preliminary list of phytophagous insects attacking wild Cynareae (Compositae) species in Europe. *Commonwealth Institute of Biological Control Technical Bulletin* 6, 81–154.

Zwölfer, H.; Harris, P. (in press) Biology and host specificity of *Rhinocyllus conicus*, (Froel.) (Col.: Curculionidae), a successful biocontrol agent of the thistle *Carduus nutans* L. *Zeitschrift für angewandte Entomologie*.

Zwölfer, H.; Preiss, M. (1983) Host selection and oviposition behaviour in west-European ecotypes of *Rhinocyllus conicus* Froel. (Col.: Curculionidae). *Zeitschrift für angewandte Entomologie* 95, 113-122.

Centaurea diffusa Lam. and *C. maculosa* Lam. *s. lat.*, Diffuse and Spotted Knapweed (Compositae)

P. HARRIS and J.H. MYERS

Pest Status

Diffuse and spotted knapweed are herbaceous plants introduced from Europe to the dry grasslands of western Canada and other parts of North America. The combination of their allelopathic properties (Fletcher & Renney 1963), low forage value, and drought adaptation have allowed these two knapweeds to displace most other herbaceous plants over large areas.

Diffuse knapweed is typically a biennial. The species was recorded in Washington State in 1907 (Howell 1959) and early Canadian herbarium specimens suggest a northward spread into British Columbia: Oyama 1936, Penticton 1939, Grandforks 1940 (Groh 1943); however, it may also have been introduced directly into the province with Turkestan alfalfa from the Caspian Sea region. Renney (1959) suggested that the weed occurred at Pritchard and Lytton, British Columbia, prior to 1930 and certainly other weeds from southern USSR were recorded at Kamloops (near Pritchard) as early as 1920 (Groh 1940). From these beginnings the weed spread by 1972 to infest 25 952 ha of dryland range and roadside in British Columbia as well as small areas in Alberta (Watson & Renney 1974). Distribution of diffuse knapweed in 1977 is shown in Fig. 4. In 1981 about 25 ha of diffuse knapweed, 8 heavily infested, were found on one farm near Morden, Manitoba (B. Todd, 1982, personal communication). Harris & Cranston (1979) predicted that the relatively high availability of summer moisture in Manitoba would put it beyond the range of the weed. It is now not clear whether the potential of the weed was seriously underestimated or the infestation merely represented a local site anomaly as it was on a light soil with a southern exposure and dry. In 1982 several kilometers of railway track were found to be infested at Walsh, Saskatchewan. This is only a slight extension of the range from the stand at Irvine, Alberta (Fig. 4); but a small stand was also found on the railway track at Colonsay, Saskatchewan, which is southeast of Saskatoon (S. McKell, personal communication). Distance spread of the weed across the prairies often seems to be associated with railway ties or treated bridge timbers from British Columbia.

It appears that diffuse knapweed will continue to spread until it dominates the herbaceous community in open uncultivated sites in the brown chernozem and brunisol soils of southern British Columbia and the brown soils of Alberta and Saskatchewan. This involves about 7.5 million ha (Harris & Cranston 1979).

Spotted knapweed is a short lived perennial. The first Canadian specimen was collected by McCoun from Victoria, British Columbia, in 1893 (Groh 1943). By 1972, it was estimated to have infested 2410 ha in British Columbia as well as several small stands in Alberta (Watson & Renney 1974); but these amounts are small compared to the estimated 800 000 ha in Montana, United States (Maddox 1979). The western Canadian distribution of the weed in 1977 is shown in Fig. 5.

It is predicted that spotted knapweed will spread to dominate the herbaceous community in open uncultivated or lightly forested sites at the dry end of the Douglas fir zone in British Columbia and in the dark brown soils of Alberta and Saskatchewan. About 3.2 million ha are vulnerable to invasion (Harris & Cranston 1979).

Assuming that the two knapweeds spread to their predicted limits of approximately 10 million ha, the loss in terms of dry native pasture species would be 2.5 million tonnes/year

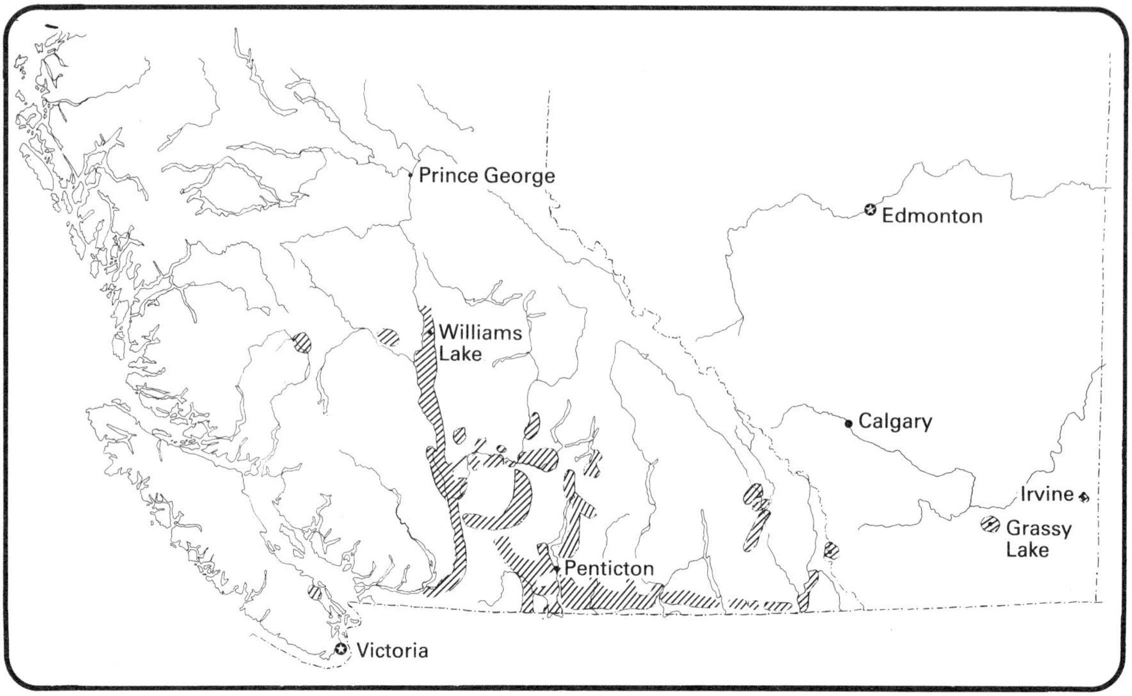

Fig. 4. Distribution of diffuse knapweed, *Centaurea diffusa* Lam., in western Canada in 1977

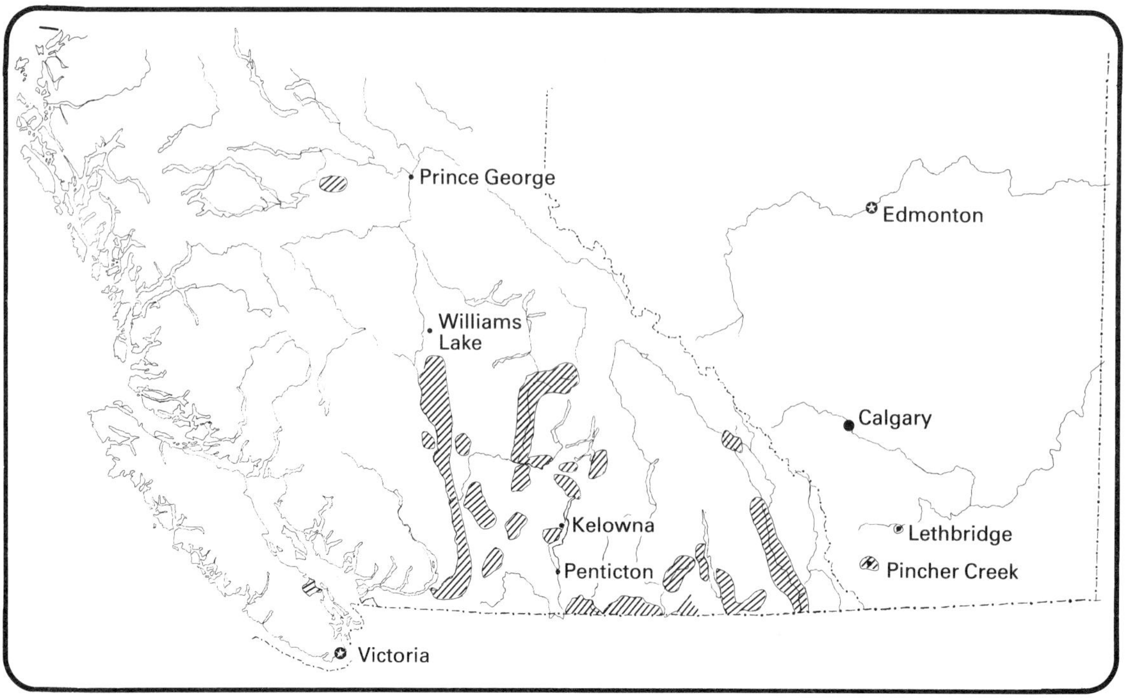

Fig. 5. Distribution of spotted knapweed, *C. maculosa* Lam., in western Canada in 1977

in western Canada. Both knapweeds can be selectively killed in grass with picloram at 0.5 kg/ha, but retreatment is necessary after about 4 years to destroy the plants returning from seed. Aerial treatment with picloram costs about $30/ha, and on good quality range is justified in terms of increased yields; however, treatment of large areas with any persistent, broad spectrum herbicide will inevitably create other problems. Spot control of small infestations with picloram costs up to $520/ha and can only be justified as a means of preventing spread to the surrounding area (Harris & Cranston 1979). Compared to these scenarios the total cost of about $2 million per weed species for biological control by inoculation is attractive (Harris 1979).

Background

The genus *Centaurea* with 530–550 species is centered in the Mediterraenean area, but species are endemic to a region that extends almost to India (Moore & Frankton 1974). Earlier treatments describe one species that is endemic to America, but it was placed in the genus *Plectocephalus* by Jeffrey (1967). Similarly, the ornamentals bachelor's button and sweet sultan were transferred to the genus *Cyanus* and *Amberboa* respectvely. The only economic *Centaurea* spp. grown in North America are all introduced ornamentals such as dusty miller, *C. cineraria* L. and *C. ragusina* L., and several other species grown for their flowers. The most economically important crop species in the same tribe (Cynareae) are safflower, *Carthamus tinctorius* L., and globe artichoke, *Cynara scolymus* L. Thus there are comparatively few closely related desirable species in North America.

There is little insect grazing of diffuse and spotted knapweed in North America apart from the species introduced for their biological control. Watson & Renney (1974) listed a number of polyphagous insects that sometimes feed on these knapweeds. Also the seeds are eaten by mice and some birds, and the pathogen *Sclerotina sclerotiorum* (Lib.) deBary was isolated from diffuse knapweed by Watson *et al.* (1974). This disease kills up to 10% of the flowering plants in some stands of the weed in British Columbia. The knapweed rust in Europe has been recovered from *C. nigra* L. in Nova Scotia (Savile 1970*b*), but not from western North America.

A good selection of potential biological control agents are available in Europe. The Commonwealth Institute of Biological Control (CIBC) has found 36 oligophagous insects that attack one or both of the knapweeds (D. Schroeder 1977 personal communication). Undoubtedly there are more species, as the center of the diffuse knapweed range, southern USSR, was not included in the CIBC surveys. Some of the species that are of particular interest are the spotted knapweed root moth, *Agapeta zoegana* (L.) (Cochylidae), and a diffuse knapweed root moth, *Pelochrista medullana* Stgr. (Tortricidae). Studies by the CIBC have shown that both moths have a narrow host range within the genus *Centaurea*, and *P. medullana* coexits with the beetle *Sphenoptera jugoslavica* Obenb. (which is already established in Canada) as it feeds on the outside and the beetle inside the root (H. Müller 1982 personal communication). The rusts, *Puccinia centaureae* D.C. and *P. jaceae* Otth., are prevalent on diffuse knapweed in Europe and normally occur together on it (Watson *et al.* 1981). According to Savile's (1970*a*) list of host records, *P. jaceae* is restricted to a few closely related knapweeds in the subgenus *Jaceae* while *P. centaureae* is found on many species in the subgenus. However, laboratory inoculations of *P. jaceae* infected a number of other genera closely related to *Centaurea: Carthamus tinctorius* (Watson & Alkoury 1981), *Amberboa moschata* (L.) D.C., *Cyanus segetum* Hill, and *Leuzea confera* (L.) D.C. Use of the rusts for biological control is further complicated by Cummins' (1977) treatment in which both the safflower rust *P. carthami* Corda (which is already present in western North America) and *P.centaureae* are treated as synonyms of *P. calcitrapae* D.C. var. *centaureae* (D.C.)

Cumm. The urediospores of *P. jaceae* and *P. centaureae* are similar in electron photomicrographs (Traquair *et al.* 1981), but are different from those of *P. carthami* (Traquair 1982 personal communication).

It has recently been realized that because of different taxonomic treatments of spotted knapweed in Europe and North America, the problem species of British Columbia has been largely missed in the surveys done in Europe for possible biological control agents. Dostal (1976) recognizes 14 European species in the section Maculosae, whereas Moore & Frankton (1974) classified all the knapweeds in the section that have been introduced to North America as *C. maculosa.* According to Dostal, *C. maculosa* is diploid (2n=18), whereas the problem knapweed of British Columbia is tetraploid (2n=36) and seems to be referable to *C. biebersteinii* D.C. *C. biebersteinii* is endemic to central and southeast Europe, but most of the survey was done on the more western species of *C. vallesiaca* (D.C.) Jordan, *C. maculosa* Lam. *s. str.*, and *C. rhenana* Bor.

Releases and Recoveries

***Urophora affinis* Frfld. (Diptera: Tephritidae)**

U. affinis oviposits into immature flower heads of diffuse and spotted knapweed. The preferred head in diffuse knapweed is 5.5 to 7.5 mm long (Berube 1980). The larva stimulates gall formation from the receptacle tissues and feeds on enlarged and modified parenchyma cells within the gall. The number of galls that can develop in a head is proportional to the receptacle area, so the heads of spotted knapweed tend to have more galls than those of diffuse knapweed (Harris 1980*a*). The mature larva overwinters in the gall in the seed-head and its emergence in the spring is synchronized with the appearance of the first flower buds which are usually all attacked. In British Columbia, the partial second generation of *U. affinis* is too small to fully utilize the diffuse knapweed flower heads produced later in the summer (Roze 1981). However, spotted knapweed heads tend to be produced concurrently in early summer so that a smaller proportion escape attack.

The releases of *U. affinis* are listed in Table 32. *U. affinis* established readily on both spotted and diffuse knapweed in British Columbia (Harris 1980*a*). It increased to a density of up to 3000 galls/m^2 and spread widely throughout British Columbia to substantially reduce knapweed seed production. In Hastings County, Ontario, some galls were found in the year following release but heavy grazing destroyed most of the knapweed heads and the site has not been surveyed since. In Alberta, there was a change in policy which resulted in the treatment of all knapweed infestations with picloram. Flies became established on at least one site but we assume no colonies currently survive. The Québec colonies released in 1980 were still present in 1981 and appear to be established (Watson 1981 personal communication).

***Urophora quadrifasciata* Mg. (Diptera: Tephritidae)**

U. quadrifasciata is bivoltine and like *U. affinis* forms galls in the flower heads of a number of knapweed species. It emerges at the same time as *U. affinis* in the spring but oviposits into slightly more mature heads (7.5–9.5 mm in diffuse knapweed (Berube 1980)). The gall is formed from the ovary wall tissues; several galls are normally found in a head and can develop in the same head as *U. affinis*. Both *U. affinis* and *U. quadrifasciata* increased rapidly after release, but at the original release sites *U. quadrifasciata* has since declined to low numbers, while *U. affinis* has stabilized at moderate levels of attack (Fig. 6). *U. quadrifasciata* has spread more widely and rapidly than *U. affinis* and is now found on even small remote stands of knapweed in British Columbia. Berube (1980) suggested that *U. quadrifasciata* has been less successful because heads

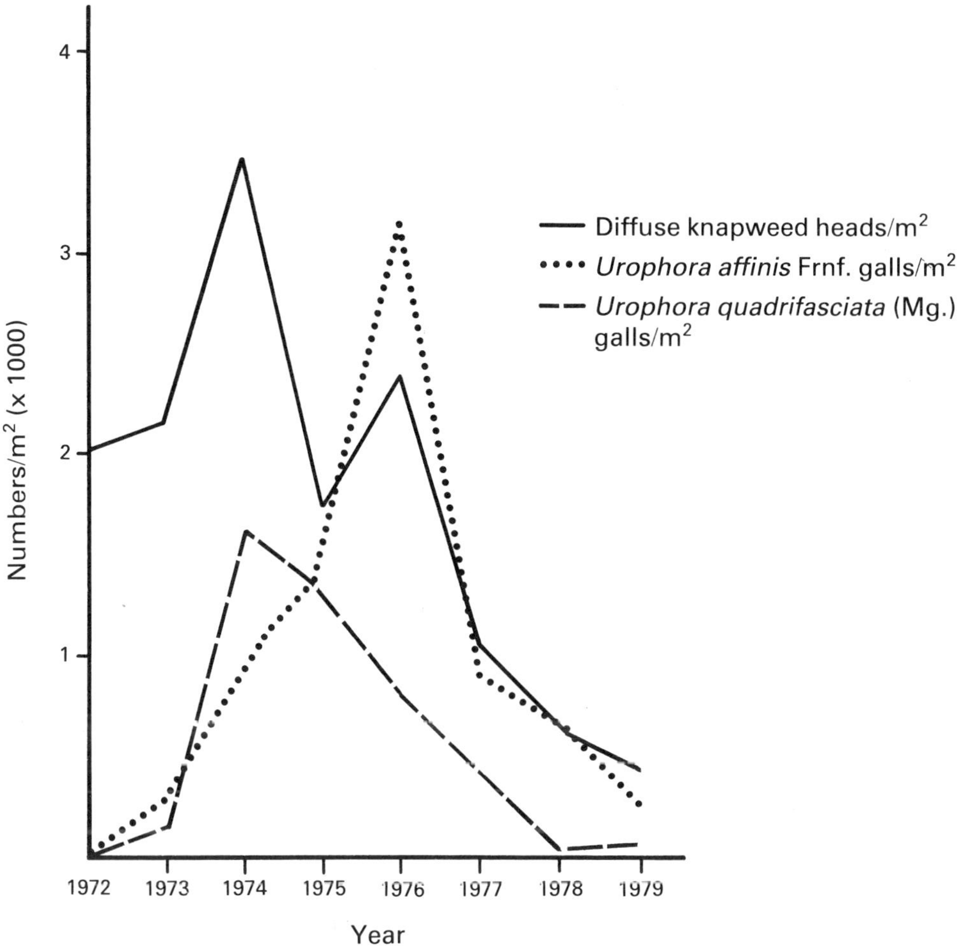

Fig. 6. Density of diffuse knapweed heads and of *U. affinis* and *U. quadrifasciata* at Ned's Creek (Kamloops, British Columbia). Data from Harris (1980*a*)

attacked by *U. affinis* elongate more slowly and hence reach the preferred size for *U quadrifasciata* after the death of most of the flies emerging in the spring.

The necessity of a second generation of *U. quadrifasciata* reduces its success, since in some years there are few buds available for oviposition late in the summer. In spite of interference between the *U. affinis* and *U. quadrifasciata*, Myers & Harris (1980) found that the presence of the two species together increased the destruction of knapweed seed. The combined level of attack of both species on diffuse knapweed has reached a plateau at 0.7–1.6 galls/head with the proportion of each species varying annually. Berube (1980) suggested that a few days difference in emergence time relative to the development of the knapweed accounted for the varying proportions of the two species with the later emergence favouring *U. quadrifasciata*. The 1980 release in Québec survived to the following year (Table 32).

The developing gall of *U. affinis* and *U. quadrifasciata* is a metabolic sink that competes for resources with other parts of the plant growing at the time. Thus the production of flower heads, of seed in unattacked heads, and of vegetative growth are affected. In diffuse knapweed, the combined attack of the two *Urophora* spp. has reduced seed production from around 25 000 to 1 500/m^2 (Harris 1980*b*).

Metzneria paucipunctella Zell. (Lepidoptera: Gelechiidae)

The life cycle of the moth on *C. vallesiaca* in the Swiss Rhone Valley (the source of the stock released in British Columbia) is reported by Englert (1971, 1972). The moth is univoltine and emerges at the end of May. It lays 60–100 eggs with a maximum of three eggs on an individual flower head base or adjacent stem. A single larva survives to develop in a head where it feeds principally on the achene while the seed coat is still soft. The mature larva overwinters in the seed head which in spotted knapweed remains standing all winter having shed its seed in the fall. Up to nine seeds are eaten by a larva and in 83 attacked heads of *C. vallesiaca*, 95% of the viable seed was destroyed. Usually one-third to one-half of the heads were attacked even though the plants were widely scattered. Approximately 20% of the eggs and 30–40% of the larvae were parasitized in Europe. Field records of the moth are restricted to knapweeds in the section Maculosae, although in the laboratory it developed on *C. diffusa*.

Adult *M. paucipunctella* were released (Table 32) in 1973 and 1974 at Castelgar airport and Westwold, British Columbia. *M. paucipunctella* has not been recovered from Castelgar; the number released was slightly less than at Westwold (73 in 1973 and 76 in 1974) and in 1974, the flowering plants were mown shortly after the release was made. At Westwold, the initial establishment was tenuous with only four larvae found at the release point in August 1973. By the following year, larvae were found up to 20 m away (Table 29) and in 1976, they had spread a radius of at least 50 m. By 1979, the population was evenly distributed within a 50 m radius and by 1981, *M. paucipunctella* had spread at least 0.5 km.

Table 29 Establishment of *M. paucipunctella* Zell. on spotted knapweed, *C. maculosa* Lam., at Westwold, British Columbia.

Year (fall)	Radius Sampled (m)	No. Heads Examined	% heads with a *M. paucipunctella* larva
1973	25	4351	0.1
1974	20	2999	0.7
1976	50	485	2.0
1977	30	1443	3.7
1978	25	1719	27.5
1979	50	531	20.3
1980	15	443	37.7

The population growth of *M. paucipunctella* has reflected the winter mortality of larvae in the previous year. For example 68% of the larvae survived the winter of 1977–78 compared with only 28% in 1978–79. The reason for the mortality is not clear. The larvae can survive colder temperatures than they experience at Westwold; survival in the mild humid winter of Vancouver in 1977–78 was 82%, and it was 78% in the dry cold winter of Walachin, British Columbia. Possibly larvae are affected by a combination of moisture and cold and the rate of onset of cold weather. Larvae absorb and lose moisture readily: 14 overwintering larvae dissected from knapweed heads and put on moist filter paper at 10°C in 3 days increased their weight by 50% (from 3.17 ± 0.56 g to 4.79 ± 1.13 g). After 16 days in a desiccator, they had returned to their original weight of 3.11 ± 0.66 g.

Less than 1% of the larvae are parasitized by an *Elachertus* sp. (Chalcidae) and some larvae are eaten by mice which feed on the seed heads particularly in the late summer.

The percentage of seed destroyed by *M. paucipunctella* in Canada is much lower than the 95% reported by Englert (1971). However, *C. biebersteinii* heads at Westwold contain an average 17 seeds compared with the 9–10 seeds for *C. vallesiaca* in the Swiss Rhone Valley. Only one *M. paucipunctella* larva develops in a head and destroys approximately four seeds or about one quarter of the production in a head at Westwold. However, at Westwold, *M. paucipunctella* does coexist in the same heads with *U. quadrifasciata* and there was no significant difference in the number of galls in heads with and without *M. paucipunctella*. Certainly *U. quadrifasciata* is adding to the seed destruction by *M. paucipunctella* but the moth larvae destroy *U. affinis* galls in the head.

***Sphenoptera jugoslavica* Obenb. (Coleoptera: Buprestidae)**

The biology and host specificity of *S. jugoslavica* were studied by Zwölfer (1976). The beetle is indigenous to the Balkans where it is found only on *C. diffusa* and *C. jurineafolia*. The adults feed on the knapweed foliage and do relatively minor damage. The female oviposits between the bases of tightly appressed rosette leaves, and if rosette growth occurs during the egg stage, the larva is unable to penetrate into the root crown. Thus the beetle is restricted to regions with a reliable summer drought during this stage. From the root crown a single larva bores into the root leaving a cylinder of cortex undamaged. The rosette continues to live, but in Europe often does not flower in the following year and so is subject to a second attack by *S. jugoslavica*. Up to 70% of the larvae in Europe are lost to parasitoids, predators, and competitors, but they were able to coexist with insects that feed on the outside of the knapweed roots. The beetle is univoltine and produces 33–65 eggs/female.

S. jugoslavica was released at Grandforks (51 adults) and White Lake, British Columbia (188 adults) in 1976 (Table 32). No survivors were found at Grandforks, but the colony at White Lake increased to infest 25–50% of the rosettes within a 250 m radius in 1981, and over half the flowering plants had been attacked (Table 30).

Table 30 Proportion of diffuse knapweed plants, *C. diffusa* L., attacked by *Sphenoptera* at the White Lake Observatory release site.

Year	Distance From Release Site									Total*		
										Stem Plants/0.25 m²		
	0	3 m	6 m	10 m	13 m	26 m	39 m	52 m	65 m	X	±S.D.	n
1977	.13 (40)**											
1978	.38 (52)	.29 (17)	.07 (41)							10.0	7.6	11
1979	.39 (18)	.80 (5)	.29 (7)	.05 (20)						3.1	3.5	15
1980	.40 (5)		.50 (34)	.46 (13)	.80 (15)	.52 (23)	.57 (21)	.29 (14)	.33 (9)	6.5	4.2	8
1981	.59 (94)in 23 quadrates 0–250 m									4.3	1.8	23

* Ungrazed area only
** Number of flowering plants

Evaluation of Control Attempts

Predictions of the eventual effects of the knapweed gall flies, *U. affinis* and *U. quadrifasciata*, are conflicting. The immediate effects of the flies on the study sites at Kamloops have been a decline in seed production from around 25 000 to 1 500/m² and a reduction in the biomass of the weed, but the number of plants per unit area has remained the same. Obviously any effects of seed destruction would be masked for several years by the seed bank in the soil which, under dense stands of knapweed, is massive. Survivorship studies by Roze (1981) indicate that 1 500 seeds/m² should be enough to maintain the density of the weed, so no decline should be expected. On the other hand, Berube & Myers (1982) suggested that the knapweed would be controlled on sites where there was a high seedling loss either from drought or in moist sites from strong competition from other vegetation. In a later study, Myers & Berube (in press) found that the number of knapweed plants surviving to the flowering stage was directly proportional to the number of seedlings over a wide range of knapweed densities. If the results of this study are generally applicable, the flies should eventually achieve a large decline in knapweed density. The difference in the two predictions may be related to the study methods. Roze (1981) achieved a low knapweed density by removing seedlings from plots of almost pure knapweed. This had the effect of temporarily reducing the competition for the remaining knapweed. In the Myers & Berube (in press) study, the knapweed was always in competition: with itself at high densities and with grass at low densities.

The effect of the beetle *S. jugoslavica* on diffuse knapweed was not studied in detail as until recently the colony was too small for destructive sampling. That the beetle tends to reduce seed production is evident from Table 31. Its effect may be much greater if it

Table 31 Effect of *S. jugoslavica* Obenb. on diffuse knapweed, *C. diffusa* L., seed production per flowering plant at White Lake, British Columbia

	Attacked			Unattacked			
Year	X	±SD	n	X	±SD	n	
1977	18.6	13.4	5	20.4	14.8	28	
1978	69.1	45.9	26	73.5	55.7	82	
1979	38.9	33.9	10	66.5	80.4	16	
1980	31.6	24.4	25	54.3	52.5	27	P 0.05 U Test

tends to prevent rosettes from bolting: approximately 20% of the rosettes in the fenced area had been attacked in the previous year, so they had passed at least two years without flowering. No previously attacked rosettes were found in the grazed area but the significance of this is not known, nor is the percentage of rosettes that normally take two years to flower. There was an indication that *S. jugsolavica* had a synergistic effect on seed reduction in plants stressed by drought or attacked by the seed-head gall flies. Both gall flies were present on the site, and in the autumn of 1981, there was an average of 1.63 ± 1.43 (S.D.) galls in the five distal heads on 23 plants; but whether a combination of *S. jugoslavica* and the flies reduce seed production below the level needed for maintenance of the knapweed stand remains to be seen.

The effect of *M. paucipunctella* on spotted knapweed was less than expected from European studies as a different species of knapweed appears to be involved in British Columbia. Also, the overwintering larvae at Westwold, British Columbia, have suffered a periodic large and unexplained winter mortality that has prevented the rapid increase and spread of the species. However, without releases in other knapweed stands, it should not be assumed that this mortality occurs throughout the spotted knapweed region of North America. The ability of *M. paucipunctella* or the seed fly *U. quadrifasciata* to survive in the colder parts of the Canadian prairies is doubtful. In the winter of 1981, spotted knapweed heads containing overwintering larvae of the three seed-head insects were tied to a stake at normal plant height at Regina. Only *U. affinis* larvae were alive in the spring. If large stands of the weed became established on the prairies, it might be possible to select cold hardy strains of the insects but in the meantime the safest course of action is to chemically treat stands of the weed as they appear and are still small.

Auld *et al.* (1979) pointed out that the best control strategy against a weed on either a farm or regional basis depends on the rate at which it spreads. The faster the rate, the greater the necessity to treat the source area while for slowly spreading weeds, containment is the most economic course of action. The seed reduction achieved by the biological control agents already established in British Columbia should have decreased the rate of knapweed invasion from headlands onto tame pasture as well as the spread to new regions within western Canada. Unfortunately, the benefits from this reduced rate of spread will not be reflected in the chemical control programme until this is done on a strictly cost/benefit basis. To reduce knapweed with biological control agents to the point where the publicly funded chemical control programme becomes clearly unnecessary will require the establishment of more biological control agents on knapweed in Canada.

Table 32 Open releases and recoveries of insects against *Centaurea diffusa* L. and *C. maculosa* Lam. *s. lat.*

Control Agent	Province Released	Year	Origin	Number	Year of Recovery
			C. maculosa		
Urophora affinis	Ontario	1970	France	694	1971
	British Columbia	1970	France	297	1971
	Ontario	1971	France	97	
	British Columbia	1971	France	377	1972
	British Columbia	1972	British Columbia	1472	
	British Columbia	1974	British Columbia	550	
	Alberta	1976	British Columbia	200	Site treated with herbicide
	Québec	1979	British Columbia	950	
	Québec	1980	British Columbia	345	
Urophora quadrifasciata	Québec	1979	British Columbia	950	
Metzneria paucipunctella	British Columbia	1973	Switzerland	197	1974
	British Columbia	1974	Switzerland	177	1975
	British Columbia	1979	British Columbia	180	Overwintering test
			C. diffusa		
Urophora affinis	British Columbia	1970	France	284	1971
	British Columbia	1971	France	209	1972
	British Columbia	1972	USSR	797	1973
	Alberta	1977	British Columbia	681	Site treated with herbicide
	Québec	1980	British Columbia		
Urophora quadrifasciata	British Columbia	1972	USSR	50	1973
Sphenoptera jugoslavica	British Columbia	1976	Greece	239	1977

Recommendations

(1) The effects of *S. jugoslavica* on diffuse knapweed in British Columbia should be determined. If the beetle adds to the seed reduction that is achieved by the seed-head gall flies, it should be established throughout the diffuse knapweed region.

(2) The susceptibility of *M. paucipunctella* to winter mortality should be determined for a number of places within the spotted knapweed region and the moth established in favorable sites.

(3) Additional species of biological control agents should be established on both spotted and diffuse knapweed in Canada as soon as they can be screened and approved

for release. In particular, studies should be completed on the root-feeding moth *Agapeta zoegana* for spotted knapweed, and *Pelochrista medullana* for diffuse and spotted knapweed.

(4) The possibility of using or enhancing pathogens already on knapweed in North America, particularly *Sclerotina sclerotiorum* and *Puccinia jaceae* should be explored. Although *S. sclerotiorum* is a major pest of vegetable crops, its presence on knapweed infested rangeland should not aggravate the problem on cultivated land.

Acknowledgements

Aspects of the biological control of knapweed have been the subject of Ph.D. studies at the University of British Columbia (U.B.C.) by Liga Roze and Peter Morrison, of undergraduate or technical studies, from U.B.C. by Barbara Rawlek, Woody Bennet, Sandy Ockenden and from Regina by Julie Soroka. Their work and ideas have been used extensively in preparing this review. Thanks are also due to Peter Morrison and Julie Soroka for reviewing and suggesting changes in the manuscript.

Literature Cited

Auld, B.A.; Coote, B.G.; Menz, K.M. (1979) Dynamics of plant spread in relation to weed control. *Proceedings of the 7th Asian Pacific Weed Science Conference*, 399–402.

Berube, D.E. (1980) Interspecific competition between *Urophora affinis* and *U. quadrifasciata* (Diptera: Tephritidae) for ovipositional sites on diffuse knapweed (*Centaurea diffusa* Compositae). *Zeitschrift für angewandte Entomologie* 90, 299–306.

Berube, D.E.; Myers, J.H. (1982) Suppression of knapweed invasion by crested wheatgrass in the dry interior of British Columbia. *Journal of Range Management* 35, 459–461.

Cummins, G.B. (1977) Nomenclature changes and new species in the Uredinales. *Mycotaxon* 5, 398–408.

Dostal, J. (1976) *Centaurea L.* p. 254–301. In: Flora europaea, vol 4, Cambridge University Press, 505 pp.

Englert, W. (1971) *Metzneria paucipunctella* Zell. (Gelechiidae, Lepidoptera): a potential insect for the biological control of *Centaurea stoebe* L. in Canada. *Commonwealth Institute of Biological Control Progress Report* 28, 12 pp.

Englert, W.D. (1972) Revision der Gattung *Metzneria* Zeller (Lepidoptera, Gelechiidae) mit Beiträgen zur Biologie der Arten. *Zeitschrift für angewandte Entomologie* 75, 381–421.

Fletcher, R.A.; Renney, A.J. (1963) A growth inhibitor found in *Centaurea* spp. *Canadian Journal of Plant Science* 43, 475–481.

Groh, H. (1940) Turkestan alfalfa as a medium of weed introduction. *Science in Agriculture* 21, 36–43.

Groh, H. (1943) Canadian weed survey. 2nd Annual Report of the Canadian Department of Agriculture, 74 pp.

Harris, P. (1979) Cost of biological control of weeds by insects in Canada. *Weed Science* 27, 242–250.

Harris, P. (1980*a*) Establishment of *Urophora affinis* Frfld. and *U. quadrifasciata* (Meig.) in Canada for the biological control of diffuse and spotted knapweed. *Zeitschrift für angewandte Entomologie* 89, 504–514.

Harris, P. (1980*b*) Effects of *Urophora affinis* Frfld. and *U. quadrifasciata* (Meig.) (Diptera: Tephritidae) on *Centaurea diffusa* Lam. and *C. maculosa* Lam. (Compositae). *Zeitschrift für angewandte Entomologie* 90, 190–210.

Harris, P.; Cranston, R. (1979) An economic evaluation of control measures for diffuse and spotted knapweed in western Canada. *Canadian Journal of Plant Science* 59, 375–382.

Howell, J.T. (1959) Distributional data on weedy thistles in western North America. *Leaflet of Western Botany* 9,17–29.

Jeffrey, C. (1967) Notes on Compositae: 111. The Cynareae in east tropical Africa. *Kew Bulletin* 22, 107–140.

Maddox, D.M. (1979) The knapweeds: their economics and biological control in the western States, U.S.A. *Rangelands* 1, 139–143.

Moore, R.J.; Frankton, C. (1974) The thistles of Canada. *Canadian Department of Agriculture Mononograph* 10, Ottawa, 111 pp.

Myers, J.H.; Berube, D.E. (in press) Diffuse knapweed invasion into rangeland in the dry interior of British Columbia. *Canadian Journal of Plant Science.*

Myers, J.H.; Harris, P. (1980) Distribution of *Urophora* galls in flower heads of diffuse and spotted knapweed in British Columbia. *Journal of Applied Ecology* 17, 359–367.

Renney, A.J. (1959) *Centaurea* spp. infestation in British Columbia. Proceedings of a Joint Meeting of the North Central Weed Control Conference 16 and the Western Canada Weed Control Conference 10, 18–19.

Roze, L. (1981) The biological control of *Centaurea diffusa* Lam. and *C. maculosa* Lam. by *Urophora affinis* Frauenfeld and *U. quadrifasciata* Meigen (Diptera: Tephritidae). Ph.D. thesis, University of British Columbia.

Savile, D.B.O. (1970*a*) Some Eurasian *Puccinia* species attacking Cardueae. *Canadian Journal of Botany* 48, 1553–1566.

Savile, D.B.O. (1970*b*) Autoecious *Puccinia* species attacking Cardueae in North America. *Canadian Journal of Botany* 48, 1567–1584.

Traquair, J.A.; Kokko, E.G.; Harris, P. (1981) Urediniospore morphology of two rust fungi on *Centaurea* (knapweed). *Abstracts of the Canadian Botanical Association Annual Meeting* 1981.

Watson, A.K.; Alkoury, I. (1981) Response of safflower cultivars to *Puccinia jaceae* collected from diffuse knapweed in eastern Europe. *Proceedings of the 5th International Symposium on the Biological Control of Weeds*, 301–305.

Watson, A.K.; Copeman, R.J.; Renney A.J. (1974) A first record of *Sclerotina sclerotiorum* and *Microsphaeropsis centaureae* on *Centaurea diffusa*. *Canadian Journal of Botany* 52, 2639–2640.

Watson, A.K.; Renney, A.J. (1974) The biology of Canadian weeds, 6. *Centaurea diffusa* and *C. maculosa*. *Canadian Journal of Plant Science* 54, 687–701.

Watson, A.K.; Schroeder, D.; Alkoury, I. (1981) Collection of *Puccinia* species from diffuse knapweed in eastern Europe. *Canadian Journal of Plant Pathology* 3, 6–8.

Zwölfer H. (1976) Investigations on *Sphenoptera* (*Chilostetha*) *jugoslavica* Obenb. (Col. Buprestidae), a possible biocontrol agent of the weed *Centaurea diffusa* Lam. (Compositae) in Canada. *Zeitschrift für angewandte Entomologie* 80, 170–190.

Cirsium arvense (L.) Scop., Canada Thistle (Compositae)

D.P. PESCHKEN

Pest Status

New data have been published on the pest status of Canada thistle, *Cirsium arvense* (L.) Scop., since the previous ten year review (Peschken 1971). Thomas (1980) summarized the abundance of Canada thistle on cultivated land in Canada. This is a particularly serious weed in the three Prairie Provinces, especially in the black soil zone and in Manitoba. In this province it ranked fifth and sixth in abundance in 1978 and 1979 respectively; ninth and twelfth in Saskatchewan from 1976 to 1979; and on the average thirteenth in Alberta from 1973 to 1979. Of their most troublesome weeds, farmers in Alberta, Saskatchewan, and Manitoba ranked Canada thistle second, fifth, and third respectively, i.e. considerably higher than the rank based on relative abundance (Thomas 1980). Peschken *et al.* (1980) estimated the loss in yield of wheat in Saskatchewan based on the sampling estimate of the density of Canada thistle in wheat fields (Thomas 1976) and estimates of the reduction in yield of wheat at various density levels of Canada thistle. Assuming the 10 year average yield of 1607 kg per ha and a price of $14.7 per 100 kg, Saskatchewan farmers lost 24 300 t of wheat or $3 600 000 in the reduction of yield of wheat alone due to Canada thistle in 1976. The costs of chemical and cultural control of Canada thistle are high. Thus the total cost of Canada thistle in terms of cost of control and losses to all crops is a multiple of the $3 600 000 estimated for wheat.

Background

Natural enemies of Canada thistle and the initial stages of the biological control work were discussed in the previous review (Peschken 1971). Work on the biological control of Canada thistle has also been done in the United States with *Altica carduorum* Guer. (Coleoptera: Chrysomelidae), *Urophora cardui* (Diptera: Tephritidae), and *Ceutorhynchus litura* (Coleoptera: Curculionidae) (Andres 1980, Schaber *et al.* 1975, Story 1980). The latter species is established in Idaho, South Dakota, and Montana (Julien, 1982). Baker *et al.* (1972) reported on releases of *A. carduorum* against Canada thistle in England. Ward & Pienkowski (1978*a*, 1978*b*), working in the state of Virginia, United States, investigated the biology and mortality of *Cassida rubiginosa* Muell. (Coleoptera: Chrysomelidae) which has been accidentally introduced into eastern North America.

One insect was screened but not released. The only confirmed host plant of *Tingis ampliata* H.-S., (Heteroptera: Tingidae) in its native habitat in Europe is *C. arvense*. However, in the laboratory it developed fertile eggs on several other plants, including the cultivated globe artichoke (*Cynara scolymus* L.) and safflower (*Carthamus tinctorius* L.). Therefore, *T. ampliata* was not recommended for release (Peschken 1977).

Releases and Recoveries

Altica carduorum Guer. (Coleoptera: Chrysomelidae)

(a) Ecology

Baker *et al.* (1972) showed that the development of the immature stages of *A. carduorum* is slightly faster at 100% relative humidity (r.h.) than at 97%. At an average temperature of 15°C, development was not completed, while at 22.5°C to 25°C, it was completed in 33–38 days. Thus high temperatures and humidities are needed for an optimal rate of development. This agrees with the coastal Mediterranean and Atlantic native distribution of the beetles (Peschken 1971).

(b) Releases

Releases up to and including 1968 were reported by Peschken (1971). The colony at Lacombe, Alberta survived until 1971. In addition, 170 were released in 1969 at Fort Vermilion, Alberta (58° 28′ N), which is the most northern and coolest of all release sites of *A. carduorum* in Canada (Table 33). In 1970, 1018 beetles were released at Essex, Ontario (42° 22′ N), which is the most southern and warmest of all release sites. Both releases were made early in the spring which prevented overdispersal in high temperatures (Peschken 1977). Predation of eggs and larvae was severe at Essex and less severe at Fort Vermilion. In the fall of the respective release years, 18 adults were found at Fort Vermilion and only one at Essex, although six times as many beetles had been released at the latter site. At Essex, the carabid beetle *Lebia viridis* Say, was implicated as a main predator of eggs and larvae. In feeding tests it consumed 0.7 eggs and larvae per hour (Peschken 1971). No *A. carduorum* were found in subsequent years.

***Ceutorhynchus litura* (F.) (Coleoptera: Curculionidae)**

(a) Ecology

The ecology of this stem-mining weevil was summarized by Peschken (1971). Since then Peschken & Beecher (1973) and Peschken & Wilkinson (1981) have published additional information. The weevil larvae feed on the parenchyma tissue of the stem and avoid the vascular bundles. In Canada, the females oviposit from early May until the beginning of June. The new adults copulate and feed heavily during sunny, late summer and early fall days with temperatures over 18°C, but do not oviposit. At this time, males outnumber females 2:1.

In spring, females outnumber the males 1.6:1. In the laboratory females laid an average of 122 eggs in temperatures varying from 17°C to 27°C during a 16.5 hour day and 7°C to 17°C during a 7.5 hour night. An average of 80% of the eggs hatched.

(b) Releases

A total of 1188 weevils was released in five provinces at 11 release sites (Table 33). The insect became established at nine sites in varying climates in four provinces of Canada. It spread slowly, 2.9 km in 12 years in Ontario, 90 m and 75 m in six and four years respectively at two release sites in Saskatchewan. The length of mines varied on the different release sites from 5 to 23 cm. The short, 5 cm, mines were found at Lacombe, Alberta where almost all of the thistles were smooth-leaved, resembling those of *C. arvense* var. *integrifolium* Wimm. & Grab. The weevil preferred dense patches of thistle.

Mined shoots were from 1.09 to 1.96 times taller than unmined ones in the fall (Peschken & Wilkinson 1981). This is explained as follows: early emerged, unattacked thistle rosettes grow into taller shoots by fall than later emerged ones (Peschken & Wilkinson 1981). The weevil oviposits into the early emerged rosettes and it is assumed that larval mining does not prevent the early emerged rosettes from growing into taller shoots. This also means that damage to the thistles is light.

The development of the weevils and thistles was monitored in detail at one release site in Ontario. The weevil was released on a dense patch of Canada thistle, and eventually it spread to other more or less contiguous thistle patches on the same permanent pasture covering about 1 ha. On the original release patch, the level of mined stems peaked at 72% in the fifth growing season with an average of 2.75 larvae per stem, and thistle density subsequently diminished to zero (Peschken & Beecher 1973, Peschken & Wilkinson 1981). However, on another patch in the same pasture, thistle density did not diminish although 77–91% of the stems were mined for four years, while on a third thistle density declined to zero per m^2 with only up to 2% of the stems being mined. Thus no consistent correlation between an increase in the weevil population and a decline of the thistles was observed. The rust *Puccinia punctiformis* (Str.) Rohl. and the thistle feeding beetles *Cleonus piger* Scop. (Coleoptera: Curculionidae) and *Cassida rubiginosa* Muell. (Coleoptera: Chrysomelidae) were also present on this release pasture, and dense thistles tend to decrease to scattered shoots behind an advancing front (Amor & Harris 1975). It is assumed that the decline of thistles on some of the patches was caused by one or a combination of these factors. Similarly, the fluctuating thistle densities at release sites in western Canada could not be associated with levels of weevil infestation.

Urophora cardui (L.) (Diptera: Tephritidae)

(a) Ecology

This gall fly occurs from France in the west (Zwölfer 1967) to near the Crimea (Dirlbeck & Dirlbeck 1964) and Siberia in the east (Leclercq 1967), and from Sweden in the north (Zetterstedt 1847) to the Mediterranean in the south (Zwölfer 1967). Areas in half shade seem to be preferred over those in full sun (Zwölfer 1967).

The fly is monophagous on Canada thistle and oviposits into the vegetative buds of the main or side shoots. In Saskatchewan, the oviposition period extends from the end of May to the third week in July. The larvae overwinter and pupate in spring inside a multi-locular lignified gall. There is a strong correlation in the number of larvae and the size of field collected or laboratory produced galls (Peschken *et al.* 1982).

(b) Releases

A total of 4932 flies and 290 galls was released in 14 locations in the four western provinces of Canada and in 10 locations in Ontario, Québec, and New Brunswick (Table 33). In western Canada, galls were usually produced in the year of the release, and on two sites it survived for two and three years respectively. However, the fly died out in western Canada except for a very small colony at Camrose, Alberta.

In contrast, *U. cardui* is established at five locations in Ontario, Québec, and New Brunswick. Near Sussex, New Brunswick, the fly had been released at two locations, 4.5 km apart. The population of one release had spread over 1000 ha by 1980, but only 6% of the shoots were galled in any one thistle patch. By 1981, populations of the two release sites had merged and spread over 3000 ha. On some thistle patches up to seven galls were found on one thistle shoot and there was an average of three on the attacked shoots (D.B. Finnamore, 1981, personal communication). In 1979, 13% of a sample of larvae was diagnosed to contain *Nosema* sp. (Microsporidia) spores and the percentage was higher in dead than in live larvae. Three percent of the larvae were parasitized by *Habrocytus elevatus* Walker (Hymenoptera: Pteromalidae). In 1980, only 3% of the larvae contained *Nosema* spores and 1.3% were parasitized by an as yet unidentified hymenopterous parasite. In Saskatchewan and Alberta, the mortality was higher: 35% of a total of 402 larvae collected in the spring of 1975, 1976, and 1977 was dead. Comparing only those thistles that emerged during the oviposition period of the fly, shoots with galls on the main shoot were 57% shorter, and those with sideshoot galls were 10% shorter than ungalled shoots in Québec when growing on an uncultivated site with competing vegetation. However, in Saskatchewan even an average of 13 galls per shoot did not reduce the height, dry weight, number of seed heads or the dry weight of new shoots produced by thistles growing in the absence of competition on heavy fertile soil with sufficient moisture (Peschken *et al.* 1982).

Lema cyanella (L.) (Coleoptera: Chrysomelidae)

(a) Ecology

This leaf-feeder is widely distributed in Europe and Asia between latitudes 35°N and 65°N and occurs about three-quarters of the range of *C. arvense* (Peschken & Johnson 1979). It is rare in Germany, Austria, Czechoslovakia, and Poland and is more frequent in western Europe (Zwölfer & Pattullo 1970, Winiarska 1973). Collection records indicate that moist habitats are preferred (Winiarska 1973, Zwölfer & Pattullo 1970, Lopatin 1960, Peschken unpublished).

The adult beetles hibernate and in Europe they appear on the thistle rosettes at about the end of April. They oviposit in May and the new generation of adults can be observed from the end of June until August. There is one generation per year.

(b) Host specificity

The only confirmed breeding host in Europe is *C. arvense* (Zwölfer & Pattullo 1970, Peschken & Johnson 1979). In the screening tests for host specificity, it did not feed on any cultivated plants but on several *Cirsium* spp. native to North America (Peschken & Johnson 1979). One of these, *Cirsium drummondii* T. & G., was even preferred in favour of Canada thistle in choice tests in the laboratory. In a field cage, however, the females laid most eggs on those thistles which were most prevalent in terms of biomass of leaves and stems (Peschken unpublished).

(c) Releases

The releases listed on Table 33 were made for the purpose of testing the host preference of *L. cyanella* in the more natural environment of a field cage, not for establishment in the field. In the course of these field cage studies, 4 out of 28 beetles survived the 1978/79 winter. In subsequent years the beetles were collected in the fall and overwintered in the laboratory. However, another adult was collected in the field in the spring of 1981. Thus *L. cyanella* can survive the winter in Saskatchewan. No larvae were found in 1981. The release of *L. cyanella* with the intent of establishment has not yet been approved out of concern for the *Cirsium* species of our native North American flora.

Table 33 Open releases and recoveries of insects against *Cirsium arvense* (L.) Scop.

Species and Province	Origin	Year	Number	Year of Recovery
Altica carduorum (Guerin-Meneville)				
British Columbia	Switzerland*	1969	5534	
Alberta	Switzerland*	1969	170	1971
Ontario	Switzerland*	1970	1018	
	France	1970	345	
TOTAL			7067	
Ceutorhynchus litura (F.)				
British Columbia	Germany, Switzerland France, Italy	1975	69	1976–80
Alberta	Germany, Switzerland France, Italy	1975	57	1976–80
	Indian Head, Saskatchewan	1978	280	1979–80
Saskatchewan	Switzerland*	1973	70	1974–80
	Switzerland*	1974	30	1975–80
	Germany, Switzerland France, Italy	1975	41	1976–80
	Switzerland*	1976	56	
	Switzerland*	1978	25	1979
	Indian Head, Saskatchewan	1979	228	1980
Ontario	Switzerland**	1966	56	
New Brunswick	Indian Head, Saskatchewan	1978	276	1981
TOTAL			1188	
Urophora cardui (L.)				
British Columbia	Germany	1974	274	
	Germany	1975	98	
	France, Austria plus lab. reared from French stock	1976	199	
Alberta	Germany	1975	100	1976
	Austria, France plus lab. reared from French stock	1976	472	
	Lab. reared from stock collected in Canada	1977	406	1978–81
Saskatchewan	Germany	1974	625	
	Germany, plus lab. reared from German stock	1975	274	1976–80
	France, Austria plus lab. reared from French stock	1976	915	1977–78
	Lab. reared from stock collected in Canada	1977	392	
Ontario	Germany	1975	81	1976–80
	France, Austria plus lab. reared from French stock	1976	52	1977–80

Table 33 continued

Species and Province	Origin	Year	Number	Year of Recovery
Québec	Germany	1975	96	
	France, Austria plus lab. reared from French stock	1976	451	1977–79
	Lab. reared from stock collected in Canada	1977	99	
	Field collected at Compton, Québec	1979	290 galls	1981
New Brunswick	France, Austria plus lab. reared from French stock	1976	191	1977–82
	Lab. reared from stock collected in Canada	1977	207	1978–82
TOTAL			4932 flies plus 290 galls	
Lema cyanella (L.)				
	Switzerland*	1978	28	1979
		1979	68	
		1980	35	1981
		1982	31	1982
TOTAL			162	

* These were laboratory reared beetles from stock collected in Switzerland.
** Not reported in Peschken 1971.

Evaluation of Control Attempts

The cool climate combined with predation prevented establishment of *A. carduorum*. The weevil, *C. litura*, is established in four provinces and probably can be colonized in all agricultural areas of Canada. However, damage to Canada thistle by larval mining is small. Furthermore, Canada thistle is a major weed in the grain belt of the three Prairie Provinces, where destruction of eggs and larvae by cultivation of thistles along field margins prevents the buildup of dense populations. *U. cardui* is not established in western Canada (except for a small colony at Camrose, Alberta), but it can probably be established anywhere in the agricultural areas of eastern Canada. The galls do cause noticeable stress to thistles growing in competition with other vegetation (Peschken *et al*. 1982). Overdispersal prevented the buildup of dense populations and reduced effectiveness so far. It is not known why *U. cardui* did not become established in western Canada. The 35% rate of mortality alone does not explain the failure of the release colonies, many of which bred well in the year of their release. Possibly severe spring frosts during the pupation period were a problem. Most release stock originated from the upper Rhine Valley which is considerably warmer and moister than the Prairie Provinces. Lalonde & Shorthouse (1982) suggest that inadequate moisture on the release sites in western Canada prevented the breakdown of the callus plug in the exit channel out of the gall, which would cause poor and unsynchronized emergence. Thus *U. cardui* may not have been preadapted to the climate in western Canada.

On some release sites of *C. litura*, Canada thistle declined to very low densities but new dense stands developed within a few paces of the old one. Density declines such as these could not be explained by the activity of the weevil and were observed also by Amor & Harris (1975) without the presence of the weevil. In Ontario, *C. litura*, the rust *Puccinia punctiformis* (Str.) Rohl., the stenophagous beetles *Cassida rubiginosa* Muell. and *Cleonus piger* Scop. (Coleoptera: Curculionidae), and autotoxicity of Canada thistle may have combined to cause the decline of Canada thistle on the release site (Ferdinandsen 1923, Bendall 1975, Wilson 1981, Peschken & Wilkinson 1982).

Additional insects or disease organisms must be established in western Canada in order to control this vigorous weed. This may be difficult to achieve with the present concern for native *Cirsium* species in the North American flora. Laboratory host ranges tend to be wider than those found in the field (Dunn 1978), and Zwölfer (1969) concluded that all beetles specialized on *Carduus* and *Cirsium* have a similar laboratory host range which includes the *Silybum*, *Carduus*, and *Cirsium* complex. Thus, *L. cyanella*, which is found almost exclusively on Canada thistle in its native range, may theoretically damage populations of one of more of the native *Cirsium* spp. It is difficult if not impossible to determine with absolute certainty by laboratory or field cage tests the potential damage an insect may do to native flora after it is liberated. This problem should be resolved before the testing of further known stenophagous *Cirsium* insects or the search for new ones is undertaken.

Recommendations

(1) Further attempts should be made to establish *U. cardui* in western Canada using stock from eastern Canada which is probably better adapted by now than that from Germany or France.

(2) The effect of the galls produced by *U. cardui* and larval mining by *Ceutorhynchus litura* on the energetics of Canada thistle should be more precisely determined.

(3) The potential danger to native flora by the establishment of biological control agents should be assessed and compared with the possible benefits of a biological control of weeds project. This problem needs to be resolved because not only Canada thistle, but also other important weeds targeted for biological control, have relatives in the native North American flora.

(4) Eastern Europe and Asia should be surveyed for additional *Cirsium* insects which would be preadapted to the climate of the three Prairie Provinces in Canada.

Literature Cited

Amor, R.L.; Harris, R.V. (1975) Seedling establishment and vegetative spread of *Cirsium arvense* (L.) Scop. in Victoria, Australia. *Weed Research* 15, 417–411.

Andres, L.A. (1980) The biological control of Canada thistle (*Cirsium arvense* (L.) Scop.) in the United States. *Proceedings of the Canada Thistle Symposium, Regina*, pp. 112–127.

Baker, C.R.B.; Blackman, R.L.; Claridge, M.F. (1972) Studies on *Haltica carduorum* Guerin (Coleoptera: Chrysomelidae), an alien beetle released in Britain as a contribution to the biological control of creeping thistle, *Cirsium arvense* (L.) Scop. *Journal of Applied Ecology* 9, 819–830.

Bendall, G.M. (1975) The allelopathic activity of California thistle (*Cirsium arvense* (L.) Scop.) in Victoria, Australia. *Weed Research* 15, 77–81.

Dirlbeck, J.; Dirlbeck, K. (1964) Notizen zum Vorkommen der Bohrfliegen (Diptera: Trypetidae) an der Schwarzmeerküste der Sowjetunion. *Sbornik Faunistických Praci Entomologického oddělini Národniho Musea v Praze* 10(87), 5–12.

Dunn, P.H. (1978) Shortcomings in the classic tests of candidate insects for the biocontrol of weeds. pp. 51–56. In: Proceedings of the 4th International Symposium on the Biological Control of Weeds, University of Florida.

Ferdinandsen, C. (1923) Biologiske Undersogelser over Tidselrust *(Puccinia suaveolens* (Pers. Rostr.)). *Beretning om Nordiske jordbrugs-forskeres forenings Kongres* 5–8, 475–487.

Julien, M.H. (Ed.) (1982) Biological control of weeds: a world catalogue of agents and their target weeds. Commonwealth Agricultural Bureaux, 108 pp.

Lalonde, R.G.; Shorthouse, J.D. (1982) Exit strategy of *Urophora cardui* (Diptera: Tephritidae) from its gall on Canada thistle. *Canadian Entomologist* 114, 873–878.

Leclercq, M. (1967) Contribution à l'étude des Trypetidae (Diptera) paléarctiques et de leurs relations avec les végétaux. *Bulletin des Recherches Agronomiques de Gembloux* (N.S.) 2, 64–105.

Lopatin, I.K. (1960) Data on the fauna and ecology of leaf-beetles (Coleoptera: Chrysomelidae) on the southern rear bank of the Dnieper River. *Entomologicheskoe Obozrenie* 39(3), 447–457.

Peschken, D.P. (1971) *Cirsium arvense* (L.) Scop., Canada Thistle. *In*: Biological control programmes against insects and weeds in Canada 1959–1968. *Commonwealth Institute of Biological Control Technical Communication* 4, 79–83.

Peschken, D.P. (1977) Host specificity of *Tingis ampliata* (Tingidae: Heteroptera): a candidate for the biological control of Canada thistle (*Cirsium arvense*). *Canadian Entomologist* 109, 669–674.

Peschken, D.P.; Beecher, R.W. (1973) *Ceutorhynchus litura* (Coleoptera: Curculionidae): biology and first releases for biological control of the weed Canada thistle (*Cirsium arvense*) in Ontario, Canada. *Canadian Entomologist* 105, 1489–1494.

Peschken, D.P.; Finnamore D.B.; Watson, A.K. (1982) Biocontrol of the weed Canada thistle (*Cirsium arvense*): releases and development of the gall fly *Urophora cardui* (Diptera: Tephritidae) in Canada. *Canadian Entomologist* 114, 349–357.

Peschken, D.P.; Hunter, J.H.; Thomas, A.G. (1980) Damage in dollars caused by Canada thistle in wheat in Saskatchewan. *Proceedings of the Canada Thistle Symposium, Regina*, p. 37–43.

Peschken, D.P.; Johnson, G.R. (1979) Host specificity and suitability of *Lema cyanella* (Coleoptera: Chrysomelida), a candidate for the biological control of Canada thistle (*Cirsium arvense*). *Canadian Entomologist* 111, 1059–1068.

Peschken, D.P.; Wilkinson, A.T.S. (1981) Biocontrol of Canada thistle (*Cirsium arvense*): releases and effectiveness of *Ceutorhynchus litura* (Coleoptera: Curculionidae) in Canada. *Canadian Entomologist* 113, 777–785.

Schaber, B.C.; Balsbaugh, E.U., Jr.; Kantack, B.H. (1975) Releases of *Altica carduorum* (Col.: Chrysomelidae) on Canada thistle (*Cirsium arvense*) in South Dakota. *Entomophaga* 20, 325–335.

Story, J.M. (1980) Biological control of Canada thistle in Montana. *Proceedings of the Canada Thistle Symposium, Regina*, p. 128–129.

Thomas, A.G. (1976) Weed survey of cultivated land in Saskatchewan. Weed survey series, Publication 76–1, 93 pp.

Thomas, A.G. (1980) Relative abundance of Canada thistle on cultivated land in Canada. *Proceedings of the Canada Thistle Symposium, Regina*, p. 167–181.

Ward, R.H.; Pienkowski, R.L. (1978*a*) Biology of *Cassida rubiginosa*, a thistle-feeding shield beetle. *Annals of the Entomological Society of America* 71(4), 585–591.

Ward, R.H.; Pienkowski, R.L. (1978*b*) Mortality and parasitism of *Cassida rubiginosa*, a thistle-feeding shield beetle accidentally introduced into North America. *Environmental Entomology* 7(4), 536–540.

Wilson, R.G. (1981) Effect of Canada thistle residue on growth of some crops, *Weed Science* 29(2), 159–164.

Winiarska, W. (1973) Observations on the biology of *Lema cyanella* L. Chrysomelidae, Criocerinae. *Polskie Pismo Entomologiczne* 43(2), 373–382 (in Polish).

*Zetterstedt, J.W. (1847) Diptera Scandinavia. V. VI. Lund, p. 2163–2580.

Zwölfer, H. (1967) Observations on *Urophora cardui* L. (Trypetidae). Weed projects for Canada. *Commonwealth Institute of Biological Control Progress Report* 14, Delémont, 11 pp.

Zwölfer, H. (1969) Experimental feeding ranges of species of Chrysomelidae (Col.) associated with Cynareae (Compositae) in Europe. *Commonwealth Institute of Biological Control Technical Bulletin* 12, 115–130.

Zwölfer, H.; Pattullo, W. (1970) Zur Lebensweise and Wirtsbindung des Distel-Blattkäfers *Lema cyanella* L. (*puncticollis* Curt.) (Col. Chrysomelidae). *Anzeiger für Schädlingskunde, Pflanzenschutz, Umweltschutz*. 43(4), 53–59

* The article has not been seen by the author.

Cirsium vulgare (Savi) Ten., Bull Thistle (Compositae)

P. HARRIS and A.T.S. WILKINSON

Pest Status

Cirsium vulgare (Savi) Ten. is a biennial of Eurasian origin that propagates itself only by seed. It is found in all regions of southern Canada but is most abundant in Québec, Ontario, and British Columbia (Moore & Frankton 1974) on recently disturbed uncultivated ground or overgrazed permanent pasture. In southern British Columbia, it occurs along roads, ditches, dykes and fences, and on pastures or range where the sward has been broken; it is often troublesome in forage crops in the second year after planting. The thistle does not withstand cultivation so it is not a pest in arable crops. It may have one to several flowering stems, usually 0.5 to 2 m high but sometimes 3 m. Plants can be up to one meter in diameter with as many as 343 seed heads. The sharp prickles on the leaves discourage livestock from grazing it.

Bull thistle can be controlled with 2,4-D or MCPA at 1 kg/ha; mowing is also effective if it is done when the plant is reaching maturity; but if cut too early the plant regrows and sets seed before frost. In spite of the ease of control, bull thistle persists as a common weed although it is a minor problem compared with the other thistles targeted for biological control in this volume.

Background

Surveys of the European insect fauna on *C. vulgare* were made by Zwölfer (1965) as part of a broad study of thistle insects that might be used for biological control in Canada. In North America, the flower buds of *C. vulgare* are attacked by the artichoke plume moth, *Platyptilia carduidactyla* (Riley) (Pterophoridae), and the vegetative parts by *Entylia carinata* (Forst.) (Membracidae) as well as by a number of more polyphagous insects. There was an absence of specialized insects such as *Urophora stylata* F. (Tephritidae) in the seed heads. As a result of studies by Zwölfer (1965), Persson (1963), and Redfern (1968), it was apparent that *U. stylata* was a promising biological control agent. Thus its screening and introduction to Canada was partly opportunistic. The lack of screening and release of other agents for the control of *C. vulgare* reflects the relatively minor pest status of the thistle.

Releases and Recoveries

Urophora stylata F. (Diptera: Tephritidae)

(a) Ecology

The biology of *U. stylata* is reported by Redfern (1968) and Zwölfer (1972). The mature larvae overwinter in the seed heads of *C. vulgare* and pupate in the spring. The univoltine adults emerge in early summer (June) and the males establish territories on bolting thistle plants. Shortly before the thistles start to bloom the female visits the plants, mates, oviposits into the flower buds at a precise stage of maturity, and then leaves for another thistle. The eggs are laid in small groups through the bracts at the top of the bud. A second-instar larva hatches in 5–8 days and bores down a corolla tube into the ovariole. The ovariole and the adjacent receptacle become enlarged and the larva feeds inside this gall. Lignified tissues which develop on the outside of the gall gradually coalesce with those of adjacent galls so that the larvae are eventually enclosed in a hard multilocular gall.

Zwölfer (1972) and Redfern (1968) reported that approximately one quarter of the *C. vulgare* heads collected in their studies in western and central Europe were attacked by *U. stylata*. This is much lower than the population reached at the best site in British Columbia. On the European continent, 37% of the larvae were parasitized (Zwölfer 1972) and 24% in Britain (Redfern 1968). No larval or pupal parasitoids were encountered in Canada. In certain regions of Europe, strains of *U. stylata* are specialized on certain other thistles, but no attack of other thistles by the strain released has been found in Canada.

(b) Releases

The releases of *U. stylata* are summarized in Table 34. The fly established itself readily at sites with a dense stand of *C. vulgare*. It has subsequently declined or disappeared where the thistle became sparse following the reestablishment of grasses and other herbaceous perennials as has happened at Danville and Ile Jésus, Québec, and Cranbrook, British Columbia. At Oka, Québec, the thistle was mowed one year after release and the colony disappeared. The release sites at Nanaimo and at Cloverdale, British Columbia, were also destroyed but the fly can be found in the surrounding area. The current status of the other release sites is unknown. The results of establishment suggest that fly dispersal is slow and they require a relatively stable and dense stand of *C. vulgare* for survival.

Table 34 Open releases and recoveries of *Urophora stylata* F. on *Cirsium vulgare* (Savi) Ten. in Canada.

Year	Location	Origin	No. released	Year of recovery
1973	Nanaimo, British Columbia	Germany - Switzerland	174 adults	1974 (site cleared and mowed in 1975)
1973	Westham Isle, British Columbia	Germany - Switzerland	462 adults	1974
1973	Cloverdale, British Columbia	Germany - Switzerland	132 adults	1974
1973	Cranbrook, British Columbia	Germany - Switzerland	459 adults	1974–1976 (no thistles or *U. stylata* in 1981)
1976	Oka, Québec	France - Austria	132 adults	Site mowed – no establishment
1977	Danville, Québec	British Columbia	157 adults	1978
1977	Ile Jésus, Québec	British Columbia	248 adults	1978
1978	New Denver, British Columbia	British Columbia	957 pupae	1979
1978	Williams Lake, British Columbia (lots 6069, 7017, 69)	British Columbia	2000 pupae	1979

The establishment and build-up of *U. stylata* populations were followed in more detail at Cranbrook, Nanaimo, and particularly Cloverdale and Ladner, British Columbia (Table 35). The Cranbrook site was along a power right-of-way through forest; Cloverdale was a permanent pasture on black muck soil in the Fraser delta on which the thistle grew where the cattle hooves broke the sward. The site was destroyed for development in 1980, but the fly survives in the area. At Ladner, also in the Fraser delta, the thistle was confined to the edge of a drainage ditch through an intensely cultivated area used largely to grow seed sugar beet and carrots. The populations were followed by sampling, at the end of the summer, all the seed heads on thistle plants growing at approximately 5 m intervals in 4 directions from the release point. The exception was the linear stand at Ladner. Samples were taken to the edge of the colony or the site, whichever was least.

Table 35 Size and spread of *U. stylata* F. colonies at release sites in British Columbia.

Location	Year	Heads/plant ± S.E.M.	No. heads sampled	% heads galled	No. larvae/gall ± S.E.M.	Colony radius m	Distance (m) to nearest thistle ± SD
Cranbook	1973	21.2 ± 6.0	525	6.7		20	0.8 ± 0.1
	1974	12.1 ± 1.8	320	2.8	2.1 ± 0.4	100	2.9 ± 2.9
	1976	6.2 ± 1.0	194	3.6	2.0 ± 0.4	80	thistle scarce
	1981	No thistles present and no *U. stylata* found on nearby roadside *C. vulgare*					
Nanaimo	1973	9.4 ± 2.1	113	3.5		30	0.7 ± 0.8
	1974	8.7 ± 1.5	180	42.8	3.0 ± 0.3	100	0.2 ±0.3
	1975	Area bulldozed, sown to grass and surrounding thistle cut					
Cloverdale	1973	–	53	30.2	3.0 ± 1.1		
	1974	29.3 ± 3.5	299	41.5	3.2 ± 0.2		
	1975	32.2 ± 5.0	1064	15.9		140	2.4 ± 3.4 (299 plants/ha)
	1976	39.7 ± 9.7	1110	65.3		140 +	36.1 ±5.2
	1977	39.8 ± 5.9	1153	84.7	9.1 ± 0.3	200 +	3.5 ± 4.1
	1978	38.4 ± 5.7	1283	95.8	13.7 ± 0.6	250 +	2.4 ± 0.4
	1979	34.7 ± 7.2	520	92.7			
	1980	Site destroyed					
(site 2)	1978	58.0 ± 19.4	580	92.2			
	1979	60.1 ± 14.5	402	83.6			
Ladner	1973	22.5 ± 4.7	449	12.2	3.1 ± 0.3	100	0.3 ± 0.3
	1974	Not sampled – all but 2 thistles mowed					
	1975	54.1 ± 9.2	865	24.4		125	1.0 ± 1.1
	1976	38.6 ± 6.9	1313	5.7		386 +	2.8 ± 6.3
	1977	80.6 ± 15.7	2338	6.0		160 +	1.7 ± 1.7
	1978	59.7 ± 10.6	1851	31.7		150 +	0.9 ± 1.3
	1979	61.2 ± 10.9	2348	46.3	14.2 ± 5.2		
	1980	41.8 ± 5.9	1918	39.9			
	1981	55.3 ± 11.8	775	26.3			
	1982	39.6 ± 9.0	912	38.8			

The results (Table 35) show that at Cloverdale the fly population increased to attack over 90% of the heads formed. The number of larvae per head increased with the percentage of heads attacked (Equation 8, Table 36). This is also apparent from Table 35: at low densities of the fly, a density of 2–3 larvae per gall was found, but at high fly densities the average increased to around 14 larvae per gall with maximum numbers of over 30 larvae. Presumably the high larval numbers are the result of multiple ovipositon. The data for Ladner were excluded from this analysis where it is suspected that the inability of the fly to attack more than half the heads resulted from their loss over the surrounding fields which were treated with insecticide during the season. Thus although the high number of larvae/gall in 1979 indicated a high density of flies in early summer, the percentage of heads attacked suggests that those forming late in the season escaped the fly.

The average distance of the sampled thistle to its neighbour (Table 35) gives an idea of density although, except in one instance, this cannot be related to the number of thistles per unit area as the distribution contagion was not measured. No effect of the fly on the distance between thistles was apparent; however as *C. vulgare* seed was found to be still viable after approximately 36 years under forest cover (Harrington 1972), it will take a number of years of reduced input into the soil seed-bank before the density of the thistle is affected.

Table 36 Evaluation of the effect of *U. stylata* F. on *C. vulgare* (Savi) Ten.

Regression			Cor.Coef.	n	P
1.	No. heads on unattacked plants	$0.40\,x + 2.0$	0.97	25	<0.001
2.	No. heads on attacked plants	$0.32\,x + 6.2$	0.81	25	<0.001
3.	No. plump seed/head from unattacked plants	$3.66\,r^2 - 18.7$	0.53	118	<0.001
4.	No. plump seed/head from attacked plants	$-0.12\,r^2 + 67.7$	0.04	291	N.S.
5.	Mean no. larvae/head in head size classes	$0.16\,r^2 - 1.2$	0.94	13	<0.001
6.	Maximum number of larvae/head in head size classes	$0.24\,r^2 + 2.8$	0.62	13	<0.05
7.	Live larval weight in mg	$-0.08\,n + 9.5$	−0.22	980	<0.001
8.	No. larvae/gall	$0.11\,z + 0.3$	0.92	7	<0.01
9.	No. larvae/head	$-0.006\,x + 9.6$	−0.07	25	N.S.
10.	Total seed production from unattacked plants	$55.51\,x + 235.2$	0.96	25	<0.001
11.	Total seed production from attacked plants	$17.69\,x + 571.4$	0.73	25	<0.001

Where x = dry weight of plant in g
r = radius of receptacle attachment to the involucre in mm
z = the percentage of heads attacked
n = no. larvae/head

Evaluation of Control Attempts

In the absence of any data on *C. vulgare* seed numbers in the soil, the best indication of the control value of *U. stylata* is its effect on seed production. The effects of gall formers on their host are complex as they sequester resources to the gall (Harris 1980); also the seed from the first heads had fallen before the last ones were mature and it was not possible to visit the stand regularly throughout the season. Thus a series of studies was done at Cloverdale to determine the effects of the fly.

Effects of *U. stylata* on the number of seed heads on *C. vulgare* plants

(a) Methods

A total of 25 unattacked and 25 attacked *C. vulgare* plants was collected at the end of September 1977 from within the *U. stylata* colony at Cloverdale and from just outside it. The plants were cut at ground level, oven dried and weighed. The number and size of the seed heads with a basal receptacle diameter of over 10 mm (productive seed heads) were recorded for each plant, as well as the number of heads attacked, and the number of *U. stylata* larvae per gall. Attack by *U. stylata* enlarges the receptacle but does not affect the base where it is attached to the involucre, so its diameter was used to compare the size of the productive heads on the attacked and unattacked plants.

(b) Results

An average of 88% of the productive heads was galled on the attacked plants. This percentage is higher than that obtained from the transect samples (Table 35), but as a range of thistle sizes was selected from near the center of the colony, it was not a random sample. The 564 galled heads contained an average of 10.3 ± 0.3 SEM larvae. This is also slightly higher than the mean in Table 35.

Plant weight accounted for most of the variation in the number of seed heads on both the attacked and unattacked plants (Equations 1 and 2, Table 36). The level of attack on individual plants ranged from an average of 1.7 larvae/head to 18.1 larvae/head, but the attack level was not affected by plant size (Equation 9, Table 36). The slopes of the regression lines (Equations 1 and 2) suggest that attack reduced the number of heads: plants of 100 g had 9.5% fewer and those of 200 g had 14.6% fewer than the unattacked plants. However, the slopes were not significantly different and the differences in the averages were partly balanced by a slightly larger head size on the attacked plants (average diameter 13.7 ± 0.07 mm compared with 12.8 ± 0.06 for the unattacked plants). It is concluded that the receptacle area in the two groups of plants was practically identical. This was in contrast to the results of the seed-head *Urophora* spp. on diffuse knapweed (Harris 1980) where the number of heads and plant size were reduced by attack.

The number of seeds per head vs. head size and the number of *U. stylata* larvae present

(a) Methods

Heads from both attacked and unattacked plants were bagged shortly before the pappus was ready to fly. Seed head size was measured as the radius of the receptacle attachment to the involucre as previously described, and the plump seed, separated with a blower, was counted for each head. Plump seed rather than germinated seed was used for regression analysis: both produced similar trends but the correlations were higher with the former. Part of the problem was a high variability of germination between heads. Anderson (1968) reported 40% germination at 20°C although this was increased to 80% with temperatures alternating between 20–30°C. We obtained 66.2 ± 2.0% germination of plump seed from unattacked plants at room temperature and 52.4 ± 4.8% from attacked plants.

(b) Results

On unattacked plants, large heads produced more plump seed than did small ones (Equation 3, Table 36) as is normal in composites. In contrast on the attacked plants, seed production decreased with head size (Equation 4, Table 36) although the slope and the correlation coefficient were so low that the production of 62 seeds for an average head of 13.7 mm diameter can be used as a constant for all heads. This means that for the plant it is now more productive to develop several small heads rather than a few large ones.

The effect of head size on seed production in the attacked plants is nullified because the number of *U. stylata* also increases with head size (Equation 5, Table 36). The mean number of larvae per head was considerably lower than the maximum number in the heads of each size (Equation 6, Table 36). Thus for a head with a diameter of 14 mm the average number from the regression was 6.6 larvae while the maximum was 14.6 larvae. Thus the larval population is well below the physical limits of the thistle heads. There was a rather small effect of crowding on larval weight. A regression of live weight of 980 larvae from heads with 1-33 larvae in a gall showed that one larva per gall averaged 9.4 mg compared with 6.7 mg for those in heads with 33 larvae. The slope of the regression is highly significant, although as indicated by the small correlation coefficient there was a large variation in larval weights at all densities (Equation 7, Table 36). For the purpose of the calculations in this paper the size and hence the effect of a larva was assumed to be the same whether it was single or crowded.

Calculation of seed production by attacked and unattacked plants

(a) Methods

Seed production for each head on the unattacked plants was determined from regression Equation 3 (Table 36). These were summed and the total for each plant regressed on plant weight so that average production could be determined for a plant of any size. The same was done for the attacked plants except that the heads were assumed to produce 62 seeds regardless of size.

(b) Results

The production of 4398 seeds for a plant of 75 g (Equation 10, Table 36), is close to the figure of 4000 seeds/plant given by Salisbury (1964) for the thistle in Britain. The attacked plant of this size produced 1898 seeds (Equation 11, Table 36), a reduction of 57%. For a plant of 150 g, the reduction was 62%.

The calculations were made for a *U. stylata* population infesting 88% of the thistle heads. This is slightly below the population plateau at Cloverdale, so the reduction in plump seed production at this site should be at least 60%. If the difference in the germination between attacked and unattacked plants is accepted, the reduction is over 65%. Under field conditions germination and seedling survival seem to depend on breaks in the sward rather than competition within the *C. vulgare* population. This means that the reduction in thistle density should be similar to the reduction of seed as soon as the soil seed-bank has reached its new equilibrium.

Recommendations

(1) *U. stylata* should be distributed to sites across Canada with stable populations of *C. vulgare*; but there is little point in releasing it against temporary outbreaks.

(2) The introduction of additional biological control agents does not appear to be warranted as most of the problems with *C. vulgare* are either temporary following the clearing of land or the result of poor farming practice.

Literature Cited

Anderson, R.N. (1968) Germination and establishment of weeds for experimental purposes. *Weed Science Society of America*, 636 pp.

Harrington, J.F. (1972) Seed storage and longevity. In: Kozlowski T.T. *(Ed.)* Seed biology vol. III, Academic Press, pp. 145–245.

Harris, P. (1980) Effects of *Urophora affinis* Frfld. and *U. quadrifasciata* (Meig.) (Diptera: Tephritidae) on *Centaurea diffusa* Lam. and *C. maculosa* Lam. (Compositae). *Zeitschrift für angewandte Entomologie* 90, 190–210.

Moore, R.J.; Frankton, C. (1974) The thistles of Canada. *Agriculture Canada Monograph* 10, 111 pp.

Persson, P.I. (1963) Studies on the biology and larval morphology of some trypetids (Dipt.). *Opuscula Entomologica* 28, 3–69.

Redfern, M. (1968) The natural history of spear thistle heads. *Field Studies* 2, 669–717.

Salisbury, E. (1964) Weeds and aliens. Collins, 384 pp.

Zwölfer, H. (1965) A preliminary list of phytophagous insects attacking wild Cynareae species in Europe. *Commonwealth Institute of Biological Control Technical Bulletin* 6, 81–154.

Zwölfer, H. (1972) Investigations on *Urophora stylata* Fabr., a possible agent for the biological control of *Cirsium vulgare* in Canada. Weed projects for Canada. *Commonwealth Institute of Biological Control Progress Report* 29, 20 pp.

Convolvulus arvensis L., Field Bindweed (Convolvulaceae)

M.G. MAW

Pest Status

Field bindweed, *Convolvulus arvensis* L., a deep rooted perennial of Eurasian origin, is a serious problem in many areas and has been classed as the worst weed of the prairie and the Great Plains States (Bakke *et al.* 1939). It is a serious weed in Ontario and Québec, especially in corn, and although it is of less importance in the western provinces some local areas are badly infested.

Losses from bindweed competition vary with the crop and cultural methods. Rye yields may be reduced by 20% while grain sorghum yields may be reduced by 78%. On bindweed free land, wheat, barley, and oats yielded respectively 350, 576, and 515 kg per ha more than on infested fields (Wiese & Phillips 1976). Fast growing crops that shade the ground may compete effectively with the weed, and winter wheat on the drier parts of the Great Plains is a good competitor because winter and spring growth occurs when bindweed is dormant. Summer crops may however be severely stunted or even killed by severe bindweed competition (Weise & Phillips 1976).

Background

Field bindweed was first noticed in America in Virginia as early as 1739, and by 1900 it was well established and recognized as a serious weed in most of the western states. In addition to a progressive spread across America, the weed may have been introduced several times. For example, it may have been introduced directly into Kansas with wheat brought from the Ukraine by immigrants about 1875 (Weise & Phillips 1976).

Bindweed is difficult to control because of its extensive root system and prolific seed production. Roots and rhizomes may cover an area up to 6 m in diameter (Frazier 1943) and extend down 9 m below the soil surface (Phillips 1978). In addition roots store a range of sugars, starches, and proteins that can support the plants over long periods even though top growth may be repeatedly destroyed (Weise & Phillips 1976). Over 11 million seeds may be produced per hectare and they may remain viable in soil for up to 30 years (Timmons 1949).

Seedlings develop deep taproots in a few weeks and in 9 to 11 weeks spread radially by lateral roots. Within one season a single plant may form a patch of more than 3 m in diameter (Weise & Phillips 1976).

Biological Control

Control of bindweed but not eradication can be achieved through an integrated programme of frequent cultivation and selected herbicides. However, with the growing concern of the overload of chemicals in the environment, their cost, the cost of energy, and costs in lost production there has been interest in biological control. Mohyuddin (1969) listed the insects from *Calystegia* spp. and *Convolvulus* spp. found by him in Ontario and by others in other parts of the world, and the University of California and the USDA jointly mounted a project to survey the Mediterranean area for natural enemies of *Convolvulus* (Rosenthal 1980). In all, 140 species of insects, three species of mites, and three fungi were found attacking *C. arvensis* and other closely related Convolvulaceae. Of these, a mite, *Eriophyes* sp.; a bruchid, *Spermophagus sericeus* Geoff.; a chrysomelid, *Galeruca rufa* Germ.; and a mildew, *Erysiphe convolvuli* DC., showed the most potential as control agents (Rosenthal 1980). None, however, was found to be strictly monophagous.

Little biological control work has been done in Canada. However, limited attempts to augment existing populations or to establish new colonies in Alberta and British Columbia with insects from Saskatchewan and Ontario (Table 37 and Table 38) were not successful.

Table 37 Collection and release in Canada of chrysomelid insects in control of *Convolvulus arvensis* L. projects

Species	Number	Year	Source	Release or Lab Study
Chirida guttata	114	1969	Canada	British Columbia (study)
(Ol.)	125	1970	Ontario	British Columbia (study)
	236	1979	Saskatchewan	Alberta (release)
Chelymorpha cassidea (F.)	97	1979	Saskatchewan	Alberta (release)
Metriona bicolor (F.)	18	1969	Canada	British Columbia (study)
	7	1970	Ontario	British Columbia (study)
Metriona purpurata Boh.	224	1979	Saskatchewan	Alberta (release)

Table 38 Collection and release in Canada of chrysomelid insects in control of *Convolvulus sepium* L. projects

Species	Number	Year	Source	Release
Chirida guttata (Ol.)	120	1971	Ontario	British Columbia
Metriona bicolor (F.)	200	1971	Ontario	British Columbia

Discussion

Bindweed competes with crop plants for nutrients and water, and thus lowers yields. It twines around plants, smothering them and interfering with harvesting, and is a haven for crop pests (Rosenthal 1980). It is considered to be the 12th most important weed in the world and the 14th most important weed in the United States (Rosenthal 1980). It is a serious weed in Ontario and Québec but a lesser problem in the western provinces. It is mainly a weed in cultivated fields and so not a good candidate for biological control.

A large number of insects are associated with bindweed in North America, but only sporadically in localized areas is appreciable damage done to the plant (Maw unpublished data). Also, it appears that most insect damage occurs when bindweed is supported by other plants or fences, rather than in the prostrate growth in fields. Attempts to move *Chirida guttata*, *Metriona bicolor* and *M. purpurata*, and *Chelymorpha cassidea* into new areas were unsuccessful. This may indicate that these species have reached their geographic limits.

Rosenthal & Buckingham (1982) found a number of European organisms that may be useful as control agents of bindweed in North America. Of these, the arthropods *Galeruca rufa*, *Spermophagus sericeus*, and *Eriophyes* sp. probably have the most potential (Rosenthal 1980) and several other species may also be useful (Rosenthal &

Buckingham 1982). None of the agents is strictly monophagous but usually they feed only on species of *Convolvulus* and *Calystegia*. As there are no commercially valuable plants in either genus, these oligophagous species are not likely to become economic pests. There are however related species in both genera that are native to North America, and this may cause some concern for native flora. In addition some species may harm the sweet potato *Ipomoea batatas*.

Recommendations

(1) Because the United States has already expended so much effort in searching for control agents and screening trials may begin in the United States, it is recommended that Canada not initiate a programme of its own but cooperate in every way with the United States.

(2) Information on the biology, ecology, economics, and status of insect fauna of the weed should be compiled.

Literature Cited

Bakke, A.L.; Gaessler, W.G.; Loomis, W.E. (1939) Relation of root reserves to the control of European bindweed, *Convolvulus arvensis* L. Research Bulletin 254, Iowa State College of Agriculture and Mechanical Arts, Ames, Iowa.

Frazier, J.C. (1943) Nature and rate of development of root systems of *Convolvulus arvensis*. Botanical Gazette 104, 417–425.

Mohyuddin, A.I. (1969) Insects from *Calystegia* spp. and *Convolvulus* spp. *Commonwealth Institute of Biological Control Technical Bulletin* 11, 93–104.

Phillips, W.M. (1978) Field bindweed: The weed and the problem. *North Central Weed Control Conference* 33, 140–141

Rosenthal, S.S. (1980) European organisms of interest for biological control of *Convolvulus arvensis* in the United States. Proceedings of the 5th International Symposium on Biological Control of Weeds, Brisbane, Australia, pp. 537–544.

Rosenthal, S.S.; Buckingham, G.R. (1982) Natural enemies of *Convolvulus arvensis* in Western Mediterranean Europe. *Hilgardia* 50(2), 19pp.

Timmons, F.L. (1949) Duration of viability of bindweed seed under field conditions and experimental results in the control of bindweed seedlings. *Agronomy Journal* 41, 130–133.

Wiese, A.F.; Phillips, W.M. (1976) Field bindweed. *Weeds Today* 7, 22–23.

Euphorbia esula-virgata complex, Leafy Spurge and *E. cyparissias* L., Cypress Spurge (Euphorbiaceae)

P. HARRIS

Pest Status

Leafy and cypress spurge, *Euphorbia esula-virgata* complex and *E. cyparissias* L., are herbaceous perennials of European origin that have become problem species in parts of North America. Leafy spurge is particularly serious on the Canadian prairies and the north central United States, while cypress spurge is most prevalent in Ontario and Québec. Both spurges are unpalatable to most grazing animals except sheep and displace other herbaceous vegetation. In two leafy spurge stands investigated in Saskatchewan in 1981, there was 65–72% less grass within the stand than just outside it. Cypress spurge on a thin limestone soil at Braeside, Ontario, comprised 56% of the herbaceous vegetation in a dry year and 26% in a wet year (Table 39). Cypress spurge is

Table 39 Density of *Hyles euphorbiae* (L.) larvae, dry weight of *E. cyparissias* L. and total herbaceous vegetation at Braeside, Ontario.

Year	No. *H. euphorbiae* larvae/m^2	Dry weight of *E. cyparissias* g/m^2	Dry weight of all vegetation/m^2
1968	6.8×10^{-3}		
1969	1.75×10^{-2}	56.6	216.0
1970	0.4	44.7	79.2
1971	1.0	73.1	147.1
1972	0.3	91.9	213.1
1974	0.6		
1975	2.3		
1976	0.4		

also of concern as extremely small amounts of its latex are intensely irritating to the eyes. The provincial and most municipal governments of the Canadian prairies have an active programme for the control of leafy spurge. For example in the early 1950s, Saskatchewan spent around \$100 000/year against persistent perennial weeds and by 1979 the amount had risen to \$150 000/year, with leafy spurge being the main target. The sum covers the cost of treatment on provincial and municipal lands and roadsides but only the cost of the chemical on private lands. Thus the actual cost of controlling leafy spurge was considerably higher. In Manitoba in 1981, \$38 649 was spent on leafy spurge control by the provincial government of which \$32 890 represented the cost of the chemical (E. Johnson, 1982, personal communication). In Saskatchewan in 1982, the full cost was \$640/ha of spurge treated as the infestations are scattered or on rough terrain.

In the first twenty years of the control programme in Saskatchewan the chemicals used were soil sterilants such as sodium chlorate and various borates either alone or in combination with other herbicides. These were both expensive and ecologically undesirable. In 1971, the treatment of 65 ha with borascu (disodium tetraborate and 5 bromo-3-sec-butyl-6-methylurate) cost $15 727 while the treatment of 86 ha with picloram (4-amino-3,5,6-trichlorpiclinic acid) cost $12 073 (Saskatchewan Production and Marketing Branch 1972). Not only was picloram cheaper but the grass cover remaining was greatly preferable to the bare patches following treatment with soil sterilants. Thus in the past ten years picloram has been used almost exclusively on uncultivated stands except in Manitoba where its use is more restricted.

The effectiveness of the chemical control programme is not impressive. Eleven Saskatchewan municipalities were resurveyed for leafy spurge after approximately 30 years of chemical control. The weed has disappeared on cereal land with normal cultivation and application of 2,4-D, while on uncultivated land the area infested with spurge has more than doubled. These two trends almost cancelled each other in the municipalities sampled.

In Manitoba, where little picloram has been used, the increase has been large: from 3236 ha in 1952 (Craik 1953) to an area estimated by Johnson (1982, personal communication) to be 46 693 ha in 1981. The difference in spurge increase between Saskatchewan and Manitoba is attributed entirely to the use of picloram; but against the benefits of the herbicide is the hazard that large scale continued use will contaminate the ground runoff water since the herbicide is subject to little microbial breakdown.

The largest cypress spurge infestation in Ontario is on thin uncultivated limestone soils at Braeside. The species appears to have been introduced in 1870 and by 1981 had spread, initially with highway construction, to about 14 000 ha on two limestone ridges. The main cause for concern is that the present infestation is only 15 km away from the 360 000 ha Smith Falls Limestone Plain.

Background

The discussion of the biological control of leafy spurge by Harris *et al.* (in press) lists many insects and pathogens that are restricted to the genus *Euphorbia*. The difficulty is finding species that will survive on North American leafy spurge which seems to be a species complex. Specialized *E. esula* insects such as the aphid *Acyrthosiphon neerlandicum* HRL and the clear-winged moth, *Chamaesphecia tenthrediniformis*, do not survive on the common North American leafy spurge. To find insects that will accept the spurge, it has been necessary to seek those that can survive on several *Euphorbia* spp. Conversely, there are several spurges that should not be damaged in North America: *E. antisyphilitica* (a source of high quality wax in Mexico), *E. pulcherrima* (the ornamental poinsettia), *E. lathyris* (a plant suggested as a source of hydrocarbon (Calvin 1978)) and native North American spurges in general. Hence the biological control of leafy spurge is not without its difficulties even though the presence of large dense stands that are almost unutilized by phytophagous organisms suggests that it is an extremely amenable target for biological control.

About three quarters of the insects attacking *Euphorbia* in Europe are restricted to the genus. This is a higher proportion than for most other herbaceous plants. For example, at least three quarters of the insects that feed on the genus *Solidago* attack other plant genera (Harris *et al.*, in press). It appears that the degree of specialization in the insect guild on spurge is determined by the presence of enzymes and toxins that make it difficult for utilization by unspecialized insects and this extends to individual spurge species. For example, Ellis & Lennox (1941) reported the proteolytic enzyme euphorbain from *E. lathyris* but that *E. peplus* contained little or none of it. That some of the compounds are extremely toxic has been long known. For example the Geoponika (6th or 7th century) records that tithymalus (possibly *E. cyparissias*) mixed in a bait could be used to control mice by making them blind. I can confirm that minute amounts of *E. cyparissias* latex in the eye are extremely painful. In the 12th century, Ibn Al Awan (Clemont-Mullet 1864) described the use of a spurge extract for repelling and killing insects. Portier & Busnel (1951) found that *E. cyparissias* latex contained a substance that produced violent muscle tetany in most insects although the specialized spurge hawkmoth larvae, *Hyles euphorbiae* (L.), recovered without ill effects. Geilser (1955) reported that the effects of the substance were similar to those of acetylcholine in that atropine was antagonistic to it.

Such differences presumably account for the specialization of some of the *Chamaesphecia* spp. moths to a single *Euphorbia* sp. For biological control purposes, this makes the precise identification of the target leafy spurges of North America an urgent matter. As this has so far proved difficult on morphological grounds perhaps a chemical or isozyme classification should be attempted.

Releases and Recoveries

Hyles euphorbiae (L.) (Lepidoptera: Sphinguidae)

(a) Ecology

H. euphorbiae, the spurge hawkmoth, is a large defoliator of *Euphorbia* species in the subgenus *esula* in south and central Europe, northern India, and central Asia. The pupae can survive cold winters (Harris & Alex 1971) and its northern distribution seems to be limited by a requirement for high summer temperatures as the larvae feed little and eventually die if kept below 14°C. Also for unknown reasons the moth is uncommon south of the Mediterranean. *H. euphorbiae* has three generations a year in the south of its range and one in the north.

The moth was selected a a biological control agent as it appeared to have the capability of defoliating spurge stands over large areas. Indeed it was a major pest of commercially grown *E. lathyris* in the USSR (Malyuta 1934). The moth was approved for release in 1965 and subsequently attempts were made to establish it in both Canada and the United States.

(b) Releases

Large numbers of *H. euphorbiae* were bred in the laboratory and released, mostly as larvae, in spurge stands across Canada (Table 40). The stock used was from scattered collecting in Switzerland, France, and Germany on *E. cyparissias* and *E. seguierana*. Other larvae presumably from these hosts were purchased from collectors in East Germany. Initial establishment was disappointing and the first evidence of breeding was 68 widely scattered nearly mature larvae recovered at one of four release sites at Braeside, Ontario in 1968. These were collected and two generations bred from them over winter in the laboratory with the result that over 5000 of their progeny were released at Braeside in the spring of 1969. The only other site where a few field bred larvae were collected was at Sidney, Ontario, but this colony has not been visited since 1972.

Most of the releases of *H. euphorbiae* larvae made in 1966 and 1967 suffered a mortality of over 95% during the subsequent two weeks. The exception was a series of releases made in September at Sidney, Ontario, and one release at Braeside, Ontario, where 23% of the larvae survived this period. The development of the larvae released late in the season was so slow that many of them failed to pupate and died. Forwood (1977) with a release of 51 larvae on leafy spurge in Nebraska found that they had all disappeared within seven days. The reason for the larval mortality was predation by

Table 40 Open releases and recoveries of *Hyles euphorbiae* (L.) against *Euphorbia* spp. in Canada.

Year	*Euphorbia* sp.	Location	No. and Stage	Year of Recovery
1965	Cypress spurge	Sidney, Ontario	206 larvae	
1966	Leafy spurge	Jameson, Saskatchewan	1200 larvae	
		Wilmer, British Columbia	300 larvae	
		Miami, Manitoba	195 larvae	
			14 pupae	
		Hartney, Manitoba	186 larvae	
	Cypress spurge	Sidney, Ontario	3606 larvae	
		Jones Falls, Ontario	550 larvae	
1967	Leafy spurge	Wilmer, British Columbia	1900 larvae	
		Kamloops, British Columbia	3000 larvae	
		Moose Jaw Creek, Saskatchewan	979 larvae	
		Balgonie, Saskatchewan	361 larvae	
		Nr. Balgonie, Saskatchewan	3000 larvae	
		Milestone, Saskatchewan	901 larvae	
		Miami, Manitoba	1100 larvae	
		Picton, Ontario	825 larvae	
1967	Cypress spurge	Braeside, Ontario	13680 larvae	1968
1968	Cypress spurge	Sidney, Ontario	1565 larvae	
		Braeside, Ontario	2801 larvae	
1969	Cypress spurge	Braeside, Ontario	5210 larvae	
1973	Cypress spurge	Hornings Mills, Ontario	600 larvae	1982
1973	Leafy spurge	Guelph, Ontario	200 larvae	no recovery
1974	Leafy spurge	Regina Beach, Saskatchewan	37 larvae	
			600 pupae	
1978	Leafy spurge	Cardston, Alberta	229 larvae	

ants, mice, carabids, wasps, pentatomids, and spiders (Harris & Alex 1971). At three sites in Ontario the percentage survival of larvae after four days decreased almost directly with the average number of ants caught in a standard series of pitfall traps during the period (Table 41).

The 68 larvae found in 1968 and the 179 found in 1979 that had not been released that summer were expressed on a m^2 basis as if all the larvae were confined to a hectare. In most of the subsequent years the survey was restricted to 100 m^2 samples taken in early August at paced intervals from a hectare within the release field but in 1975 the samples were taken from the adjacent roadside which usually supported a higher density of larvae.

Table 41 Relationship between the number of ants in pitfall traps and the survival of *H. euphorbiae* (L.) larvae in a four day period

Place	No. ants/trap	% *H. euphorbiae* larvae recovered
Braeside	5.4	34.5
Sidney	35.9	20.5
Picton	58.7	11.8

An *H. euphorbiae* larva requires 2.8 g (dry weight) of *E. cyparissias* foliage for development (New 1971) and in Nebraska they consumed 17.4 g (fresh weight) of leafy spurge (Forwood 1977). On cypress spurge about 70% by weight of the stem and foliage can be consumed by the larva, so at Braeside 11 larvae/m^2 would be needed for defoliation in dry years and 23 larvae/m^2 in wet years. Thus the population at Braeside has stabilized at a level that is using less than 5% of the available *E. cyparissias* foliage. The insect is obviously of little biological control value. The moth has, however, spread well; in 1971 it was found 3 km away from the release site and by 1974 larvae were found in all the spurge stands examined in the 80 km^2 area including the sites where it did not become established initially.

Releases made at Hornings Mills, Ontario, are more encouraging as far as an impact on cypress spurge is concerned (J. Alex, 1982, personal communication). In 1973, 600 larvae collected from Braeside were released (some in field cages) at three sites 3–5 km apart. The only evidence of survival the following summer was one adult in an overturned cage. In 1975, two adults were seen near one of the sites but no moths or larvae were found in 1976. However, in 1982 larvae were found 9 km from the nearest release site. There were 27 larvae/m^2 at one place, 17/m^2 at another, and smaller numbers scattered over a 5 x 8 km^2 region. The higher larval densities resulted in defoliation of the spurge. As the moth has only recently been established and the area along the Niagara limestone escarpment infested with cypress spurge is large, it is likely that the moth will become more numerous in the region over the next few years.

The failure of the releases on leafy spurge at Guelph and elsewhere may indicate that there are host strains of the moth, although in the laboratory it can be reared on both spurges. The only instance of establishment on leafy spurge is in Montana, United States, where large releases (circa 10 000 larvae) of stock originating from Braeside were made betwen 1966–74 (J. Story, 1982, personal communication). Hopefully the Montana colony will provide a strain of leafy spurge accepting moth that can be established widely in western North America. The moth is also established on cypress spurge in New York (S. Batra, 1982, personal communication).

***Chamaesphecia empiformis* (Esp.) and *C. tenthrediniformis* (Denis & Schiff) (Lepidoptera: Sesiidae)**

(a) Ecology

Chamaesphecia is a genus of root-boring moths that have largely specialized on *Euphorbia* spp. It was originally thought that a single species attacked both *E. esula* and *E. cyparissias* but in the course of host specificity studies (Schroeder 1969*a*, 1969*b*, 1970), it was found that there were two sympatric species that differed in their host plants and various aspects of their biology. Nauman & Schroeder (1980) were unable to distinguish them with certainty by adult morphology but they found differences in egg size and in the number of crotchets on the larval prolegs. These authors defined the species on *E. cyparissias* as *C. empiformis* and that on *E. esula* as *C. tenthrediniformis*.

On *E. esula*, larval hatch is confined to a two week period at the start of seed formation (16–30 June in Austria in 1970 (Zwölfer 1970)). On *E. cyparissias*, it occurs after most of the seeds are mature and continues for a longer period: the first eggs were collected 25 June and hatch continued until 12 August. In leafy spurge, the larvae mine into the root directly or through the crown, while on cypress spurge they initially mine the stem and then enter the root in the fourth or fifth instar. The change in mining from stems to the roots on *E. cyparissias* follows changes in nutrient storage. At Belleville, Ontario, seed maturity was marked by an increase in the carbohydrate reserves of the roots and then a decline associated with a second flush of shoot growth (Fig. 7). Thus stem mining would occur during the period of carbohydrate export from the roots to the stems and root mining during the period of high reserve accumulation in the roots.

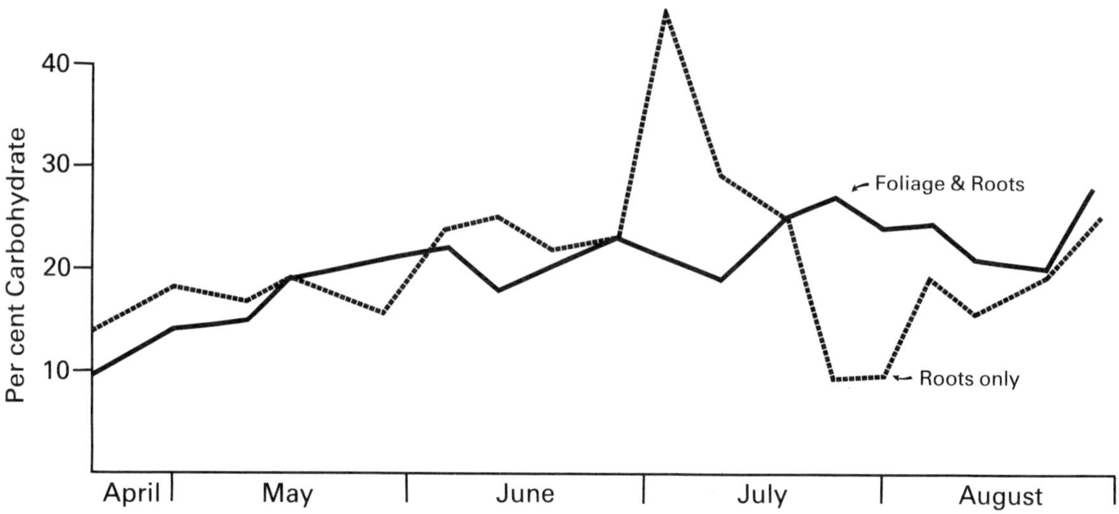

Fig. 7. Seasonal change of available carbohydrate as a percentage of dry weight in the roots and whole plant of *Euphorbia cyparissias* L. at Belleville, Ontario in 1971.

In Austria, *C. tenthrediniformis* was the main insect controlling *E. esula* and the attack level varied from 10–100% of the plants in a stand; however, *E. virgata* even when growing in the same stand as *E. esula* was never found to be attacked. In Switzerland, 10–20% of *E. cyparissias* were attacked by *C. empiformis*. Its effect was to destroy about half of the upper root. Attacked plants regrew from dormant buds below the damage but the shoots produced were considerably smaller and less vigorous. The upper damage threshold inflicted is limited by the antagonism of the larvae so that two can survive only in extremely large plants.

(b) Releases

The *Chamaesphecia* spp. were approved for release in 1970 and 1971 but in spite of several attempts neither species became established (Table 42).

In laboratory tests, newly hatched *C. tenthrediniformis* failed to bore into either the stems or roots of potted Canadian leafy spurge and if pushed into an artificially made hole, they backed out or became paralyzed on contact with a small amount of latex. A few larger larvae did continue boring when placed in holes made in roots that had been treated with maleic hydrazide to eliminate the latex but none completed development. It appears that Canadian leafy spurge is not a suitable host for this insect and the fact that it does not attack either *E. virgata* or *E. cyparissias* in Europe is an indication of an extremely narrow host range. In light of the study by Nauman & Schroeder (1980), it remains to be determined whether the moth from *E. velenovskyi* is *C. tenthrediniformis* or a third practically identical species with another host plant.

Table 42 Open releases and recoveries of *Chamaesphecia empiformis* (Esp.) and *C. tenthrediniformis* (Denis & Schiff) on *Euphorbia* spp. in Canada.

Year	No. and stage released	Source	Release	Recovery
C. empiformis				
1970	200 newly hatched larvae	*E. cyparissias* ex Switzerland	*E. cyparissias* Belleville, Ontario	
1971	1832 eggs	*E. cyparissias* ex Switzerland	*E. cyparissias* Sidney,Ontario	
C. tenthrediniformis				
1971	800 eggs	*E. esula* ex Austria	*E. esula-virgata* complex Picton, Ontario	
1973	44 moths	*E. velenovskyi* Greece	*E. esula-virgata* complex Jameson, Saskatchewan	
1974	54 moths	*E. velenovskyi* Greece	*E. esula-virgata* complex Jameson, Saskatchewan	

It is not known why the releases of *C. empiformis* failed on *E. cyparissias* in Ontario. The eggs were collected from a tetraploid *E. cyparissias* in the Swiss Jura and released on the sterile diploid form at Belleville and Sidney, Ontario. Possibly there are chemical differences between the diploid and tetraploid forms but it is also obvious that transfers of eggs and first instar larvae result in a large loss. It was later found that the third instar larva could, in the laboratory, be readily established in the tetraploid cypress spurge by placing it in a hole made in the pith of a cut stem. This would seem to justify another attempt to get establishment on the *E. cyparissias* stand at Braeside, Ontario.

Oberea erythrocephala (Schr.) (Coleoptera: Cerambycidae)

(a) Ecology

The biology and host range of *O. erythrocephala* were described by Schroeder (1980). The beetle is found in southern and central Europe, Asia Minor, and southern Siberia where it breeds on several perennial spurges in the sections *Esula, Helioscopa*, and *Paralias*, but not on species of economic importance or on other genera of plants. Main beetle emergence occurs in June and the adult lives for 3–8 weeks with the first two weeks spent feeding on tender parts of the spurge foliage. According to Schroeder (1980) there is no long range attraction of the male to the female, so the field encounter between the sexes seems to largely depend on the use of spurge as a rendez-vous. Mating occurs frequently through the oviposition period during which the female produces about 60 eggs which are typically laid singly in the upper part of flowering spurge stems. Just beneath the oviposition site the stem is girdled with 1–4 narrow grooves. The eggs hatch in 7–10 days and the larvae feed on the pith reaching the root in the fourth or fifth instar at the end of July or early August. Subsequent feeding is done in the crown and upper part of the root where pupation occurs the following spring

Most spurge plants resprout after attack by *O. erythrocephala* although small plants are killed. The main effect is to reduce the number and vigor of the stems produced in the following year: on *E. cyparissias* when 90% of the plants were attacked, the number of flowering stems was reduced by 85%. Population densities of the beetle are high, particularly in the eastern part of its range. Thus Schroeder (1980) noted that 5–50% of the *E. cyparissias* plants in the Swiss Valais were attacked and 50–100% of the *E. esula* in East Austria with 3–5 larvae per plant. Some field mortality was noted: at one site on *E. cyparissias*, 31% of the eggs failed to hatch although they developed embryos and 9% of the larvae died in the early instars. The population in part seems to be controlled by oviposition behaviour as multiple oviposition into a stem is rare and when it does occur only one larva survives to enter the root. In choice tests *O. erythrocephala* attacked Canadian leafy spurge more often than *E. esula*, but there was no indication that association with a particular spurge species in the field had formed a host race.

(b) Releases

Authority to release *O. erythrocephala* was obtained in October 1979. The stock was field collected as beetles in Switzerland on *E. seguierana* and *E. cyparissias*, and in Austria on *E. cyparissias* and *E. esula* (Table 43). The sexes were separated on collection and then released together at the release site in Canada later.

Table 43 Open releases and recoveries of *Oberea erythrocephala* (Schr.) on *E. esula-virgata* complex in Canada.

Release Date	No. and Stage	Source	Release Site	Recovery
15 Oct 79	30 larvae	Switzerland	Caronport, Saskatchewan	1 of 12 alive May 80
27 Jun 80	70 males 60 females	Switzerland	Caronport, Saskatchewan	15 larvae Sep 80 31 larvae Sep 81
3 Jul 80	70 males 53 females	Switzerland	Cardston, Alberta	17 larvae Sep 80 0 larvae Aug 81
14 Jul 80	49 males 57 females	Switzerland	Jameson, Saskatchewan	1 larva Sep 80 0 larvae Aug 81
26 Jun 81	55 males	Switzerland	Weyburn, Saskatchewan	14 larvae Aug 81 (8 on plants attacked in lab)
30 Jun 81	59 males 36 females	Switzerland	Lauder, Manitoba	12 larvae Aug 81
10 Jul 81	90 adults	Switzerland & Austria	Manitou Lake, Saskatchewan	13 larvae Aug 81
5 Aug 81	17 larvae	Austria	Caronport, Saskatchewan	

In 1979, 30 leafy spurge plants that had been attacked in the laboratory were transplanted in mid-October to a spurge stand in Saskatchewan (Table 43) to check on the ability of the larvae to survive the winter. The plants did not have time to establish themselves before winter and some were killed. In the following May one larva was found in 12 crowns examined and subsequently some oviposition girdling was found at the site but no eggs or larvae.

Field collected beetles were released on three spurge stands in 1980 (Table 43) and during the summer, stems with oviposition scars were tagged and then harvested in September to determine the number of larvae that had tunnelled into the roots before winter (Table 44).

The Cardston site was extremely fertile and the spurge so tall and dense that many stems with oviposition scars were certainly missed but it had by far the best ratio of oviposition scars to larvae. Unfortunately the site was burnt just before winter and in the following June the spurge was sprayed with picloram. Thus it was hardly surprising that no sign of the beetle was found in August 1981. In contrast at Caronport winter survival

Table 44 Oviposition attempts and success by *O. erythrocephala* (Schr.) at three sites

Site	No. stems with oviposition scars	No. larvae in roots by September	No. larvae in stems in September
Jameson, Saskatchewan	25	1	3
Caronport, Saskatchewan	93	15	2
Cardston, Alberta	29	17	2

was good in spite of a poor snow cover as live larvae were found in three out of four plants examined in May. About twice as many larvae were found at the end of the summer as in the previous year including two plants (untagged the year before) with larvae developing over a two year period. The colony had a radius of about 25 m in the fall of 1981. No progeny were found at the Jameson release site.

In 1981, beetles were released at three widely separated sites but as in the previous year breeding success was poor (Table 43). The table indicates that the date of release had little influence on breeding success so the problem does not appear to be related to the synchrony of the plant and larval development.

The larvae developed readily in the stems of spurge plants from the Jameson area grown in pots but these plants tended to have larger stems than most of those in the field. The colony at Caronport was still present in 1982 and for the first time one adult was found (a fertilized female). A search made in mid-September produced one larva which was still in the stem only a few centimeters from the oviposition point. As most of the spurge stems were dead by October it had little prospect of survival. The problem is apparently not cool summer temperatures as the Cardston site had fewer heat units than the other release sites. The stem diameter at Caronport was, however, considerably less than at Cardston (the soil is sand and too light to be cultivated) so it appears that the beetle is only likely to thrive on spurge plants growing on fertile soils.

Evaluation of Control Attempts

The biological control of spurge in Canada has so far been a failure. The one insect established on cypress spurge has had little impact and the possible impact of *O. erythrocephala* on leafy spurge will not be apparent until its population has reached its plateau. Other insects tested have failed to survive on Canadian leafy spurge although some like *C. tenthrediniformis* appear to have a major impact on *E. esula* in Europe. The initial difficulty arose from equating the *E. esula* of Europe with the plant called *E. esula* in North America. Some difficulties still remain as it is not possible to search Europe for the Canadian spp. of leafy spurge for possible agents since the precise origin and identity of this spurge is not clear. Indeed, some of it may have arisen in North America through hybridization. Thus agents have to be found by testing the acceptability of North American leafy spurge to organisms found attacking closely related European spurges (those in the section *Esula*). A number of these are already known and can be subjected to the usual screening tests to establish their specificity. A few species have already been screened and effort should be made to establish them as widely as possible.

The difference in the density of spurge in European and North American stands appears to be related to the cropping pressure by a complex of insects and pathogens in Europe, although it would be nice to demonstrate this by removing the insects from a European stand. One possible explanation for the richness of the complex in Europe is that like *H. euphorbiae* in Ontario each insect species is only able to utilize a small proportion of the total resource. The establishment of a complex of insects in North America large enough to reduce spurge density seems to be largely a matter of time and effort. Meantime any insect that can be established in reasonable numbers should be regarded as a success as it contributes to the cropping pressure on the weed in North America.

Recommendations

The priorities in the biological control of leafy and cypress spurge for the immediate future are recognized as follows:

(1) To complete host specificity studies on additional biological control agents in order to obtain approval for their release in North America. A reasonable target is the establishment of 10 agents in Canada since several species of spurge are probably involved.

The root feeding beetles in the genus *Aphthona* (Chrysomelidae) are of particular interest as they are widespread and common on *Euphorbia* spp. in Europe. Investigations by Maw (1981) showed that *A. cyparissiae* Koch, *A. flava* Guill., and *A. czwalinae* Weise would accept Canadian leafy spurge although the first two species are most common on *E. cyparissias* and the latter on *E. esula*. *A. cyparissiae* is likely to be the most useful for the Canadian prairies: it has the broadest ecological range of the three species, but prefers dry open sites with dense spurge and the larvae are reasonably cold tolerant. *A. czwalinae* will probably be restricted to moist sites such as along streams and in well wooded areas. *A. flava* prefers dry open sites but is less cold tolerant than *A. cyparissias* and so may be most applicable for southern Ontario or the United States.

The following insects have been bred on Canadian leafy spurge in the laboratory and warrant further investigation as biological control agents. The leaf-tying moth *Lobesia euphorbiana* (Ferr.) (Tortricidae), the defoliating moths *Minoa murinata* Scop. and *Biston fiducarius* Anker (Geometridae): both the former two moths in feeding tests were restricted to a few species of *Euphorbia* and they appear suitable for use as biological control agents. The gall midges *Bayeria capitigena* Bremi, *Dasineura capsulae* Kiett., and *D. loewi* Mik. (Cecidomyiidae); the aphids *Acyrthosiphum cyparissiae* Koch and *Aphis euphorbiae* Kltb. (Aphididae) are also worthy of testing.

(2) The root-feeding moth, *Chamaesphecia empiformis* Esp. (Sesiidae), can almost certainly be established on the tetraploid *E. cyparissias* in Ontario and Québec. Renewed efforts should be made to do this.

(3) *Chamaesphecia tenthrediniformis* (D. & S.) should be established on *E. esula* s. str. if this spurge is a problem anywhere in North America. It is the most effective agent on *E. esula* in Austria.

(4) The poor survival of *Oberea erythrocephala* (Cerambycidae) on the Canadian prairies may be related to the low fertility of most of the release sites. Releases should be tried again at Cardston, Alberta, and on cypress spurge in Ontario. Possibly the colony at Caronport can be made to flourish and hence be a source for distribution if the site is fertilized with nitrogen.

(5) The populations of the spurge hawkmoth, *Hyles euphorbiae* (L.) (Sphingidae), established on cypress spurge near Mulmur, Ontario, suggest that the species may have value in other places in Ontario and Québec. The strain imported from Europe appears to have a strong ovipositional preference for cypress spurge over leafy spurge. Stock from the colony established on leafy spurge in Montana should be tried again on leafy spurge in western Canada. As summer temperature may be limiting on much of the Canadian prairies, the releases should be made first in the regions receiving the greatest number of Corn Heat Units such as around Medicine Hat, Alberta, Estevan, Saskatchewan, and southcentral Manitoba.

(6) There is a need for a detailed taxonomic study of North American leafy spurge to determine the regions of Europe with the closest counterparts as these are the most promising source for biological control agents.

Literature Cited

Calvin, M. (1978) Chemistry, population, resources. *Pure and Applied Chemistry* 50, 407–425

Craig, H.A. (1953) Leafy spurge, *Euphorbia esula* L. 3–7. In: Committee reports, National Weed Committee (Western Section). Ottawa; Department of Agriculture, 80 pp.

Ellis, W.J.; Lennox, F.G. (1941) A proteolytic enzyme in the latex of the weed, *Euphorbia lathyris* (Caper spurge). *Australian Journal of Science* 4, 187–188.

Forwood, J.R. (1977) Spurge hawkmoth (*Celerio euphorbiae* L.) as a biocontrol agent for leafy spurge in Nebraska. M.Sc. thesis, University of Nebraska, 52 pp.

Geisler, M. (1955) Active principle of cypress spurge. *Pharmazie* 10, 677–678.

Geoponika of Cassianus Bassus, translated by T. Owen, London 1805.

Harris, P.; Alex, J. (1971) *Euphorbia esula* L., leafy spurge and *E. cyparissias* L.. cypress spurge (Euphorbiaceae) In: Biological control programmes against insects and weeds in Canada 1959–1968. *Commonwealth Institute of Biological Control Technical Communication* 4, 83–88.

Harris, P.; Dunn, P.H.; Schroeder, D.; Vonmoos, R. (in press) Biological control of leafy spurge in North America. In: Watson A.K. *(Ed.)* Monograph on leafy spurge, *Weed Science Society of America*.

Ibn Al Awan. (12 Century) Le livre de l'agriculture d'Ibn Al Awan (2 vols). Translated by J.J. Clement-Mullett, Paris 1864.

Malyuta, D.A. (1934) Sphingids as pests of spurge. *All-Union Institute of Scientific Research on Oil Cultivation* 7, 81–84 (in Russian).

Maw, E. (1981) The biology of some *Aphthona* spp. (Col: Chrysomelidae) feeding on *Euphorbia* spp. (Euphorbiaceae), with special reference to leafy spurge (*Euphorbia* sp. near *esula*). M.Sc. thesis, University of Alberta, Edmonton, 258 pp.

Naumann, C.M.; Schroeder, D. (1980) Ein weiteres Zwillingsarten-parr Mitteleuropaischer Sesiiden: *Chamaesphecia tenthrediniformis* (Denis & Schiffermuller, 1975) und *Chamaesphecia empiformis* (Esper, 1783) (Lepidoptera, Sesiidae). *Zeitschrift Arbeitsgemeinschaft Osterreicher Entomologie* 32, 29–46.

New, T.R. (1971) The consumption of *Euphorbia cyparissias* (Euphorbiaceae) by larvae of *Celerio euphorbiae* (Lepidoptera: Sphingidae). *Canadian Entomologist* 103, 59–66.

Portier, P.; Busnel, R.G. (1951) Imminite relative de la chenille de *Deilephila euphorbiae* L. au poison des euphorbes; sensibilité du papilon à ce même poison. *Comptes Rendus de l'Acadamie des Séances, Paris* 233, 897–898.

Saskatchewan Production and Marketing Branch (1972) Fifth annual report, 247–310.

Selleck, G.W.; Coupland, R.T.; Frankton, C. (1962) Leafy spurge in Saskatchewan. *Ecological Monographs* 32, 1–29.

Schroeder, D. (1969*a*) Studies on phytophagous insects of *Euphorbia* spp. *Chamaesphecia empiformis* (Esp.). Weed projects for Canada. *Commonwealth Institute of Biological Control Progress Report* 22, 21 pp.

Schroeder, D. (1969*b*) Studies on phytophagous insects of *Euphorbia* spp. *Chamaesphecia empiformis* (Esp.). Weed projects for Canada. *Commonwealth Institute of Biological Control Progress Report* 23, 23 pp.

Schroeder, D. (1970) Investigations on insects attacking *Euphorbia* spp. Weed projects for Canada. *Commonwealth Institute of Biological Control Progress Report* 26, 6 pp.

Schroeder, D. (1980) Investigations on *Oberea erythrocephala* (Schrank) (Col.: Cerambycidae), a possible biocontrol agent of leafy spurge, *Euphorbia* spp. (Euphorbiaceae) in Canada. *Zeitschrift für angewandte Entomologie* 90, 237–254.

Zwölfer, H. (1970) Investigations on insects attacking *Euphorbia* spp. (Additional screening tests with *Chamaesphecia;* notes on other *Euphorbia* insects.). Weeds projects for Canada. *Commonwealth Institute of Biological Control Progress Report* 36, Delémont, Switzerland.

Hypericum perforatum L., St. John's-wort (Hypericaceae)

P. HARRIS and M. MAW

Pest Status

In the early 1950s, St. John's-wort was forming large stands on the grasslands in the interior of British Columbia where it became the dominant herbaceous species. A number of biological control agents were introduced (McLeod 1962, Harris & Peschken 1971) and in most areas the weed has been reduced to scattered plants or patches up to 2 m in diameter and usually 100 to 200 m apart. The use of herbicides against the weed on public and private lands practically ceased. The main agent responsible for this reduction has been the beetle *Chrysolina quadrigemina* (Suffr.). In the few areas, such as the eutric brunisols on fluvial glacial gravel and morainal deposits near Cranbrook, British Columbia, where the beetle population has been patchy, the weed has continued to spread. In this area, dense stands of St. John's-wort occur on about 30 km^2 of lightly grazed Douglas fir, *Pseudotsuga menziesii*, forest. At present, the weed is causing little monetary loss, but if it continues to spread, extensive wildlife and cattle grazing areas in the east Kootenays are threatened.

In Ontario, St. John's-wort continues to be a common weed on poor quality grasslands, on thin dry soils, on gravel, limestone, or rock; but it was removed from the noxious weed list when the Weed Control Act was amended in 1976. If the availability of an effective biological control agent contributed to its removal from the list, biological control can be considered to have been a success. The weed is localized in New Brunswick and Nova Scotia, but it is capable of forming dense stands and appears to be spreading.

Background

The background to the biological control of St. John's-wort in Canada was described by McLeod (1962) (Harris & Peschken 1971). In these accounts, the beetle *C. quadrigemina* was most effective in the semiarid to dry subhumid regions (Harris *et al.* 1969) as they had an obligatory aestival diapause which they could not enter under continuously wet conditions (Huffaker 1967). The smaller beetle, *C. hyperici*, was most effective in the moist subhumid to humid regions. This assessment must now be revised with the spread of *C. quadrigemina* onto the moist coastal plain at Agassiz, British Columbia. In the years immediately following the release of the beetles, eggs and larvae were only found on the procumbent foliage from September to May. In 1981 in a moist region of British Columbia, eggs and first instar larvae were found on 23 June on the upright stems just prior to bloom while in the drier regions the beetle was in the adult pre-aestival feeding stage.

The colour forms of *C. quadrigemina* vary markedly between release sites in British Columbia. Peschken (1972) found that the prevalence of the bronze forms was correlated with the mildness of the winter. He also found several behavioural adaptations between beetles in British Columbia and those in California which were the source of the British Columbia stock.

Releases and Recoveries

***Agrilus hyperici* (Crotch) (Coleoptera: Buprestidae)**

Renewed attempts to establish the root boring beetle *A. hyperici* in the Cranbrook, British Columbia, area (Table 45) were not successful. This beetle is established at Mt. Shasta, California, and the climate of southern British Columbia is within the range of the species in Europe. The problem seems to be related to shipping stress in the newly emerged beetles, two shipments arrived dead at Cranbrook airport and no eggs or larvae were found from the release of live beetles. If further attempts are made to establish this insect, it is suggested that it should be done by transferring newly hatched larvae to plants growing at the release site.

***Anaitis plagiata* (L.) (Lepidoptera: Geometridae)**

Two releases of *A. plagiata* were made in the Cranbrook British Columbia area and one in New Brunswick. No establishment has been reported from New Brunswick. In British Columbia, two moths were caught in September 1980, about 3 km from the release site and on 23 June 1981, when the first flowers had opened, an egg, a first instar larva, and several moths were seen. The moth was also found in 1981 near Grandforks, an area where it had been released in 1967 and not previously recovered. The moths were flying actively between the widely scattered clumps of the weed and had survived in spite of strong populations of *C. quadrigemina*. Many larvae were found at Grandforks in late July and a few near Cranbrook in late July and early October (K. Williams 1981 personal communication). The colony at Agassiz, British Columbia, in 1981 caused browning and defoliation of *H. perforatum* in patches (L. Sigurgeirson 1981 personal communication).

The *A. plagiata* larvae released in 1977 at North Tay Creek, New Brunswick, may have been killed by insecticide drift from aerial spray of the adjacent forest for spruce budworm (D. Finnamore 1981 personal communication).

***Aphis chloris* (Koch) (Homoptera: Aphididae)**

(a)Ecology

The sexual generation of the aphid *A. chloris* appears in the fall (October in Cranbrook) in response to lower temperatures and shortened day length. During this period and in early November, an average of four (maximum of eight spread over a month) yellow eggs, which turn shiny black, are laid per female at the base of the flowering stems or on the procumbent foliage. Hatching occurs in April or May giving rise to a generation of apterous females. These mature in about a month and several generations of apterous viviparous females are produced. During midsummer, when colonies become crowded, alate viviparous females develop and disperse to found new colonies.

A. chloris has a wide geographic range. It is found in Sweden near the northern limit of St. John's-wort (Johansson 1962), in Poland (Czechowskii 1975), in Israel (Neser, 1973, personal communication), in Britian in cool and damp conditions, and in the arid south of France, from sea level to 1500 m in the mountains (Wilson 1943).

In the south of France and at Cranbrook, the aphids in summer are found near the tip of the shoots or beneath the flowers. In contrast, in England, northern France, and in the Alps, they are normally found at the base of the stems. Wilson (1943) suggested that the feeding site depends on conditions of heat and moisture. If so, the ability of the aphid to adjust in this manner bodes well for its survival in the varied conditions at Cranbrook, British Columbia.

In France, *A. chloris* can be heavily parasitized by *Aphidius cardui* Marsh. (Braconidae) and *Aphidencyrtus aphidiverus* Mayr. (Encyrtidae), but its reproductive capacity is such that it remains a common species (Wilson 1943). The stocks of *A. chloris* received from the Rhine Valley and Alsace contained *Lysiphlebus faborum (*Marsh.) (Braconidae). This is an aggressive parasitoid and if left unchecked can destroy a laboratory colony.

Although *A. chloris* will attack various species of *Hypericum*, the damage done to the plant or its acceptance by the aphid varies with the species. For example, at the end of 5 days the number of offspring from two mature apterous females on *H. perforatum*, *H. rhodopaeum*, *H. calycinum*, and *H. densiflorum* were 37, 10, 7, and 0, respectively. Large colonies of *A. chloris* consistently killed vigorous plants of *H. perforatum* in about a month, while *H. rhodopaeum* lived for over seven months and put on new growth in spite of a heavy aphid population.

(b) Releases

Two releases of *A. chloris* were made in the Cranbrook, British Columbia, area (Table 45). Infested *H. perforatum* plants were transplanted into the field. In 1979 this was done during a period of drought: the transplanted plants died and most of the surrounding *H. perforatum* had lost their foliage. No evidence of aphid survival was found. In 1980 infested plants were transplanted earlier in the spring and the aphids moved readily onto the surrounding plants. By early November all of the rosettes sampled in a 20 m radius at the two release sites had a few fundatrigeniae and/or eggs. By late June 1981 following a wet and cool spring, scattered plants within a radius of 50 m of the release point had small colonies of *A. chloris* near the flower buds. By far the strongest colony of the aphid was on the site least favoured by *C. quadrigemina*.

Chrysolina hyperici (Forester) (Coleoptera: Chrysomelidae)

British Columbia was used as a source for obtaining stock for release in Ontario, Nova Scotia, and New Brunswick (Table 45).

In Ontario, *C. hyperici* became established at the release site near Picton. Overwintering survival was 63% for adults and 31% for eggs compared to 19% and 6% respectively, for *C. quadrigemina* (Harris & Peschken 1974). Nevertheless *C. quadrigemina* rapidly became the most common species, although both species were still present in 1979 (13 ♂♂ and 4 ♀♀ *C. quadrigemina:* 4 ♂♂ and 3 ♀♀ *C. hyperici* in the sample collected).

The releases in Nova Scotia contained a few individuals of *C. quadrigemina* collected with the *C. hyperici* from Fruitvale, British Columbia. Both species became established but by 1979 *C. hyperici* had become numerous enough at one site in the Annapolis Valley to generate one query about the identity of the beetles sitting on the side of a house (H. Specht, 1979, personal communication). This was about 30 km from the release site. According to M. Neary (1979, personal communication), the original release site has been cleared of the weed and it has been substantially reduced in the Kingston-Auburn area. *C. hyperici* is also established in the Dundurn and Fenwick regions. At Dundurn between 1970–72 the number of flowering stems declined from 5.8 to 3.4/m^2. This is a more gradual decline than has been usual with *C. quadrigemina* in British Columbia.

C. hyperici released at Tay Creek, New Brunswick, were well established in 1977, but were not numerous enough to control the weed. A problem with the release made in 1975 is that all the individuals examined were female. Apparently the females continued their

Table 45 Open releases and recoveries of insects against *Hypericum perforatum* L.

Release site	Year	Origin	Number	Year of Recovery
Chrysolina hyperici				
Pictou Co., Nova Scotia	1969	British Columbia	2 100 adults	1970
Prince Edward Co., Ontario	1969	British Columbia	70 adults	1970
Annapolis Co., Nova Scotia	1969	British Columbia	1 320 adults	1979
Prince Edward Co., Ontario	1970	British Columbia	182 adults	1970
Tay Creek, New Brunswick	1971	British Columbia	1 150 adults	1972
Lime Kiln, New Brunswick	1971	British Columbia	1 150 adults	
Tay Creek, New Brunswick	1975	British Columbia	496 adults	
Chrysolina quadrigemina				
Prince Edward Co., Ontario	1969	British Columbia	182 adults	1970
Sidney Twp., Ontario	1971	British Columbia	15 adults	
	1971	California	139 adults	
Tay Creek, New Brunswick	1975	British Columbia	128 adults	Not found in 1981
Tay Creek, New Brunswick	1977	British Columbia	55 adults	
Guelph, Ontario	1979	Ontario	40 adults*	7 000 collected in 1981
4 sites near Guelph, Ontario	1979	Ontario	560 adults	No recovery
34 sites in and around Guelph, Ontario	1981	Ontario	7 000 adults	
Agrilus hyperici				
Elko, British Columbia	1977	California	56 adults	No recovery up to 1981
Aphis chloris				
Elko, British Columbia	1979	Germany	300 plus	Did not survive
Elko, British Columbia	1980	Germany	150 000	1981
Anaitis plagiata				
Grandforks, British Columbia	1967	Europe	875 larvae	1981
Westbridge, British Columbia	1967	Europe	500 larvae	
Tay Creek, New Brunswick	1977	Switzerland	715 adults	Not found in 1978
Elko, British Columbia	1977	Switzerland	3 648 larvae	
Agassiz, British Columbia	1977	Switzerland	476 larvae	1981
Elko, British Columbia	1980	France	169 larvae	1981**

* Sample examined contained 17 *C. quadrigemina*: 7 *C. hyperici*
** In view of the small population in 1981, it is assumed that establishment occurred from the 1980 releases.

pre-aestival feeding on the flowering plants for longer than the males, so that mass collecting produced a badly distorted sex ratio. It is not known whether other releases were adversely affected in this manner. The release site at North Tay Creek, New Brunswick, has been subject to insecticide drift from aerial treatments of the forest against spruce budworm. In 1978, 1979, and 1980, a buffer strip of two miles was left between the treated forest and the release field and beetle density increased noticeably in the first two of these years. In 1980, the area burned in the spring and few beetles were found (D. Finnamore, 1981, personal communication). In 1981, 28 beetles (11 ♂♂ and 17 ♀♀) collected in mid-August were all *C. hyperici,* so apparently *C. quadrigemina* has not survived at this site.

***Chrysolina quadrigemina* (Suffr.) (Coleoptera: Chrysomelidae)**

C. quadrigemina from British Columbia were used as stock for release in Ontario and New Brunswick.

In an ungrazed meadow near Picton, Ontario, the beetles increased rapidly to defoliate the St. John's-wort. The main effects, reported by Harris & Peschken (1974) and Dougan (1975), were the decline of the weed to a negligible amount; a large but short lived increase in other weed species and an approximately 300% increase of forage species. The total yield recovered by the third year to its original value and by the fifth year was nearly double as a result of a wet summer in 1974 (Table 46), but the amount of St. John's-wort did not increase with the moisture at the beetle site although it did on the control sites (Dougan 1975). The beetles have spread approximately 11 km from the release site (Alex 1981 personal communication).

In New Brunswick, *C. quadrigemina* from Agassiz, British Columbia, a moist area, were released at Tay Creek in 1977, but apparently did not survive.

Table 46 Effect of *Chrysolina* spp. on *H. perforatum* L. in an ungrazed meadow near Picton, Ontario.

	Dry weight (g/m^2) of foliage in the first week of July			
Year	*H. perforatum*	Forage spp.	Weed spp.	Total
1970	91.4	65.7	16.5	173.6
1971	12.0	60.6	30.3	102.9
1972	0.2	99.0	66.0	165.2
1973*	0.8	101.1	114.0	215.9
1974*	0.2	265.2	38.1	303.5

* Data from Dougan (1975)

Evaluation of Control Attempts

The biological control of *H. perforatum* has been a success. In a number of widely separated grassland areas in the interior of British Columbia where the weed had been accidentally introduced, it had become the dominant herbaceous species. It appeared likely to spread to all but the driest grassland habitats in the region. The introduction of the beetle *C. quadrigemina* has reduced the weed in most areas to less than 2% of its former density (Harris *et al.* 1969). At this density, it seems to be causing no problems to cattle and the chemical control programme conducted by the British Columbia Departments of Forests and of Agriculture for the containment of the weed has been discontinued. Even small remote stands of the weed that periodically appear are being colonized by the beetle.

Initially, *C. quadrigemina* was effective only in the drier *H. perforatum* sites and a related beetle, *C. hyperici*, was needed to control the weed in moist sites. *C. quadrigemina* has now become numerous at the moist sites of Fruitvale and Agassiz, British Columbia, although *C. hyperici* has persisted and spread to other sites in the province (K. Williams 1981 personal communication). However, in spite of the release of pioneer moist-adapted individuals of *C. quadrigemina* with *C. hyperici* in the maritimes, the latter species has become dominant and *C. quadrigemina* was not present in the samples examined. *C. hyperici* is only slightly smaller than *C. quadrigemina* and the two species feed in a similar manner on *H. perforatum*, so it is surprising that they coexist in the same habitats. It is not known what occurred to allow *C. quadrigemina* to colonize moist habitats and invade the thriving colony of *C. hyperici* at Fruitvale. The adaption has not extended to the maritimes and again the reasons are unknown. From a purely pragmatic purpose of controlling *H. perforatum* the important thing is that one species or another will reduce the weed to a low density in most places in Canada. For example, at the release site in Ontario, the weed declined to only 0.2% of its former density.

A strong colony of *C. quadrigemina* with some *C. hyperici* was established near Guelph, Ontario, with stock from the Picton, Ontario, colony. A total of 7000 beetles collected from the Guelph colony before its destruction in 1981 was distributed to 34 other sites in and around the city. By 1983 there should be massive populations of the beetle in the Guelph area. These could be collected at little cost and released at other *H. perforatum* sites throughout Ontario. Some of the releases would fail (two out of five release sites of 1979 at Guelph were destroyed) but the beetle would distribute itself from the survivors. The pilot studies at Picton, Ontario as well as those in British Columbia indicate that the prevalence of *H. perforatum* would be greatly decreased by such a programme with a corresponding increase in forage for livestock or wildlife.

The apparently few sites where *Chrysolina* spp. have been ineffective are on nutrient-poor soils and in insecticide-drift areas. No insects are likely to be effective for biological control if they receive an insecticide spray while they are exposed on their host plant. It might be possible to find a root-boring insect that was inside the plant during the spruce budworm spraying season; however, in view of the public outcry, the spray programme is likely to become increasingly circumspect, so the problem for *C. hyperici* should diminish.

More investigations are needed to determine whether the effectiveness of the *Chrysolina* spp. can be increased on nutrient-poor soils; however, as the moth *A. plagiata* is increasing on some of the areas not favored by the beetle, we already may have part of the solution. *Aphis chloris* was chosen as a biological control agent as the presence of symbionts in aphids often enables them to survive on plants that lack one or more essential amino acids or B vitamins. However, its effectiveness as a biological control agent for *H. perforatum* remains to be seen.

Recommendations

(1) *C. quadrigemina* and/or *C. hyperici* have shown that they are able to reduce *H. perforatum* to well below the economic threshold in most regions of Canada. Any province with *H. perforatum* on their noxious weed list should distribute the beetles or, if the importance of the weed does not justify this small effort, remove the weed from the list.

(2) Provinces where the weed is being satisfactorily controlled should remove *H. perforatum* from their noxious weed list as mowing or spraying operations against the weed are a waste and not likely to assist its biological control.

(3) Further study is required to determine the role of *Anaitis plagiata* and/or *Aphis chloris*.

(4) If further attempts are made to establish *Agrilus hyperici*, these should be done by transferring the young larvae to plants in the field.

(5) The potential importance of *H. perforatum* in the Cranbrook area should be determined. It may be possible to relate the weed to a soil type and use this to predict its future spread.

(6) The possibility of solving the *H. perforatum* problem by supplying nutrients that increase the density of biological control agents should be explored.

Acknowledgements

Most of the releases of *Hypericum* insects in British Columbia were made by the British Columbia Forestry Service and they provided stocks for release in other parts of the country. Nello Angerilli released *Anaitis plagiata* and *Aphis chloris* at Elko, British Columbia, and was the first person to recover the moth in the region. Releases in New Brunswick were made by Dave Finnamore. Linda Sigurgeirson and Kathy Williams read the text and contributed observations about *Hypericum* insects in British Columbia and Jack Alex did this for Ontario. We would like to thank all these people and others who have contributed at various times.

Literature Cited

Czechowskii, W. (1975) Bionomics of *Formica* (*Coptoformica*) *pressilabris* Nyl. (Hymenoptera, Formicidae) *Annals of Zoology* 33, 103–125.

Dougan, J.M. (1975) The ecological and biological control of *Hypericum perforatum* L. (Hypericaceae) in Ontario. M.Sc. thesis, University of Guelph, 123 pp.

Harris, P.; Peschken, D.; Milroy, J. (1969) The status of biological control of the weed *Hypericum perforatum* in British Columbia. *Canadian Entomologist* 101: 1–15.

Harris, P.; Peschken, D.P. (1971) *Hypericum perforatum* L., St. John's-wort (Hypericaceae). In: Biological control programmes against insects and weeds in Canada 1959–1968. *Commonwealth Institute of Biological Control Technical Communication* 4, 89–94.

Harris, P.; Peschken, D.P. (1974) Biological control of St. John's-wort. *Canadian Agriculture* 19, 13–16.

Huffaker, C.B. (1967) A comparison of the status of biological control of St. John's-wort in California and Australia. *Mushi* 39: 51–73.

Johansson, S. (1962) Insects associated with *Hypericum* L. 2. Lepidoptera, Diptera, Hymenoptera and general remarks. *Opuscula Entomologica* 27: 175–192.

McLeod, J.H. (1962) Biological control of weeds. *In*: A review of the biological control attempts against insects and weeds in Canada. *Commonwealth Institute of Biological Control Technical Communication* 2, 20–21.

Peschken, D.P. (1972) *Chrysolina quadrigemina* (Coleoptera: Chrysomelidae) introduced from California to British Columbia against the weed *Hypericum perforatum*: comparison of behaviour, physiology and colour in association with post-colonization adaptation. *Canadian Entomologist* 104: 1689–1698.

Wilson, F. (1943) The entomological control of St. John's-wort (*Hypericum perforatum* L.) with particular reference to the insect enemies of the weed in southern France. *Council of Scientific and Industrial Research, Melbourne, Bulletin* 169, 87 pp.

Linaria vulgaris Miller, Yellow Toadflax and *L. dalmatica* (L.) Mill., Broad-leaved Toadflax (Scrophulariaceae)

P. HARRIS

Pest Status

The importance of *Linaria vulgaris* Mill. in western Canada has declined since the early 1950s so that now, as in eastern Canada, it is a conspicuous but relatively unimportant weed. In a study in the Peace River area of Alberta, Darwent *et al.* (1975) found that in

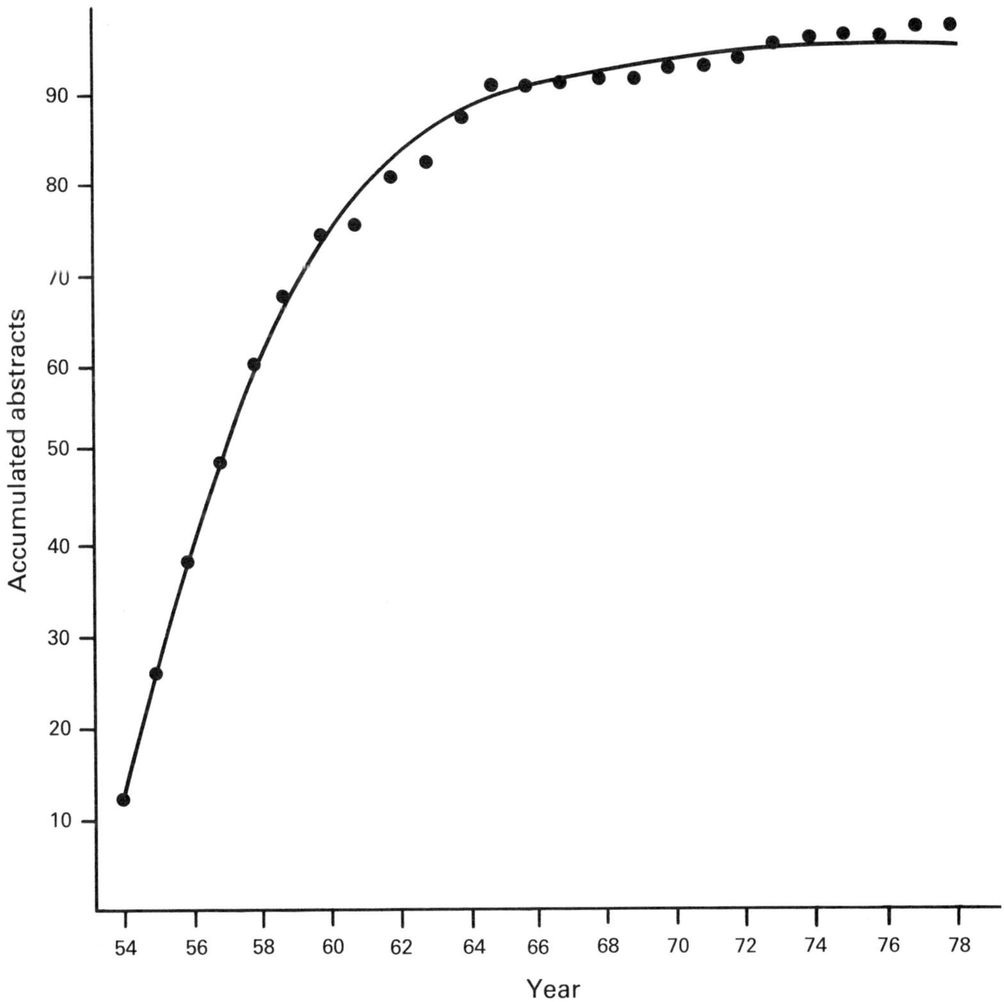

Fig. 8. Cumulative number of abstracts on the control of toadflax with herbides in the Research Report of the Expert Committee on Weeds (Canada Western Section)

most areas it was less abundant in the early 1970s than it had been in 1956 and that toadflax was not a serious problem in cereal crops. In creeping red fescue, 180 toadflax stems/m^2 reduced the yield by a third; but in most areas there were 20 or less stems/m^2 with no measurable effect on yield. Most of the dense stands of the weed were on sites that had been cleared and abandoned or were cropped infrequently.

In Saskatchewan, the weed is still dense on uncultivated sandy soils from North Battleford west to Alberta. In other regions of the province it has disappeared or greatly declined since the 1950s.

The decrease in the importance of the weed is reflected by the decline in the number of abstracts relating to its chemical control in the Research Report of the Expert Committee on Weeds (Canada Western Section) (Fig. 8). The decrease of the weed has not however changed its status under noxious weed legislation. For example in Saskatchewan provincial subsidies are still paid to municipalities for its control.

L. dalmatica (L.) is considerably more persistent that *L. vulgaris*, and it appears to exclude most other herbaceous species from the stand. Stands are local and the total area occupied does not appear to be large. Thus in the absence of documented evidence it does not appear to offer a serious threat.

Background

A more detailed history of the biological control of toadflax is given by Harris & Carder (1971). The decline of the weed in western Canada coincided with the appearance of the flower-feeding beetle *Brachypterolus pulicarius* (L.). This European beetle was present in eastern Canada and spread to the west. In addition the seed-feeding weevil *Gymnaetron antirrhinii* (Payk.), another European insect present in the east, was released in western Canada in 1957. Both insects are now present in regions of western Canada where toadflax is common.

In 1962, a European defoliating moth, *Calophasia lunula* (Hfn.), was first released in Canada, and it became established on the railway right-of-way at Belleville, Ontario in 1966. The moth has two overlapping generations and lays single well distributed eggs. The first two larval instars are inconspicuous and frequently feed inside the flowers, but the third to fifth instars feed externally on the foliage and are easily seen. Samples taken along the railway right-of-way (Table 47) indicate that larvae were present from late May to early September, usually at a density of 2–6 larvae/100 stems regardless of the toadflax density. The density of the toadflax increased greatly following soil disturbance in the previous year. The overall level of damage was not impressive with less than 20% of the stems being defoliated.

Table 47 Density of *C. lunula* and toadflax, *L. vulgaris* Mill. and *L. dalmatica* (L.) Mill., along the railway at Belleville, Ontario.

Date	No. samples*	No. Toadflax stems/m^2	No. 3rd–5th instar *C. lunula* larvae/m^2
2 Jun 1968	26	85.7	2.0
8 Aug 1968	19	89.1	3.8
3 Sep 1969	158	6.6	0.05
18 Jun 1970	52	7.2	0.31
13 Aug 1970	205		0.98
15 Aug 1971	31	7.7	0.13

* (0.25 m^2) samples taken at 5 pace intervals

The area has not been surveyed since 1972, but the moth has continued to spread; it was recovered from light traps at Ottawa, Ontario in 1977, a distance of over 300 km from the release site (E. Monroe, 1978, personal communication). Thus *C. lunula* is now probably present in most of southern Ontario. According to I. Jones (1978, personal communication), larvae were scarce in the Ottawa area in 1980: less than half a dozen were seen in several organized and many casual searches. In 1981, there was a large increase in the population, larvae were seen from early June to mid-October and along the railway track 47 to 57 larvae could be collected in an hour. Defoliated stems were evident, and some small stands were completely defoliated. Both the larvae and defoliation were associated with stunted toadflax plants and they rarely occurred in adjacent lush stands.

Table 48 Open releases and recoveries of *Calophasia lunula* against *L. vulgaris* Mill. and *L. dalmatica* (L.) Mill.

Weed Species	Province	Year	Origin	Number	Year of Recovery
L. vulgaris	New Brunswick	1971	Ontario	1565 larvae & 136 pupae	
L. vulgaris	Saskatchewan	1973	Ontario	2725 larvae	
L. dalmatica	British Columbia	1969	Ontario	200 pupae	
L. dalmatica	British Columbia	1971	Ontario	1180 larvae	

Releases and Recoveries

Four attempts were made to establish *Calophasia lunula* in regions of Canada outside Ontario and all were unsuccessful (Table 48). The release site in Saskatchewan was flooded in the year following the release. The larvae released on Dalmation toadflax in British Columbia disappeared within days, apparently from ant predation. Survival was not followed closely in New Brunswick but no evidence of establishment was found in the following year.

Evaluation of Control Attempts

The moth, *C. lunula*, is obviously well established in Ontario and it should be possible to establish it in the rest of Canada. *C. lunula* has not defoliated enough *L. vulgaris* to control the weed on its own, but it has added to the stress inflicted by the now ubiquitous beetles, *G. antirrhinii* and *B. pulicarius*. The possibility that it may sometimes inflict more damage is indicated by the populations in Ottawa in 1981 which seem to be about ten times the density at Belleville prior to 1972.

C. lunula favors the stunted toadflax stands that are found on dry soils. It is these stands that are persistent as edaphic conditions that permit the formation of tall lush stands also lead to the displacement of toadflax by grass and other vegetation. Thus *C. lunula* concentrates its attack where the weed is most persistent.

The combination of the three toadflax insects can control *L. vulgaris* at below the economic threshold in most parts of Canada. Indeed in most regions this has been achieved by the two beetles alone. An exception is North Battleford, Saskatchewan, where there are extensive dense stands of toadflax. Based on the results from Ontario, it is doubtful whether the establishment of *C. lunula* in the area would solve the problem. Whether it is worth screening and establishing a fourth agent for the control of toadflax around North Battleford needs to be determined by a cost-benefit study as opinions vary about the importance of toadflax to the region.

Recommendations

Surveys for the biological control agents of toadflax and their host specificity testing should not be done by Canada unless it can be shown that the potential benefits are large enough to justify the costs of biological control. However, the costs to Canada for biological control would be minimal if the tests were done by other countries. In this event the control agent should be imported as there would be benefits to individual land holders from the biological control of Dalmation and common toadflax. The moth *C. lunula* can almost certainly be established outside Ontario.

Acknowledgements

I am grateful to Mr Ian Jones for the observations on *C. lunula* in the Ottawa area in 1980 and 1981.

Literature Cited

Darwent, A.L.; Lobay, W.; Yarish, W.; Harris, P. (1975) Distribution and importance in northwestern Alberta of toadflax and its insect enemies. *Canadian Journal of Plant Science* 55, 157–162.

Harris, P.; Carder, A.C. (1971) *Linaria vulgaris* Mill., yellow toadflax and *L. dalmatica* (L.) Mill., broad-leaved toadflax (Scrophulariaceae). In: Biological control programmes against insects and weeds in Canada 1959–1968. *Commonwealth Institute of Biological Control Technical Communication* 4, 94–97.

Opuntia polyacantha Haworth, Plains Prickly-pear Cactus (Cactaceae)

M.G. MAW

There are three species of cacti on the Canadian Prairies: the prickly-pears, *Opuntia polyacantha* Haw. and *O. fragilis* (Nutt.) Haw., and the pincushion or ball cactus, *Mammillaria vivipara* (Nutt.) Haw. *Mammillaria* is not considered to be a problem.

In the western provinces cacti are found on dry hillsides and flatlands throughout the more arid parts of the brown soils zone. These areas have some grazing capacity, but have limitations that make improvement and cultivation impractical. For example, when the surface is disturbed the soils are likely to erode and blow easily.

Periodically, especially during droughts, prickly-pear becomes a concern. Ranchers then worry that it will displace valuable grass, use scarce moisture, and cause injury to stock. It is generally thought that overgrazing increases cactus growth and spread. Certainly, severely overgrazed grasslands greatly facilitate the spread of cacti for when soils have been denuded, cactus pads broken from clumps come into contact with bare soil, root, and establish new clumps. On the other hand, Bement's (1968) study of heavy, moderate, and light pasture use showed that while populations of cactus doubled in area in 28 years, it never exceeded 2.4% of land area. Removal of cactus did not increase forage production significantly but did make it more available for grazing. In lightly grazed pastures, grasses tended to mask cactus while under heavy grazing, cactus was more conspicuous. Farmers in the Avonlea area of Saskatchewan maintain that cactus has not changed much in the past 20 years and where it has increased, it was through careless efforts to break up the clumps with harrows (personal interviews).

The value of cactus in the North American ecosystem is not always appreciated. During drought and under heavy grazing, cactus reduces soil erosion. Clumps of cactus conserve moisture by holding snow, and after summer showers the soil within the clumps remains moist for several hours longer than soil outside them. Clumps also protect desirable plants from grazing so that seeds can be produced to repopulate the range after drought or heavy grazing (Houston 1963).

Prickly-pear is a food for at least 44 species of birds and mammals. For example, seeds may account for 65% of the diet of the Harris ground squirrel and up to 5.0% of the browse of deer and antelope (Martin *et al.* 1961).

The earliest record of an insect attacking cactus dates back several hundred years, and there is indication that cultivation of the cochineal, *Dactylopius coccus* Costa, was conducted in Mexico and Central America for many centuries before the voyage of Columbus (Mann 1969).

The first comprehensive study of insects of Cactaceae in the United States was done by Hunter *et al.* (1912). This work described the life history and habits of various species, summarized previous information, and provided an extensive and useful bibliography.

Surveys for insects associated with *O. polyacantha* and *O. fragilis* in Saskatchewan and Alberta were made from 1974 to 1979 (Maw & Molloy 1980). Although only 20 species were found, three, *Chelinidea vittiger* Uhler, *Dactylopius confusus* (Ckll.), and *Melitara dentata* (Grote), were numerous enough to exert considerable pressure on the cactus. *D. confusus* and *M. dentata* can be very destructive and each year destroy about 10% of the pads (Maw & Molloy 1980)

Two species of Lepidoptera, *Cactoblastis doddi* Heinrich and *C. bucyrus* Dyar, were obtained from the high altitude valleys of Jujuy province in Northern Argentina for screening at Regina. The two species fed on local prickly-pear, but it was necessary to injure the pads before the larvae would enter to feed. Once inside, the larvae completely hollowed the pads, but then would not leave them to attack new ones. Either the local cactus was unsuitable or the insects were diseased for both species became weak and the colonies died out after a few generations in the laboratory.

Discussion

No introductions of cactus insects are considered, for any harm or inconvenience caused by our prairie cactus is balanced by benefits from it. The native biotic agents exert considerable pressure on the cactus and our native grasses, when not overgrazed are well able to compete.

Any release of biological control agents to control prickly-pear would be vigorously opposed by both the United States and Mexico, because *Opuntia* spp. are used as emergency fodder for cattle and as human food. In addition, at least one species of *Opuntia* is listed as endangered and 12 as threatened and still others are valued as ornamentals throughout the American southwest.

Recommendations

(1) Introduction of biological control agents for *Opuntia* spp. in Canada should not be pursued.

(2) Any control of cactus in Canada should be approached as a range management problem.

Literature Cited

Bement, R.E. (1968) Plains prickly-pear; relation to grazing intensity and blue grama yield on Central Great Plains. *Journal of Range Management* 19, 83–86.

Houston, W.R. (1963) Plains prickly-pear, weather, and grazing in the Northern Great Plains. *Ecology* 44, 569–574.

Hunter, W.D.; Pratt, F.C.; Mitchell, J.D. (1912) The principle cactus insects of the United States. *USDA Bureau of Entomology* 113, 1–71.

Mann, J. (1969) Cactus-feeding insects and mites. *Smithsonian Institute Bulletin* 256, 158 pp.

Martin, A.C.; Zim, H.S.; Nelson, A.L. (1961) American wildlife and plants. A Guide to wildlife food habits. New York, Dover Publications Inc., 500 pp.

Maw, M.G.; Molloy, M.M. (1980) Prickly-pear cactus on the Canadian prairies. *Blue Jay* 38, 208–211.

Rhamnus cathartica L., Common or European Buckthorn (Rhamnaceae)

M.G. MAW

Pest Status

Common or European buckthorn, *Rhamnus cathartica* L., is the alternate host of crown rust or leaf rust of oats, *Puccinia coronata* Cda. f. sp. *avenae* Eriks.

Considerable amounts of urediospore inoculum of crown rust is blown northwards every year from the United States and has nothing to do with the presence of buckthorn in Canada. Moreover, in some years, crown rust is so widespread and severe, such as in 1980, that the presence or absence of buckthorn is probably not important (R.V. Clark, 1980, personal communication) (Hanson & Grau 1979, Peturson 1954). Nevertheless, buckthorn is responsible for virtually all of the early infection by crown rust in the northern United States and in Canada (Hanson & Grau 1979), infecting oats from two weeks (Harder & McKenzie 1974) to a month (Harder 1975) before the arrival of the urediospores from the south.

Buckthorn leaves are the site of spermagonial and the aecial stages of crown rust of oats. This is a potential source of new pathogenic variation because hybridization and recombination occur in the pathogen on buckthorn (Simons *et al.* 1979). The greater diversity of rare races in the east, as opposed to the west, may be attributed to the presence of buckthorn (Fleischmann *et al.* 1963). Although there are a number of species of *Rhamnus*, both native and introduced, in North America, of those susceptible to *P. coronata* f. sp. *avenae*, only *R. cathartica* occurs in any significant numbers in Canada (Peturson 1954).

It is difficult to assess overall losses due to crown rust of oats. Unless there is a general epidemic of the rust, the infection is generally extremely variable. In Ontario, in 1977, infections ranged from a trace in some fields to nearly complete destruction of the oat crop in others (Harder 1978). Yield reductions are negligible if crown rust development does not precede heading, especially in relatively dry seasons. A late attack can, however, cause losses of 763–1029 kg per ha if the season is delayed by cool and wet weather (Fleischmann 1968). When there are few late fields of oats, severe infections of 30–90% may cause little damage (Fleischmann 1963).

In Manitoba, estimates of combined crown and stem rust damage in 1970 indicated a loss in excess of 154 000 tonnes but less than half of it was attributed to crown rust (Martens 1971). In 1969, in Manitoba, losses to crown rust were estimated at 108 000 tonnes and losses in individual fields ranged from 5–30% (Fleischmann 1969).

It is only when there is a minimal urediospore shower that the losses due to crown rust associated with buckthorn can be determined. Buckthorn is responsible for local and persistent infestations of the rust wherever buckthorn and oats are in close proximity and a single large buckthorn is sufficient to generate a moderately severe infection in an adjacent field (Harder 1975, 1978). The distribution of crown rust in Ontario in 1977 indicates that buckthorn was responsible in that year for most of the rust infection of oats. The very light infection occurring in the absence of buckthorn suggests there was little influx of spores from elsewhere (Harder 1978).

Crown rust adversely affects yield, quality, weight, and protein content of oat seed. Light infections cause minor reductions, but infections of 68–80% cause yield reductions of 19.2–27.8%, kernal weight of 12.3–16.1%, and a grade reduction of about one grade (Peturson 1952).

Background

R. cathartica was introduced into North America as a hedge and shelterbelt shrub or as an ornamental in parks and gardens. It varies considerably in size, ranging upwards of 10 m high under favourable conditions but is more usually less than half that height. It is very bushy, often many stemmed, and the shoots and branches terminate in a sharp spine. The leaves are alternate, smooth, and elliptical with toothed margins. The flowers are in the axils of the leaves, small, greenish, and dioecious. The mature fruit is a bluish to nearly black drupe, bitter to the taste, and about 8 mm in diameter. The seed is 3–4 grooved.

Reproduction is entirely by seeds which germinate 90–100% when the fruit is just ripe. The plant regenerates quickly from cut stems and after burning and grazing (Montgomery 1956, Mulligan 1952, Rydberg 1932).

(a) Habitat

In eastern Canada, buckthorn is found along fence rows, along roadsides, in open woods, and along the gently sloping banks of lakes, rivers, and streams, it is often seen in solid masses along every fence row in many locations, but apparently is not found to any extent in pastures. Plants of about 2.5 m can be found in wood openings with many smaller plants and seedlings in the more shaded locations. Where the trees are removed, these smaller plants grow rapidly forming a forest of buckthorn. It tends to be more commonly on soils of high moisture content than on light, dry soils (Mulligan 1952).

In Manitoba, there are relatively few areas outside of the larger towns and cities where buckthorn is found (Peturson 1954). It is concentrated in Winnipeg, Brandon, and near Macdonald, generally as park and hedgerow plantings (Bassett 1958), and in a sheltered ravine near Morden (Harder 1975).

In England, it is found chiefly on alkaline peat and limestone soils (Godwin 1943).

(b) Geographic range

In England, it is found in the midland, southern and eastern counties, but is absent from Cornwall, North Devon, Northumberland, and western counties of Wales.

In Ireland, it is present in a wide belt across the central Irish plain, but absent from the southern counties and absent or rare in the northern.

It is doubtfully native in Scotland, except possibly in Dumfries.

Its range extends through the greater part of Europe to 60° 48′ in Norway and 61° 41′ in Sweden, in southern Finland, Esthonia, and across Russia into western Asia, Afghanistan, and Turkestan.

It is present in southern Europe to middle and eastern Spain and Macedonia, and sparingly at high altitudes in Morocco and Algeria, but absent in the Balearics, Corsica, and Sardinia, also from western Siberia, Transcaucasia to Altai (Godwin 1943).

In North America, *R. cathartica* is found in the New England states, Virginia, westward to Illinois, Missouri, Wisconsin, Minnesota (Fernald 1950), Québec, Ontario Manitoba, and to a limited extent in Saskatchewan, Alberta, and the Maritimes.

(c) Uses

R. cathartica is a useful hedge and shelterbelt shrub and as such has been extensively planted in Ontario and Québec and to a lesser extent in other provinces. It is also used as an ornamental in gardens and parks.

The fruit is eaten by birds but usually only as a last resort. For example, robins will take them sparingly when no other food is available in the early spring. It is claimed that emodin in the green fruits prevents premature predation by birds (Trial & Dimond 1979). Perhaps late retention of fruit well into the spring attests to this.

Several species of birds are cited as feeding on the fruit in Germany and England, and mice have been seen to eat the seed from dried fallen fruit.

(d) Control

Buckthorn is so prevalent in Ontario, especially in the eastern counties, and the bush is of such a large size that it is an almost impossible task to eradicate it by spraying, cutting, or bulldozing. So very little, if anything, is being done to control it.

Herbicides give unsatisfactory results and 2,4,5-T which was recommended as a treatment of freshly cut stumps is now outlawed in many states and provinces (Hanson & Grau 1979). In any event, combinations of 2,4,-D and 2,4,5-T had little permanent effect and treated bushes regrew from the base and many of the test buckthorn leafed out one year after treatment (Switzer 1961). Ammate-X (ammonium sulfamate, duPont) is recommended as either a foliar spray or a stump treatment in Wisconsin (Hanson & Grau 1979) but early tests in Ontario showed that Ammate-X and fenuron (3-phenyl-1, 1-dimethylurea) had little effect in that most of the test plants over four feet high continued to grow (Switzer *et al.* 1960, Switzer 1961).

Cutting buckthorn gives at best temporary control because the stumps send up vigorous shoots in a relatively short time. However, cutting of individual bushes can reduce the incidence of crown rust in nearby fields.

(e) Biological control

Because chemical control of buckthorn is at best temporary, biological control is an attractive alternative and so at the request of Canada a survey for biological control agents was made by the Commonwealth Institute of Biological Control in 1964 (Malicky *et al.* 1970). It was found that practically all the phytophagous insects specific to or closely associated with Rhamnaceae are in the orders Lepidoptera and Hemiptera. Many Coleoptera were found on Rhamnaceae but none was specific (Malicky *et al.* 1970),

Investigations, particularly starvation tests suggest that the geometrids *Scotosia vetulata* Schiff., *Triphosa dubitata* L., and the lycaenid *Thecla spini* L. may be possible species for the biological control of *R. cathartica* in Canada. The geometrids combine a high degree of host specificity with a potential to defoliate their host, however, *T. dubitata* overwinters as an adult which may prove difficult in Canadian situations. *T. spini* may have too wide a host range to be an acceptable agent, but there is some doubt as to the correctness of the host records (Malicky *et al.* 1970).

Discussion

R. cathartica is only one of several species of Rhamnaceae capable of being an alternate host to crown rust of oats (Dietz 1926). It is, however, the only one occurring in any significant numbers in Canada (Peturson 1954). Eradication of buckthorn would not erase the problems of crown rust but would remove the site of the aecial or sexual spores of the rust and thus the source of pathogenic variation through hybridization and recombination of genetic material (Simons *et al.* 1979). Hybridization and recombination, however, may not be as important as once thought. New races appear first in the main north-south rust population, and only later become established in populations cycling on buckthorn. (Simmons M.D. personal communication 1983).

Infections of rust due to buckthorn are usually very local but can result in complete loss of the crop near buckthorn. Infection levels usually decline linearly with distance from the buckthorn (Harder 1975) and a few hundred meters from the source is sufficient to make a significant difference in infection incidence. Growers must either avoid planting oats near buckthorn or eradicate it from their farms. Buckthorn is so widespread and persistent that complete eradication is not feasible.

Biological control of buckthorn appears to be the only course to lessen the impact of crown rust on oat production. It is not expected that introduced biological agents will eradicate buckthorn, but they might add to the stress of climate and soil type to slow the spread of the shrub. Also, it is expected that any reduction in leaf surface will aid in diminishing rust inoculum early in the season when the greatest injury is caused to the oat crop. Not many insects find buckthorn an attractive food source because of the feeding deterrent chemical, emodin (Trial & Dimond 1979). The original sources of the North American populations of buckthorn are not known and there could be different biotypes containing varying amounts of emodin. This may make selection of effective biological control agents difficult.

Surveys for phytophagous insects in Europe found no species specific to *R. cathartica*, although the Lepidoptera *Scotosia vetulata* and *Thecla spini* exhibit a degree of specificity to Rhamnaceae.

Thus when attempting to control buckthorn biologically, one has to consider that species other than *R. cathartica* will be attacked. The degree to which these other species will be damaged must be determined in the screening tests, and decisions made as to whether damage and possible loss of other species of Rhamnaceae can be tolerated.

Recommendations

Even though *R. cathartica* is the source of early infection of oats by crown rust and is the site of the aecial stage of the rust, its complete eradication would at best result in only a reduced incidence of the disease and the appearance of new rust races. Therefore, it is recommended that biological control work on *R. cathartica* be discontinued.

Literature Cited

Bassett, I.J. (1958) A survey of the European buckthorn and common barberry in southern Manitoba. Science Service, Division of Botany and Plant Pathology, Canadian Department of Agriculture Ottawa, 22 pp.

Dietz, S.M. (1926) The alternate hosts of crown rust, *Puccinia coronata* Corda. *Journal of Agricultural Research* 33, 953–970.

Fernald, M.L. (1950) Gray's manual of botany. American Book Corp., New York, 1632 pp.

Fleischmann, G. (1963) Crown rust of oats in Canada in 1963. *Canadian Plant Disease Survey* 43, 168–172.

Fleischmann, G. (1968) Crown rust of oats in Canada in 1968. *Canadian Plant Disease Survey* 48, 99-101.

Fleischmann, G. (1969) Crown rust of oats in Canada in 1969. *Canadian Plant Disease Survey* 49, 91–94.

Fleischmann, G.; Samborski, D.J.; Peturson, B. (1963) The distribution and frequency of occurrence of physiologic races of *Puccinia coronata* Corda. f. sp. *avenae* Erikas. in Canada. *Canadian Journal of Botany* 41, 481–487.

Godwin, H. (1943) Rhamnaceae. *Journal of Ecology* 31, 66-92.

Hanson, E.W.; Grau, G.R. (1979) The buckthorn menace to oat production. *University of Wisconsin Extension Bulletin* A 2860, 2 pp.

Harder, D.E. (1975) Crown rust of oats in Canada in 1974. *Canadian Plant Disease Survey* 55, 63–65.

Harder, D.E. (1978) Crown rust of oats in Canada in 1977. *Canadian Plant Disease Survey* 58, 39–43.

Harder, D.E.; McKenzie, R.I.H. (1974) Crown rust of oats in Canada in 1973. *Canadian Plant Disease Survey* 54, 16–20.

Malicky, H.; Sobhian, R.; Zwölfer, H. (1970) Investigations on the possibilities of a biological control of *Rhamnus cathartica* L. in Canada. Host ranges, feeding sites, and phenology of insects associated with European Rhamnaceae. *Zeitschrift für angewandte Entomologie* 65, 77–97.

Martens, J.W. (1971) Stem rust of oats in Canada in 1970. *Canadian Plant Disease Survey* 51, 11–13.

Montgomery, F.H. (1956) Common weeds of Ontario. *Ontario Department of Agriculture Bulletin* 505, Toronto, 104 pp.

Mulligan, G.A. (1952) Survey of common barberry (*Berberis vulgaris* L.) and the European buckthorn (*Rhamnus cathartica* L.) in eastern Ontario. Science Service, Division of Botany and Plant Pathology, Canadian Department of Agriculture, Ottawa, 16 pp.

Peturson, B. (1952) Studies on specialization and pathogenicity in *Puccinia coronata* Corda. Dissertation (Publication 4872), University of Minnesota, 78 pp.

Peturson, B. (1954) The relative prevalence of specialized forms of *Puccinia coronata* that occur on *Rhamnus cathartica* in Canada. *Canadian Journal Botany* 32, 40–47.

Rydberg, A. (1932) Flora of the prairies and plains of Central North America. Lancaster, Pennsylvania; Science Press Printing Co., 969 pp.

Simons, M.D.; Rothman, P.G.; Michel, L.J. (1979) Pathogenicity of *Puccinia coronata* from buckthorn and from oats adjacent to and distant from buckthorn. *Phytopathology* 69, 156–158.

Switzer, C.M. (1961) The effects of 2,4,-D, 2,4,5-T, ammate-X, and fenuron on bush one year after treatment. *Research Report National Weed Committee, Eastern Section, Canadian Department of Agriculture*, Ottawa, pp. 99–100.

Switzer, C.M.; Appleton, A.R.; Malcomson, H. (1960) Brush control with 2,4,-D, 2,4,5-T, ammate-X, and fenuron. *Research Report, National Weed Committee, Eastern Section, Canadian Department of Acriculture*, Ottawa, pp. 88–89.

Trial, H.; Dimond, J.B. (1979) Emodin in buckthorn: a feeding deterrent to phytophagous insects. *Canadian Entomologist* 111, 207–212.

Salsola pestifer A. Nels., Russian Thistle (Chenopodiaceae)

P. HARRIS

Pest Status

Russian thistle, *Salsola pestifer* A. Nels. (= *S. iberica* Sennen & Pau; = *S. kali* var. *tenuifolia* Tausch.) is an annual, native to southern Russia and western Siberia. It was introduced as a flax seed contaminant into South Dakota in 1873 (Robbins *et al.* 1951), but it was probably also present in the many introductions of Turkestan alfalfa imported into Canada and the United States. The reason for suspecting this source was that Wilcox & Stevenson (1909) found it in 51 out of 201 samples of alfalfa seed in Nebraska. In Canada, it became particularly prevalent in the 1930s: Moynan (1939) reported Russian thistle constantly invading new territory in Manitoba, and that between 1935–38 it became established on illustration stations at Roblin, Gilbert Plains, and Ste. Rose. It is now present in all provinces except Newfoundland, although in eastern Canada it is almost exclusively a railway and roadside weed (Frankton & Mulligan 1970). It is, however, abundant in the drier parts of the prairies and British Columbia on both cultivated land and over-grazed or disturbed pasture. In Saskatchewan, Russian thistle varied between the fifth to the seventh most abundant weed on cultivated land and in the southwest of the province it was present in 64% to 100% of the 0.25 m^2 samples surveyed (Thomas 1976, 1977, 1978, 1979). In winter the plants become tumble weeds that collect on fences and other obstructions.

In western United States, Russian thistle is the favored alternate host of the beet leafhopper, *Circulifer tenellus* (Baker), which is the vector of the curly top virus of sugar beet and other crops (Severin 1933). The losses caused by this disease have been a major impetus for the biological control of the weed in the United States, however, the disease does not occur in the sugar beet areas of either Manitoba (Robertson 1968) or Alberta (R.J. Howard, 1981, personal communication).

Background

The insect guild on Russian thistle in North America is typical of an introduced weed. Goeden & Ricker (1968) found 91 species of insects feeding on it, but none of them did appreciable damage and all but two were polyphagous. In contrast, in the palaearctic the plant is attacked by a number of specialized insects (Hawkes 1969, Goeden 1973). Two of these have been cleared by workers in the United States. The Canadian programme against Russian thistle has been based on this work.

The stem boring moth, *Coleophora parthenica* Meyr., was released in the United States in 1973 following host specificity studies by Hawkes & Mayfield (1976). The species has a high temperature threshold: a minimum of 15.5°C is required for development and about 500 degree-days above this temperature for maturation; however, about twice this number of degree-days are required for a population increase (Hawkes, 1976, personal communication). The moth is well established and has attained high population densities at many places in southwestern United States where it normally has three generations a year (Hawkes & Mayfield 1978*b*). At Indio, California, Goeden & Ricker (1979) found an average of one larva per 7.4 cm of stem. This level of attack had no effect on plant size, mortality, or seed production so the moth will have to be supplemented with other biological control agents to achieve beneficial results.

The case bearer, *Coleophora klimeschiella* Toll., was released in California in 1977 following screening studies by Hawkes & Mayfield (1978*a*). The biology of this insect is reported by Khan & Baloch (1976): the first two instars are leafminers; in the last three, the larva lives outside the leaf in a case of hollowed leaves, and feeds by protruding its head and forelegs through a hole in the leaf so that it continues to feed inside the leaf. The insect overwinters as a larva inside the case attached to *Salsola* stems or stones. There are one to three generations a year in Pakistan, depending on elevation. *C. klimeschiella* is established in California (Julien 1982).

Releases and Recoveries

The releases of both *C. parthenica* and *C. klimeschiella* are listed in Table 49.

C. parthenica was released on a number of occasions in July and August in stands of Russian thistle at the Regina and Swift Current Research Stations. The moths remained on the plants where they were released for several days, but no eggs or larvae were found in the vicinity. Larvae in the stems of potted plants that were placed outside failed to complete development. The problem was presumed to be the cool summer temperatures of the Canadian prairies. At Regina in July and August, the minimum temperature was below the threshold for development for 50 days in July and August, and on 16 August, the temperature fell to 1.3°C. At all sites on the Canadian prairies, there is normally less than the 500 degree-days required by the insect.

C. klimeschiella was released in June at the Regina Research Station and bred in the summer of release and the following summer. The population was less than one larva per Russian thistle plant and they did no appreciable damage. In the spring of 1979, the release site was flooded for several weeks in the spring and the colony was destroyed.

Table 49 Open releases and recoveries against *Salsola pestifer* A. Nels.

Year	No. of moths released	Source	Release site	Recovery
Coleophora parthenica				
1975	273	California	Swift Current, Saskatchewan	
1975	66	California	Regina, Saskatchewan	
1975	33	Pakistan	Regina, Saskatchewan	
Coleophora klimeschiella				
1977	91	California ex Pakistan	Regina, Saskatchewan	bred in 1977 and 1978 but not in 1979

Evaluation of Control Attempts

Summer temperatures on the Canadian prairies are too low for *C. parthenica*, and there is no purpose in making additional releases unless a strain with a low temperature threshold can be found. Even if such a strain can be established in Canada, on the basis of results in the United States, the density of Russian thistle is unlikely to be affected.

C. klimeschiella can survive in Saskatchewan and hence presumably elsewhere in southern Canada. Insects released for the biological control of a weed often require several years for the selection of a strain adapted to the region, so the lack of a rapid increase in the population should not be the sole reason for discontinuing further attempts. More discouraging is the absence of any data on the losses caused by this extremely common weed on the prairies and the present general lack of interest in a biological control solution for Russian thistle.

Recommendations

Biological control of *S. pestifer* should be discontinued until its use can be justified in economic terms.

Acknowledgements

I am grateful to R.B. Hawkes for reviewing the manuscript.

Literature Cited

Frankton, C.; Mulligan, G.A. (1970) Weeds of Canada. Canadian Department of Agriculture, 217 pp.

Goeden, R.D. (1973) Phytophagous insects found on *Salsola* in Turkey during exploration for biological weed control agents for California. *Entomophaga* 18, 439–448.

Goeden, R.D.; Ricker, D.W. (1968) The phytophagous insect fauna of Russian thistle (*Salsola kali* var. *tenuifolia*) in southern California. *Annals of the Entomological Society of America* 61, 67–72.

Goeden, R.D.; Ricker, D.W. (1979) Field analysis of *Coleophora parthenica* (Lep.: Coleophoridae) as an imported natural enemy of Russian thistle, *Salsola iberica*, in the Coachella valley of southern California. *Environmental Entomology* 8, 1099–1101.

Hawkes, R.B. (1969) Insects recorded from *Salsola* spp. and *Halogeton* spp. Report of the Entomology Research Division, USDA, ARS, Albany, California, 55 pp.

Hawkes, R.B.; Mayfield, A. (1976) Host specificity and biological studies of *Coleophora parthenica* Meyrick, an insect for the biological control of Russian thistle. Commemorative Volume in Entomology, Department of Entomology, University of Idaho, 37–43.

Hawkes, R.B.; Mayfield, A. (1978*a*) *Coleophora klimeschiella*, biological control agent for Russian thistle: host specificity testing. *Environmental Entomology* 7, 257–261.

Hawkes, R.B.; Mayfield, A. (1978*b*) *Coleophora* spp. as biological control agents against Russian thistle. *Proceedings of the 4th International Symposium on Biological Control of Weeds*, 113–116.

Julien, M.H. (1982) Biological control of weeds: a world catalogue of agents and their target weeds. Commonwealth Institute of Biological Control, 108 pp.

Khan, A.G.; Baloch, G.M. (1976) *Coleophora klimeschiella* (Lep: Coleophoridae) a promising biocontrol agent for Russian thistles, *Salsola* spp. *Entomophaga* 21, 425–428.

Moynan, J.C. (1939) Division of illustration stations. Progress report 1934 to 1938. Part II. Department of Agriculture, Canada, 130 pp.

Robertson, H. (1968) Sugar farmers of Manitoba. The Manitoba Beet Growers Association, 176 pp.

Robbins, W.W.; Bellue, M.K.; Ball, W.S. (1951) Weeds of California. Sacramento; California State Printing Office, 547 pp.

Severin, H.H.P. (1933) Field observations on the beet leafhopper, *Eutettix tennellus*, in California. *Hilgardia* 7, 281–360.

Thomas, A.G. (1976) Weed survey of cultivated land in Saskatchewan. First annual report, Agricultural Canada, 93 pp.

Thomas, A.G. (1977) Weed survey of cultivated land in Saskatchewan. Second annual report, Agriculture Canada, 103 pp.

Thomas, A.G. (1978) Weed survey of cultivated land in Saskatchewan. Third annual report, Agriculture Canada, 113 pp.

Thomas, A.G. (1979) Weed survey of cultivated land in Saskatchewan. Fourth annual report, Agriculture Canada, 141 pp.

Wilcox, E.M.; Stevenson, N, (1909) Report of the Nebraska seed laboratory. *Bulletin of the Agricultural Experiment Station, Nebraska* 110, 29 pp.

Senecio jacobaea L., Tansy Ragwort (Compositae)

P. HARRIS, A.T.S. WILKINSON and J.H. MYERS

Pest Status

The status of tansy ragwort, *Senecio jacobaea* L., in Canada was described by Harris *et al.* (1971). Since that time the distribution has changed slightly. A small stand of the weed that had existed at Guelph, Ontario, for many years recently started to spread (J. Alex, 1981, personal communication). A small infestation in the Gaspé Peninsula, Québec, recorded prior to 1900 could not be found by Watson & Muirhead (1979). The weed has continued to spread in the lower Fraser Valley in British Columbia, and it remains a serious socio-economic problem in the maritimes where there are many small mixed farms with a few cattle. The small field size and lack of specialized equipment makes chemical and mechanical control difficult and expensive. On the other hand, the death of one or two cattle from ragwort poisoning is a serious loss to the family food and income. There are no provincially compiled statistics of cattle deaths from ragwort as diagnosis is difficult and many cattle are sent for early slaughter if ragwort poisoning is suspected. However, in 1981, in Prince Edward Island the Montague Veterinary Clinic reported 12–20 ragwort-related cattle deaths; the Kensington Clinic had 20 but many more were suspected; the Provincial Veterinary Pathology Laboratory reported 10–12 but suspected more (L. Thompson, 1981, personal communication). Thus the total cattle mortality from ragwort is similar to the 40–65 deaths estimated to have occurred in 1968 (Harris *et al.* 1971).

Background

The cinnabar moth, *Tyria jacobaeae* (L.), was established in the Canadian maritimes in 1964 and in British Columbia in 1965 for the biological control of tansy ragwort. Despite difficulties in getting the moth established (Harris *et al.* 1975), it is now present throughout the ragwort infested regions of Canada, including Newfoundland where no releases were made (Larson & Jackson 1980).

The effect of the moth on the weed has varied widely. At Prince Charles, Prince Edward Island, and Durham, Nova Scotia, the weed has practically disappeared from the release sites although it has persisted in the disturbed ground along the roadsides (Harris *et al.* 1978). At Durham it was largely replaced by goldenrod *Solidago* sp. At Sussex, New Brunswick, the ragwort stand collapsed after defoliation by cinnabar larvae while the Canada thistle, *Cirsium arvense* (L.) Scop., increased. Tansy ragwort has since returned to the site and the density of cinnabar reached circa three larvae/stem in 1980 with scattered but numerous larvae in 1981 (D. Finnamore, 1981, personal communication). Near Nanaimo, British Columbia, over the past ten years an average of 38% of the ragwort was defoliated with another 18% partially defoliated. This has had little effect on the density of the flowering plants but they are smaller. In California, Hawkes & Johnson (1978) found that, with cinnabar moth attack, the number and size of the ragwort flowering stems decreased and the number of rosettes increased. Not all the declines were the result of the cinnabar moth as Myers (1980) found that two of five stands in Nova Scotia declined even though the cinnabar population was not large enough to cause extensive defoliation.

Many papers have been written in Canada and elsewhere on the causes of the diverse effects of the cinnabar moth on tansy ragwort. Harris *et al.* (1978*a*) related the collapse of some maritime stands of the weed following cinnabar defoliation to the frost sensitivity of the regenerating plants. Several studies (Green 1974, Campbell 1975, Myers 1976, Myers & Campbell 1976*a*, 1976*b*) have investigated the importance of ragwort spacing on cinnabar egg distribution and the success of larval transfer between plants. They showed that one requirement for explosive increases of the moth was dense stands of the weed. Dempster (1971) reported a higher mortality of cinnabar eggs and larvae on ragwort rosettes than on flowering stems and Myers (1980) found that cinnabar populations in areas with a high density of closely spaced rosettes tended to be less stable than on those with low rosette density. A high nitrogen content in the plant increased both moth fecundity and egg mass size in the following generation. Myers & Post (1981) argued that this would tend to destabilize the cinnabar population by over-exploitation of the food resource. In a study by Lakhani & Dempster (1981), the number of eggs had a rather small effect on the subsequent moth numbers because high egg density was often associated with larval starvation. In a dune habitat, Meijden (1979) described the cinnabar population as "walking on ice floes" because small stands of ragwort became extinct one or two years after attack by the moth and then reappeared after the moth emigrated to other stands. Predators were important in some instances (Wilkinson 1965, Myers & Campbell 1976*c*, Meijden 1979). In 1980 at Nanaimo, British Columbia, most of the pupae were destroyed resulting in a low population in 1981 in which only 3% of the plants were attacked.

Philogène (1975) found that the day length and temperature during larval development did not affect the obligatory pupal diapause. However, the moth has adapted its temperature threshold for emergence in the spring so that in the various regions of North America larval feeding remains synchronized with ragwort flowering (Myers 1979). Richards & Myers (1980) found that maternal moth size and temperature requirements for moth emergence were heritable and Myers (1978) found that the present populations of the moth have regional differences in their enzyme systems that are irrespective of their origin. Thus in a few years since release, the moth has adapted to its new habitat with the result that its behaviour now is not necessarily the same as that of the stock released.

A model by Lakhani & Dempster (1981) showed that for both Nanaimo, British Columbia, and Weeting Heath in Britain, the changes in the density of the weed closely conformed to the amount of spring and early summer rainfall. The model showed that at these sites the cinnabar population merely tracked the changes in the density of the weed. Myers (1980) also concluded that population fluctuations after an initial reduction of plant size by introduced cinnabar moths have a large environmental component. One indication that different factors are important in different sites is that the model did not fit the results from a more moist site in Oregon. In a more generalized model Roff & Myers (unpublished) found that depending on the degree of larval dispersal, the cinnabar population could be cyclic, stable, or chaotic. The gamut of these conditions occurs in nature. There are several ragwort habitats that are not utilized by the moth: the weed tends to be avoided when growing in partial shade; the cinnabar pupae are not able to overwinter in wet sites (Dempster 1971) so the moth tends to be absent in Europe from pastures on river flood plains, which often have dense ragwort stands; the moth is rare in the Swiss Jura on the widely separated plants or small clumps of plants. It is a weak flier and has dispersed less rapidly in California than the beetle *Longitarsus jacobaeae* (Waterhouse).

In many habitats the cinnabar moth has not reduced the density of tansy ragwort, but plants defoliated annually tend to be smaller so there has been some reduction in ragwort biomass. Also for about two months in the summer there is little ragwort foliage in the pastures or hay fields so that availability of the toxic foliage to cattle is reduced. The level of control is not satisfactory as a high density of the weed remains on many sites for much of the year, so the cinnabar moth needs to be supplemented by additional agents.

Releases and Recoveries

***Hylemya seneciella* Meade (Diptera: Muscidae)**

(a) Ecology

H. seneciella larvae reduce seed production by feeding in ragwort flower heads. The biology was described by Miller (1970), and an unsuccessful attempt to establish it in British Columbia and Prince Edward Island in 1968 was reported by Harris *et al.* (1971). The fly has been established in New Zealand where it infested up to 77% of the heads at peak flowering but the effect of the reduced seed production on ragwort density was not determined. The fly was established in California and probably in Oregon and Washington (Frick 1969*a*) but the California site was subsequently destroyed (Julien 1981).

(b) Releases

A summary of releases in the current review period is shown in Table 50; none of them became established. At least part of the problem was poor synchrony between emergence of the flies and ragwort flowering: in 1970 the *H. seneciella* were received in May and placed outside in a field cage in Prince Edward Island. The flies emerged between 8 and 17 June but the host plant did not flower until August. In 1971 emergence was delayed until the second half of July but still did not result in establishment (L. Thompson, 1981, personal communication). The results from British Columbia were similar.

Establishment in the United States was obtained with much larger releases of circa 2000 flies. As egg viability was circa 25% (Frick 1969*b*) a larger release than that used in Canada is necessary. The problem with the synchronism of fly emergence can probably be overcome by releasing the mature larvae at the end of the summer; this could be done with the Oregon or Washington stock providing that it is free of parasites. The value of establishing the fly is less clear since most ragwort reproduction in pastures is by vegetative propagation from existing plants. It is unlikely that the fly would survive in stands defoliated by cinnabar larvae which consume the flowers preferentially.

***Tyria jacobaeae* (L.) (Lepidoptera: Arctiidae)**

Releases

Field collected larvae from Nova Scotia were released at two sites in New Brunswick in 1970 and at seven sites in Ontario in 1979 and 1981 (Table 50). These were the only releases since those reported by Harris *et al.* (1971). Establishment occurred at both sites in New Brunswick resulting in widespread defoliation of tansy ragwort but on a long-term basis this is a less than desirable level of control. The insect did not become established in Ontario, possibly because of predation by ants (Alex, 1981, personal communication). About 20% of the Ontario shipment died in transit indicating that the population was heavily stressed and probably diseased.

Longitarsus jacobaeae (Waterhouse) (Coleoptera: Chrysomelidae)

(a) Ecology

L. jacobaeae is one of six species of *Longitarsus* known from *Senecio* in Europe although some of them are general feeders (Freude *et al.* 1966). The adults feed on the foliage, making holes in the leaves but the damage is minor. The main damage is done by the larvae which mine the cork-like periderm on the outside of the main root and the parenchyma cells in the center of the root. They also feed externally on the lateral roots producing long scarred grooves (Hawkes & Johnson 1978).

L. jacobaeae is distributed from Ireland east to Siberia and Tibet, but in a region south of latitude 52°N in Britain and east of longitude 5°E on the continent of Europe, it is replaced by *L. flavicornis* Steph., a species with which it has been frequently confused (Shute 1975). There are several biotypes of *L. jacobaeae* distinguished by their life cycle. Frick & Johnson (1973) reported that beetles from Italy entered aestivation under long days that delayed oviposition for an average of 88 days. In effect this means that, although the beetles emerge in the spring, they do not oviposit until the fall. Once laid, the eggs hatched in two to three weeks. In contrast, beetles from west Switzerland started oviposition in spring 13 days after emergence but the eggs took 1 to 9 months to hatch (Frick 1971). The end result is that egg hatch occurs in the fall when root reserves are high: in Prince Edward Island carbohydrate reserves in the roots of ragwort rosettes peaked in September-October (Harris *et al.* 1978*a*). In Britain, Newton (1933) found that adult emergence did not start until July compared to May for the other two biotypes. Frick (1971) suggested that the Swiss biotype would not survive in a climate with a dry summer as the eggs are vulnerable to desiccation and certainly in Britain the beetles are typically found in sites that are moist in summer. On the other hand, the Italian biotype did poorly in the cool-moist summer of Vancouver, British Columbia (Table 50).

Over 1000 eggs/female may be produced over a period of several months. Four years after their release in California they had practically eliminated ragwort from 5 ha and had spread at least 2 km from the release site (Hawkes & Johnson 1978). By 1979, ten years after release, they had spread 30 km.

(b) Releases

The initial releases of *L. jacobaeae* were made in British Columbia with stock from California that originated near Rome, Italy (Table 50). It was poorly adapted to British Columbia as only three pupae were recovered from the release of 100 beetles in the artificial ragwort stand at the University of British Columbia: they produced 200 adults for transfer to Abbotsford in the following year. In 1975 two beetles were found at Abbotsford in a one hour search, 20/hour in 1976, and 800 beetles were collected for release near Chilliwack in 1978 before the site was mined for gravel. In 1981 in a small stand of ragwort around the gravel pit 96% of the plants were attacked with an average of 13.1 larvae per plant.

The colony at Nanaimo, British Columbia, was established from 92 beetles from England that were released in 1974 (Table 50). By 1976 the population had increased to a count of 14 beetles/hour. A large number of beetles from Oregon (originally from Italy) were added to the colony in 1976. The extent of their contribution is not known but by 1981, 97% of the plants examined at the release site were attacked with an average of five larvae per ragwort plant.

L. jacobaeae was recovered at the two Prince Edward Island sites in 1982 and 1983 although numbers were low (L.S. Thompson personal communication 1983). Also the possibility of some root damage from larval feeding was noted in New Brunswick (D. Finnamore personal communication 1981). Prince Edward Island has a more even distribution of summer rain than occurs on the south west coast of British Columbia, so possibly the summers are too moist for the biotype released. Frick & Johnson (1972) found that the life cycle of the Swiss biotype adapted in the laboratory by a decrease in the length of the egg diapause. With the beetle's rapid adaptability and the probable wide genetic variation of the British Columbia population (stock from three sources has been released there), it should be possible to select an adapted biotype in a field cage and increase the survivors in the laboratory if the field population falls critically low. The other alternative is to import beetles from a climate with a similar summer rainfall to Prince Edward Island, but even so some adaptation would probably be necessary.

Recommendations

(1) The cinnabar moth has reduced the seriousness of tansy ragwort in many areas; however additional species of biological control agents are needed if a fully satisfactory level of ragwort control is to be achieved.

(2) The most obvious candidate as a second biological control agent is the beetle *Longitarsus jacobaeae*. A biotype of this beetle is now adapted to the southwest coast of British Columbia and should be distributed to ragwort infestations in the region. Estab-

Table 50 Open releases and recoveries against *Senecio jacobaea* L.

Species and release site	Year	Origin	No.	Recovered
Longitarsus jacobaeae (Waterhouse)				
British Columbia				
Vancouver	1971	California ex Italy	100 adults	1974
Vancouver	1973	Switzerland	38 adults	
Abbotsford	1971	California ex Italy	400 adults	
	1972	Lab reared ex California and England	400 adults	
	1972	California and England	250 adults	
	1974	UBC site ex Switzerland	200 adults	1975
Nanaimo	1974	England	92 adults	1976
	1976	Oregon ex Italy	1000 adults	
Chilliwack	1978	Abbotsford	500 adults	1979
Prince Edward Island				
Mt. Herbert	1978	British Columbia	300 adults	1982
Charlottetown	1981	British Columbia	185 adults	1982
New Brunswick				
Fredericton	1981	British Columbia	180 adults	
Tyria jacobaeae (L.)				
New Brunswick				
Sussex	1970	Nova Scotia	1000 larvae	1971
Bathurst	1970	Nova Scotia	1000 larvae	1971
Ontario				
Guelph	1979	Nova Scotia	4500 larvae	Not recovered
Guelph	1981	Nova Scotia	4500 larvae	Not recovered
Hylemya seneciella (Meade)				
British Columbia				
Abbotsford	1970	Switzerland	65 adults	
			453 puparia	Not recovered
Prince Edward Island				
Charlottetown	1970	Switzerland	115 adults	Not recovered
Charlottetown	1971	United States	170 adults	Not recovered

lishment of *L. jacobaeae* in the Canadian maritimes should be attempted by increasing the survivors in the laboratory if the population falls to critically low levels. If this fails, a stock should be imported from a region with a similar summer climate.

(3) Annual sampling of tansy ragwort, the cinnabar moth, and *L. jacobaeae* should be continued to determine their single and combined effects.

(4) If there are large areas where ragwort flower production is not controlled by the cinnabar moth and/or *L. jacobaeae*, another attempt should be made to establish the seed-head fly *Hylemya seneciella*. This should be done by releasing the mature larvae in the fall, so that there is opportunity for fly emergence in the following year to be synchronized with ragwort flower production.

(5) Preliminary host specificity studies by the CIBC indicate that *Ceutorhyncus atlanticus* Dieckmann and *Cochylis atricapitana* (Stephens) are restricted to the genus *Senecio*. Tests on these species should be completed.

Literature Cited

Campbell, B.J. (1975) Food plant spacing and the dispersal tendency of the cinnabar moth larva. M.Sc. thesis, University of British Columbia, 94 pp.

Dempster, J.P. (1971) The population ecology of the cinnabar moth *Tyria jacobaeae* L. (Lepidoptera, Arctiidae). *Oecologia* 7: 26–67.

Freude, H.; Harde, K.W.; Lohse, G.A. (1966) Die Käfer Mitteleuropas 9. Cerambycidae; Chrysomelidae. Drefeld; Goecke and Evers, 299 pp.

Frick, K.E. (1969*a*) Tansy ragwort control aided by the establishment of seedfly from Paris. *California Agriculture* 12: 10–11.

Frick, K.E. (1969*b*) Attempt to establish the ragwort seed fly in the United States. *Journal of Economic Entomology* 62: 1135–1138.

Frick, K.E. (1971) *Longitarsus jacobaeae* (Coleoptera: Chrysomelidae), a flea beetle for the biological control of tansy ragwort. II. Life history of a Swiss biotype. *Annals of the Entomological Society of America* 64, 834–840.

Frick, K.E.; Johnson, G.R. (1972) *Longitarsus jacobaeae* (Coleoptera: Chrysomelidae), flea beetle for the biological control of tansy ragwort. 3. Comparison of the biologies of the egg stage of Swiss and Italian biotypes. *Annals of the Entomological Society of America* 65: 406–410.

Frick, K.E.; Johnson, G.R. (1973) *Longitarsus jacobaeae* (Coleoptera: Chrysomelidae), a flea beetle for the biological control of tansy report. 4. Life history and aestivation of an Italian biotype, *Annals of the Entomological Society of America* 66: 358–367.

Green, W.Q. (1974) An antagonistic insect/plant system: the problem of persistence. Ph.D. thesis, University of British Columbia, 247 pp.

Harris, P.; Wilkinson, A.T.S.; Neary, M.E.; Thompson, L.S. (1971) *Senecio jacobaea* L., tansy ragwort (Compositae). *In*: Biological control programmes against insects and weeds in Canada 1959–1968. *Commonwealth Institute of Biological Control Technical Communication* 4, 97–104.

Harris, P.; Wilkinson, A.T.S.; Neary, M.E.; Thompson, L.S.; Finnamore, D. (1975) Establishment in Canada of the cinnabar moth, *Tyria jacobaeae* (Lepidoptera: Arctiidae) for controlling the weed *Senecio jacobaea. Canadian Entomologist* 107, 913–917.

Harris, P.; Thompson, L.S.; Wilkinson, A.T.S.; Neary, M.E. (1978*a*) Reproductive biology of tansy ragwort, climate and biological control by the cinnabar moth in Canada. *Proceedings of the 4th International Symposium on Biological Control of Weeds*, 163–173.

Harris, P., Wilkinson, A.T.S.; Thompson, L.S.; Neary, M. (1978*b*) Interaction between the cinnabar moth, *Tyria jacobaeae* L. (Lep: Arctiidae) and ragwort, *Senecio jacobaea* L. (Compositae) in Canada. *Proceedings of the 4th International Symposium on Biological Control of Weeds*, 174–180

Hawkes, R.B.; Johnson, G.R. (1978) *Longitarsus jacobaeae* aids moth in the biological control of tansy ragwort. *Proceedings of the 4th International Symposium on Biological Control of Weeds*, 193–196.

Julien, M.H. (1982) Biological control of weeds: a world catalogue of agents and their target weeds. Commonwealth Institute of Biological Control, 108 pp.

Lakhani, K.H.; Dempster, J.P. (1981) Cinnabar moth and its food plant, ragwort: further analysis of a simple interaction model. *Journal of Animal Ecology* 50, 231–249.

Larson, D.J.; Jackson, B.S. (1980) First record of a cinnabar moth, *Tyria jacobaeae*, in Newfoundland. *Canadian Field-Naturalist* 94, 343–344.

Meijden, E. van der (1979) Herbivore exploitation of a fugitive plant species: local survival and extinction of the cinnabar moth and ragwort in a heterogenous environment. *Oecologia* 42, 307–325.

Miller, D. (1970) Biological control of weeds in New Zealand 1927–48. *New Zealand Department of Scientific Industrial Research Information Service* 74, 104 pp.

Myers, J.H. (1976) Distribution and dispersal in populations. A simulation model. *Oecologia* 23, 255–269.

Myers, J.H. (1978) Biological control introduction as grandiose field experiments: adaptations of the cinnabar moth to new surroundings. *Proceedings of the 4th International Symposium on Biological Control of Weeds*, 181–188.

Myers, J.H. (1979) The effects of food quantity and quality on emergence time in the cinnabar moth. *Canadian Journal of Zoology* 57, 1150–1156.

Myers, J.H. (1980) Is the insect or the plant the driving force in the cinnabar moth-tansy ragwort system? *Oecologia* 47, 16–21.

Myers, J.H.; Campbell, B.J. (1976*a*) Indirect measures of larval dispersal in the cinnabar moth, *Tyria jacobaeae* (Lepidoptera: Arctiidae). *Canadian Entomologist* 108, 967–972.

Myers, J.H.; Campbell, B.J. (1976*b*) Distribution and dispersal in populations capable of resource depletion. A field study on cinnabar moths. *Oecologia* 24, 7–20.

Myers, J.H.; Campbell, B.J. (1976*c*) Predation by carpenter ants: a deterrent to the spread of cinnabar moth. *Journal of the Entomological Society of British Columbia* 73, 7–9.

Myers, J.H.; Post, B.J. (1981) Plant nitrogen and fluctuations of insect populations: a test with the cinnabar moth-tansy ragwort system. *Oecologia* 48, 151–156.

Newton, H.C.F. (1933) On the biology of some species of *Longitarsus* living on ragwort. *Bulletin of Entomological Research* 24, 511–520.

Philogène, B.J.R. (1975) Responses of the cinnabar moth, *Hypocrita jacobaeae* to various temperature/photoperiod regimes. *Journal of Insect Physiology* 21, 1415-1417.

Richards, L.J.; Myers, J.H. (1980) Maternal influences on size and emergence time of the cinnabar moth. *Canadian Journal of Zoology* 58, 1452–1457.

Roff, D.; Myers, J.H. Simulating the dynamics of the cinnabar moth populations: a study of complexity in a deterministic world (unpublished manuscript).

Shute, S.L. (1975) *Longitarsus jacobaeae* Waterhouse (Col., Chrysomelidae): identity and distribution. *Entomologists' Monthly Magazine*, 111 (1328–13300), 33–39.

Watson, A.K.; Muirhead, L. (1979) Exploration of the Gaspé region of Québec for tansy ragwort. p. 411. In: Research report, Expert Committee on Weeds (Eastern Canada), 412 pp.

Wilkinson, A.T.S. (1965) Releases of cinnabar moth, *Hypocrita jacobaeae* (L.) (Lepidoptera: Arctiidae) on tansy ragwort in British Columbia. *Proceedings of the Entomological Society of British Columbia* 62, 10–13.

Silene cucubalus Wibel, Bladder Campion (Caryophyllaceae)

M.G. MAW

Pest Status

Silene cucubalus Wibel is a deep rooted perennial of Eurasian origin. It is spread by seed (as many as 20 000 per plant) as well as vegetatively by parts of the root crown severed by implements. It is resistant to the common herbicides 2,4-D (2,4-dichlorophenoxy-acetic acid) and MCPA (4-chloro-2-methylphenoxyacetic acid) at rates which can be safely used in grain crops. The weed grows in waste places, roadsides, and railyards, and although it is an infrequent weed in cultivated ground (Boivin 1968), it can be troublesome in well drained pastures, cereal crops, and hay. It is found in the north-eastern and central United States and in every province of Canada to latitude 54°N (Scoggan 1978) and is more common in the eastern than in the western parts of its range (Frankton *et al.* 1970).

Although the weed is generally not a great problem throughout its Canadian range, there are localized areas where it causes concern. For example, in the Ethelbert district of Manitoba, a survey of weeds in cultivated fields (Thomas 1978) showed it to be present in 25% of the samples in 50% of the fields surveyed with a mean density of 5.9 plants per square meter. In contrast, the weed was reported in only one of the remaining 38 Manitoba districts surveyed, and there it was found only in 0.4% of the samples in 8.3% of the fields at a density of 0.6 plants per square meter.

Although it would seem that there has been little change in the status of the weed on cultivated lands since the 1966 survey (Alex 1966), agricultural practices have altered the situation in areas such as in southeastern Manitoba. Here a shift has been away from annual tillage to establishment of pastures and forage crops and forage seed production. Where bladder campion was not a problem as recently as three years ago, it now is a concern in over 4200 hectares and only limited control is being provided by herbicides and cultural methods.

Background

Surveys of the European insect fauna on species of *Silene* and *Melandrium* were made by the Commonwealth Institute of Biological Control (Miotk 1973). Sixty-six percent of the phytophagous insects on the two genera were oligophagous and the supposedly stenophagous species were either weevils or noctuids.

The chrysomelid, *Cassida azurea* Fab. (mistakenly identified as *Cassida hemisphaerica* Hbst.) (Maw & Steinhausen 1980*a*, 1980*b*), was collected in Switzerland and screened in Canada (Maw 1976). The Cassidinae are generally specialized in their feeding habits and species in a subfamily are usually restricted to a limited group of plants. This was found to be so with *C. azurea.*

First instar larvae fed on all the plants in the tribes Sileneae and Alsineae tested but they developed only on *S. cucubalus, S. cserei, S. glauca, S. noctiflora, S. maritima, S. alba, S. acaulis, Gypsophila repens, Dianthus chinensis,,* and *D. plumarius*. Damage by young larvae is confined to the epidermis of the leaf, while older larvae and adults eat large holes in the leaf blade. Flowers of *Silene* are eaten by both larvae and adults.

Discussion

Although starvation tests give the impression that *C. azurea* has a wide host range, only plants in the family Caryophyllaceae supported development and in most instances only the younger tender leaves were eaten. Older and harder leaves and those with waxy coatings resist attack from young larvae and older larvae fare little better. It is expected that under field conditions, *C. azurea* would direct its attack to *S. cucubalus*. Other *Silene* would at best be marginal hosts. *Dianthus* spp. might be eaten when *Silene* spp. are unavailable. This would, however, be very restricted because the hard texture and waxy coating of the leaves make sustained feeding difficult for older larvae and adults and almost impossible for young larvae.

With the main concern being the spread of *S. cucubalus* into forage, hay, and forage seed production fields, leaf feeding biological control agents may not be too successful. They will however be useful in situations where bladder campion is not cut, such as in waste areas. Root feeders should be of more use and so should be actively sought for biological control of *S. cucubalus* in hay and forage crops.

Recommendations

(1) A biological control programme for *S. cucubalus* be continued.

(2) *C. azurea* be reconsidered as a control agent.

(3) Root feeders such as *Hadena luteago* Schiff. and *H. andalusica* Stgr. be studied and their potential as control agents be assessed.

Literature Cited

Alex, J.F. (1966) Survey of weeds of cultivated land in the prairie provinces. Regina, Saskatchewan; Canada Agriculture, Research Branch, 68 pp.

Boivin, B. (1968) Flora of the prairie provinces. Part 2. Provancherea 3. Herbier Louis-Marie. Québec; Université Laval, 185 pp.

Frankton, C.; Mulligan, G.A.; Wright, W.H.; Steins, I. (1970) Weeds of Canada. Ottawa; Canadian Department of Agriculture, 217 pp.

Maw, M.G. (1976) Biology of the tortoise beetle *Cassida hemisphaerica* (Coleoptera: Chrysomelidae), a possible biological control agent for bladder campion, *Silene cucubalus* (Caryophyllaceae), in Canada. *Canadian Entomologist* 108, 945–954.

Maw, M.G.; Steinhausen, W.R. (1980*a*) Corrigendum for "Biology of the tortoise beetle *Cassida hemisphaerica* (Coleoptera: Chrysomelidae), a possible biological control agent for bladder campion *Silene cucubalus* (Caryophyllaceae), in Canada. *Canadian Entomologist* 112, 639.

Maw, M.G.; Steinhausen, W.R. (1980*b*) *Cassida azurea* (Coleoptera: Chrysomelidae) - not *C. hemisphaerica* - as a possible biological control agent of bladder campion, *Silene cucubalus* (Caryophyllaceae) in Canada. *Zeitschrift für angewandte Entomologie* 90, 420–422.

Miotk, P. (1973) Phytophagous insects associated with weeds in central Europe. Part II. Weed Projects for Canada. Progress Report 32, Commonwealth Institute of Biological Control, Delémont, Switzerland, 29 pp.

Scoggan, H.J. (1978) The flora of Canada. Part 3. Ottawa; National Museum of Natural Sciences, pp. 547–1115.

Thomas, A.G. (1978) The 1978 weed survey of cultivated land in Manitoba. Regina; Agriculture Canada Research Station, 109 pp.

Sonchus arvensis L., Perennial Sow-thistle, *S. oleraceus* L., Annual Sow-thistle and *S. asper* (L.) Hill, Spiny Annual Sow-thistle (Compositae)

D.P. PESCHKEN

Pest Status

Four species in the genus *Sonchus* have been introduced into North America from Europe and Asia. Three of them are weeds. Perennial sow-thistle, *Sonchus arvensis* L., is a common weed in all provinces of Canada but it is particularly abundant in Ontario, Manitoba, and the northern agricultural areas of Québec, Ontario, and the Prairie Provinces (Frankton & Mulligan 1970). It includes the varieties *S. arvensis* L. s. str. and *S. arvensis* var. *glabrescens* Guenth., Grab & Wimm. (Boulos 1961). It reproduces by seed and an extensive underground root system. About 9750 achenes are produced per flowering shoot (Stevens 1932) and 59 500 per m^2 (Stevens 1924). The seeds are dispersed by means of a pappus. Stevens (1924) noted that most seeds are not carried far and that there is a difference in the persistence of the pappus in different lots. In our experiments only one of 26 seeds which were released into the air at wind speeds averaging 7 km per hour and gusting to 22 km per hour became detached from its pappus. Eighteen out of 20 seeds flew out of sight at average windspeeds of 15 km per hour and gusting to 38. The pappus of one seed became entangled on vegetation and fluttered in the wind for 15 minutes, but did not separate from the seed (Peschken, unpublished).

Seedlings establish more readily where there is litter, and the moisture in low habitats may help seedling establishment in pond and ditch margins (Stevens 1926). This is the reason why *S. arvensis* is particularly serious in the irrigated area around Outlook, Saskatchewan (W.J. King, 1980, personal communication). Perennial sow-thistle thrives on low heavy soils and on lighter soils with a good moisture supply (Stevens 1924).

Fifty-nine shoots per m^2 reduced the yield of oats by 25% (Friesen & Shebeski 1960). Shaskov *et al.* (1977) report reduction in wheat yield by 4.5% to 27% in Kazakh SSR, at 3–15 plants per m^2. In Saskatchewan five shoots per m^2 reduced the yield of rapeseed by 12% and 10 shoots by 18% (Peschken unpublished).

Control of perennial sow-thistle growing in grain is possible with herbicides, but difficult in broad-leaved crops. Cultivation has to be repeated every three to four weeks for at least one year to deplete root reserves (Derscheid & Parker 1972).

The annual sow-thistle, *S. oleraceus* L., an annual or rarely biennial weed, reproduces by seeds and has a strong tap root which may reach to a depth of 1 m and may spread to 1.29 m (Kutschera 1960). It grows up to 2 m tall and it produces about 6000 seeds per plant (Salisbury 1961). It is a weed of cultivated fields, lawns, gardens, grainfields, vineyards and occurs in all provinces of Canada. In Canada, it may harbor virus diseases such as tobacco streak, alfalfa mosaic, cucumber mosaic, and the mycoplasma disease, aster yellows (Gayed 1978, Holm *et al.* 1977, Smith 1972). The spiny annual sow-thistle, *S. asper* (L.) Hill, is an erect annual or sometimes biennial weed up to 2 m tall, with a strong taproot reaching down to 2 m (Kutschera 1960). Salisbury (1961) estimates seed production to be about 18 000 per plant.

Both annual sow-thistles occur throughout Canada, but are more abundant in Ontario, Québec, and British Columbia than in other provinces (Frankton & Mulligan 1970). *S. tenerrimus* is only reported from California (Shetler & Skog 1978).

Background

No biological control of sow-thistles has been reported from other parts of the world. In Canada, only one as yet unidentified Lepidoptera larva has been found feeding on the flower heads or destroying seeds, and no monophagous insects have been reported feeding on any other part of the perennial sow-thistle (M. Maw, 1982, personal communication). Conners (1967) lists three disease organisms specialized on *S. arvensis*: *Marssonina sonchi* Dearn. & Bisby, *Septoria sonchi-arvensis* Dearn. & Bisby, and *S. sonchifolia* Cke. However, these do not control their host. The survey in eastern Austria and the Swiss Jura, and of the literature by Schröder (1974), and in Iran, Pakistan, and Japan by Zwölfer (1973) on *Sonchus* spp. produced only six insect species which appeared to be sufficiently host specific to warrant further study. Five of these are endophytic in the flower heads and one is a leaf gall fly. Two of these, *Tephritis dilacerata* Loew and *Cystiphora sonchi* (Bremi) have been studied in detail and will be discussed below. In addition, the moth *Celypha rosaceana* (Schläger) is reported from *Sonchus* spp. and *Taraxacum officinale* Weber only and should be studied further (Bradley *et al.* 1979).

Tephritis dilacerata Lw. (Diptera: Tephritidae)

Ecology

The fly deposits clutches of six to seven eggs into flower buds and in the laboratory the oviposition period may last up to ten weeks (Berube 1978*a*,1978*b*). The larvae transform the flower bud into a gall in which they pupate when mature. The adult fly enters diapause when exposed to a short day or cool temperatures or both and it overwinters as an adult. There is one generation per year (Berube 1978*a*). *T. dilacerata* is widely distributed in the palaearctic covering about three quarters of the range of *S. arvensis* (Berube 1978*a*).

Cystiphora sonchi (Bremi) (Diptera: Cecidomyiidae)

Ecology

C. sonchi is recorded from Sweden (Sylvén 1975) and Denmark in the north (Henriksen 1944) to Italy and Romania (Skuhravá *et al.* 1972) in the south, and from France (Kieffer 1899) in the west to the European part of the USSR in the east (Peschken, in press).

C. sonchi females lay their eggs on the underside of leaves. Galls are produced and become about 5 mm in diameter when mature. The larvae pupate in the gall or in the ground. The adults live up to 16 hours in the laboratory. Up to 721 galls were produced on one perennial sow-thistle rosette by six females. There are three generations per year in the field (Skuhravá & Skuhravý 1973).

The fly is host specific to *Sonchus* spp. (Peschken in press).

Releases and Recoveries

T. dilacerata has been released on a total of five sites in Saskatchewan in 1979, 1980, and 1981, and one site each in Alberta, Québec, and Prince Edward Island (Table 51). Breeding occurred in 1979 when the flies were released in large field cages (180 cm high, 180 x 180 cm). At Estevan three galls were found in August and 250 at Regina. From the latter an estimated 1150 flies emerged in the cage at Regina. Many of these were allowed to escape so that they could search for overwintering sites but no survivors were found in 1980. In 1981, only three galls were found in Outlook and 22 at Wishart, Saskatchewen. However, 98% of the flies were released for overwintering in the fall of 1980 and 97% in 1981. All fall releases were made with flies that emerged from galls collected in Austria. In 1980, the flies emerged in the quarantine laboratory in a 16 hour day except for 1000, which had emerged in an eight hour day at 14°C and were released into a large field cage at Regina, Saskatchewan in September 1980. A log pile covered loosely with straw was provided as a possible overwintering site. The cage was removed in late fall to allow additional insulation by snow and replaced on 15 April 1981. No flies emerged.

Subsequently it was learned that flies emerging at room temperature in a 16 hour day do not enter diapause. Releases made in the fall of 1981 were all made with flies conditioned for overwintering.

C. sonchi was released on four sites in Saskatchewan in 1981, where it survived the first winter on two sites and produced two generations in 1982 (Table 51). The colony in Alberta produced galls in 1981 but did not survive the winter. No galls were found from the 1982 release in Alberta nor from the 1981 and 1982 releases in Manitoba and Québec.

Table 51 Open releases and recoveries against *Sonchus arvensis* L.

Species and Province	Year	Number	Year of Recovery
Tephritis dilacerata Lw.*			
Saskatchewan	1979	810	1979
	1980	11 486	
	1981	1 093	
Alberta	1981	2 000	
Québec	1981	1 947	
Prince Edward Island	1981	2 000	
Total		19 336	
Cystiphora sonchi (Bremi)*			
Saskatchewan	1981	21 900	1982
Alberta	1981	5 000	1981
	1982	4 500	
Manitoba	1982	2 000	
Québec	1981	5 000	
	1982	2 100	
Total		40 500	

* All insects were originally collected in Austria. All *C. sonchi* and 6% of the *T. dilacerata* releases were made with laboratory reared stock, and the remaining *T. dilacerata* releases were made with flies that emerged from pupae collected in the field in Austria.

Evaluation of Control Attempts

So far no *T. dilacerata* overwintered in Saskatchewan. Saskatchewan, with its thin or often non-existent snow-cover, constitutes a harsh climate for overwintering. Also suitable food for the adult flies may be scarce in spring and fall on the prairies.

Recommendations

(1) Releases of *Cystiphora sonchi* and *T. dilacerata* should be made and monitored in areas where *S. arvensis* is frequent, i.e. in Ontario and Manitoba, and the northerly areas of Ontario, Québec, and the Prairie Provinces, and in irrigated areas where *S. arvensis* is a problem.

(2) A study of the overwintering habits of *T. dilacerata* should be made.

(3) The role of seeds of *S. arvensis* in the propagation of existing stands and the establishment of new stands should be investigated to determine whether the establishment of seed-destroying biological control agents should be pursued.

(4) The moth *Celypha rosaceana* (Schläger) (Lepidoptera: Tortricidae) and the seed-head fly *Contarinia schlechtendaliana* Rübs. (= *sonchi* Kieffer) appear to have a narrow host range and warrant screening as biological control agents.

(5) The survey for *S. arvensis* insects should be resumed and concentrated in the northern agricultural areas of Europe, such as Sweden, Finland, and the northern areas of European and Asian USSR, where *S. arvensis* is frequent.

Literature Cited

Berube, D.E. (1978*a*) Larval descriptions and biology of *Tephritis dilacerata* (Dipt.: Tephritidae) a candidate for biocontrol of *Sonchus arvensis* in Canada. *Entomophaga* 23, 69–82.

Berube, D.E. (1978*b*) The basis of host plant specificity in *Tephritis dilacerata* and *T. formosa* (Dipt.: Tephritidae). *Entomophaga* 23, 331–337.

Boulos, L. (1961) Cytotaxonomic studies in the genus *Sonchus*. 3. On the cytotaxonomy and distribution of *Sonchus arvensis* L. *Botaniska Notiser* 114(1), 57–64.

Bradley, J.D.; Tremewan, W.G.; Smith, A. (1979) British tortricoid moths. Tortricidae: Olethreutinae. Vol. II. London; The Ray Society and British Museum, 320 pp.

Conners, I.L. (1967) An annotated index of plant diseases in Canada, and fungi recorded on plants in Alaska, Canada and Greenland. *Canadian Department of Agriculture Publication* 1251, 381 pp.

Derscheid, L.A.; Parker, R. (1972) Thistles. Canada thistle, perennial sow-thistle. South Dakota State University Extension Fact Sheet 450, 4 pp.

Frankton, D.; Mulligan, G.A. (1970) Weeds of Canada. *Canadian Department of Agriculture Publication* 948, 217 pp.

Friesen, G.; Shebeski, L.H. (1960) Economic losses caused by weed competition in Manitoba grain fields. I. Weed species, their relative abundance and their effect on crop yields. *Canadian Journal of Plant Science* 40, 457–467.

Gayed, S.K. (1978) Tobacco diseases. *Canadian Department of Agriculture Publication* 1641, 57 pp.

Henriksen, K.L. (1944) Fortegnelse over de danske galler (zoodcecidier). *Spolia Zoologica Musei Hauniensis* VI, 1–212.

Holm, L.G.; Plucknett, D.L.; Pancho, J.V.; Herberger, J.P. (1977) The world's worst weeds. Honolulu; University Press of Hawaii, 609 pp.

Kieffer, J.J. (1899) Enumération des cécidies recueillies aux Petites-Dalles (Seine-Inférieure) avec description de deux Cécidomyies nouvelles. *Bulletin de la Société des Amis des Sciences Naturelles de Rouen*, 89–105.

Kutschera, L. (1960) Wurzelatlas mitteleuropäischer Ackerunkräuter und Kulturpflanzen. Frankfurt; DLG, 574 pp.

Peschken, D.P. (in press) Host specificity and biology of *Cystiphora sonchi* (Loew) (Diptera: Cecidomyiidae), a candidate for the biological control of *Sonchus* species. *Entomophaga*.

Salisbury, E. (1976) Weeds and aliens. London; Collins, 384 pp.

Schröder, D. (1974) The phytophagous insects attacking *Sonchus* spp. (Compositae) in Europe. Proceedings of the 3rd International Symposium on Biological Control of Weeds, Montpellier, France. *Commonwealth Institute of Biological Control Miscellaneous Publication* 8, 89–96.

Shaskov, V.P.; Kokmakov, P.P.; Volkov, E.E.; Trifanova, L.F. (1977) The influence of rhizomatous weeds in spring wheat crops on the utilization of nitrogen, phosphoros and potassium. *Agrokhimiya* 14(3), 57–59.

Shetler, S.G.; Skog, L.E. (1978) A provisional checklist of species for Flora North America (Revised). Flora North America Report 84, Missouri Botanical Garden, 199 pp.

Skuhravá, M.; Skuhravý, V. (1973) Gallmücken and ihre Gallen auf Wildpflanzen. Wittenberg Lutherstadt; Ziemsen, 118 pp.

Skuhravá , M.; Skuhravý, V.: Neacsu, P. (1972) Verbreitung der Gallmücken in Rumänien. *Deutsche entomologische Zeitschrift* 19, 375–392.

Smith, K.M. (1972) A textbook of plant virus diseases. 3rd ed., New York; Academic Press, 684 pp.

Stevens, O.A. (1924) Perennial sow-thistle, growth and reproduction. *North Dakota Agricultural Experiment Station Bulletin* 181, 44 pp.

Stevens, O.A. (1926) The sow thistle. *North Dakota Experiment Station Circular* 32, 16 pp.

Stevens, O.A. (1932) The number and weight of seeds produced by weeds. *American Journal of Botany* 19, 784–794.

Sylvén, E. (1975) Study on relationships between habits and external structures in Oligotrophidi larvae. (Diptera, Cecidomyiidae). *Zoologica Scripta* 4, 55–92.

Zwölfer, H. (1973) A survey for weed insects in Japan, Iran and Pakistan. Weed projects for Canada. *Commonwealth Institute of Biological Control Progress Report* 30, 27 pp.

Verbascum thapsus L., Common Mullein (Scrophulariaceae)

M.G. MAW

Pest Status

Verbascum thapsus L. is a biennial or, rarely, an annual of Eurasian origin. It was probably introduced into North America several times as a medicinal herb (Gross & Werner 1978). It is found over most of the United States and southern Canada, but is rare on the Canadian prairies, North Dakota, and Montana (Gross & Werner 1978, Maw 1980). It is most vigorous on well drained gravelly or stony soils in rough pastures, fence rows, roadsides, railway yards, and waste places. It is not a major weed, not allergenic, not poisonous to humans, and not a problem on cultivated land. It does, however, harbor insects that may be vectors of diseases of economic plants. For example, the mullein leaf bug, *Campylomma verbasci* Meyer, whose normal plant food is mullein, may cause damage to apple or pear fruit and potato leaves (Pickett 1939) and may also be a vector of fire blight, *Bacillus amylovorus* (Burr.) Trev. (Leonard 1965).

Biological Control Studies

The most effective control agents of mullein in Europe appear to be a weevil, *Gymnaetron tetrum* F., and the mullein shark moth, *Cucullia verbasci* L. (Noctuidae) (Miotk 1973). The weevil is already well established in Ontario, British Columbia, and the United States from New England to Arkansas and the Pacific states (Blatchley & Leng 1916, Hatch 1971). The adults eat mullein leaves, and the larvae and adults destroy up to 50% of the seeds (Gross & Werner 1978).

C. verbasci was screened (Maw 1980) and although there was some feeding on all Scrophulariaceae tested, and there was some nibbling on tomato, *Nicotiana* sp. and *Brassica napus*, sustained feeding and development occurred only on *Verbascum* spp.

Discussion

V. thapsus is an early colonizer on abandoned agricultural fields with the tall flowering stocks usually appearing the second year after abandonment. Populations are usually scattered but locally abundant, and the weed may cover entire fields where the ground has been left undisturbed. This situation is, however, relatively short-lived. In a study by Gross & Werner (1978) it was found that in a barley field that had been abandoned for three years, the density of *V. thapsus* was 1.0 plants per m^2. In an adjacent field that had been abandoned for 12 years, the density of *V. thapsus* had declined to 0.17/m^2. Since seeds can be dormant but viable for 35 years, the typical pattern seems to be one of ephemeral adult populations and long-lived seed pools (Gross & Werner 1978).

Although the mullein leaf bug, *Campylomma verbasci*, may damage apple and pear fruit, the bug is only partially phytophagous. It cannot complete its life cycle without insect prey and feeds on several species of mites and insects that attack the fruit crop (Gross & Werner 1978).

The mullein moth, *Cucullia verbasci*, is considered to be a safe biological control agent to release. However, the inherent risk in releasing any organism into new environmental conditions remains and until the economic value in controlling the weed is determined, no release should be made.

Recommendations

Until the loss caused by mullein has been determined, no releases of biological control agents should be made.

Literature Cited

Blatchley, W.S.; Leng, C.W. (1916) Rhynchophora or weevils of North Eastern America. Indianapolis; Nature Publishing Co., 682 pp.

Gross, K.L.; Werner, P.A. (1978) The biology of Canadian weeds. 28. *Verbascum thapsus* L. and *V. blattaria* L. *Canadian Journal of Plant Science* 58, 401–413.

Hatch, M.H. (1971) The Beetles of the Pacific Northwest. Part V. University of Washington Press, 662 pp.

Leonard, D.E. (1965) *Atractotomus mali* and *Campylomma verbasci* (Heteroptera: Miridae) on apples in Connecticut. *Journal of Economic Entomology* 58(5), 1031.

Maw, M.G. (1980) *Cucullia verbasci*, an agent for the biological control of common mullein (*Verbascum thapsus*). *Weed Science* 28, 27–30.

Miotk, P. (1973) Phytophagous insects associated with weeds in central Europe. Part II. Weed Projects for Canada. *Commonwealth Institute of Biological Control Progress Report* 32, 29 pp.

Pickett, A.D. (1939) The mullein leaf bug, *Campylomma verbasci* Meyer, as a pest of apples in Nova Scotia. *Entomological Society of Ontario Annual Report* 69, 105–106.

PART III

BIOLOGICAL CONTROL OF FOREST INSECT PESTS IN CANADA 1969–80

Biological Control of Forest Insect Pests in Canada 1969-1980: Retrospect and Prospect

M.A. HULME and G.W. GREEN

This contribution is the third in a series of reviews on biological control of forest insect pests in Canada. The first review (McGugan & Coppel 1962), covering the half century up to 1958, described control attempts against 35 pests, and although progress was made in both theoretical and practical aspects, many studies did not progress beyond exploratory work and little attempt was made to assess control success. Only 12 attempts were systematically planned for control of a target pest. Many of the remaining attempts, including those against several sawflies, were not continued beyond this first review period and simply involved the release of stocks of parasitoids left over from other control attempts, or the release of small numbers of parasitoids without any follow-up plans for monitoring. McGugan & Coppel (1962) tentatively assessed the more rigorously planned and promising attempts as follows: problems associated with *Gilpinia hercyniae* (Htg.), *Leucoma salicis* (L.), *Coleophora laricella* (Hb.) and *Lecanium tiliae* L. were considered to be "markedly reduced" following parasitoid introductions; and "significant results" were obtained against *Euproctis chrysorrhoea* (L.), *Pristiphora erichsonii* (Htg.), *Adelges piceae* (Ratz.), and *Diprion similis* (Htg.). In most of the other serious attempts, more time was considered necessary before any judgement could be made on the level of control obtained.

The second review (Reeks & Cameron 1971), covering the period 1959–68, concentrated on 12 pests: 11 were subjects of the earlier review and one new sawfly was added — *Neodiprion lecontei* (Fitch). An effort was made in this review to evaluate control success in terms of regulation of the pest population, and five of the control attempts were considered successful in widespread pest regulation. These included *G. hercyniae, L. salicis*, and *C. laricella*, all mentioned in the preceding paragraph, plus *N. lecontei* and *Operophtera brumata* (L.). The success with *N. lecontei* resulted from the first major control attempt to make sole use of a forest insect entomopathogen, in this case a nuclear polyhedrosis virus (NPV). Three further attempts, those against *P. erichsonii, Neodiprion sertifer* (Geoff.), and *Neodiprion swainei* Midd., were all considered to have achieved local pest control and all involved work continued from the previous review period. Two other continuing attempts – those against *Rhyacionia buoliana* (Schiff.) and *Malacosoma disstria* Hb. — resulted in slight pest reduction but the assessments were still considered premature. On the negative side, continued work with parasitoids for *A. piceae* showed no control success, and parasitoid work on *Choristoneura fumiferana* (Clem.), begun during the previous review period, was not continued. During this review period, more emphasis was given to entomopathogens as control agents, particularly using NPV against *N. lecontei*, as noted above, and the bacterium *Bacillus thuringiensis* Berliner (*B. t.*) against *C. fumiferana*. Most results from these initial tests were equivocal, and Reeks & Cameron (1971) summarized the entomopathogen work by stating that "success in control has been somewhat variable".

During the current period of review, work on all 12 pests from the previous review was continued and 9 further pests were examined as potential candidates for biological control. The additions are a western Canadian species of spruce budworm, *Choristoneura occidentalis* Freeman; a casebearer, *Coleophora serratella* (L.), and a leafminer, *Fenusa pusilla* (Lep.), both of which feed on birch; the gypsy moth, *Lymantria dispar* (L.); the balsam fir sawfly, *Neodiprion abietis* (Harr.), which, in the first review, was only examined using surplus parasitoid stock; Bruce spanworm, *Operophtera bruceata* (Hulst); two tussock moths, *Orgyia leucostigma* (J.E. Smith) and *Orgyia pseudotsugata* (McDunnough); and the mountain-ash sawfly, *Pristiphora geniculata* (Htg.). Effort was almost equally divided between treatments with entomopathogens and releases of predators or parasitoids. The current review thus contains evaluations of control attempts against 21 insect pests, although in some cases this simply entails updating assessments of control attempts made during the previous review period.

Despite the increased number of pests covered in the current review, total effort devoted to biological control of forest pests declined as the Canadian Forestry Service went through a period of austerity. Work with predators and parasitoids suffered particularly severe reductions of resources; and certain aspects of all programmes, such as the population dynamics of the pest and of the control agent, were almost totally neglected. These cutbacks are reflected in the lack of depth with which many of our investigations were undertaken and inevitably reduced our chances of finding successful controls. Despite these difficulties significant progress has been made, as detailed later.

The definition of biological control of forest insect pests used in this review follows that of the previous review – it is confined to the regulation of pest populations by the introduction of parasitoids, predators, or entomopathogens. Population regulation is considered acceptable if damage is controlled at tolerable levels in economic and/or social terms. The methods of introduction of the control agent cover all strategies from inoculative releases of an organism, which then spreads through the pest population, to augmenting agents already present in the pest population, or to inundation of the pest's environment with a control agent that may or may not already be present (Knipling 1979).

The chapters that follow this overview are organized in alphabetical generic sequence, and each chapter is written by the scientists responsible for conducting or assessing the control attempts. An outline is given of the present status of the pest, background is given where needed on reasons for the choice of control agents, and the authors' evaluation is given of each attempt. Each chapter concludes with specific suggestions for future work.

In describing the pest status, and in many of the assessments of control effectiveness reference is made to the Forest Insect and Disease Survey (FIDS), an organization within the Canadian Forestry Service that carries out major surveys of pest populations and pest damage across the country. In earlier times surveys included routine rearings to measure parasitoid levels in pest populations, but due to staff cutbacks such measurements are rarely performed now and surveys by FIDS are generally confined to measurements of pest numbers and pest damage. This in turn has seriously impeded the effectiveness of the biological control research applied to forest pests.

For taxonomic and other reasons, many changes in the names of insects or the spelling of insect names have taken place during the current review period. In order to standardize some of these changes, forestry contributions follow the nomenclature of insect pests given in the "Common names of insects and related organisms" published by the Entomological Society of America in 1980.

Selection of Biological Control Agents

Inoculative releases of parasitoids and predators

When planning this strategy of biological control it has become customary to first establish whether the target pest is a native or an introduced species. Of the 21 pests included in this review, 9 are considered native: *C. fumiferana*, spruce budworm; *C. occidentalis*, western spruce budworm; *M. disstria*, forest tent caterpillar; *N. abietis*, balsam fir sawfly; *N. lecontei*, redheaded pine sawfly; *N. swainei*, Swaine jack pine sawfly; *Operophtera bruceata,* Bruce spanworm; *Orgyia leucostigma,* whitemarked tussock moth; and *O. pseudotsugata,* Douglas-fir tussock moth. There is disagreement on whether *P. erichsonii,* larch sawfly, is native or introduced, although most favour the latter view (A. Giroux[1] 1981, personal communication; Pschorn-Walcher 1977). Following this division into introduced and native species it has been customary to concentrate on introduced pests as being promising candidates for biological control by inoculative introduction of predators or parasitoids obtained from the pest's native habitat. Thus, in the present review, all but two such control attempts are against introduced pests: the exceptions are attempts against *C. fumiferana,* where mostly limited and unpromising cage studies were formed with parasitoids, and attempts against *N. swainei,* where trials were made to introduce *Formica* spp. of ants as predators. The former attempts are best described as preliminary investigations rather than control attempts, and the latter trials as introductions of polyphagous predators aimed at control of several forest pests rather than just the sawfly. In general, the number of parasitoids or predators released in each control attempt depended on numbers obtained from field collections. Because of limited resources of both people and material, little rearing was carried out to increase insect numbers and few of the female parasitoids were deliberately mated before release. Reeks & Cameron (1971) feel that a minimum of 500 mated females should be released but this was rarely achieved in the work reported here. Indeed, Beirne (1975) regards releasing inadequate numbers and other faulty procedures as important reasons for failure in control attempts.

The question of whether to release one or several species of parasitoids or predators continues to evoke controversy (e.g. DeBach *et al.* 1976). Both release strategies were practised during the current review period without either showing obvious superiority in terms of control success. In co-operation with CIBC Delémont and other shipping stations, precautions were taken to screen out unwanted parasitoids, such as cleptoparasitoids and hyperparasitoids, before releases were attempted. A further difficult question is whether to release monophagous or polyphagous parasitoids. In the case-studies reported here both types shared in the control successes. In all, 31 species of parasitoids and predators were released against 11 target pests and a further 3 species of parasitoids were tested in laboratory and field cage studies against *C. fumiferana*. Well over half of the introductions were from Europe, three were from Japan, and one from Argentina; the remainder were relocations within Canada. All but three of the species belonged to the order of Hymenoptera; the three exceptions, all Diptera, were a laboratory culture of *Agria housei* Shewell tested unsuccessfully against *C. fumiferana*, *Lypha dubia* Fallén, released against *R. buoliana* but not known to be established, and *Cyzenis albicans* (Fall.) successfully established and helping to control *O. brumata*. Approximately half of the released parasitoids and predators are known to be established and the list may be extended as more time is allowed for establishment to become evident. A summary of all free field releases is shown in Table 52. Three parasitoids tested only in cages are also included to complete the list of those examined during the review period.

Application of entomopathogens

Entomopathogens were used for attempted control against all the native insect pests covered in this review, and similar controls were explored against the introduced pests *N. sertifer* and *L. dispar*. The main entomopathogens used here were the bacterium *B.t.* and a number of baculoviruses.

[1] Biosystematics Research Institute, Ottawa

Table 52 Free field releases of parasitoids and predators against forest insect pests between 1969 and 1980

Host	Parasitoids/predators	Total number released
Choristoneura fumiferana	*Agria housei*	2 800 *
	Cephaloglypta laricis	**
	Cephaloglypta murinanae	**
	Lissonota sp.	59
	Trichogramma sp.	20 000 **
Coleophora laricella	*Agathis pumila*	2 633
	Chrysocharis laricinellae	530
	Diadegma laricinellum	101
	Dicladocerus japonicus	292
Coleophora serratella	*Apanteles coleophorae* }	
	Apanteles mesoxanthus }	337
	Apanteles corvinus }	
	Campoplex borealis }	
	Campoplex sp. }	1 480
Fenusa pusilla	*Grypocentrus albipes*	1 753
	Lathrolestes nigricollis	4 994
Lymantria dispar	*Anastatus disparis*	1 404
	Ooencyrtus kuvanae	25 000
Neodiprion sertifer	*Dipriocampe diprioni*	1 189
	Exenterus abruptorius	4 072
	Lophyroplectus luteator	2 389
	Pleolophus basizonus	1 032
Neodiprion swainei	*Formica lugubris*	1 000 000
	Formica obscuripes	5 000 000
Operophtera brumata	*Agrypon flaveolatum*	820
	Cyzenis albicans	1 238
Pristiphora erichsonii	*Mesoleius tenthredinis*	467
	Olesicampe benefactor	8 588
Pristiphora geniculata	*Olesicampe geniculatae*	912
	Rhorus sp.	63
Rhyacionia buoliana	*Lypha dubia*	496
	Orgilus obscurator	560
	Parasierola nigrifemus	595
	Agathis binominata	85

* Laboratory culture.
** Cage release only.

B.t. has been tried against a variety of pests, all belonging to the Lepidoptera, because the pathogenic effects of many strains of the bacterium are promoted by the alkaline digestive conditions typical of this order of insects. Much of the pioneering work on the bacterium was carried out by the Canadian Forestry Service and work on the mode of action of the entomopathogen is continuing. *B.t.* is now used operationally by forest managers: the control agent is fully registered as a pesticide in Canada and is commercially produced in a number of different formulations for use against *C. fumiferana*, *L. dispar*, and several other lepidopterous defoliators. Methods of applying the bacterium have varied. In a few cases the bacterium was applied from ground sprays on localized areas of pest outbreaks (examples are treatments to control *M. disstria* and *L. dispar*), whereas in most other cases the bacterium was applied with aerial sprays (examples here include attempted control of *C. fumiferana*, *C. occidentalis*, and *O. pseudotsugata*). Most development of aerial spraying of *B.t.* was conducted with *C. fumiferana*. Both water-based and oil-based formulations were used and the spray aircraft included fixed-wing aircraft equipped with from one to four engines, and various sizes of helicopters. Areas treated ranged as high as 90 000 ha, and in all cases results to a large extent depended on the application techniques that were employed. The time period during which *B.t.* can be applied during the insects' development is shorter than that permissible with chemical insecticides because the bacterium must be ingested to be effective whereas the standard chemical insecticides are neurotoxins and exhibit contact toxicity. An interesting international aspect of this work involved co-operative trials between Canada and the United States under the 6-year CANUSA Spruce Budworms Research Programme that started in 1977. The two countries also co-operated in one trial using *B.t.* against *O. pseudotsugata* in British Columbia.

The use of viruses, particularly baculoviruses, for control of forest insect pests has progressed rapidly since the last review, and Morris (1980) provides a convenient tabulaton of all field tests where either ground or aerial applications of virus were used. Most attention has been focused on the neodiprionid sawflies because NPVs that attack these sawflies were found to be particularly virulent. Many sawflies also feed colonially, which facilitates cross infection, and some of the viruses spread from the point of infection, thus allowing spot introductions of virus to be used rather than complete coverage of the infected areas. Predacious and scavenging insects are thought to be the main disseminators of the viruses. In some cases strips of the infected area were sprayed, and in the case of *N. swainei* spot introductions of laboratory-infected pupae were used. Other viruses, again mainly NPVs, with good virulence against a number of Lepidoptera, have also been evaluated. Examples are tests against the tussock moths, *O. leucostigma*, and *O. pseudotsugata*. Some less virulent viruses were also applied aerially against, for example, *C. fumiferana* and *C. occidentalis*; others were applied by ground sprays, for example, those against *M. disstria* and *L. dispar*.

The remaining entomopathogens covered in this review are essentially at the laboratory stage of development. Fungi are well known to cause epizootics, such as the one believed to have helped collapse a recent outbreak of hemlock looper, *Lambdina fiscellaria fiscellaria* (Guenée), in Newfoundland (Otvos 1973). Other North American forest pests in which fungal epizootics have been recorded include spruce budworm *C. fumiferana*, blackheaded budworm, *Acleris variana* (Fern), and forest tent caterpillar, *M. disstria*. *Entomophthora* spp. of fungi are generally responsible and these are the fungi on which the Canadian Forestry Service concentrates its effort. Much of the biology of these fungi is now elucidated; before direct applications are attempted, efforts continue to devise techniques for mass production of resting spores, to devise practical application systems, and to induce spore germination, particularly under field conditions. One study, not reported elsewhere in this review, is an examination of control possibilities of mountain pine beetle, *Dendroctonus ponderosae* Hopkins, with *Beauveria bassiana* Vuillemin. This work can be summarized by stating that although the fungus is lethal to the insect in laboratory trials, no satisfactory way has yet been found to infect insects in the field (S. Whitney[2] 1981, personal communication).

Work with protozoa roughly parallels that with fungi and emphasis is now on epizootiological studies. Protozoa are dominant entomopathogens of insects such as spruce budworm, *C. fumiferana*, but their effects are to debilitate rather than to kill the insects, by affecting larval and pupal vigour and by reducing adult longevity and fecundity. Because these entomopathogens are ubiquitous, work continues to elucidate their interaction with other control agents.

[2] Pacific Forest Research Centre, Victoria, British Columbia.

Virtually no work has been carried out with nematodes during the review period, although recent laboratory studies have shown that *Heterorhabditis heliothidis* (Khan *et al.*) is lethal to *C. fumiferana* (G. Finney[3] 1981 personal communication). Work with rickettsia has been deliberately avoided because of potential human health problems.

Evaluation of Control Attempts

We have attempted to summarize the authors' evaluations of control success under the headings of predators and parasitoids, and of entomopathogens. As previous reviews in this series have been devoted largely to predators and parasitoids, we have also summarized these earlier evaluations of the major control attempts to show progress through the review periods. Simmonds' classification of control success has been used to provide continuity with the previous review (Simmonds 1969). Practical control using entomopathogens received relatively little attention until the current review period, except for work with NPVs against *G. hercyniae* and *N. lecontei*, hence few comparisons can be made with earlier work. Furthermore, Simmonds' classification of control success does not adequately describe the control strategy attempted with many entomopathogens and a separate rating system has thus been devised for entomopathogens that better reflects the behaviour of these control agents.

The summary of control attempts given in Tables 53 and 54 shows that two-thirds of the forest insect pests covered in this review were considered to be successfully

Table 53 Evaluation of biological control attempts against forest insect pests using predators and parasitoids

Pest	Degree of success*		
	1910–58	1958–68	1969–80
Adelges piceae	promising	–	–
Coleophora laricella	good	++++	++++
Coleophora serratella			–
Fenusa pusilla			+
Gilpinia hercyniae	good	++++**	++++**
Leucoma salicis	good	+++	+++
Lymantria dispar			–
Neodiprion sertifer		++	++
Operophtera brumata		++++	++++
Pristiphora erichsonii	promising	++	+++
Pristiphora geniculata			++
Rhyacionia buoliana		+	+++

* Based on McGugan & Coppel's assessment up to 1958, and on Reeks & Cameron's evaluation for 1959–68. Simmonds' (1969) classification of control success is used in the final two columns, viz.

–	No control.
+	Slight pest reduction or too early for evaluation of control.
++	Local control; distribution restricted or not fully investigated.
+++	Control widespread but local damage occurs.
++++	Control complete

** Acting in conjunction with a virus.

[3] Memorial University of Newfoundland, St. Johns, Newfoundland.

Table 54 Evaluation of biological control attempts against forest insect pests using entomopathogens*

Pest	Entomopathogens	
	Type	Degree of success**
Choristoneura fumiferana	NPV-CPV-GV-EV***	slight
	B.t.	good
Choristoneura occidentalis	NPV	none
	B.t.	none
Gilpinia hercyniae	NPV	excellent****
Lymantria dispar	NPV	slight
	B.t.	slight
Malacosoma disstria	NPV	none
	B.t.	good
Neodiprion abietis	NPV	unknown
Neodiprion lecontei	NPV	excellent
Neodiprion sertifer	NPV	excellent
Neodiprion swainei	NPV	good
Operophtera bruceata	NPV	unknown
Orgyia leucostigma	NPV	good
Orgyia pseudotsugata	NPV	good
	B.t.	slight

* Covering the period 1969–80, except for *N. lecontei* and *G. hercyniae* where work was undertaken before 1969.

** The rating system is as follows:

slight - decreased pest population but little damage prevented.
good - decreased pest population and damage reduced to economically acceptable levels
excellent - pest populations and damage reduced to almost zero.

*** NPV nuclear polyhedrosis virus
CPV cytoplasmic polyhedrosis virus
GV granulosis virus
EV entomopox virus
B.t. *Bacillus thuringiensis*

**** Acting in conjunction with parasitoids.

regulated by using biological control agents and that success was equally divided between parasitoids and entomopathogens. A number of evaluations cannot yet be made due to lack of data, although in some cases, as mentioned later, prospects for success are promising. Brief details of control evaluations with parasitoids and predators, and with entomopathogens are as follows.

Parasitoids and predators

Many of the following chapters report work continued from the previous review period. Most evaluations have been directed to technical success in control and little further can be added to the earlier tentative economic assessments of Reeks & Cameron (1971). *C.*

laricella, *G. hercyniae*, *L. salicis*, *O. brumata*, and *P. erichsonii* continue to be controlled by introduced parasitoids (Table 53), as does *R. buoliana*, provided temperatures favour parasitoids by synchronising their development with nectar and pollen supplies. These supplies are provided by encouraging the growth of flowering plants such as buckwheat (*Fagopyrum esculentum* Moench), wild carrot (*Daucus carota* L.), and milkweed (*Asclepias syriaca* L.). This effective and novel approach emphasizes the importance of examining several alternative tactics for inoculating and establishing the control agent in the pest population. The geographic range of control of some of the above insect pests does not yet cover the entire country and attempts are now being made to control, for example, *C. laricella* in western Canada by introducing a braconid, an ichneumonid and two eulophids similar to those that are successful in eastern Canada.

Several new subjects were chosen for control attempts. Small introductions of braconids and ichneumonids were made to control *C. serratella*, so far without success. However, prospects still seem good because these parasitoids are important regulators of populations in Europe. Limited introductions were made against *F. pusilla* and at least one of the parasitoids has become established, although it is too early to evaluate control potential. *P. geniculata* was chosen as a promising candidate for control because it is so heavily parasitized in its native Europe that once sufficient numbers of the ichneumonid were received from Europe good local control was obtained, underlining the importance of making adequate releases. Attempts against *C. fumiferana* are not evaluated here because they essentially did not progress beyond cage studies. Attempts against *N. swainei* are also not evaluated because they involved release of a general predator of many forest pests.

Entomopathogens

Entomopathogens have been used for most of our current biological control attempts against native pests, mainly by attempting to inundate the environment of the pest insect. In general, entomopathogens must be applied early in the insect's development if foliage is to be saved in the year of application. Early application is particularly important when viruses are used.

Remarkable progress has been made with *B.t.* during the review period. Improved strains of *B.t.* (Dulmage 1970) followed by improvements in formulation and application technology have now reached the point where many forest managers find *B.t.* can be as effective as chemical insecticides against *C. fumiferana* (e.g. Dorais *et al.* 1980); costs are three to five times higher. It is, of course, essential that the material be applied at the correct time, in adequate dosage, and with good application techniques that include suitable formulation of the spray ingredients. Many of these developments with *B.t.* sprays have not yet been adequately tested on other Lepidoptera but there are no *a priori* reasons why success should not be similar.

Recent success with viruses has been outstanding. Effective viruses were found against all the tested neodiprionid sawflies, with the possible exception of *N. abietis*, where sufficient data are not yet available to make an assessment. The NPV of *N. lecontei* has been particularly successful in spreading from the point of application (apparently via predacious and scavenging insects) and has allowed spot introductions to be used. Observations suggest that many applied viruses continue to exert control in the years following application, although it is not yet clear in many cases how this carry-over is achieved. The virus may infect offspring transovarially, for example, or the virus may simply contaminate the environment by residing in cadavers, twig crotches, and other protected locations. Further research is required to elucidate the specific mechanisms. Effective viruses were also found that could control certain Lepidoptera, notably the NPVs for *O. leucostigma* and *O. pseudotsugata*. Success against *C. fumiferana* and *L. dispar* has been only marginal because of the lower virulence of the viruses tested.

The estimated cost for virus material largely depends on production methods. Thus NPV from field-infected sawflies costs about $2.50/ha, whereas the tussock moth NPV produced *in vivo* in the laboratory costs about $50/ha.

There have been no control successes with either fungi or protozoa during the review period because, with the exception of trials to infect *D. ponderosae* with *B. bassiana* mentioned earlier, direct control has not been attempted. Basic questions in epizootiology remain to be answered first. Protozoa are, in any case, not generally envisaged as satisfactory control agents by themselves and emphasis is directed to control strategies in which they act synergistically with other agents.

Future Considerations

Like earlier reviews, the present reports cover many approaches to biological control. These case histories provide a valuable base on which to build future work, whether it be from the standpoint of assessing development with specific control agents or of examining control opportunities with specific insect pests. Each of these aspects will be considered separately in the following discussion.

Predators and parasitoids as control agents

There are two schools of thought on whether to emphasize introduced pests in inoculative control attempts with exotic parasitoids or predators. Although it is agreed that this approach has been fruitful for many introduced pests, some favour a continuation of this planning strategy, whereas others such as Munroe (1971) point out that this concentration of effort on introduced pests has no theoretical basis to support it. Indeed, Pimentel (1963) lists a number of successes against native pests. One interesting recent addition to such lists is *Oxydia trychiata* Gn., a native forest defoliator in Colombia that was controlled by *Telenomus alsophilae* Vier., an egg parasitoid of *Alsophila pometaria* (Harris) imported from Virginia in the United States. Not only does this example show control of a native pest by a parasitoid from a host of a different genus but it also demonstrates a successful transfer from temperate to tropical latitudes (Bustillo & Drooz 1977). The example serves to underline the earlier dictum of Huffaker *et al.* (1971) that "native as well as exotic pests are suitable subjects for biological control importations". Indeed we hope that inoculative introductions of exotic parasitoids and predators against native forest pests will be given more attention than in the past.

A further point that perhaps needs to be reconsidered is the method by which parasitoid species are selected. Disproportionate attention has been devoted to the Hymenoptera during the review period, yet among other orders the Diptera, in particular, contain numerous species of successful parasitoids that deserve closer examination. Diptera can be more difficult to rear and the adults are sometimes considered more difficult to identify than Hymenoptera, but these factors should not prejudice the choice of candidate parasitoids unless all other comparative factors are judged equal. Munroe (1971) pointed out that during the previous review period the establishment rate for introduced Diptera was actually higher than that for introduced Hymenoptera although it is, of course, recognized that control does not always follow establishment.

In terms of the number of species released we concur with the views of Reeks & Cameron (1971) that either single or multiple species' releases are acceptable when supporting work shows that reasons are sound. Case histories suggest even where competitive displacement occurs, the resulting control may well be improved (e.g. Caltagirone 1981). Precautions should, of course, be taken to screen out undesirable parasitoids, especially hyperparasitoids and cleptoparasitoids (e.g. Schroeder 1974). Desirable parasitoids should have good numerical and/or functional response to host population changes through attributes such as high fecundity, efficient host searching ability, and good synchronization of reproduction with the host (Pschorn-Walcher 1977). Inbreeding that restricts genetic variability should be limited. It is always advisable to try to ensure genetic variation between released individuals by collecting breeding stock from several locations (Bennett 1974, Mackauer 1976).

Selection of collecting locations for exotic parasitoids and predators has rightly been focused on the native habitat of the pest when introduced insects are the subject of control attempts and this has inevitably led to a concentration of collection efforts in Europe. Case histories, however, show that successful control agents have also been found on different hosts and in different parts of the world, and to allow for this possibility, and thus increase the probability of success, a proportion of effort should be devoted to collecting from areas other than the native habitat of the pest (Pschorn-Walcher 1977). This approach is, of course, essential where native pests are the subject of control attempts by inoculative releases.

Although the above remarks are aimed principally at so-called classical biological control, they are not intended to preclude attention to other release strategies. Augmentative or inundative releases of parasitoids or predators should perhaps be considered in some cases. These techniques remain essentially untested in Canadian forestry applications despite successes claimed in Russia and China (Tropin *et al.* 1980, McFadden *et al.* 1981). For example, in Russia release of *Trichogramma evanescens* Westw. against eggs of *R. buoliana* and *Petrona resinella* (L.) reduced the number of damaged pine buds by over 50% (Beglyarov & Smetnik 1977); and in China *Trichogramma dendrolimi* Mats. reduced infestation by *Dendrolimus sibericus* Tschetw. from 63% of trees in 1956 to 1% of trees in 1970 (Hussey 1977). Attention could also be given to the use of behavioural chemicals such as kairomones (e.g. Gross Jr. 1981, Weseloh 1981, Vinson 1977); these semiochemicals should improve control by augmenting the ability of natural enemies to locate hosts.

Finally, two fundamental research needs in predator and parasitoid introductions have been raised repeatedly in this overview. The first need covers rearing techniques to ensure that adequate numbers of biological control organisms (mated females) are available for whatever release strategy is employed. The second fundamental need is for better knowledge of the population dynamics of the pest insect and of the candidate predators or parasitoids. Many feel that neglecting this aspect in favour of an *ad hoc* approach inevitably reduces chances of obtaining successful control. It is of course recognized, as Pschorn-Walcher (1977) states, that "predictions derived from multivariate analysis or population models are no guarantee that the biological control agents selected will be effective." Judgement must be exercised in deciding how much background information is desirable before releases are attempted. As Simmonds (1972) points out, "The introduction . . . is the crucial experiment. Promising biological control agents have failed to live up to expectations whereas apparently unlikely species have been very successful".

Entomopathogens as control agents

Bacteria and viruses have dominated control successes with entomopathogens and will probably continue to do so in the immediate future as improved methods are found for producing and applying these organisms. With *B.t.*, for example, efficacy against *C. fumiferana* has been widely demonstrated and future research emphasis will be on reducing costs. This can be approached in a number of ways. Development of more concentrated formulations to reduce the volume of carrier liquid will reduce both shipping and application costs. Development of even more potent strains of *B.t.* will lead to similar reductions in costs. Finally, development of better application methods will increase the probability that larvae will ingest a lethal dose from a given quantity of applied *B.t.* by optimising the distribution of *B.t.* on the leaf surface. These methods entail all aspects of application technology: generating optimum-sized droplets that contain the optimum amount of *B.t.*; reducing evaporation of the carrier liquid; reducing unwanted drift of the spray cloud; and ensuring that the material is not quickly inactivated on the leaf surface by ultra-violet light, phytotoxins, or other processes.

The above spectrum of research needs is not confined to *B.t.* It can also be applied to several of the viruses whose efficacy is already well established (Table 54). Production costs in particular need attention; many viruses can only be propagated *in vivo*, often by growing them in small insects that are handled manually. Automation of handling, utilizaton of larger host insects, and *in vitro* culturing are all potential ways to reduce costs. Many sawfly viruses can be produced inexpensively by infecting natural populations in the field and collecting the cadavers. The virus is usually virulent but quality control is a problem that may need addressing if widespread use of the virus is planned.

Viruses, in common with other entomopathogens, must be federally registered as safe and effective insecticides before widespread operational use is permitted. This means that besides efficacy data, extensive data are required on the environmental safety of the formulations. Many of these latter data have already been collected but the scope of the data required remains in doubt pending establishment of firm testing protocols for viral insecticides by the federal registration authorities. Indeed, this clarification of data requirements for registration remains an urgent need for all entomopathogens other than *B.t.* Meanwhile, the first petition for Canadian registration of a viral insecticide will be submitted in March 1981 for an NPV against *N. lecontei* and petitions will follow for NPVs of *Orgyia pseudotsugata, L. dispar,* and *N. sertifer*. The viruses of *O. pseudotsugata* and *L. dispar* are already registered in the United States and the NPV for *N. sertifer* is now before their registration authorities. Not all the viruses listed in Table 54 are at the stage of development where registration is being contemplated. Where efficacy is only marginal, the most important need is to obtain strains with improved virulence, either from even more widespread searches for candidate strains, or perhaps eventually by genetic manipulation of viruses already in culture (see later).

Control using protozoa or fungi poses a variety of research needs. Now that many of the common organisms found in pest populations have been characterized, the immediate need is for more epizootiological studies to elucidate the natural processes of infection. Studies on production methods will, of course, eventually be needed to obtain an efficacious product at reasonable cost, and much of the research on application technology described earlier will also need focusing on these entomopathogens.

Nematodes are a group of entomopathogens that have been neglected in disease studies of Canadian forest insect pests. These potential control agents deserve closer attention, first in laboratory studies of their pathogenicity and later in field applications. Two species have already been produced commercially in the United States for use against *Anopheles* spp. of mosquitoes (Poinar 1979).

One exciting research area that has potential application to many entomopathogens is genetic manipulation to improve virulence. In the case of *B.t.*, attempts are imminent to clone the plasmid thought to control endotoxin production from *B.t.* into *Escherichia coli* (Migula) Castellani & Chalmers, and to develop monoclonal antibodies for more precise measurement of toxicity. It is difficult to judge the practical importance of these techniques for pest control during the next few years because, although advances in genetic engineering have been rapid, much basic work remains to be carried out before field work can be contemplated. A watching brief on advances seems the most prudent approach at the time of writing.

An important aspect of the work that should not be overlooked is mass production of the control agent. In all work with entomopathogens, and some of the treatment strategies used for predators and parasitoids, ultimate operational use of the control agent depends on being able to produce adequate quantities of material economically. Biological control agents have a narrower spectrum of target pests than chemical insecticides and in some cases they are self-replicating. For these and other reasons, production of biological control agents is seen as a less attractive commercial proposition, and it is thus important that government organizations maintain an interest in providing opportunities for large-scale production of biological control agents.

Economic aspects of various biological control strategies have been briefly mentioned in this chapter. These aspects need addressing in more detail to supplement the few cost estimates developed here by biologists. Analysis by economists could provide a more comprehensive picture of costs and benefits: cost figures should separate production from application; and benefit figures should include carry-over effects following the year of application.

Target pests for control

Judging by progress reported in this and other reviews of biological control there seems little reason to rule out any forest pest on *a priori* grounds as a suitable candidate for biological control. Each decision to attempt control must, of course, take into account the specific situation of pest and host tree(s) and the range of control options that might be applied. However, in considering biological control there seems good reason to give more attention to research opportunities with the key economic pests in Canada's forests, which would include *C. fumiferana* in eastern Canada and *D. ponderosae* and *D. rufipennis* (Kby.) in western Canada

In the case of *C. fumiferana*, a good short-term option using biological control is now available commercially. Longer term control is desirable but biological control attempts have met with limited success. There is, of course, scope for further work and, although all attempts will be difficult, we suggest that searches for more virulent viruses and for exotic parasitoids be given special emphasis. Candidate viruses to date exhibit marginal virulence against *C. fumiferana* but some control has nevertheless been obtained and further work is warranted. Parasitoid introductions as a biological control option have been given little more than a cursory examination, perhaps because *C. fumiferana* is a native pest. Many of the releases were conducted in unsuitable conditions and a more systematic study of this control option would allow a more factual comparison of its potential with other control options. The most common candidate hosts in Europe that have been suggested for parasitoid collections, based on the similarity of their life cycles to *C. fumiferana*, are *C. murinana* Hb. and *Epinotia nigricana* L. on *Abies* spp., *Dichelia histrionana* Froel. on *Picea* spp., and *Archips oporana* L. on *Larix* spp., *Pinus* spp., and *Picea* spp. Several *Zeiraphera* spp. offer more remote possibilities. Collection activities should not, however, be confined to Europe. China, for example, has a *Choristoneura* sp. that attacks *Picea* spp. and appears to be well controlled by natural factors (Macdonald & Pollard 1976).

With *Dendroctonus* spp. of bark beetles, particularly *D. ponderosae* and *D. rufipennis* that are major pests in western Canada perhaps the whole range of biological control agents needs closer examination to see which, if any, have practical potential for control. Certainly a wealth of information exists on related scolytids such as the southern pine bark beetle, *D. frontalis* Zimmerman (Thatcher *et al.* 1978), some of which could provide a useful starting point in considering options with other *Dendroctonus* spp. More field examinations of the natural enemy complex of *Dendroctonus* spp. in Canada is obviously required. Multiphagous parasitoids of bark beetles that attack hardwoods should not be overlooked, as some also parasitize bark beetles that attack softwoods.

Other particularly destructive pests that are not included in this review are the hemlock looper, *L. fiscellaria fiscellaria*, and the western hemlock looper, *L. fiscellaria lugubrosa* (Hulst); populations are now at endemic levels and irruptions are likely within the next few years. For the present, however, the most logical approach is to give priority to biological control research, mainly in accordance with the current importance of the pest, as the scope of research opportunities is limited not by lack of suitable problems but by the resources we are able to devote to biological control. There seems little doubt that interest in this method of pest control is growing and more resources are likely to be committed in the future. This increased effort should in turn lead to progress that better reflects the potential of biological control as a viable pest management option.

Literature Cited

Beglyarov, G.A.; Smetnik, A.J. (1977) Seasonal colonization of entomophages in the USSR. In: Ridgway, R.L.; Vinson, S.B. *(Eds.)* Biological control by augmentation of natural enemies. New York; Plenum, pp. 283–328.

Beirne, B.P. (1975) Biological control attempts by introductions against pest insects in the field in Canada. *Canadian Entomologist* 107, 225–236.

Bennett, F.D. (1974) Criteria for determination of candidate hosts and for selection of biotic agents. In: Maxwell, F.G.; Harris, F.A. *(Eds.)* Proceedings of a Summer Institute on Biological Control of Plant Insects and Diseases. Jackson, Mississippi; University Press of Mississippi, pp. 87–96.

Bustillo, A.E.; Drooz, A.T. (1977) Comparative establishment of a Virginia (USA) strain of *Telenomus alsophilae* on *Oxydia trychiata* in Columbia. *Journal of Economic Entomology* 70, 767–770.

Caltagirone, L.E. (1981) Landmark examples of classical biological control. *Annual Review of Entomology* 26, 213–232.

DeBach, P.; Huffaker, C.B.; MacPhee, A.W. (1976) Evaluation of the impact of natural enemies. In: Huffaker, C.B.; Messenger, P.S. (*Eds.*) Theory and practice of biological control. London, New York; Academic Press, pp. 255–285.

Dorais, L.; Pelletier, M.; Smirnoff, W.A. (1980) Pulvérisations aériennes de *Bacillus thuringiensis* Berliner realisées au Québec de 1971 á 1979 contre la tordeuse des bourgeons de l'épinette, *Choristoneura fumiferana* (Clem.). [Abstract] Rapport présenté au VIe Congrès International de l'Aviation Agricole à Turin, Italie 22-26 Sept. 1980, unpaginated.

Dulmage, H.T. (1970) Insectidical activity of HD-1, a new isolate of *Bacillus thuringiensis* var. *alesti*. *Journal of Invertebrate Pathology* 15, 232–239.

Gross, H.R. Jr. (1981) Employment of kairomones in the management of parasitoids. In: Norlund, D.A.; Jones, R.L.; Lewis, W.J. (*Eds.*) Semiochemicals and their role in pest control. New York; John Wiley, pp. 137–152.

Huffaker, C.B.; Messenger, P.S.; DeBach, P. (1971) The natural enemy component in natural control and the theory of biological control. In: Huffaker, C.B. (*Ed.*) Biological control. New York, London; Plenum, pp. 253–293

Hussey, N.W. (1977) Biological control techniques. In: Rishbeth, R.S. (*leader*). The Royal Society delegation on biological control to China, Aug. 15 – Sept. 3, London; Royal Society, 38 pp.

Knipling, E.F. (1979) The basic principles of insect population, suppression, and management. *US Department of Agriculture Handbook* 512, 659 pp.

Macdonald, D.R.; Pollard, D.F.W. (1976) Report on the scientific exchange in forestry with the People's Republic of China. *Canadian Forestry Service Report Pacific Forestry Research Centre*, 32 pp.

McFadden, M.W.; Dahlsten, D.L.; Berisford, C.W.; Knight, F.B.; Metterhouse, M.W. (1981) Integrated pest management in China's forests. *Journal of Forestry* 79, 723–726.

McGugan, B.M.; Coppel, H.C. (1962) Biological control of forest insects 1910–1958. In: A review of the biological control attempts against insects and weeds in Canada. *Commonwealth Institute of Biological Control Technical Communication* 2, 35–127.

Mackauer, M. (1976) Genetic problems in the production of biological control agents. *Annual Review of Entomology* 21, 369–385.

Morris, O.N. (1980) Entomopathogenic viruses: strategies for use in for use in forest insect pest management. *Canadian Entomologist* 112, 573–584.

Munroe, E.G. (1971) Status and potential of biological control in Canada. In: Biological control programme against insects and weeds in Canada. *Commonwealth Institute of Biological Control Technical Communication* 4, 213–255.

Otvos, I. (1973) Biological control agents and their role in the population fluctuation of the eastern hemlock looper in Newfoundland.*Canadian Forestry Service Information Report* N-X-102, 34 pp.

Pimentel, D. (1963) Introducing parasites and predators to control native pests. *Canadian Entomologist* 95, 785–792.

Poinar, G.O. Jr. (1979) Nematodes for biological control of insects. Boca Raton, Florida; Chemical Rubber Company, 220 pp.

Pschorn-Walcher, H. (1977) Biological control of forest insects. *Annual Review of Entomology* 22, 1–22.

Reeks, W.A.; Cameron, J.M. (1971) Current approaches to biological control of forest insects 1959–1968. *In*: Biological control programmes against insects and weeds in Canada. *Commonwealth Institute of Biological Control Technical Communication* 4, 105–112.

Schroeder, D. (1974) A study of the interaction between internal larval parasites of *Rhyacionia buoliana*. *Entomophaga* 19, 145–171.

Simmonds, F.J. (1969) Brief resumé of activities and recent successes achieved. *Commonwealth Institute of Biological Control Report*, Trinidad, 16 pp.

Simmonds, F.J. (1972) Approaches to biological control problems. *Entomophaga* 17, 251–264.

Thatcher, R.C.; Searcy, J.L.; Coster, J.E.; Hertel, G.D. (1978) The southern pine beetle. *US Department of Agriculture Technical Bulletin* 1631, 266 pp.

Tropin, I.V.; Vedernikov, N.M.; Kranguaz, R.A.; Maslov, A.D.; Zubov, P.A.; Khramtsov, N.N.; Andreeva, G.I.;Lyashenko, L.I. (1980) A manual on the protection of the forest against pests and diseases. Moscow; Leanaya promyshelnnost, 500 pp.

Vinson, S.B. (1977) Behavioural chemicals in the augmentation of natural enemies. In: Ridgway, R.L.; Vinson, S.B. (*Eds.*) Biological control by augmentation of natural enemies. New York; Plenum, pp. 237–265.

Weseloh, R.M. (1981) Host location by parasitoids. In: Nordlund, D.A.; Jones, R.L.; Lewis, W.J. (*Eds.*) Semiochemicals and their role in pest control. New York; John Wiley, pp. 79–96.

Adelges piceae (Ratz.), Balsam Woolly Adelgid (Homoptera: Adelgidae)

H.O. SCHOOLEY, J.W.E. HARRIS and B. PENDREL

Pest Status

The balsam woolly adelgid, *Adelges piceae* (Ratz.), is a destructive pest of firs, *Abies* spp. In recent years, however, populations have been low and dispersal has been limited. The distribution of the adelgid remains about the same as in 1968 (Clark *et al.* 1971). In eastern Canada, expansion of the infestation has been largely into those small areas that had been bypassed as the initial infestation developed (Schooley 1980). The adelgid is now present throughout most of the fir forests of the Maritime Provinces, Newfoundland, and the eastern tip of the Gaspé Peninsula. A total area of about 103 600 km^2 is affected. In western Canada there have been only minor increases in the size of the infestation, which now covers about 10 360 km^2 in southwestern British Columbia. The infestations in Canada extend southward into the United States on both the east and west coasts (Mitchell *et al.* 1970). Climate seems to be the main factor limiting further spread of this insect in Canada (Greenbank 1970).

The balsam woolly adelgid attacks all species of firs in Canada but some are more resistant to injury than others. In eastern Canada, mortality of balsam fir, *Abies balsamea* (L.) Mill., has been common. In western Canada, subalpine fir, *A. lasiocarpa* (Hook.) Nutt., growing at low elevations, is the most vulnerable species, but it rarely occurs within the present infestation boundaries (Harris 1968). Pacific silver or amabilis fir, *A. amabilis* (Dougl.) Forb., is the second most vulnerable western host and considerable mortality of this species has occurred. Grand fir, *A. grandis* (Dougl.) Lindl., is also deformed and killed by the adelgid, primarily on southern Vancouver Island. The adelgid is also threatening to destroy Fraser fir, *A. fraseri* (Pursh.) Poir., in the few remaining spruce–fir stands of the southern Appalachian Mountains of eastern United States (Johnson 1980).

Some of the causes of variation in the response to adelgid infestation among *Abies* host species have been identified. Puritch (1973), for example, has shown that the onset and intensity of adelgid attack caused premature heartwood formation and a resulting abnormal water stress. Thus, ability to withstand this stress among fir species corresponds to their susceptibility to adelgid damage. It has also been observed that adelgid attack has little effect on European species because a layer of dead bark tissue develops during the wound healing process and protects the trees from further attack for several years (Oechssler 1962). This healing process is delayed or absent in balsam fir that is under continuous adelgid attack.

Background and Evaluation of Control Attempts

Clark *et al.* (1971) stated that there was little scope for additional study in the control of this adelgid by introduced predators. Field trials in eastern Canada tested all the apparently suitable insects over a 35-year period and failed to find a predator or predator complex that would control the adelgid. Only eight species of introduced predators have become established, at least temporarily, but none has been effective (Table 55) in reducing damage to economically tolerable levels. Consequently, it was recommended that the importation and release of predators be discontinued. This recommendation has been accepted and no introductions or releases have been made since 1969. The search for surviving introduced predators was discontinued in 1969, 1971, and 1978 in both New Brunswick and Nova Scotia (D.O. Greenbank, personal communication), in Newfoundland (Bryant 1971), and in British Columbia (Harris & Dawson 1979). The history of predator releases and recoveries tabulated by Clark *et al.* (1971) has been updated

Table 55 Open releases and recoveries of predators against *Adelges piceae* (Ratz.)

Species and province	Year	Origin	Number	Year of Recovery
Adalia luteopicta Mulsant				
Newfoundland	1960	India	159	
Adalia ronina (Lewis)				
Newfoundland	1961	Japan	67	
Nova Scotia	1963	Japan	290	
New Brunswick	1960	Japan	25	1960
	1962	Japan	37	1962
	1963	Japan	585	1963
Adalia tetraspilota (Hope)				
Newfoundland	1960	India	33	
Aphidecta obliterata (L.)				
Newfoundland	1959	Czechoslovakia	735	
	1960	Germany	934	
	1962	Czechoslovakia	848	
	1963	Germany	989	
	1964	Germany	1 187	
	1965	Czechoslovakia	185	
		Germany	267	
	1966	Czechoslovakia	509	
	1966	Austria	1 380	1966
	1967	Austria	4 288	1967
	1968	Germany and Austria	1 096	1968
Nova Scotia	1964	Germany	1 773	1967
	1966	Austria	370	
		Norway	47	
	1967	Norway	422	
New Brunswick	1962	Germany	1 827	1962
	1963	Germany	3 157	1963
British Columbia	1960	Germany	1 050	
	1961	Germany	1 141	
	1962	Germany	796	
	1963	Germany	1 997	
	1965	Czechoslovakia	660	
	1968	Germany	1 069	
	1969	Germany	420	
Aphidoletes thompsoni (Möhn)				
Newfoundland	1959	Czechoslovakia	3 124	
		Germany	25 952	
	1962	Germany	270	1961
	1963	Germany	5 201	

Table 55 continued

Species and province	Year	Origin	Number	Year of recovery
	1965	Germany	826	1967
	1966	Germany	35 119	
	1968	Germany	7 248	
Nova Scotia	1965	Germany	450	1965
	1966	Germany	37 110	1966–67
New Brunswick	1959	Germany	36 131	1959–61
	1965	Germany	551	1965
	1966	Germany	6 300	1966
British Columbia	1962	Germany	280	
	1963	Germany	516	
	1965	Germany	1 080	1967
	1966	Germany	6 845	
Balaustium sp.				
Québec	1967	Pakistan	126	
Ballia eucharis Mulsant				
Newfoundland	1960	India	32	
New Brunswick	1959	Pakistan	79	
	1960	India	55	
Coccinella septempunctata L.				
New Brunswick	1959	India	28	
	1960	India	22	
Cremifania nigrocellulata Czerney				
Newfoundland	1959	Germany	198	1971
	1961	Germany	17	
Nova Scotia	1966	Germany	169	1966
New Brunswick	1963	Germany	48	1964–65
	1966	Germany	17	1966–68
British Columbia	1966	Germany	137	
	1968	Germany	706	1969
Exochomus lituratus Gorham				
Newfoundland	1960	Pakistan	110	
Nova Scotia	1963	Pakistan	991	
New Brunswick	1963	Pakistan	209	
Exochomus uropygialis Mulsant				
Newfoundland	1960	India	226	
	1960	Pakistan	2 839	
Nova Scotia	1963	Pakistan	8 782	
New Brunswick	1959	Pakistan	2 550	
	1963	Pakistan	247	
	1964	Pakistan	238	
Harmonia breiti Mader				
Newfoundland	1960	Pakistan	88	
New Brunswick	1959	India	85	
Laricobius erichsonii Rosenhauer				
Newfoundland	1959	Czechoslovakia	1 043	1959
		Germany	2 159	1960

Table 55 continued

Species and province	Year	Origin	Number	Year of recovery
	1960	Germany	8 228	1961
	1961	Germany	1 118	1963
	1962	Germany	7 940	1964
	1963	Germany	1 869	1965
		Newfoundland via Europe	1 984	1966
	1964	Germany	1 819	1968
		Newfoundland via Europe	300	
	1965	Germany	152	
Nova Scotia	1966	Germany	2 575	1967
New Brunswick	1960	Germany	14 266	1958–68
	1966	Germany	3 497	1958–68
British Columbia	1960	Germany	800	1962–65
	1961	Germany	1 432	1978
	1963	Germany	4 871	1966,74
	1965	Germany	612	
	1968	Germany	3 164	1969,71,74
Leucopis (Leucopis) n. sp. nr. *'melanopus'* Tanasijtshuk*				
Newfoundland	1959	Germany	160	1959–62
	1968	Germany	1 040	
New Brunswick	1960	Germany	166	1958–68
	1966	Germany	241	
British Columbia	1968	Germany	2 273	
Leucopis (Neoleucopis) obscura Holiday				
Newfoundland	1965	Austria	24	1959–68
Leucopis (Neoleucopis) atralula Ratz.				
New Brunswick	1965	Germany	385	
Pullus impexus (Mulsant)				
Newfoundland	1959	Germany	9 500	
	1960	Germany	1 146	1960
	1961	Germany	131	1961
	1966	Germany	18 036	
Nova Scotia	1963	Germany	1 000	1965–67
	1964	Germany	2 031	1965–67
	1966	Germany	4 700	1965–67
New Brunswick	1962	Germany	685	1958–59
	1963	Germany	397	1958–59
	1964	Germany	200	1958–59
	1966	Germany	26 550	1958–59
British Columbia	1960	Germany	1 240	1961
	1963	Germany	1 400	
	1965	Germany	2 417	
	1966	Germany	18 513	
	1968	Germany	2 079	1969,71,74,78

Table 55 continued

Species and province	Year	Origin	Number	Year of recovery
Scymnus pumilio (Weise)				
Newfoundland	1960	Australia	9 687	
New Brunswick	1959	Australia	5 590	
	1960	Australia	7 286	
British Columbia	1960	Australia	2 930	
Tetraphleps abdulghani Ghauri				
Nova Scotia	1965	Pakistan	949	
New Brunswick	1962	Pakistan	1 972	
	1963	Pakistan	1 157	
	1964	Pakistan	45	
	1965	India	278	
		Pakistan	2 921	
British Columbia	1965	India	19	
		Pakistan	1 257	
Tetraphleps raoi Ghauri				
Nova Scotia	1965	India	59	
New Brunswick	1965	India	59	

* Species 'M'.

Attempts to control the adelgid with contact or systemic insecticides have been unsuccessful. Several insecticides tested proved suitable for ornamental and nursery applications where it was possible to saturate infested trees with chemical solutions. However, not one of the 51 insecticides used in laboratory and greenhouse experiments caused acceptable levels of adelgid mortality under field conditions (Nigam 1972, 1976). Recently, some success has been obtained in controlling adelgids on individual trees with insecticidal soaps (Puritch 1975). These chemicals are expected to be useful only for ornamental and nursery applications.

There are no known insect parasites of the balsam woolly adelgid but several fungal diseases have been reported, including *Fusarium larvarum* Forbel and *Cephalosporium coccorum* Petch found in the Gaspé region of Québec (Smirnoff 1970), *Cephalosporium* sp. and *Penicillium* sp. found in British Columbia (Harris *et al.* 1966), and *F. nivale* (Fries) Cesati found on adelgids in North Carolina, USA (Fedde 1971). Greenhouse and field experiments with the diseases from Québec have been unsuccessful. The potential of organisms from other locations as natural control agents remains unknown.

Failure to control the balsam woolly adelgid is viewed with special concern in Atlantic Canada. In Nova Scotia and Prince Edward Island, balsam fir stands are usually damaged before reaching a marketable size. On the island of Newfoundland, spectacular and unprecedented mortality of balsam fir has occurred on an accelerated scale in stands where adelgid attack has been followed by severe spruce budworm defoliation (Schooley 1981). In affected areas, stand conversion, i.e. the replacement of balsam fir with other tree species, is being conducted on a large scale. Black spruce, *Picea mariana* (Mill.) B.S.P., is the favoured alternate species but white spruce, *P. glauca* (Moench) Voss, eastern larch, *Larix laricina* (Du Roi) K. Koch, and hybrid larch species are being considered.

Recommendations

As previously recommended by Clark *et al.* (1971) no further introduction of predators should be made into eastern Canada until new species that might effectively control the adelgids are identified. Suitable predators may be found attacking two closely related fir shoot adelgids, *Adelges merkeris* Eichhorn and *Adelges nusslini* C.B., in Europe. The milder winter temperatures of coastal British Columbia could favour a species that failed in the east. Also, some species already tested on the west coast may have failed because too few individuals were released or because of unfavourable weather conditions at the time of release. Therefore, further introductions may be considered for western Canada. Any predator successfully introduced and proved to be effective in western Canada should be considered for testing in the mild coastal areas of eastern Canada.

Population dynamics studies should be conducted regularly in all regions where the adelgid is present. These studies should include observations that will identify any factors that successfully control mortality. Financial support should also be made available for periodic reappraisal of the effect that natural insect and disease organisms have in controlling the balsam woolly adelgid in other countries.

Literature Cited

Bryant, D.G. (1971) Forest protection: balsam woolly aphid, *Adelges piceae* (Ratz.). *Annual Report Newfoundland Forest Protection Association 1971*, 43 pp.

Clark, R.C.; Greenbank, D.O.; Bryant, D.G.; Harris, J.W.E. (1971) *Adelges piceae* (Ratz.), balsam woolly aphid (Homoptera: Adelgidae). In: Biological control programmes against insects and weeds in Canada. *Commonwealth Institute of Biological Control Technical Communication* 4, 113–127.

Fedde, G.F. (1971) A parasitic fungus disease of *Adelges piceae* (Homoptera: Phylloxeridae) in North Carolina. *Annals of the Entomological Society of America* 64, 749–750

Greenbank, D.O. (1970) Climate and the ecology of the balsam woolly aphid. *Canadian Entomologist* 102, 546–578.

Harris, J.W.E. (1968) Balsam woolly aphid in British Columbia. *Canadian Department of Forestry and Rural Development Forestry Research Forest Pest Leaflet* B.C.-1, 5 pp.

Harris, J.W.E.; Dawson, A.F. (1979) Predator release program for balsam woolly aphid, *Adelges piceae* (Homoptera: Adelgidae), in British Columbia, 1960–1969. *Journal of the Entomological Society of British Columbia* 76, 21–26.

Harris, J.W.E.; Allen, S.J.; Collis, D.G.; Harvey, E.G. (1966) Studies of the balsam woolly aphid, *Adelges piceae* (Ratz.), in British Columbia. *Canadian Department of Forestry Information* Report BC-X-5, 5 pp.

Johnson, K. (1980) Fraser fir and balsam woolly aphid. Summary of information. *Southern Appalachian Reserve Resources Management Committee Report*, 62 pp.

Mitchell, R.G.; Amman, D.G.; Waters, W.E. (1970) Balsam woolly aphid. *US Department of Agriculture Forestry Service, Forest Pest Leaflet* 118, 10 pp.

Nigam, P.C. (1972) Summary of toxicity of insecticides and chemical control studies against balsam woolly aphid. *Canadian Forestry Service Information Report* CC-X-26, 7 pp.

Nigam, P.C. (1976) Summary of chemical control studies against balsam woolly aphid in British Columbia, 1973–74.*Canadian Forestry Service Information Report* CC-X-133, 6 pp.

Oechssler, G. (1962) Sucking injuries to the tissue of native and exotic firs caused by central European fir aphids. *Zeitschrift für angewandte Entomologie* 50, 418–459 (*Canadian Department of Forestry Translation* 57).

Puritch, G.S. (1973) The effects of water stress on photosynthesis, respiration and transpiration of four *Abies* species. *Canadian Journal of Forest Research* 3, 293—298.

Puritch, G.S. (1975) The toxic effects of fatty acids and their salts on the balsam woolly aphid, *Adelges piceae* (Ratz.). *Canadian Journal of Forest Research* 5, 515–522.

Schooley, H.O. (1980) The introduction, spread and occasional resurgence of the balsam woolly aphid in Newfoundland. *Proceedings of the International Union Forest Research Organizations Conference on Dispersal of Forest Insects, August 27–31, 1979*. Washington State University, Pullman, Washington, pp. 116–127.

Schooley, H.O. (1981) An evaluation of the hazard rating system for balsam woolly aphid damage in Newfoundland. *Proceedings of a Symposium on Hazard Rating Systems in Forest Insect Pest Management, July 31 – August 1, 1980*, Athens, Georgia, *US Department of Agriculture Forestry Service General Technical Report* WO–27.

Smirnoff, W.A. (1970) Fungus diseases affecting *Adelges piceae* in the fir forest of the Gaspé Peninsula, Québec. *Canadian Entomologist* 102, 799–805.

Choristoneura fumiferana (Clemens), Spruce Budworm (Lepidoptera: Tortricidae)

C.A. MILLER

Pest Status

The spruce budworm, *Choristoneura fumiferana* (Clem.), is a native defoliator of the spruce, *Picea* spp., and balsam fir, *Abies balsamea* (L.) Mill., forests of North America, which extend from northern Alberta to the Atlantic seaboard. It is the most important economic pest in forests of eastern Canada because of the frequency of epidemics and the intensity and extent of fibre losses. Budworm populations periodically irrupt to epidemic proportions and cause extensive tree mortality and reduced growth in surviving trees. There have been three major epidemics in this century. The first occurred around 1915 and covered most of the eastern Québec – Atlantic Region; the second irrupted in Ontario – western Québec in the late 1930s and appeared in the Atlantic Region in the late 1940s, but had little impact in Nova Scotia and Newfoundland; and the current outbreak irrupted in the late 1960s. In sequence, these three outbreaks of the 20th century have tended to increase in size and frequency.

The review period of this publication, 1969–80 largely coincides with the third major epidemic. The prolonged second epidemic collapsed in the 1960s, except for a persistent infestation covering less than a million hectares in central New Brunswick. During the 4 years 1967–70, spruce budworm population increases were recorded in all provinces from Ontario to Newfoundland, and by 1975 the epidemic reached its peak, covering about 54 x 10^6 ha, or about half of the land area in eastern Canada classified as productive forest. Much of the remainder is not occupied by spruce and fir.

In general, the intensity of the infestation, as measured by egg-mass densities across eastern Canada, declined in the latter part of the 1970s but the density of feeding larvae relative to the decreased quality of foliage on previously attacked trees remained high enough to cause severe damage. Thus in 1980 the epidemic still caused moderate to severe defoliation over 36 x 10^6 ha and it is expected that extensive defoliation will continue into the mid-1980s.

Maturing balsam fir trees in unprotected forests usually die after 3 or 4 successive years of severe spruce budworm attack. Spruce trees, particularly black spruce, *Picea mariana* (Mill.) B.S.P., are more tolerant of defoliation and may withstand up to 6 years of attack. In the current outbreak, tree mortality was first recorded in northeastern Ontario and western Québec in 1972, about 5 years after the beginning of the epidemic. In Nova Scotia (Cape Breton Highlands) and Newfoundland, mortality was first recorded between 1975 and 1976. Throughout the whole region, the area of vulnerable forest containing dying and dead trees increased each year and by 1980 was estimated at about 19 x 10^6 ha (Table 56). Although 'zones of heavy defoliation' and 'zones of tree mortality' were delineated, it has proved difficult to translate these spatial data to a volumetric estimate of fibre loss. Inventory data on spruce and fir before the outbreak are meagre; moreover, the rapid detection of dying stands of trees by remote sensing is not yet operational. However, some estimates of fibre loss in mortality zones are available (Table 56). Generally, in eastern Canada in the late 1970s there was (1) continuing loss of radial and height growth in severely defoliated stands; (2) tree mortality ranging from 10 to 80% at the stand level in a mortality zone of about 19 x 10^6 ha; (3) a loss in fibre quality and volume due to secondary pest attack in dead trees that were salvaged; and (4) evidence that the future marketable volume of young stands that recover from the epidemic may be significantly lower than the site potential (Baskerville & MacLean 1979).

Table 56 Estimates of tree mortality in eastern Canada during the spruce budworm, *Choristoneura fumiferana* (Clem.), outbreak of the 1970s*

Province	Date of first tree mortality	Dead and dying trees, 1980: Area (x 10^6 ha)	Dead and dying trees, 1980: Volume (x 10^6 m^3)
Ontario	1972	8.4	?
Québec	1972	9.3	?
New Brunswick	– **	0.7	30
Nova Scotia	1976		16
Newfoundland	1975	0.4	40

* Data extracted from: Forest Insect and Disease Conditions in Canada 1980. Canadian Forestry Service, Ottawa, 1981.

** Persistent infestation in central New Brunswick from the mid-1950s, with some tree mortality.

No comprehensive policy was developed to deal with the spruce budworm problem in the 1970s in eastern Canada. The main reason is that each province administers its own forest resource and each gave a different priority to the social, economic, and political problems caused by spruce budworm. For example, in New Brunswick (Table 57) where the forest industry is of great importance to the provincial economy and the rate of utilization of spruce and fir is high, the policy was to protect Crown forests and designated private holdings where there was risk of severe tree damage and imminent tree mortality (Miller & Kettela 1975). Extensive annual use of chemical insecticides was the principal tactic in implementing this policy. In contrast, Ontario adopted a policy limiting chemical

Table 57 Softwood utilization relative to spruce, *Picea* spp., and balsam fir, *Abies balsamea* (L.) Mill., supply in New Brunswick and Ontario

	Pre-outbreak 1973: Estimated spruce and fir volume* (x 10^6 m^3)	Pre-outbreak 1973: Estimated annual growth** (x 10^6 m^3)	Average softwood*** harvest 1979: Allowable (x 10^6 m^3)	Average softwood*** harvest 1979: Actual (x 10^6 m^3)	Potential wood supply (D−C)	Average annual area of sprayed forest (ha) 1975–79
	A	B	C	D		
New Brunswick	348	8.7	7.1	7.4	Deficit	2.3 x 10^6
Ontario	1744	44****	27.1	19.3	Excess****	4.0 x 10^4

* Statistics Canada, 1973.

** Broad-scale estimate.

*** F.L.C. Reed and Associates Ltd., Canadian Forestry Congress, 1980.

**** No correction for accessibility.

control programmes to high investment areas such as parks, nurseries, and high value timber stands, because in that province wood supply problems are less pressing and fir is a minor component of utilized softwood. In all provinces, the issue of protection kindled debate on the use of chemicals. Various citizen groups argued that the use of chemical insecticides in forest spraying created unacceptable environmental and human health risks; they called for biological control and other pest management options. All provincial administration reacted to these concerns, but in Nova Scotia and Newfoundland the reaction resulted in curtailment of chemical spraying in forests that, with no intervention, had already suffered severe defoliation and increasing mortality.

Although there has been research into many potential spruce budworm control techniques, operational control tactics in the face of a widespread epidemic are largely limited to removal of vulnerable stands and the suppression of feeding larval populations with chemical insecticides or with *Bacillus thuringiensis* Berliner. The following sections summarize the progress made with each of the biological control methods.

Literature Cited

Baskerville, G.L.; MacLean, D.A. (1979) Budworm caused mortality and 20-year recovery in immature balsam fir stands. *Canadian Forestry Service, Maritime Forest Research Centre Information Report* M-X-102.

Miller, C.A.; Kettela, E.G. (1975) Aerial control operations against the spruce budworm in New Brunswick, 1952–1973. In: Prebble M.L. (*Ed.*) Aerial control of forest insects in Canada. Ottawa, Ontario; Thorn Press, pp. 95–112.

A. Field Development of *Bacillus thuringiensis* Berliner in Eastern Canada, 1970–80

W.A. SMIRNOFF and O.N. MORRIS

Background

Bacillus thuringiensis var. *kurstaki*, (*B.t.*) is a bacterium registered in Canada for control of agricultural and forest pests. The registered preparations, derived from cultures of the serotype 3a, 3b, are a mixture of dormant endospores and endotoxin crystals suspended in a liquid carrier.

Forest protection agencies have found spraying to be more expensive with *B.t.* than with chemical insecticides, yet *B.t.* remains a desirable alternative because its toxicity is confined to lepidopterous larvae and its use rarely arouses public antagonism. The Canadian experience of *B.t.* aerially applied against spruce budworm, *Choristoneura fumiferana* (Clem.) is that its efficacy is variable. Field trials have indicated differences among commercial products, varying efficacies among field formulations, uncertainties over dosages, technical difficulties in spraying methods, weaknesses in means of measuring deposits, and inadequacies in techniques of assessing the agent's effectiveness in reducing larval populations and protecting foliage.

Research into these problems was conducted in the 1970s, in two main areas, by two laboratories of the Canadian Forestry Service (C.F.S.): (a) in Québec, where the Laurentian Forest Research Centre collaborated with the Québec Department of Lands and Forests, and (b) in Ontario and the Atlantic Provinces, where the Forest Pest Management Institute participated in provincial-federal field trials. This paper collates the results from both laboratories.

Mode of action

B. thuringiensis serotype 3a, 3b is a bacterium that is pathogenic to lepidopterous larvae. On spruce budworm and other tortricids the commercial formulations cause enterotoxicosis by the action of the crystal-like parasporal inclusions (= delta endotoxin), followed by septicaemia from the action of their spores (Smirnoff & Valéro 1979, Fast 1981). The mode of action of the crystal toxin is described by Cooksey (1971).

The pathogenic action of *B.t.* on spruce budworm depends on the dose ingested, the age and physiological state of the larvae, the temperature, and possibly on the presence of natural infections by microsporidia. The optimal temperature for this action is 20–22°C (Smirnoff 1967). Young larvae are more susceptible to *B.t.* than fifth- and sixth-instar larvae (Morris 1973).

Formulations

Initially *B.t.* was tested against spruce budworm in New Brunswick in 1960. The material used was Thuricide® SO-75 (Bioferm Corp.), based on the Berliner serotype. Next, a small-scale aerial application of Thuricide® 90T was made in 1969, using both suspensions in water and water-in-oil emulsion (Morris *et al.* 1975). Assessment of population density reduction and foliage protection was inconclusive owing to clogging of the spray nozzles and poor distribution of the active ingredient.

Improved commercial preparations of *B.t.* became available for testing between 1970 and 1980, namely Thuricide® 24B, Thuricide® 32B, Dipel® 36B (Abbot Ltd.), and Novabac® 32B (Cyanamid Canada) or Novabac® wettable powder titrating 80 000

I.U./mg. These concentrates included various additives designed to improve deposit efficiency and adherence to foliage (rain fastness), and to prolong residual activity of the biological agent.

Three field formulations were developed at the Laurentian Forest Research Centre, and have been used experimentally in Québec. The first formulation, based on an aqueous solution of sodium dihydrogen phosphate (NaH_2PO_4), and the second, based on an aqueous solution of sorbitol, incorporate chitinase and Chevron® sticker (Chevron Chemical (Canada) Ltd.), and have been used since 1973 in low-volume sprays (4.7 l/ha). Sorbitol is a humectant, whereas sodium dihydrogen phosphate is added to increase the specific gravity of the liquid *B.t.* formulation. The third formulation, named Futura by the Laurentian Forest Research Centre, also uses sorbitol, chitinase, and sticker but incorporates less aqueous carrier and has been used for reduced-volume sprays (2.5 l/ha) since 1978 (Smirnoff 1980*a*, 1981*a*). Formulations used in Ontario and the Atlantic Provinces were similar, but did not include chitinase, sorbitol, or sodium dihydrogen phosphate (Morris 1980).

Smirnoff (1973, 1977) concluded after several years of experiments that the addition of minute quantities of chitinase increases the efficacy of *B.t.* against spruce budworm; in his view, it enhances penetration of ingested ingredients through the gut lining, and rapidly potentiates symptoms of infection (lethargy, weight loss) even at low temperatures. Some other additives have been tested for compatibility. Smirnoff (1981*b*) found that either mineral oil in large quantities or various dyes inhibited spore development.

During 1970–79, Thuricide® 16B (Sandoz. Inc.) was the only product fully registered for spruce budworm control under the Canadian Pest Control Products Act; around 1980 Dipel® 88 and Novabac® (Cyanamid Canada) have been fully registered.

Spray delivery systems and measurement of deposit

Sprays were applied by various aircraft ranging from four-engined freight planes to single-engined biplanes and helicopters. They included the Lockheed Constellation L-749, Douglas DC-6B, Canadair CL-215, Grumman Avenger TBM, Grumman Ag-cat, Boeing Stearman, Piper Pawnee, Cessna Agtruck, and the Sikorsky S55-T helicopter. Several kinds of spray emission hardware were tried, including Beecomist® (Beeco Products Co., USA), Micronair® (Micronair (Aerial) Ltd., U.K.) boom and flat fan Teejet® nozzles, and boom and open nozzles. Spray aircraft were calibrated to ensure appropriate emission rates for prescribed swath widths. Swath width varied from 305 m for a Constellation to 38 m for an Ag-cat (Smirnoff 1979*a*, Smirnoff & Juneau 1982).

There is no general agreement on the optimal droplet size or optimal foliage coverage. Spray coverage is measured as the number of droplets per square centimetre on Kromekote® cards, or as the number of colonies per square centimetre collected on agar plates or Millipore® filters, placed at ground level in forest clearings. The number of spores deposited per unit area on glass plates at ground level has also been measured (Morris 1980). In 1979, Smirnoff & Valéro proposed a new method whereby a given volume of peptonized water was exposed to the spray, so that the number of spores deposited per unit area could be calculated. For *B.t.* treatments, this value is more important than the total volume of liquid deposited (Smirnoff 1980*a*). Field tests have shown the necessity for calibration of the spray system: assessment based on the viable spore count deposited per unit area provides an adequate calibration.

The concentration of *B.t.* within the proprietary product, the properties of the adjuvants added at the airstrip mixing plant, and the composition of the tank concentrate all affect the rate of deposition and the success of operations (Smirnoff 1980*b*).

Operators used various kinds of spray hardware designed for chemical insecticides. Beecomist® spray heads were satisfactory on the Sikorsky helicopter. However, a boom and nozzle spray system composed of Teejet® 8004 flat fan nozzles gave better results with a Grumman Ag-cat fixed-wing aircraft (Smirnoff 1979*a*). A larger aircraft such as the DC-6B is used with open type nozzles, provided that the pumping system is able to produce a 413 kPa to 482 kPa pressure in the booms.

In summary, a variety of means was used in the development of *B.t.* application technology: field trials using new formulations; improved hardware and nozzle systems; atomization and distribution of *B.t.*; knowledge of meteorological parameters; swath paths; improved deposit analysis; dosage rates; application timing and frequency; methods of measuring treatment effectiveness; and knowledge of the spruce budworm population densities most amenable to effective control (Morris 1980, Morris *et al.* 1981, Smirnoff 1979*a*, 1979*b*, 1980*a*, Auger *et al.* 1981).

Assessment of efficacy

The assessment of efficacy continues to be a major problem because both criteria – foliage saved and number of larvae killed – present sampling difficulties. The estimate of foliage saved by treatment is derived from comparisons of foliage persisting in untreated check plots with foliage retained in the sprayed plot. An important element of the comparison is the production of new buds as a measure of foliage potential in the year following treatment. Quantification of foliage saved is based on counts of shoots, but the method lacks sensitivity because all shoots do not develop consistently. Since 1978 an improved method has been developed at the Laurentian Forest Research Centre based on weight of new foliage produced and retained; it compares the weight of foliage produced by a defoliated tree with the expected weight of foliage produced by a healthy tree of similar branch area.

The estimation of larval population reduction has usually been based on a comparison of the survival rate in the treated plot with that in an untreated check plot, using Abbott's formula to isolate the effect of the spray. Most workers recognize that this technique has major problems of sensitivity owing to the variability encountered in sampling.

Non-lethal residual effects from *B.t.* have been observed by Morris (1973), Smirnoff (1978), and Morris *et al.* (1981); the influence of these effects on the dynamics of treated budworm populations cannot yet be assessed, but should not be ignored. Smirnoff detected changes in the "vitality" of survivors in treated populations, calculated by measurements of pupal weight, total lipids, calcium, and various enzymes. In a comparison of pupal populations in stands treated with the chemical insecticide fenitrothion, stands treated with *B.t.*, and untreated stands, the survivors exposed to *B.t.* had the lowest vitality assessed by these biochemical criteria (Smirnoff 1979*c*).

Field Trials

Province of Québec

The field trials (Table 58) had various objectives, but were especially designed to test formulations and to measure deposits, larval mortality, and foliage protection under operational and experimental circumstances. The first trials in 1971 and 1972 encouraged the expectation that spruce budworm mortality and defoliation control could be achieved by adequate deposits of *B.t.* (Smirnoff *et al.* 1973*a*, 1973*b*). Small trials in 1973 showed that *B.t.* sprays might be ineffective where spruce budworm populations were very high and stands were already weakened (Smirnoff *et al.* 1974). In 1974 and 1975, four-engined aircraft were used on a large scale for the first time, but the results were inconclusive because of improper calibration of the spray system (Smirnoff *et al.* 1976, Smirnoff &

Table 58 Large scale tests with *Bacillus thuringiensis* Berliner against the spruce budworm, *Choristoneura fumiferana* (Clem.), on balsam fir, *Abies balsamea* (L.) Mill., in Québec

Formulation	Year	Area (ha)	Dosage (I.U. x 10^9 /ha)	Volume (l/ha)	Air-craft	Ground deposit	Pre-spray larval density per 45-cm branch	Corrected larval mortality (%)	Percentage defoliation		
									Check	Treatment	Saved
Thuricide®	1972	4 000			TBM						
Thuricide® 16B	1974	3 250	16.7	4.6	DC-6B CL-215	5.7 *	68.7	88.7		82	
Thuricide® 16B, 26B, Dipel® 36B (Average of treatments)	1975	91 000	16.7	2.3 4.6	DC-6B	3.0 *	44.0	91.0	86	80	6
Thuricide® 32B	1978	6 000	39.54	9.36	DC-6B	43**	12.0	93.0	73	30	43
Novabac® 32B Dipel® 36B Thuricide® 32B	1978	7 000			Ag-Cat		(Inconclusive results)				
Thuricide® 32B	1979		29.65		L-749	34.2 **	15.0	95.0	73	57	16
Thuricide® 32B		17 030	19.77		L-749	30.6 **	10.2	95.0	73	58	15
Novabac® 32B			29.65		L-749	37.6 **	11.1	84.0	73	48	25
Dipel® 32B			19.77		L-749	13.3 **	11.0	91.0	73	34	39
Dipel® 88	1980	1 209	19.54		L-749	16.2 **	6.4	87.0	43	17	61
Thuricide®		20 971	29.64	7.015	L-749	26.16**	9.7	82.0	55	26	29

* Drops/cm^2.
** Colonies/cm^2.

Juneau 1979). The use of a helicopter with Beecomist® heads in 1976 reaffirmed that good deposition, resulting in satisfactory control of larvae and defoliation, was attainable with low-volume formulations (Smirnoff & Hardy 1978). In 1977 and 1978, apparently moribund stands were sprayed by Ag-cat aircraft for two successive years and recovered their vigour (Smirnoff 1979*a*). Operational spraying in 1978 demonstrated good dispersion of a sorbitol formulation and good foliage protection by use of four-engined aircraft (DC-6B). An experimental test of the sodium dihydrogen phosphate formulation was inconclusive owing to navigational problems. In 1979, the dispersion capability of the Futura formulation was tested; this led to a further field trial in 1980. In both 1979 and 1980, four-engined L-749 aircraft with boom and nozzle spray equipment were employed in extensive operations, in which *B.t.* treatments compared favourably with chemical insecticide treatments.

In 1980, the Québec Department of Energy and Resources concluded that "after 8 years of research and development with *B.t.* against the spruce budworm in the forest, it seems that this product can be used on an operational basis in Québec with the same chance of success as chemical insecticides provided that some constraints, relevant to the fact that *B.t.* has to be ingested to work, are respected" (Dorais *et al.* 1980).

Ontario and the Atlantic Provinces

The effectiveness of *B.t.* showed marked improvement during the 1970s, yet results were inconsistent, due partly to inconsistent deposits of the water-based formulations.

If the retention of 50% of a year's growth on budworm-infested trees is taken as acceptable protection, then five out of eight treatments on balsam fir, *Abies balsamea* (L.) Mill., (Table 59) and seven out of nine treatments on white spruce, *Picea glauca*

Table 59 Tests of *Bacillus thuringiensis* Berliner aerially applied against the spruce budworm, *Choristoneura fumiferana* (Clem.), on balsam fir, *Abies balsamea* (L.) Mill., in Ontario and the Atlantic Provinces 1970–80*

Formulation**	Year	Dosage applied (I.U. x 10^9/ha)	Ground deposit rate (drops/cm^2)	Pre-spray larval density per 45-cm branch	Percentage defoliation Expected	Percentage defoliation Actual	Percentage foliage saved
			Ontario				
Dipel® WP + molasses	1973	1 x 10	15	52	98	85	13
Thuricide® 16B	1973	1 x 10	37	28	90	79	0
Dipel® WP + molasses	1974	1 x 20	25	11	55	28	27
Thuricide® 16B	1974	2 x 20	Good	29	80	44	36
Thuricide® 16B	1974	2 x 10	Good	67	100	11	89
Dipel® 36B + Sorbitol	1975	1 x 20	12	11	55	57	0
Dipel® 36B + carboxylmethyl-cellulose	1976	1 x 20	36	6	42	3	39
Thuricide® 16B	1976	1 x 10	Good	13	53	18	35
Thuricide® 16B	1979	2 x 10	NR***	16	62	55	7
Acephate followed by Novabac®	1979	280 g + 1 x 20	29	25	75	86	0
Dipel® 88	1980	1 x 20	2 I.U. x 10^9/ha	11	55	45	10
Dipel® 88	1980	1 x 20	31	18	69	21	48
Dipel® 88	1980	1 x 20	15	24	74	48	26
Thuricide® 24B	1980	1 x 20	38	5	40	22	18
Thuricide® 16B	1980	1 x 10	NR	13	53	7	46
Thuricide® 16B	1980	1 x 20	40	13	53	20	33
Thuricide® 24B	1980	1 x 20	25	49	96	62	34
			New Brunswick				
Dipel® 88	1979	1 x 40	20	21	70	22	48
Dipel® 45B + Sorbitol	1979	1 x 33	NR	26	76	22	54
Thuricide® 16	1979	1 x 20	12–33	20	70	58	12
			Nova Scotia				
Thuricide® 16B	1979	1 x 20	26–90	24	74	19	55
Thuricide® 16B	1979	2 x 20	34–56	35	88	10	78
Thuricide® 16B	1980	1 x 20	36	24	74	40	34
Thuricide® 16B	1980	1 x 20	19	27	75	26	49
Thuricide® 16B	1980	1 x 20	24	42	95	12	83
Thuricide® 16B	1980	1 x 20	23	56	100	43	57
			Newfoundland				
Thuricide® 16B	1979	1 x 20	NR	9	50	53	0
Thuricide® 24BA	1979	1 x 20	NR	9	50	42	6
Thuricide® 24BC	1979	1 x 20	NR	11	55	26	29
Novabac® 45B	1979	1 x 20	NR	5	40	20	20
Dipel® 88	1980	2 x 20	5	30	83	66	17
Dipel® 88	1980	2 x 20	NR	91	100	100	0
Dipel® 88	1980	2 x 20	NR	100	100	96	4
Thuricide® 16B	1980	1 x 20	NR	8	49	46	3
Thuricide® 16B	1980	1 x 20	NR	8	49	35	14
Thuricide® 16B	1980	1 x 20	NR	10	54	37	17
Thuricide® 16B	1980	1 x 20	NR	14	60	36	24
Thuricide® 16B	1980	1 x 20	NR	27	75	64	11
Thuricide® 16B	1980	1 x 20	NR	38	90	27	63
Thuricide® 32B	1980	1 x 20	9	40	93	93	0
Thuricide® 24B	1980	1 x 20	15	42	95	91	4
Thuricide¶ 16B	1980	1 x 20	NR	47	96	99	0
Novabac® -3	1980	1 x 20	12	11	55	36	19
Novabac® -3	1980	1 x 20	NR	21	70	90	0
Novabac® -45B	1980	1 x 20	8	27	75	96	0

* Includes only tests that reported pre-spray larval densities.
** Includes only formulations without chemical additives, e.g. chemical pesticides or enzymes.
*** NR, not reported.

(Moench) Voss, (Table 60) were successful during 1973–76. In these tables, and in Table 61 expected defoliation was calculated from measurements of larval density per 45-cm branch plotted against defoliation of that year's growth on 850–1700 budworm-infested branch tips from 17 unsprayed plots in Ontario (Morris & Moore 1975). No correlation was evident between pre-spray larval density or deposit rate and percentage of foliage saved, indicating that *B.t.* effectiveness is influenced by a complex of interrelated factors.

During 1978–80, 23 out of 38 treatments (61%) were successful in retaining 50% of a year's growth on firs, *Abies* spp. (Table 59). Success rates on white spruce and red spruce, *Picea rubens* Sarg., were 15/23 (65%) and 8/8 (100%), respectively (Tables 60 and 61). The high rate of success on red spruce is not surprising, because most of the experimental stands were evidently not at the risk of higher than 50% defoliation; assessed on the basis of percentage of foliage saved, *B.t.* efficacy on red spruce was unexceptional.

In 1979 and 1980, a total of 51 093 ha of forest was sprayed with *B.t.* in eastern Canada (Table 62). Success rates, based on the percentage of the sprayed areas satisfactorily

Table 60 Tests of *Bacillus thuringiensis* Berliner aerially applied against the spruce budworm, *Choristoneura fumiferana* (Clem.), on white spruce, *Picea glauca* (Moench) Voss, in Ontario and the Atlantic Provinces 1970–80*

Formulation**	Year	Dosage applied (I.U. x 10^9/ha)	Ground deposit rate (drops/cm^2)	Pre-spray density per 45-cm branch	Percentage defoliation Expected	Percentage defoliation Actual	Percentage foliage saved
			Ontario				
Dipel® WP + molasses	1974	1 x 20	25	19	51	52	0
Dipel® WP + molasses	1974	1 x 20	34	84	92	67	25
Dipel® WP + molasses	1974	1 x 10	28	90	94	45	49
Thuricide® 16B	1974	2 x 10	NR***	108	93	18	75
Thuricide® 16B	1974	2 x 20	NR	137	95	42	53
Dipel® 36B + Sorbitol	1975	1 x 20	12	16	46	48	0
Dipel® 36B + carboxymethyl-cellulose	1976	1 x 20	36	11	36	10	26
Thuricide® 16B	1976	1 x 10	NR	13	36	7	29
Thuricide® 16B	1976	1 x 10	NR	26	63	20	43
Thuricide® 16B	1979	1 x 10	NR	9	32	12	20
Thuricide® 16B	1979	2 x 10	NR	27	65	12-38	› 27
Thuricide® 16B	1979	1 x 10	NR	29	67	63	4
Thuricide® 16B	1979	2 x 10	NR	42	73	91	0
Novabac® 32B	1979	2 x 20	NR	34-55	70-82	47-92	0-23
Acephate followed by Novabac® 32B	1979	280 g + 1 x 20	29	47	78	69	9
Dipel® 88	1980	1 x 20	15	29	67	50	17
Dipel® 88	1980	1 x 20	2 I.U. x 10^9/ha	35	70	69	1
Dipel® 88	1980	1 x 20	31	44	75	31	44
Thuricide® 24B	1980	1 x 20	38	21	58	25	33
Thuricide® 16B	1980	1 x 20	40	30	68	22	46
Thuricide® 16B	1980	1 x 20	25	67	87	70	17
			New Brunswick				
Dipel® 88	1979	1 x 40	20	38	73	33	40
Thuricide® 16B	1979	1 x 20	12-33	44	76	49	27

* Includes only tests reporting pre-spray larval densities.
** Includes only formulations without chemical additives, e.g. chemical pesticides and enzymes.
*** NR, not reported.

Table 61 Tests of *Bacillus thuringiensis* Berliner aerially applied against the spruce budworm, *Choristoneura fumiferana* (Clem.), on red spruce *Picea rubens* Sarg., in Nova Scotia and New Brunswick 1979–80*

Formulation**	Year	Dosage applied (I.U. x 10^9/ha)	Ground deposit rate (drops/cm²)	Pre-spray larval density per 45-cm branch	Percentage defoliation Expected	Actual	Percentage foliage saved
			Nova Scotia				
Thuricide® 16B	1979	1 x 20	38–56	22	60	14	46
Thuricide® 16B	1979	1 x 20	51	14	38	9	28
Thuricide® 16B	1979	2 x 20	27–31	12	36	14	22
Thuricide® 16B	1980	1 x 20	24	7	25	0.2	25
Thuricide® 16B	1980	1 x 20	30	7	25	14	11
Thuricide® 16B	1980	1 x 20	19	6	25	12	13
			New Brunswick				
Dipel® 88	1979	1 x 40	20	17	46	23	23
Thuricide® 16B	1979	1 x 20	12–33	19	50	28	22

* Includes only tests reporting pre-spray larval densities.
** Includes only formulations without chemical additives, e.g. chemical pesticides and enzymes.

Table 62 Extent of acceptable protection with *Bacillus thuringiensis* Berliner in eastern Canada 1979–80*

Province	Area treated (ha)	Mean no. of droplets/cm² (range)	Percentage of area acceptably protected**
Ontario	4 846	22 (11–30)	74
– FPMI (1980 only)	320	34 (24–40)	100
New Brunswick (1979 only)	360	23 (2–53)	100
Nova Scotia	30 188	51 (9–111)	92
Newfoundland and Labrador	15 699	9 (5–16)	48
Québec	33 174	43 (10–87)	89

* Based on efficacy data provided by provincial departments or agencies.
** Acceptable protection is less than 50% defoliation of current year's growth.

protected, ranged from 74 to 100% except for Newfoundland and Labrador, where the success rate was only 48%. The poor efficacy in Newfoundland was apparently due to inadequate spray deposits.

Evaluation of Control Attempts

The encouraging results with *B.t.* up to 1978 prompted the C.F.S. to formulate technical guidelines for the use of this product against the spruce budworm. A preliminary C.F.S. policy statement recommended "the operational testing of *B.t.* in environmentally sensitive areas or in sensitive areas where the forest manager either chooses not to use or is forbidden to use conventional chemical insecticides for the control of the spruce budworm". The main technical recommendations were as follows:

Objective – primarily to limit the average annual defoliation to less than 50% and secondarily to reduce the numbers of spruce budworm by 75–85%.
Formulation – Thuricide® 16B, water (not chlorinated), Chevron® sticker (0.1%). Optional for experimental trials: Thuricide® 32B, sorbitol, Chevron® sticker, chitinase (9880 nephelometric units per hectare), water, and Erio Acid Red (experimental tracer). Note that dyes may inhibit *B.t.* efficacy.
Dosage – 20 x 10^9 I.U. of *B.t.* in a minimum of 4.7 l/ha. In mixed spruce-fir stands, two applications of 10–20 x 10^9 I.U./ha each.
Operational constraints – pre-spray population density of 25–30 larvae per 45-cm branch tip. Application should start at shoot flare and continue no later than the date when 30% of the spruce budworm have developed to the fifth instar.
Deposit specifications – agar plates or Millipore® filters should be used to collect ground level deposits, aiming for a density of 25 droplets/cm² or more.

In an attempt to reduce the inconsistencies of results in previous years, the *B.t.* applications using these guidelines were co-ordinated in 1979 and 1980 (Morris 1980, 1981). In general, the guidelines were followed unless extenuating circumstances were encountered. Clearly, *B.t.* is an effective alternative to the use of chemical pesticides for the protection of foliage. An analysis of the results of these applications (Morris 1980) showed that successful treatments had the following common characteristics:

1. Pre-spray densities were usually less than 28 larvae per 45-cm branch tip
2. Larval development at spray time was around fourth instar
3. Bud flushing was 80–100% complete (except on red spruce)
4. I.U. applied per hectare were 20–40 x 10^9 in single or double applications
5. Ground level droplet density was greater than 25/cm²
6. Spray time relative humidity was mostly higher than 65%
7. Good weather followed spraying.

These conditions are now generally accepted by researchers as desirable for *B.t.* application in the forest.

Smirnoff (1980*c*) advocates further conditions for reducing costs and increasing success;

a) The required *B.t.* dosage should be emitted in a formulation dispensible at 2.5–3.0 l/ha; or (in the case or Thuricide® 32B with sorbitol in water) at 4.7 l/ha;
b) formulations should include Chevron® sticker at 1:1600; and chitinase at 10 000 nephelometric units per hectare;
c) spore viability in the concentrate drum should be checked;
d) large capacity aircraft equipped with boom and nozzle are more cost-effective than small aircraft. Spray output must be calibrated and checked by counts of viable spores at ground level; the ground deposit should attain at least 60% of the emission rate;
e) *B.t.* treatments should be carried out when most of the larvae are third instar, and buds are flushing (Smirnoff 1980*c*, Auger *et al.* 1981).

Recommendations

Improvements in commercial formulations and application technology during the past decade have brought *B.t.* to a point where it is now considered a clear alternative to chemical insecticides for use against spruce budworm, especially in environmentally sensitive areas. It is almost as effective for tree protection as the available chemical pesticides. Success or failure cannot yet be totally explained or predicted because relationships between foliage protection and dosage per volume emitted are not firmly established. Moreover, estimates of ground deposits have borne little practical relationship to deposits on foliage. The technical guidelines prepared in 1978 (see earlier) continue to be valid, but require updating. *B.t.* spray operations are expensive compared with the use of chemical insecticides (2.5–5 times higher around 1980). As much of the cost is in material transport, more concentrated proprietary products should be devised to lower the shipping and application costs.

Technical development of formulations, spray applications, and efficacy assessment should be continued: the search for more efficaceous formulations, droplet spectra, and timings should be continued; field dosage and volume/response relationships should be established, to provide the pest manager with cost/benefit options; and a quantitative method of measuring deposits at the feeding site is needed, to relate deposit rate of the active ingredient to its effectiveness.

Literature Cited

Auger, M.; Chabot, M.; Pelletier, M.; Bordeleau, C. (1981) Expertises entomologiques reliées aux pulvérisations aériennes contre la tordeuse des bourgeons de l'épinette au Québec en 1980. Québec; Ministére de l'Energie et des Ressources, 107 pp.

Cooksey, K.E. (1971) The protein crystal toxin of *Bacillus thuringiensis*: biochemistry and mode of action. In: Burges, H.D.; Hussey, N.W. (*Eds.*) Microbial control of insects and mites. New York; Academic Press, pp. 247–274.

Dorais, L.; Pelletier, M.; Smirnoff, W.A. (1980) Pulvérisations aériennes de *Bacillus thuringiensis* Berliner réalisées au Québec de 1971 à 1979 contre la tordeuse des bourgeons de l'épinette, *Choristoneura fumiferana* (Clem.). Rapport présenté au VIe Congrès International de l'Aviation Agricole à Turin, Italie, Septembre 1980.

Fast, P.G. (1981) The crystal toxin of *Bacillus thuringiensis*. *In*: Burges, H.D. (Ed.). Microbial control of pests and plant diseases 1970–1980. New York; Academic Press, pp. 223–248.

Morris, O.N. (1973) Dosage-mortality studies with commercial *Bacillus thuringiensis* sprayed in a modified Potter's Tower against some forest insects. *Journal of Invertebrate Pathology* 22,108–114.

Morris, O.N. (1980) Report of the 1979 CANUSA Cooperative *Bacillus thuringiensis (B.t.)* spray trials. *Canadian Forestry Service Information Report* FPM-X-40, 160 pp.

Morris, O.N. (1981) Report on the 1980 Cooperative *Bacillus thuringiensis (B.t.)* spray trials. *Canadian Forestry Service Information Report* FMP-X-48, 80 pp.

Morris, O.N.; Moore, A. (1975) Studies on the protection of insect pathogens from sunlight protection. II. Preliminary field trials. *Canadian Forestry Service Information Report* CC-X-113.

Morris, O.N.; Angus, T.A.; Smirnoff, W. (1975) In: Prebble, M.L. (*Ed.*). Aerial control of forest insects in Canada. Ottawa; Thorn Press, pp. 129–133.

Morris, O.N.; Hildebrand, M.J.; Moore, A. (1981) Comparative effectiveness of three commercial formulations of *Bacillus thuringiensis* for control of the spruce budworm, *Choristoneura fumiferana* (Clem.). *Canadian Forestry Service Information Report* FMP-X-47, 31 pp.

Smirnoff, W.A. (1967) Influence of temperature on the rate of development of six varieties of *Bacillus cereus* group. Insect pathology and microbial control. Amsterdam; North-Holland, pp. 125–130.

Smirnoff, W.A. (1973) Results of tests with *Bacillus thuringiensis* and chitinase on larvae of the spruce budworm. *Journal of Invertebrate Pathology* 21,116–118.

Smirnoff, W.A. (1977) Confirmations expérimentales du potentiel du complexe *Bacillus thuringiensis* et chitinase pour la répression de la tordeuse des bourgeons de l'épinette *Choristoneura fumiferana* (Lepidoptera: Tortricidae). *Canadian Entomologist* 109,351–358.

Smirnoff, W.A. (1978) Impact des traitements bactériologiques et des traitements chimiques sur une épidémie de la tordeuse des bourgeons de l'épinette. Résumé des communications, 46e congrès ACFAS, p. 68.

Smirnoff, W.A. (1979*a*) Results of spraying *Bacillus thuringiensis* two consecutive years over balsam fir stands damaged by spruce budworm. *Canadian Journal of Forest Research* 9,509–513.

Smirnoff, W.A. (1979*b*) Result of a 3-phase research program related to the biological control of the spruce budworm in 1979. LFRC, Report to the Candian Forest Pest Control Forum, Nov. 27–28, 1979, Ottawa, Ontario.

Smirnoff, W.A. (1979*c*) Changes in metabolites in insects during starvation, infections or poisoning. Abstracts of papers, 9th International Congress of Plant Protection and 71st Annual Meeting of the American Phytopathological Society, Washington, D.C., No. 928.

Smirnoff, W.A. (1980*a*) Deposit assessment of *Bacillus thuringiensis* formulations applied from an aircraft. *Canadian Journal of Microbiology* 26,1364–1366.

Smirnoff, W.A. (1980*b*) Calibration tests with various *Bacillus thuringiensis* preparations. *Canadian Forestry Service Bi-monthly Research Notes* 36,30–31.

Smirnoff, W.A. (1980*c*) Technological requirements for a successful *Bacillus thuringiensis* application against spruce budworm. *Canadian Forestry Service Bi-monthly Research Notes* 36,29–30.

Smirnoff, W.A. (1981*a*) Développement d'une préparation compacte et économique de *Bacillus thuringiensis* pour la répression de la tordeuse des bourgeons de l'épinette *Choristoneura fumiferana. Annales de l'ACFAS* 48,203.

Smirnoff, W.A. (1981*b*) Les insecticides biologiques doivent rester biologiques. *Annales de l'ACFAS* 48,203.

Smirnoff, W.A.; Hardy, Y.J. (1978) An assay of sequential application of fenitrothion and *Bacillus thuringiensis* for an integrated control of the spruce budworm. *Canadian Journal of Forest Research* 8,300–305.

Smirnoff, W.A.; Juneau, A. (1979) Results of aerial spraying with *Bacillus thuringiensis* against the spruce budworm in Québec in 1975. *Annals of the Entomological Society of Québec* 24,131–138.

Smirnoff, W.A.; Juneau, A. (1982). Analyse physique de la dispersion de préparations de *Bacillus thuringiensis* utilisées pour la répression de la tordeuse des bourgeons de l'épinette. *Canadian Forestry Service Information Report* LAU-X-55.

Smirnoff, W.A.; Valéro, J.R. (1979) Mode d'action de *Bacillus thuringiensis* chez *Choristoneura fumiferana* (Lepidoptera: Tortricidae), importance des spores. *Canadian Entomologist* 111,305–308.

Smirnoff, W.A.; Fettes, J.J.; Desaulniers, R. (1973*a*) Aerial spraying of a *Bacillus thuringiensis* – chitinase formulation for the control of spruce budworm. *Canadian Entomologist* 105,1535–1544.

Smirnoff, W.A.; Randall, A.P.; Martineau, R.; Haliburton, W.; Juneau, A. (1973*b*) Field test of the effectiveness of chitinase additive to *Bacillus thuringiensis* Berliner against *Choristoneura fumiferana* (Clem.). *Canadian Journal of Forest Research* 3,288–236.

Smirnoff, W.A.; Larson, L.V.; Juneau, A.; Valéro, J.R. (1974) Test with a highly concentrated low-volume formulation of *Bacillus thuringiensis* against spruce budworm. *Canadian Forestry Service Bi-monthly Research Notes* 30,9.

Smirnoff, W.A.; Desaulniers, R.; Juneau, A.; Valéro, J.; Pelletier, M. (1976) Résultats de la dispersion aérienne d'une formulation concentrée de *Bacillus thuringiensis* avec l'addition de chitinase contre la tordeuse des bourgeons de l'épinette en 1974. *Annals of the Entomological Society of Québec* 21,26–34.

B. Viruses: Application and Assessment

J.C. CUNNINGHAM and G.M. HOWSE

Background

Five different types of viruses have been found to infect larvae of the spruce budworm, *Choristoneura fumiferana* (Clem.). Four of them share one feature in common, that is, virus particles are contained within proteinaceous inclusion bodies. These infectious particles are released when the inclusion bodies are ingested by larvae and the inclusion body protein dissolves in their alkaline gut juices. The four viruses comprise a nuclear polyhedrosis virus (NPV), a granulosis virus (GV), a cytoplasmic polyhedrosis virus (CPV), and an entomopoxvirus (EPV). Between 1971 and 1980, all four viruses have been field tested either alone or in various combinations. The fifth virus, which has been found to infect spruce budworm, is cricket paralysis virus. It has no inclusion body, has a very wide host range, has little potential as a biological control agent, and will not be discussed further. Collapse of a spruce budworm population due to a naturally occurring virus epizootic has never been observed. Both NPV and CPV are occasionally detected in eastern spruce budworm populations in Ontario, but generally less than 1% of larvae are infected.

The first field trials of spruce budworm viruses were conducted in 1959 and 1960 by applying NPV and GV on single small trees (Stairs & Bird 1962). In 1969, different dosages of NPV were applied on single trees at 2- to 3-day intervals throughout the larval period (Bird & McPhee 1970). In 1970, an EPV was found in 2-year cycle spruce budworm, *C. biennis* Freeman (Bird *et al.* 1971). In the laboratory very low dosages of this EPV (in comparison to other viruses) caused high larval mortality and, although it was slow acting, there was high hope that it would prove to be an effective biological control agent. By 1970, the use of artificial diet had greatly improved methods of rearing spruce budworm (McMorran 1965), and sufficient larvae could be reared in order to produce viruses for aerial spray trials (Grisdale 1970).

The first aerial spray trial was conducted in 1971 and trials have been carried out each season since then. Sixty plots were treated with viruses or combinations of chemical and viral insecticides in Ontario between 1971 and 1980. Such factors as dosage, timing of application, tank mixtures, spray equipment, impact on different species of host trees, and persistence of viruses from one year to the next have been evaluated. The following viruses and combinations of viruses were tested during this period: NPV, EPV, GV, EPV plus NPV plus CPV, and NPV plus CPV. In addition, NPV and EPV plus NPV have been applied following chemical insecticide sprays.

Field Trials

Assessment of the impact of treatments

The same methods used to evaluate chemical insecticide treatments in Ontario were used to assess virus spray trials. In addition to carrying out larval population studies and defoliation estimates, attempts were made to determine the levels of infection in the spruce budworm population, to assess the effect on successful pupal emergence and, on some plots, to continue these studies for one or more years after the initial application.

Monitoring the deposit

In most treated plots, Kromekote® spray deposit cards were placed along roads or in open areas in a line at right angles to the flight lines. The number of droplets per square centimetre was determined and the size distribution of the droplets was tabulated, as measured on the cards. This gave an approximate comparison of different treatments. More refined deposit assessments were not undertaken.

Estimating spruce budworm populations

Pre-spray and post-spray larval and pupal counts were made on 46-cm branch tips collected at mid-crown from treated and check plots. In most cases, when plots contained only one tree species, 50 samples were examined; when two species were studied, 25 samples were taken from each plot. When treatments were made on the second-instar larvae, whole branch samples were cut up and placed in special boxes in the laboratory. Larvae emerged, wandered to a glass tube at the top, and were counted daily. When second-instar larvae were treated, the first post-spray count was made of the fourth-instar larvae. Final post-spray samples for all treatments were taken after pupation but before adult emergence. Abbott's formula (Abbott 1925) was used to calculate reductions of population caused by treatment.

Determining levels of virus infection

Random samples of ten 46-cm branch tips were collected at mid-crown from treated and check plots at various time intervals following treatment. Larvae were removed from the foliage, and squash preparations of gut and fat body were made on microscope slides. About 100–200 larvae were examined on each sampling date from each treated or check plot. These preparations were examined microscopically, using phase contrast optics, and the presence of viruses and other pathogens was recorded. Only those viruses actually seen under the light microscope were recorded; recently infected larvae containing no inclusion bodies and larvae that had already died and disappeared from the foliage were not included. Some dead larvae do remain on the foliage in webbed buds, but others fall off or are removed by scavenging insects. Percentages obtained by this method are therefore always lower than population reduction percentages. In the tables only the highest percentage infection observed in a series of samples is entered.

Recently, an alternative method has been used to determine levels of virus infection. This method involved collecting larvae from treated and check plots about a week after treatment or, for follow-up studies, when they reached the fourth and fifth instars in the year or years following application. Larvae were reared individiually on artificial diet in the laboratory until adult emergence. Percentage mortality was then recorded and cadavers were examined microscopically for the presence of pathogens. The figures obtained by this method do not take into account secondary infection of larvae that ingest inclusion bodies released from virus-killed insects. Although this infection figure is generally higher than that obtained by microscopic examination, it still does not reflect the full impact of the treatment.

Estimating adult emergence

The impact of virus applications on successful adult emergence was assessed by obtaining pupae from post-spray collections from treated and check plots, pupae obtained from laboratory rearing and, in some cases, special pupal collections. They were maintained at room temperature and the adult moths and dead pupae counted.

Estimating current year's defoliation

The percentage of the current year's defoliation was obtained by detailed examination of the 46-cm branch tips collected for the post-spray sample from the treated and check plots. These samples were collected after all surviving larvae had pupated and feeding had ceased . In the tables, the percentage of foliage saved was calculated using Abbott's formula.

Follow-up studies

Many plots were re-examined the following year and some plots were studied for several years because carry-over of viruses from one year to the next and the impact on the spruce budworm population in the years following the year of treatment are considered important factors in evaluating these biological control agents. The same observations were recorded as in the year of treatment.

Tank mixes for virus applications

Several tank mixes have been used for spruce budworm virus applications and are described in the tables. The viral product in all tests was a finely ground lyophilized preparation of virus-infected larvae. A few plots were treated with this product in water alone, but without a tracer dye no deposit assessment could be made. In many of the tests IMC 90-001 (International Minerals Corp.) UV protectant was added. It is a water-soluble wood byproduct that has a dark colour; it can also be used as a tracer dye. This material was renamed Shade® (Sandoz Inc.), but its production was discontinued in 1979.

Molasses has been used in most of the tank mixes. It is an anti-evaporant that enhances deposit of aqueous sprays, may give some UV protection, and may be a feeding attractant to spruce budworm larvae. The results from single-tree ground spray tests indicated that molasses in the mix had a marked beneficial effect. A commercial molasses preparation, Sandoz adjuvant V® (Sandoz Inc.), was used in 1975.

Many wetting and sticking agents have been tested in combination with Shade®, with or without molasses. These agents include Biofilm® (Colloidal Products Corp.), Chevron® spray sticker (Chevron Chemical Co.), Triton® X-100 and Triton® B-1956 (Rohm and Haas Co. of Canada Ltd.). All passed a laboratory test to determine their compatibility with virus at a concentration of 1%.

Two other preparations tested were:Sandoz San 285® wettable powder, which is a feeding attractant for the virus of cotton bollworm, *Heliothis zea* (Boddie), incorporated in a polymer-bound carbon formulation prepared by the Southwest Research Institute, San Antonio, Texas; and an emulsifiable oil, Sunspray® 11E (Sunoco Inc.). The process used to produce the polymer-bound carbon formulation inactivated most of the virus and the feeding attractant did not enhance virus infection levels in spruce budworm (Cunningham *et al.* 1978). The emulsifiable oil used at 66% had no adverse effect on the virus and its use would be beneficial if sprays were applied when the humidity was low.

Entomopoxvirus plus nuclear polyhedrosis virus plus cytoplasmic polyhedrosis virus

In 1971 a helicopter was used to treat six 1.07-ha plots with EPV (Table 63). When smears of larvae from treated plots were examined microscopically it was discovered that the EPV preparation was contaminated with both NPV and CPV. All three viruses gave consistently higher levels of infection in larvae on white spruce, *Picea glauca* (Moench) Voss, hosts than on balsam fir, *Abies balsamea* (L.) Mill., hosts (Bird *et al.* 1972, Howse *et al.* 1973).

Table 63 Application of EPV contaminated with NPV and CPV on six 1.07-ha plots using a helicopter fitted with boom and nozzle equipment and calibrated to deliver 28.2 l/ha*

Plot	Dosage, PIB/ha (x 10^9)	Pre-dominant instar at time of application	Deposit card, droplets/ cm^2	Highest virus infection (%) EPV bF**	EPV wS	NPV bF	NPV wS	CPV bF	CPV wS	Population reduction due to treatment (%) bF	wS	Current year's foliage saved (%) bF	wS
1	750	2	10	9	38	1	10	0	7	0	40	8	15
2	75	2	9	3	16	0	12	1	4	0	61	3	0
3	7.5	2	9	1	12	0	2	0	0	0	25	0	0
4	750	3 & 4	29	6	5	4	7	1	4	30	79	1	0
5	75	3 & 4	31	2	2	0	1	0	0	48	57	0	0
6	7.5	3 & 4	ND***	4	6	0	0	0	0	0	59	1	0

* Aqueous tank mix contained 2.5% IMC 90-001 sunlight protectant.
** bF = balsam fir, *Abies balsamea* (L.) Mill.; wS = white spruce, *Picea glauca* (Moench) Voss.
*** ND = Not determined.

Follow-up studies were undertaken between 1972 and 1975 on the two plots treated with the highest dosage (Table 64). It can be seen that, although there was little foliage protection in 1971 (Table 63), there was significant foliage protection in the subsequent 3 years. Also, the NPV, which was a minor contaminant in the EPV preparation, was considerably more prevalent in the following years than the EPV (Cunningham *et al.* 1975*a*).

Table 64 Follow-up studies on two plots treated with EPV contaminated with NPV and CPV at a dosage of 750 x 10^9 inclusion bodies/ha in 1971*

Plot	Year	Highest virus infection (%) EPV bF**	EPV wS	NPV bF	NPV wS	CPV bF	CPV wS	Population reduction due to virus (%) bF	wS	Current year's foliage saved (%) bF	wS
1	1972	0	5	7	19	2	5	64	82	72	59
	1973	2	5	3	13	0	0	3	60	27	22
	1974	1	2	1	4	0	0	0	5	29	49
	1975	1	6	5	16	1	1	0	78	22	2
4	1972	0	0	17	20	10	4	16	68	44	49
	1973	0	1	4	16	2	0	15	83	26	22
	1974	0	1	3	11	1	0	0	48	25	40
	1975	0	1	17	11	0	0	47	26	1	1

* See Table 63 for details and results in 1971
** bF = balsam fir, *Abies balsamea* (L.) Mill.; wS = white spruce, *Picea glauca* (Moench) Voss.

Entomopoxvirus

In 1972, pure EPV was tested and three plots with a total area of 512 ha were treated. The timing of the application was on second- and third-instar larvae, but an unseasonably late snow fall destroyed most of the spruce budworm larvae and killed the new growth on the trees. Population studies and defoliation estimates were abandoned and only levels of virus infection were recorded (Table 65) (Cunningham & McPhee 1973). By the end of the 1972 field season, follow-up studies on the plots treated in 1971 indicated that NPV had greater potential than EPV as a biological control agent and further testing of EPV was suspended.

Table 65 Application of EPV in 1972 on second- and third-instar larvae using a dosage of 100 x 10^9 inclusion bodies/ha delivered from a fixed-wing aircraft fitted with Micronair® equipment

Plot	Area (ha)	Emitted volume (l/ha)	Tank mix additives	Deposit card, droplets/ cm^2	Maximum EPV infection (%)	
					bF*	wS
P	256	18.8	2.5% IMC 90-001	41	6	20
Q	168	18.8	10% molasses 2.5% IMC 90-001 0.2% Biofilm	93	2	4
R	84	9.4	As plot Q	21	2	3

* bF = balsam fir, *Abies balsamea* (L.) Mill.; Ws = white spruce, *Picea glauca* (Moench) Voss.

Nuclear polyhedrosis virus plus cytoplasmic polyhedrosis virus

In 1971, two 2.3-ha mature white spruce plantations were treated with NPV, which was later found to be contaminated with a small amount of CPV. The ratio of NPV:CPV was 400:1 (W.J. Kaupp 1980 personal communication). In spite of this trace of CPV, significant

Table 66 Application of NPV contaminated with CPV on spruce budworm, *Choristoneura fumiferana* (Clem.), in 1971 from a helicopter fitted with boom nozzle equipment calibrated to deliver 28.2 l/ha, and applications in 1979 and 1980 from a fixed wing aircraft fitted with Micronair® equipment calibrated to deliver 9.4 l/ha.

Year	Plot	Area (ha)	Dosage, PIB/ha (x 10^9)	NPB:CPV	Tank mix additives	Predominant instar treated	Deposit card, droplets/ cm^2	Highest virus infection % NPV		Highest virus infection % CPV		Population reduction due to treatment (%)		Current year's foliage saved (%)	
								bF	wS	bF	wS	bF	wS	bF	wS
1971	G	2.3	750	400:1	Virus in water alone	2	ND**		21		28		69		0
	H	2.3	750	400:1	Virus in water alone	3 & 4	ND		46		35		80		0
1979	A	2	750	178:1	25% molasses 6% Shade® 1% Chevron® sticker	2	193	9	9	39	24	85	50	78	68
	B	8	750(x2)	178:1	As plot A	2	103 & 16	2	13	27	33	50	81	43	39
	C	12	1500	178:1	As plot A	4, 5 & 6	16	11	31	24	11	0	55	34	13
	D	20	750	178:1	As plot A	5 & 6	9	5	25	11	7	48	0	0	0
	E	12	750	178:1	As plot A	5 & 6	26	10	15	19	9	78	58	0	0
	F	20	750	178:1	66% Sunspray® 11E oil	5 & 6	16	8	9	4	4	61	76	0	0
1980	H	20	750	300:1	As plot A	2	71	5	7	5	5	71	45	87	75
	J	2	750	300:1	6% Shade®	2	23		11		4		0		68
	L	20	750	178:1	As plot A	2	76	5	8	2	6	52	66	26	38

* bF = balsam fir, *Abies balsamea* (L.) Mill.; wS = white spruce *Picea glauca* (Moench) Voss.
** ND = not determined.

levels of CPV as well as of NPV infection were recorded (Table 66) (Bird *et al.* 1972, Howse *et al.* 1973). Follow-up studies were undertaken on these plots between 1972 and 1977 (Table 67). The incidence of NPV was higher than CPV during this period. Although not spectacular, some saving of foliage was recorded between 1972 and 1975 (Cunningham *et al.* 1975*a*).

Table 67 Follow-up studies on two white spruce, *Picea glauca* (Moench) Voss, plantations treated with NPV plus CPV (400:1) in 1971*

Plot	Year	Highest virus infection %		Population reduction due to virus carry-over (%)	Current year's foliage saved (%)
		NPV	CPV		
G	1972	28	8	20	0
	1973	17	5	82	22
	1974	6	1	56	27
	1975	21	0	74	0
	1976	8	0	ND**	ND
	1977	3	0	ND	ND
H	1972	24	3	65	63
	1973	14	0	68	20
	1974	6	0	62	44
	1975	9	0	48	0
	1976	0	0	ND	ND
	1977	2	0	ND	ND

* See Table 66 for details and results in 1971.
** ND = not determined.

Rigorous quality control facilitated production of pure NPV between 1972 and 1978, but in 1979 a CPV contaminant was again found and attempts to eliminate it were unsuccessful both that year and in 1980. In 1979, the ratio of NPV:CPV was 178:1, and in 1980 the same preparation and one with an NPV:CPV ratio of 300:1 were applied (W.J. Kaupp 1980, personal communication) (Table 66). In 1979, the viruses were applied either as second-instar larvae emerged from hibernacula or on later instars following budflush. Following the early application, high levels of virus infection and acceptable foliage protection were recorded. On the remaining plots reasonable levels of virus infection and population reduction but little or no foliage protection were recorded. In 1980, an attempt was made to repeat the early applications made in 1979. However, when the spray was applied about 90% of the larvae had mined into needles and buds. Infection levels were lower than in 1979, although moderate levels of both population reduction and foliage protection were recorded.

Nuclear polyhedrosis virus

Tests were conducted with pure NPV in 1972 (Cunningham & McPhee 1973), in 1973 (Cunningham *et al.* 1974) in 1974 (Cunningham *et al.* 1975*b*), in 1975 (Cunningham *et al.* 1975*c*), in 1976 (Kaupp *et al.* 1978), in 1977 (Cunningham *et al.* 1978), and in 1978 (Cunningham *et al.* 1979). During that 6-year period, 33 plots with a total area of 1741 ha were treated (Table 68). In 1972 the trials were a failure because of the same late snow fall that affected the EPV trials that year. In subsequent years, tests were made with lower dosages than those used in the original 1971 test, a variety of tank mixes containing

Table 68 Applications of NPV on spruce budworm, *Choristoneura fumiferana* (Clem.), in Ontario using a fixed-wing aircraft fitted with either boom and nozzle (B and N) or Micronair® (M) equipment*

Year	Plot	Area (ha)	Dosage, PIB/ha ($x10^9$)	Spray equipment	Tank mix additives	Predominant instar treated	Deposit card, droplets/cm^2	Highest virus infection (%)		Population reduction due to treatment (%)		Current year's foliage saved (%)	
								bF**	wS	bF	wS	bF	wS
1972	1	168	100	M	2.5% IMC 90-001 10% molasses 0.2% Biofilm	II & III	98	28	26	ND***	ND	ND	ND
	2	84	100	M	As plot 1 (1972)	II & III	131	15	24	ND	ND	ND	ND
1973	1	40	25	M	2.5% IMC 90-001	II & III	16	2	8	33	17	0	0
	2	28	62.5	M	As plot 1 (1973)	II & III	17	5	17	37	50	0	0
	3	56	125	M	As plot 1 (1973)	II & III	23	12	37	14	59	0	14
	4	30	50(x2)	M	As plot 1 (1973)	II & III	7 & 14	3	11	47	74		
	5	44	62.5	M	As plot 1 (1973)	IV, V & VI	ND	2	10	21	0	9	0
	6	32	125	M	As plot 1 (1973)	IV, V & VI	14	10	13	45	43	2	4
	7	32	125	M	As plot 1 (1973)	IV, V & VI	39	17	26	ND	ND	0	2
					0.13% Chevron® sticker	IV, V & VI	39	17	26	ND	ND	0	2
	8	20	125	M	2.5% IMC 90-001 0.2% Biofilm®	IV, V & VI	32	7	13	ND	ND	ND	ND
	9	13.2	125	M	Water alone	IV, V & VI	ND	6	7	ND	ND	ND	ND
1974	1	520	250	M	3% IMC 90-001 0.13% Chevron® sticker	IV	48	19	12	58	5	0	4
	2	6	250	M	As plot 1 (1974)	IV & V	72		15		46		17
1975	1	120	125	M	50% Sandoz adjuvant V®	IV & V	167	12	13	21	91	0	11
	2	40	250	M	3% Shade® 0.13% Chevron® sticker	IV, V & VI	38	4	6	31	18	73	0
	3	80	125	B & N	6% Shade® 0.13% Chevron® sticker	IV & V	32	17	40	53	91	0	2
	4	60	125	B & N	50% Sandoz adjuvant V®	IV & V	22	31	17	65	84	0	10
	5	18	44	B & N	50% Sandoz adjuvant V®	IV & V	29	15	10	0	32	0	1
1976	1	40	125(x2)	B & N	25% molasses 6% Shade® 0.13% Chevron® sticker	II	23 & 23		12		52		0
	2	40	125	B & N	As plot 1 (1976)	II	17		5		34		0
	3	40	125(x2)	B & N	As plot 1 (1976)	II	11 & 8		9		0		13
1977	1	40	750	B & N	As plot 1 (1976)	V & VI	22	43	61		92		0
	2	20	250	B & N	Polymer bound carbon and virus 0.1% Triton® X-100	IV & V	14	1	2	45	11	3	0
	3	8	250	B & N	1 kg l San® 285 WP	IV & V	39	4	12	0	0	31	0
	4	40	250	B & N	25% molasses 6% Shade® 0.13% Chevron® sticker	IV & V	40	25	19	58	22	89	67
	5	40	250	B & N	As plot 4 (1977)	IV & V	50	16	30	60	81	10	0
1978	1	20	750	B & N	25% molasses 6% Shade® 0.05% Triton® B-1956	V & VI	47		28		33		20
	2	20	750	B & N	As plot 1 (1978)	V & VI	66	21	24	76	57	0	0
	3	20	750	B & N	As plot 1 (1978)	IV, V & VI	47		35		92		0
	4	20	750	B & N	As plot 1 (1978)	IV, V & VI	53	16	57	37	89	16	0
	5	20	750	B & N	As plot 1 (1978)	IV, V & VI	73		38		92		0
	6	20	750	B & N	As plot 1 (1978)	IV, V & VI	37	26	46	64	83	0	0
	7	14	750	B & N	66% Sunspray® 11E oil	VI	28	26	17	41	33	0	0

* Emission rate on all plots was 9.4 l/ha except on 1972 plot 1 where it was 18.8 l/ha.
** bF = balsam fir, *Abies balsamea* (L.) Mill.; wS = white spruce, *Picea glauca* (Moench) Voss.
*** ND = not determined.

ingredients to enhance the deposit and increase the leaf life of the virus, single and double applications, and applications at different stages of larval development; in addition, applications with boom and nozzle and Micronair® (Micronair (Aerial) Ltd. U.K.) equipment were compared. Mediocre results were obtained until 1977, when the same dosage used in 1971 (750 x 10^9 polyhedral inclusion bodies (PIB) per hectare) was retested. Applied on later-instar larvae, this gave an excellent level of virus infection and good population reduction, but no foliage protection (Cunningham *et al.* 1978). Between 1978 and 1980 the highest percentage of NPV infection in larvae from this plot declined from 9 to 2%. However, the spruce budworm population remained fairly low and little defoliation was recorded during this period.

Following the encouraging results with the 750 x 10^9 PIB/ha dosage used in 1977, this treatment was replicated on six plots in 1978. Levels of virus infection were not as high as in 1977 and population reductions ranged from 37 to 76% on balsam fir hosts and 33 to 92% on white spruce hosts. Again little or no foliage protection was attained in the year of application. Also, in 1978, an oil-based formulation of NPV was tested for the first time on spruce budworm. Under the high humidity conditions at the time of application, the results were no better than with aqueous formulations but this test did demonstrate that application of an oil-based formulation is feasible (Cunningham *et al.* 1979).

Granulosis virus

One 6-ha plot was treated with GV in 1979 and one 10-ha plot in 1980. It is difficult to recognize GV using a light microscope, therefore samples of larvae were reared in the laboratory to assess the impact of the treatment, in addition to population and defoliation studies (Table 69). Neither of these trials was considered to be an adequate test of this virus. A fungal epizootic decimated the spruce budworm population in the plot treated in 1979; in 1980 the treatment on second-instar larvae was applied after 90% had mined into needles and buds, and not as they emerged from hibernacula as had been intended.

Table 69 Application of GV on white spruce, *Picea glauca* (Moench) Voss, plantations using a fixed-wing aircraft fitted with Micronair® equipment and calibrated to deliver 9.4 l/ha. The tank mix contained 25% molasses, 6% Shade®, and 1% Chevron® sticker

Year	Area (ha)	Dosage, capsules/ ha	Predominant instar treated	Deposit card, droplets/ cm^2	Mortality in treated larvae reared in laboratory (%)	Population reduction due to treatment (%)	Current year's foliage saved (%)
1979	6	2 x 10^{14}	5 & 6	16	86	74	17
1980	10	6.4 x 10^{13}	2	29	27	0	49

Applications of chemical insecticides followed by viruses

In 1972, a comparison was made between a low dosage of the chemical insecticide fenitrothion alone, NPV alone, EPV plus NPV alone, fenitrothion followed by NPV, and fenitrothion followed by EPV plus NPV (Table 70) (Morris *et al.* 1972, 1974). Generally better results were obtained with the chemical–virus combinations than with either the viruses or chemical alone. More foliage was saved in the year of application in

Table 70 Application of fenitrothion at 18 g/ha in 444 ml Hysol solvent/ha followed by applications of EPV contaminated with NPV (24 h later) and NPV alone (14 h later) at 9.4 l/ha on third-, fourth-, and fifth-instar spruce budworm, *Choristoneura fumiferana* (Clem.) using a fixed wing aircraft fitted with Micronair® equipment*

				Highest levels of virus infection (%)				Population reduction due to treatment (%)		Current year's foliage saved in 1972 (%)		Current year's foliage saved in 1973 (%)	
	Area		PIB/ha	NPV		EPV							
Plot	(ha)	Treatment	(x 10^9)	bF**	wS	bF	wS	bF	wS	bF	wS	bF	wS
1	16	EPV + NPV	175	8	15	5	1	0	22	28	17	44	13
2	16	Fenitrothion followed by EPV + NPV	175	2	4	17	29	50	39	77	32	56	28
3	16	NPV	720	22	74	0	0	54	76	36	14	64	42
4	16	Fenitrothion followed by NPV	720	27	26	0	0	78	72	82	38	61	57
5	48	Fenitrothion	0	0	0	0	0	51	35	15	4	0	0

* The viruses were incorporated in a tank mix containing 10% molasses, 2.5% IMC 90-001, and 0.2% Biofilm®.

** bF = balsam fir, *Abies balsamea* (L.) Mill.; wS = white spruce, *Picea glauca* (Moench) Voss.

the plots treated with virus alone or the chemical–virus combination than with the chemical alone, and this saving of foliage was also recorded the following year.

In 1979, a 40-ha plantation containing both black spruce, *P. mariana* (Mill.) B.S.P., and white spruce was treated with acephate when larvae were in the fourth instar, and with NPV 8 days later when they were in the fifth and sixth instars. The two treatments were evaluated separately and together (Table 71). Significant foliage protection was obtained in the year of application and also in the following year.

Table 71 Application of acephate at 560 g/ha on fourth-instar spruce budworm, *Choristoneura fumiferana* (Clem.), larvae followed 8 days later by application of NPV at 750 x 10^9 PIB/ha on fifth- and sixth-instar larvae using a fixed-wing aircraft fitted with Micronair® equipment on a 40-ha black spruce, *Picea mariana* (Mill.) B.S.P., and white spruce, *Picea glauca* (Moench) Voss, plantation in 1979, and an assessment of the impact of the treatment in 1980*

	Highest NPV (%)		Population reduction due to acephate (%)		Population reduction due to NPV (%)		Combined population reduction %		Current year's foliage saved %	
Year	bS**	wS	bS	wS	bS	wS	bS	wS	bS	wS
1978	36	37	72	85	0	47	45	88	73	73
1980	ND***	ND					12	43	21	69

* The viruses were incorporated in a tank mix containing 25% molasses, 6% Shade® and 1% Chevron® sticker.

** bS = black spruce, *Picea mariana* (Mill.) B.S.P.; wS = white spruce, *Picea glauca* (Moench) Voss.

*** ND = not determined.

Evaluation of Control Attempts

After 10 years of field trials, it has become obvious that use of the viruses now available as biological control agents for spruce budworm offers no simple solution to the spruce budworm problem in eastern Canada, although a great deal of useful information has been assimilated during this period. None of the four viruses initiated an epizootic causing population collapse, and disease did not spread from treated to untreated areas. This confinement is probably a factor of the host-parasite relationship and the biology of the adult female budworm. There is no evidence of transovum or transovarial transmission of viruses by female moths. Persistence of viruses from one year to the next is thought to be by contamination of foliage with virus inoculum leaching from virus-killed larvae and pupae that overwinter webbed in nests on the foliage. Such a mode of transmission is not nearly as effective as some form of ovarial transmission. It results in little or no spread of the virus and is density-dependent. As the insect population density declines, this source of inoculum is lost.

Results of field trials with spruce budworm viruses are often difficult to interpret and anomalies can occur due to sampling errors and differing effects of natural control factors, including weather, on treated and check plots. Such anomalies are not unique to the assessment of virus treatments and often occur when assessing chemical insecticide treatments. In the tables it can be seen that low levels of virus infection sometimes give high population reduction and vice versa. When foliage protection is also considered, the results can be even more confusing although, when viruses are applied on fifth- and sixth-instar larvae, no foliage protection is anticipated in the year of application. The virus that kills the quickest, NPV, takes at least 14 days to cause mortality under the temperature conditions normally encountered in Ontario. In general, better results are obtained on white spruce hosts than balsam fir hosts and this again is probably a density-dependent factor with higher spruce budworm populations being encountered on white spruce. Although not documented in the tables, fairly consistent results have been obtained in pupal emergence studies. Pupal mortality following application of viruses is not dramatic, but successful emergence is usually from 10 to 30% less than from pupae collected from untreated check plots.

As infestations of spruce budworm age in a locality, the level of a microsporidian parasite, *Nosema fumiferanae* (Thompson), increases. In new outbreaks, fewer than 10% of larvae carry this parasite and as the population ages, levels often build up to 70 or 80% (Wilson 1981, G.G. Wilson 1981, personal communication). The level of microsporidian infection in larvae from all the plots treated with virus, and corresponding check plots, was recorded and in most cases more than 50% of the larvae were infected. Preliminary experiments have indicated that the presence of microsporidia is antagonistic to NPV infection and replication in spruce budworm larvae, but further experimentation is required to verify these results. In laboratory tests, microsporidia-free western spruce budworm, *Choristoneura occidentalis* Freeman, larvae are considerably more susceptible to both NPV and GV than microsporidia-free *C. fumiferana.* Microsporidia in populations of *C. occidentalis* are rare or totally absent, thus making it a more attractive candidate for population regulation with viruses than *C. fumiferana.*

Spruce budworm NPV has been propagated in cell cultures (Arif *et al.* 1976), but this method is at present not feasible for mass production; hence all the virus material used in field trails is grown in either spruce budworm larvae or western spruce budworm larvae. This procedure is labour intensive and expensive. Spruce budworm larvae are small in size compared to many other lepidopterous larvae in which viruses are produced on a large scale. A carefully selected spruce budworm sixth-instar larva killed with NPV yields 5×10^8 PIB, but on a production line basis this figure drops to 1×10^8. Therefore, to produce a 1-ha dosage of 750×10^9 PIB it is necessary to rear, infect, harvest, and process 7500 budworm larvae. Obviously, at the present high cost NPV cannot be applied over large areas and its use will probably be restricted to high value stands and environmentally sensitive areas. If it can be demonstrated conclusively that treatment with a virus gives foliage protection for several years following the year of application, then the high initial cost of the treatment becomes more acceptable.

The application of contaminated virus preparations in 1971 was accidental. The presence of a CPV contaminant in the NPV used in 1979 and 1980 was known, but the material was still applied on the basis of the 1971 results. The minute amounts of CPV gave dramatic infection levels in the spruce budworm population that were often higher than the levels of NPV. This was surprising because the ratios of NPV:CPV were 400:1, 300:1, and 178:1. It appears that there is a synergistic effect between NPV and CPV, but this would require detailed laboratory investigations to prove conclusively.

Foliage protection in the year of application was demonstrated in 1979 following treatment with NPV plus CPV on second-instar larvae as they emerged from hibernacula. Treatment of second-instar larvae has many benefits, although timing is extremely critical and it would only be practical to treat small, carefully monitored areas. Second-instar larvae are much more susceptible to infection with NPV than later instars and the LD_{50} in terms of numbers of PIB has been calculated for different instars as follows: second, 25; third, 204; fourth, 462; fifth, 2 514; and sixth, 3 643 (F.T. Bird 1978, unpublished data). At the time of emergence from hibernacula, leaves on the hardwood overstorey, frequently found in spruce–fir stands, have not flushed and a good aerial spray deposit can be obtained on the target species.

To date, the tank mix considered most suitable for spruce budworm virus applications was an aqueous suspension containing 25% molasses, 6% Shade® and 1% Chevron® sticker. It has been widely used with the NPVs of both Douglas-fir tussock moth, *Orgyia pseudotsugata* (McDunnough), and gypsy moth, *Lymantria dispar* (L.). Since Shade® is no longer commercially available, it will be necessary to find a new screening agent against ultra-violet light or a totally new formulation.

Safety tests on spruce budworm NPV conducted at Ontario Veterinary College showed that this virus presents no hazard to birds and mammals (Valli *et al.* 1976). Fish tests, although indicating no hazard, were inconclusive and should be repeated (Savan *et al.* 1979). There are only limited safety testing data available for EPV (Buckner & Cunningham 1972), and no CPV and GV safety testing data are available. These would be required before any of these three viruses could be registered by Canadian authorities.

Recommendations

The use of any of the four viruses with inclusion bodies that infect spruce budworm is still at the experimental stage. In the near future only treatment of small areas can be considered. Applications of NPV or NPV plus CPV on second-instar larvae as they emerge from hibernacula should be retested in an effort to find if, under closely monitored conditions, this is a feasible and effective strategy.

An improved tank mix that will keep the virus in a viable state on the foliage for more than a week would greatly increase the effectiveness of treatments and possibly allow lower dosages. Research on formulations is being conducted by both commercial and government agencies in the United States and formulations developed for other pathogens, such as bacteria and fungi, will probably have the characteristics suitable for use with viruses.

The search for new viruses or strains of viruses that infect spruce budworm should be continued. There are several isolates of NPV , GV, and EPV from different *Choristoneura* species, but none has been found that is more virulent than the strains used in field trials. The search for larger, easily reared host larva for spruce budworm virus production has also proved unsuccessful, but should be continued.

A long-term goal in the use of viruses for the regulation of spruce budworm population is to study the nucleic acid of spruce budworm NPV with a view to constructing a genetic map. Once this is accomplished, techniques of genetic manipulation can be used in an attempt either to increase the virulence of spruce budworm NPV or to modify some highly virulent, easily produced insect virus in such a way as to render it capable of infecting spruce budworm.

Literature Cited

Abbott, W.S. (1925) A method of computing the effectiveness of an insecticide. *Journal of Economic Entomology* 18,265–267.

Arif, B.M.; Sohi, S.S.; Krywienczyk, J. (1976) Replication of alkali-released NPV in a cell line of *Choristoneura fumiferana*. *Proceedings of the 1st International Colloquium on Invertebrate Pathology Kingston, Ontario,* pp. 108–110.

Bird, F.T.; McPhee, J.R. (1970) Susceptibility of spruce budworm to pure nuclear polyhedrosis virus (NPV) sprays. *Canadian Department of Fisheries and Forestry Bi-monthly Research Notes* 26,35.

Bird, F.T.; Sanders, C.J.; Burke, J.M. (1971) A newly discovered virus disease of the spruce budworm, *Choristoneura biennis* (Lepidoptera: Tortricidae). *Journal of Invertebrate Pathology* 18,159–161.

Bird, F.T.; Cunningham, J.C.; Howse, G.M. (1972) Possible use of viruses in the control of spruce budworm. *Proceedings of the Entomological Society of Ontario* 103,69–75.

Buckner, C.H.; Cunningham, J.C. (1972) The effect of the poxvirus of the spruce budworm, *Choristoneura fumiferana* (Lepidoptera: Tortricidae), on mammals and birds. *Canadian Entomologist* 104,1333–1342.

Cunningham, J.C.; McPhee, J.R. (1973) Aerial application of entomopoxvirus and nuclear polyhedrosis virus against spruce budworm at Chapleau, Ontario, 1972. *Canadian Forestry Service, Sault Ste. Marie, Information Report* IP-X-3, 27 pp.

Cunningham, J.C.; McPhee, J.R.; Howse, G.M.; Harnden, A.A. (1974) Aerial application of a nuclear polyhedrosis virus against spruce budworm near Massey and Aubrey Falls, Ontario in 1973 and a survey of the impact of the virus in 1974. *Canadian Forestry Service, Sault Ste. Marie, Information Report* IP-X-4, 45 pp.

Cunningham, J.C.; Howse, G.M.; Kaupp, W.J.; McPhee, J.R.; Harnden, A.A.; White, M.B.E. (1975*a*) The persistence of nuclear polyhedrosis virus in populations of spruce budworm. *Canadian Forestry Service, Sault Ste. Marie, Information Report* IP-X-10, 32 pp.

Cunningham, J.C.; Kaupp, W.J.; McPhee, J.R.; Howse, G.M.; Harnden, A.A. (1975*b*) Aerial application of nuclear polyhedrosis virus against spruce budworm on Manitoulin Island, Ontario in 1974 and a survey of the impact of the virus in 1975. *Canadian Forestry Service, Sault Ste. Marie, Information Report* IP-X-8, 30 pp.

Cunningham, J.C.; Kaupp, W.J.; McPhee, J.R.; Howse, G.M.; Harnden, A.A. (1975*c*) Aerial application of nuclear polyhedrosis virus on spruce budworm on Manitoulin Island, Ontario in 1975. *Canadian Forestry Service, Sault Ste. Marie, Information Report* IP-X-9, 35 pp.

Cunningham, J.C.; Kaupp, W.J.; Howse, G.M.; McPhee, J.R.; de Groot, P. (1978) Aerial application of spruce budworm baculovirus: tests of virus strains, dosages and formulations in 1977. *Canadian Forestry Service, Sault Ste. Marie, Information Report* FPM-X-3, 37 pp.

Cunningham, J.C.; Howse, G.M.; McPhee, J.R.; de Groot, P.; White, M.B.E. (1979) Aerial application of spruce budworm baculovirus: replicated tests with an aqueous formulation and a trial using an oil formulation in 1978. *Canadian Forestry Service, Sault Ste. Marie, Information Report* FPM-X-21, 19 pp.

Grisdale, D. (1970) An improved method for rearing large numbers of spruce budworm, *Choristoneura fumiferana* (Lepidoptera: Tortricidae). *Canadian Entomologist* 102,1111–1117.

Howse, G.M.; Sanders, C.J.; Harnden, A.A.; Cunningham, J.C.; Bird, F.T.; McPhee, J.R. (1973) Aerial application of viruses against spruce budworm, 1971. Part A. Impact in the year of application (1971). Part B. Impact in the year following application (1972). *Canadian Forestry Service, Sault Ste. Marie, Information Report* 0-X-189, 62pp.

Kaupp, W.J. Cunningham,; J.C.; Howse, G.M.; McPhee, J.R.; de Groot, P. (1978) Aerial application of spruce budworm baculovirus: tests on second instar larvae in 1976. *Canadian Forestry Service, Sault Ste. Marie, Information Report* FPM-X-2, 20 pp.

McMorran, A. (1965) A synthetic diet for the spruce budworm, *Choristoneura fumiferana* (Clem.) (Lepidoptera: Tortricidae). *Canadian Entomologist* 97,58–62.

Morris, O.N.; Armstrong, J.A.; Hildebrand, M.J.; Howse, G.M.; Cunningham, J.C.; McPhee, J.R. (1972) Aerial applications of virus-insecticide combinations against spruce budworm *Choristoneura fumiferana* (Clem.) (Tortricidae: Lepidoptera) at Rankin, Ontario, 1972. *Canadian Forestry Service, Ottawa, Information Report* CC-X-37, 65 pp.

Morris, O.N.; Armstrong, J.A.; Howse, G.M.; Cunningham, J.C. (1974) A 2-year study of virus-chemical insecticide combination in the integrated control of the spruce budworm, *Choristoneura fumiferana* (Tortricidae: Lepidoptera). *Canadian Entomologist* 106,813–824.

Savan, M.; Budd, J.; Reno, P.W.; Darley, S. (1979) A study of two species of fish inoculated with spruce budworm nuclear polyhedrosis virus. *Journal of Wildlife Diseases* 15,331–334.

Stairs, G.R.; Bird, F.T. (1962) Dissemination of viruses against the spruce budworm, *Choristoneura fumiferana* (Clemens). *Canadian Entomologist* 94,966–969.

Valli, V.E.; Cunningham, J.C.; Arif, B.M. (1976) Tests demonstrating the safety of a baculovirus of the spruce budworm to mammals and birds. *Proceedings of the 1st International Colloquium on Invertebrate Pathology, Kingston, Ontario,* pp. 445–446.

Wilson, G.G. (1981) *Nosema fumiferanae*, a natural pathogen of a forest pest: potential for pest management. In: Burges, H.D. (*Ed.*) Microbial control of pests and plant diseases 1970–1980. London; Academic Press, pp. 595–601.

C. Application of Microsporidia and Fungi, and of Genetic Manipulation

G.G.WILSON, D. TYRRELL and T.J. ENNIS

Background

Microsporidia

Microsporidian parasites have been recognized as important factors in the natural regulation of the densities of certain insects but only recently have a few been investigated as possible control agents of pest insects. Microsporidia generally produce chronic rather than acute diseases and will therefore be most useful in long-term control programmes. Species currently being considered in North America include *Nosema locustae* Canning against grasshoppers, *Nosema pyraustae* (Paillot) against European corn borers, *Nosema fumiferanae* (Thompson) against the spruce budworm, *Vairimorpha necatrix* (Kramer), and *Pleistophora schubergi* Zwölfer against various insects. For a more extensive review of these microsporidian parasites the reader is referred to Burges (1981).

Microsporidia are common in many insects. In order to develop the parasite for use in pest management a complete understanding of the biology and host-parasite relationships of microsporidia is essential. Furthermore, methods of evaluating the incidence and effects of microsporidiosis in natural populations of insects are required, and methods for economical propagation and efficacious spraying are needed.

N. fumiferanae is the most common microsporidian parasite of the spruce budworm, *Choristoneura fumiferana* (Clem.), and may also infect other *Choristoneura* spp. The jack pine budworm, *C. pinus* Freeman, is infected by a microsporidium that is very similar to *N. fumiferanae*. *P. schubergi* is less frequently detected in spruce budworm, but it has been identified from other lepidopterous pests including *Hyphantria cunea* (Drury), *Archips cerasivoranus* (Fitch), and others (Wilson 1981). A third microsporidium has been encountered in field populations of the spruce budworm and was tentatively assigned to the genus *Thelohania*. This microsporidium occurred in a low percentage of the spruce budworm larvae examined (0.5%) and attempts to infect healthy spruce budworm larvae with this parasite were not successful.

Observations of laboratory and wild stock of spruce budworm show that *N. fumiferanae* retards larval and pupal development, and reduces pupal weight and fecundity (Table 72) (Thomson 1958, Wilson 1977*a*).

Table 72 Effects of natural infection of *Nosema fumiferanae* (Thompson) on pupal weight and egg production of *Choristoneura fumiferana* (Clem.)

	Lab-reared stock (Wilson 1977a)			Field-collected stock (Thomson 1958)		
	Pupal weight (mg)		Egg	Pupal weight (mg)		Egg
	Male	Female	complement	Male	Female	complement
Non-infected	70(70)*	108(70)	153(70)	57(33)	102(28)**	169
Infected	63(30)*	82(30)**	126(30)***	58(17)	78(13)**	153***

* Number in parentheses indicates size of sample.
** Significantly different from non-infected sample at 0.01 level of probability.
*** Significantly different from non-infected sample at 0.05 level of probability.

The natural levels of microsporidian infection in spruce budworm populations tend to increase with the age of the infestation. Examination of spruce budworm larvae collected in the Uxbridge forest of southern Ontario over a 6-year period indicated that the infection rate from *N. fumiferanae* increased from 36% in 1973 to 69% in 1978 (Wilson 1977*b*). The Uxbridge forest was in fact an old infestation; severe defoliation was recorded in 1952 and has persisted at fluctuating levels ever since.

Fungi

Steady progress was made during the 1970s in documenting the effects of fungi in natural epizootics and developing various species for use in biological control, with the result that *Hirsutella thompsoni* Fisher, the first fungus to be registered for mite control in North America, is now available under the trade name Mycar® (Abbott Laboratories) and is also used against citrus rust mites.

The occurrence of entomophthoraceous fungi on spruce budworm has been documented throughout much of the insect's range in the last 10 years. Outbreaks of fungal disease occurred in 1974 and 1979 in northern Ontario; in the former year the outbreak was of *Zoophthora radicans* (Brefeld) A. Bakto (= *Entomophthora sphaerosperma* Frasenius) (Harvey & Burke 1974), in the latter it was of the same fungus together with *E. egressa* MacLeod (Tyrrell, unpublished). In Newfoundland, the same two fungi were first recorded from the spruce budworm in 1972. The disease was widespread in western Newfoundland in 1973, and subsequently spread to the whole of the Island (Otvos & Moody 1978). Mortality varied from 10–40%, depending on weather conditions, and occasionally reached much higher values in localized areas. Both fungal species, together with an unidentified *Conidiobolus* species, were responsible for spuce budworm mortality in northern Maine, where disease prevalence varied between 4 and 7% during 1975–77 (Vandenberg & Soper 1978).

In northern Ontario, mortality was higher in white spruce than in balsam fir in 1974, when fungal incidence was high (Harvey & Burke 1974), but no difference was observed between tree species in 1980, when fungal incidence was much lower (Tyrrell, unpublished). In Maine, fungal prevalence was found to be both density-dependent and significantly higher in the lower crown area of the tree, and appeared to be enhanced by periods of cool, wet weather (Vandenberg & Soper 1978). The link between moist conditions and increased incidence of fungus has also been observed in Newfoundland (Otvos & Moody 1978).

Fungal infection occurs in the later stages of larval development. Incidence of *Z. radicans* peaked between the fifth and sixth instars, *E. egressa* between the sixth instar and pre-pupa, and *Conidiobolus* sp. between the pre-pupal and pupal stages, respectively, in Maine (Vandenberg & Soper 1978); while in northern Ontario peak fungal infection coincided with the peak development of sixth-instar larvae (Tyrrell unpublished). Earlier instars however, are not resistant to the disease, and can readily be infected with *Z. radicans* under laboratory conditions (Kenneth 1978, Perry unpublished).

Both *Z. radicans* and *E. egressa* can be isolated and grown in pure culture, and a number of different isolates of both species are now available. *Z. radicans* grows readily on a variety of solid and liquid media, and the mycelial stage can also be grown in small fermentors. *E. egressa* grows slowly on complex media such as coagulated egg yolk, but can be grown as a protoplast in osmotically stabilized media.

Z. radicans can sporulate on artificial media to give both conidia and resting spores. The effects of environmental conditions, including temperature, photophase, and nutrient, on conidial germination have been determined (van Roermund, personal communication). Maturation of laboratory-produced resting spores under controlled conditions has been shown to be a key factor in potentiating their subsequent germination (Perry & Tyrrell in press), and the effects of environmental conditions on their germination are currently being determined. A bioassay procedure for conidia of *Z. radicans* has been developed (Vandenberg & Soper 1979), and subsequent improvements (Soper unpublished) have allowed the comparison of the infectiveness of different isolates.

Genetic manipulation

The release of sterilized or genetically altered insects to reduce reproductive success and survival of their own species has been successful against some agricultural pests and has replaced chemical insecticides in these cases (Lachance 1979). The technique employs chemical- or radiation-induced sterilization of mass-produced males without weakening their mating capability, followed by inundative release so that the reared population displaces the wild population of males in coition.

Outram (1969) first discussed the theoretical potential for genetic control of the spruce budworm following the approach that had been so successful against the screwworm.

Field Trials

Microsporidia

In order to conduct field trials, microsporidian spores must be mass produced. Propagation of the microsporidia begins 4 months before field spraying. Second-instar larvae are placed in 28.4 ml plastic cups filled with 20 ml of synthetic diet sprayed with spores of

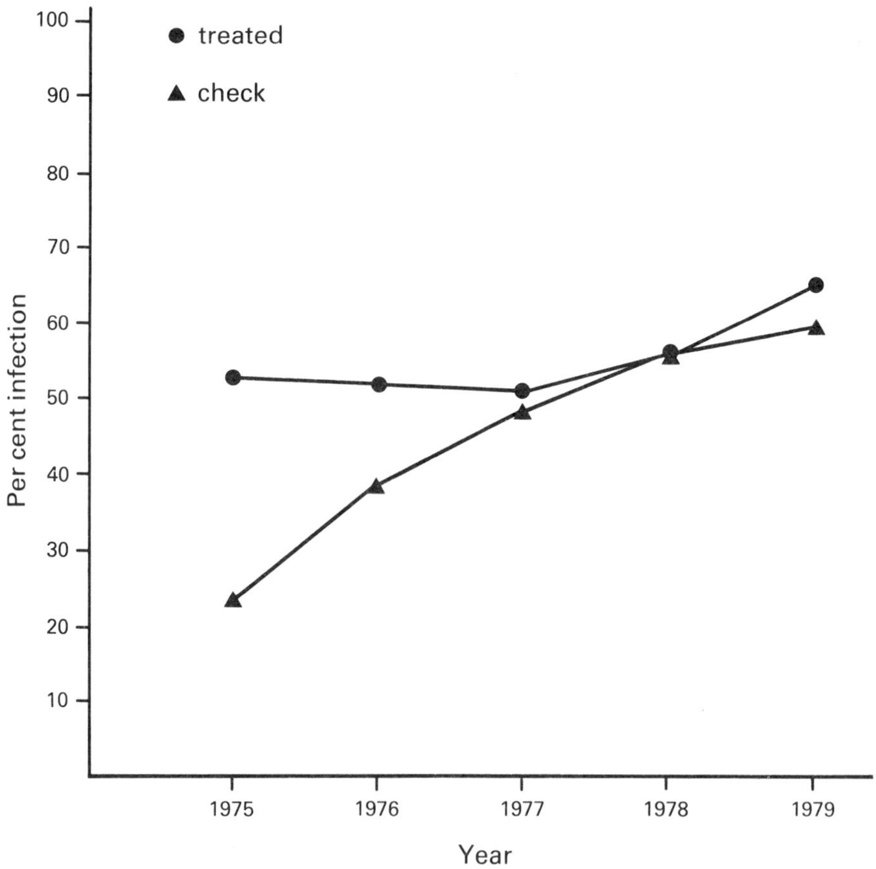

Fig. 9. Incidence of *Nosema fumiferanae* (Thomps.) in living spruce budworm, *Choristoneura fumiferana* (Clem.), larvae collected from white spruce, *Picea glauca* (Moench) Voss, trees sprayed 2 June 1975 with 1.8 x 10^{11} spores/tree

N. fumiferanae in distilled water. Shortly before pupation, the contaminated larvae are homogenized in water and strained through two layers of cheesecloth to remove debris. The filtrate containing the spores of *N. fumiferanae* is collected and stored at 4°C. Most of the supernatant can be decanted after 2–3 days, when the spores have settled. The concentration of pooled mixed batches is determined by counting with a haemacytometer.

Spray trials of the microsporidian parasites, *N. fumiferanae* and *P. schubergi*, were conducted against spruce budworm in 1975 and 1976, and progress of the infection was monitored until 1979 (Wilson 1980). Vigorous spruce budworm-infested white spruce, *Picea glauca* (Moench) Voss, about 6 m tall were selected for the 1975 trials and white spruce and balsam fir, *Abies balsamea* (L.) Mill., for the 1976 trials. Sample trees were sprayed by backpack mist blower with 1500 ml of an aqueous formulation containing 25% (v/v) molasses, 30 g/l of IMC 90-001 (International Minerals Corp.) sunlight protectant and the desired spore concentration to yield a specified application rate per tree. Application rates were 1.8 x 10^{11} spores per tree in 1975 and 5 x 10^{10} in 1976.

The results for 1975 showed that 25 days after treatment 53% of the spruce budworm collected from treated white spruce were infected with *N. fumiferanae* whereas only 23% of the larvae from the check area were infected. Levels of infection in 1976, 19 days after treatment with *N. fumiferanae*, were 44% on white spruce and 66% on balsam fir compared with 27 and 39% respectively for the checks. The higher levels of infection persisted in the treated trees for about 3 years (Figs. 9 and 10). Application of 5.0 x 10^{10} *P.*

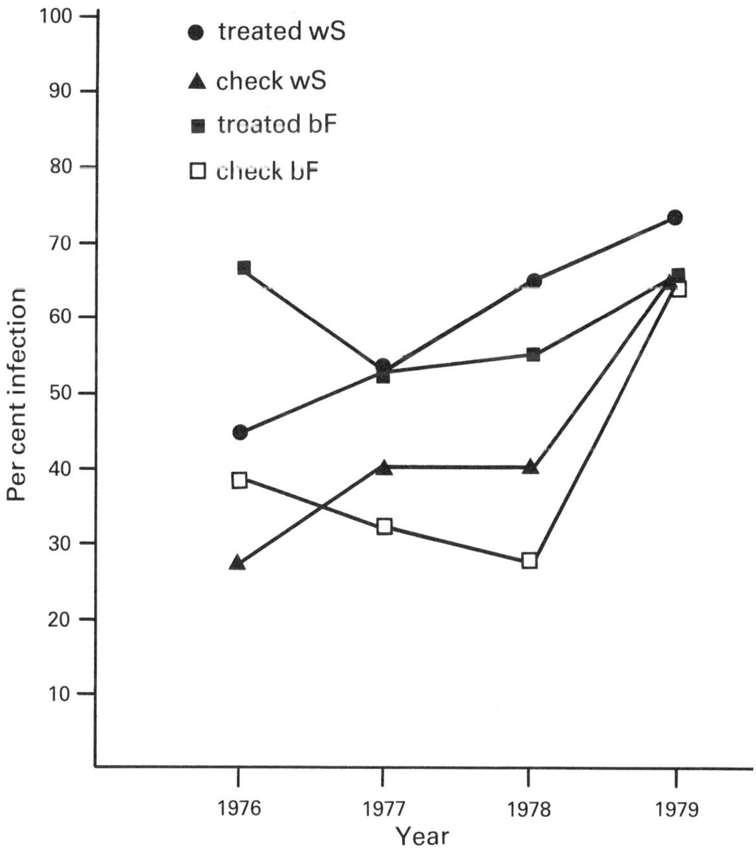

Fig 10 Incidence of *Nosema fumiferanae* (Thomps.) in spruce budworm, *Choristoneura fumiferana* (Clem.), larvae collected from white spruce, *Picea glauca* (Moench) Voss, (wS) and balsam fir, *Albies balsamea* (L.) Mill., (bF) trees sprayed 4 June 1976 with 5.0 x 10^{10} spores/tree

schubergi spores resulted in infection rates of 64.8% for white spruce and 96.3% for balsam fir 19 days after spraying. There was no infection in larvae from the check area and no carry-over of the microsporidium to the following year. All samples showed higher levels of infection in spruce budworm taken from balsam fir.

Based on laboratory results there is a relationship between levels of infection and mortality. As an example, spore concentrations of 10^9 spores/ml fed to second-instar larvae resulted in 100% mortality in up to 24 days, with 50% mortality occurring about 14 days after infection (balsam fir buds were dipped in the spore suspension). As a comparison, spore concentration of 10^6/ml resulted in 61% mortality over a period of 6–28 days. For fifth-instar larvae 10^9 spores/ml gave 75% mortality over 8–33 days, and 10^6 spores/ml resulted in only 5.6% mortality, which was similar to the controls.

Fungi

E. egressa does not sporulate well on artificial media and conidia are normally obtained from infected insect cadavers. These are easily produced by injection of insect larvae with protoplasts, which have been shown to be part of the natural life cycle of this fungus (Tyrrell 1977). Transmission of *E. egressa* from laboratory-infected to natural populations of spruce budworm under field conditions using conidia produced in this manner has been demonstrated (Lim *et al.* unpublished).

Genetic manipulation

Topical application of thiotepa or 250–300 Sv of Co^{60} irradiation induced complete sterility in male spruce budworm (Retnakaran 1970, 1971). Males sterilized with thiotepa remained fully competitive and could suppress fertility of caged laboratory populations by up to 90% (Ennis unpublished data). However, release of chemosterilized males into caged field populations produced no detectable reduction in population, mainly because of adverse effects on adult male mating behaviour. High level irradiation of male pupae or adults reduces fitness and mating competitiveness, militating against their effective use. However, at 20–50 Sv, induced chromosome aberrations have a heritable effect on fertility of subsequent generations, with depression of fertility persisting for up to three generations in the laboratory (Ennis 1979*a*). The potential of inherited semi-sterility has not yet been tested in field populations. Dispersal and behaviour of irradiated males have also been studied using pheromone re-trapping techniques (Ennis 1979*b*).

Recommendations

Few microsporidian parasites have been investigated as possible control agents of pest insects. If we take the amount of research on *Bacillus thuringiensis* Berliner (*B.t.*) before it was accepted as an insecticide as being representative in the development of a biological control agent, then it is obvious that we have barely begun the task with microsporidia. However, there are indications that many microsporidia do have good potential as biological control agents that, given appropriate research and development, may one day be realized. There is still much research needed on the development of microsporidia in a control plan for the spruce budworm. Thorough studies are required on the host-parasite relationships and the incidence and effects of the microsporidia in their natural setting. Although these parasites can be introduced into a population of the spruce budworm, their effect on insect numbers has not been quantified. The use of these parasites in integrated control should be investigated, not only in conjunction with chemicals, but also with other spruce budworm pathogens.

The level of research effort on fungi for control of spruce budworm has been low compared to the effort on such biological agents as the viruses and *B.t.*, but the results obtained to date suggest that fungi play a significant role in the regulation of spruce budworm populations, and the recent demonstration that infectious diseases can in theory drive population cycles (Anderson & May 1980) will provide further impetus for studies designed to delineate the role of fungi in the spruce budworm ecosystem. To this end, extensive epizootiological investigations on the biotic and abiotic factors affecting the fungus-budworm interaction should be carried out. These, coupled with continuing laboratory studies on the production of fungal material suitable for field dissemination, and on the selection and mode of action of the most virulent strains of fungi, will, in the long term, determine how and under what circumstances fungi may be used in the manipulation of spruce budworm populations.

At present it is difficult to draw definite conclusions on the potential of microsporidia and fungi in the management of the spruce budworm. However, because of their potential in a control programme, any comprehensive model of spruce budworm population-dynamics must embrace a thorough understanding of these pathogens.

Research into the genetic manipulation of the spruce budworm to date has supplied the basic level of information required for further testing: methods for mass sterilization, effects on fertility of caged population, and dispersal and detection of released insects. However, the large numbers of insects and extensive areas of forest affected by epidemics of this pest suggest caution in trying to develop genetic manipulation for control of outbreak populations. The logistics of application, in terms of numbers and area, and the uncertainty of success indicate that its greatest potential lies in controlling endemic population levels, where the released insects' ability to detect and mate with females even when they are relatively rare provides one of the greatest advantages of the genetic approach. Future research should be aimed at such low level populations, and should be conducted with the objective of reducing reproduction potential to a level below that necessary to support a transition from endemic to epidemic levels.

Literature Cited

Anderson, R.M.; May, R.M. (1980) Infectious diseases and population cycles of forest insects. *Science* 210,658–661.

Burges, H.D. (Ed.) (1981) Microbial control of pests and plant diseases, 1970–1980. London; Academic Press, 949 pp.

Ennis, T.J. (1979*a*) Detection and isolation of induced chromosome aberrations in Lepidoptera. *Experientia* 35,1153–1154.

Ennis, T.J. (1979*b*) A release-recapture experiment with normal and irradiated spruce budworm males. *Canadian Forestry Service Bi-monthly Research Notes* 35,9–10.

Harvey, G.T.; Burke, J.M. (1974) Mortality of the spruce budworm on white spruce caused by *Entomophthora sphaerosperma*. *Canadian Forestry Service Bi-monthly Research Notes* 30,23–24.

Kenneth, R.G. (1978) *Entomophthora sphaerosperma* as an insect pathogen for spruce budworm control. *Abstracts of the Acadian Entomological Society's 38th Annual Meeting*, pp. 35–36.

Lachance, L.E. (1979) Genetic strategies affecting the success and economy of the sterile insect release method. In: Hoy, M.A.; McKelvey, J.J. (*Eds.*) Genetics in relation to insect management. Rockefeller Foundation, USA.

Otvos, I.S.; Moody, B.H. (1978) The spruce budworm in Newfoundland: history, status and control. *Canadian Foresty Service, Newfoundland Forest Research Centre, Information Report* N-X-150.

Outram, J. (1969) Potential use of the sterility principle for spruce budworm control in eastern Canada. *Canadian Department of Fisheries and Forestry Internal Report* M–45, Fredericton, N.B.

Perry, D.; Tyrrell, D. (in press) Resting spore germination in *Zoophthora radicans*. *Mycologia*.

Retnakaran, A. (1970) Preliminary results of radiation induced sterility of the male spruce budworm. *Canadian Department of Fisheries and Forestry Bi-monthly Research Notes* 26,13–14.

Retnakaran, A. (1971) Thiotepa as an effective agent for mass sterilizing the spruce budworm, *Choristoneura fumiferana* (Lepidoptera: Tortricidae). *Canadian Entomologist* 103,1753–1756.

Thomson, H.M. (1958) The effect of a microsporidian parasite on the development, reproduction and mortality of the spruce budworm, *Choristoneura fumiferana* (Clem.). *Canadian Journal of Zoology* 36,499–511.

Tyrrell, D. (1977) Occurrence of protoplasts in the natural life cycle of *Entomophthora egressa*. *Experimental Mycology* 1,259–263.

Vandenberg, J.D.; Soper, R.S. (1978) Prevalence of entomophthorales mycoses in populations of the spruce budworm, *Choristoneura fumiferana*. *Environmental Entomology* 7,847–853.

Vandenberg, J.D.; Soper, R.S. (1979) A bioassay technique for *Entomophthora sphaerosperma* on the spruce budworm, *Choristoneura fumiferana*. *Journal of Invertebrate Pathology* 33,148–154.

Wilson, G.G. (1977*a*) The effects of feeding microsporidian (*Nosema fumiferanae*) spores to naturally infected spruce budworm (*Choristoneura fumiferana*). *Canadian Journal of Zoology* 55,249–250.

Wilson, G.G. (1977*b*) Observations on the incidence rates of *Nosema fumiferanae* (microsporidia) in a spruce budworm, *Choristoneura fumiferana*, (Lepidoptera: Tortricidae) population. *Proceedings of the Entomological Society of Ontario* 108,144–145.

Wilson, G.G. (1980) Persistence of microsporidia in populations of the spruce budworm and forest tent caterpillar. *Canadian Forestry Service, Sault Ste. Marie, Information Report* FPM-X-39.

Wilson, G.G. (1981) The potential of *Pleistophora schubergi* in microbial control of forest insects. *Canadian Forestry Service, Sault Ste. Marie, Information Report* FPM-X-49.

D. Testing of Parasitoids

I.W. VARTY

Background

The two strategies for biological control of the spruce budworm, *Choristoneura fumiferana* (Clem.), by parasitoids are the introduction of exotic species and the manipulation of native ones. Most entomologists agree that the best prospects for introduction are those species that parasitize *Choristoneura* spp. (or near relatives) in spruce (*Picea* spp.) and fir (*Abies* spp.) forests in similar climatic zones abroad. The problem is to be ready for opportunities in those few years when pest outbreaks overseas permit sizeable collections of candidate parasitoids. The manipulation of native parasitoids offers two other avenues for action: inundative release of laboratory-reared stock, and management of forest habitats to favour the dynamics of parasitoid response to available hosts.

Efforts to establish exotic parasitoids in spruce budworm populations of eastern Canada have not succeeded. In 1944–53, large numbers of four species (two tachinids, one sarcophagid, and one ichneumonoid) from unidentified western spruce budworms (British Columbia) were released at various sites from Manitoba to Newfoundland. In 1948–56, 12 ichneumonoid species from European hosts, especially *Choristoneura murinana* (Hb.) (France, Germany, Czechoslovakia), were released in small numbers in northwestern Ontario and Québec. No evidence of the establishment of any introduced species has been produced (McGugan & Coppel 1962), and there were no explanations of the causes of failure. Interest in further introductions was inhibited by these failures and by lack of sources of parasitoids abroad. However, during the 1960s unsuccessful efforts were made to obtain three ichneumonoids specific to *C. murinana* (Miller & Angus 1971).

Interest was renewed when an outbreak of *Choristoneura diversana* (Hb.) on Japanese (todo) fir (*Abies firma* Sieb & Zucc.) in Hokkaido, Japan, occurred in the late 1960s. Collaboration among four institutes (Hokkaido Forest Experimental Station, Maritimes Forest Research Centre, Commonwealth Institute of Biological Control, and the Agriculture Canada Research Station at Belleville, Ontario) resulted in the shipment of parasitoids by air from Japan to New Brunswick in 1970–75 (Tables 73 and 74). Further, a small collection of parasitoids from *C. murinana* in Czechoslovakia was shipped to New Brunswick in 1972. These introductions are detailed below.

In the early 1970s, the theory that spruce budworm epidemics may start in isolated local epicentres led to interest in methods of quelling incipient outbreaks over small areas. At the Laurentian Forest Research Centre this interest developed into a research study of the native egg parasitoid, *Trichogramma minutum* Ril., as a candidate for inundative release, now briefly reported. As pressure to find a biological alternative to

Table 73 Open releases and recoveries of parasitoids against *Choristoneura fumiferana* (Clem.)

Species and province	Year	Origin	Number	Year of recovery
Agria housei Shewell New Brunswick	1971	Ontario (laboratory culture)	2800	1971 (14 adults) Nil in 1972–73
Lissonota sp. New Brunswick	1973	Japan	59	Nil in 1974

Table 74 Laboratory and field cage studies of parasitoids against spruce budworm, *Choristoneura fumiferana* (Clem.)

Species and province	Year	Origin	Number	Year of recovery
Agria housei Shewell New Brunswick	1971	Ontario (laboratory culture)	76*	No recovery sought
Cephaloglypta laricis Momoi New Brunswick	1970	Japan	111**	1971 (2 adults)
	1971	Japan	933*,**	1972 (13 adults)
	1972	Japan	552*,**	1973 1 – 2 adults 2 – 4 adults
	1973	Japan	176*,**	1974 1 – no recovery 2 – 1 adult
	1975	Japan	47** 34*	Nil in 1976 Nil in 1976
Cephaloglypta murinanae Bauer New Brunswick	1972	Austria	22**	1973 (2 mature larvae)
Lissonota sp. New Brunswick	1972	Japan	102**	1973 (1 mature larva)
	1973	Japan	66* 455**	Nil in 1974
	1975	Japan	350* 71**	Nil in 1976
Trichogramma minutum Riley Québec	1971	USA	20 000*	No recovery

* Field cage studies.
** Laboratory or insectary studies.

chemical insecticides mounted in the 1970s, research on means to enhance the effectiveness of other native parasitoids by habitat management and inundative release was initiated at the Maritimes Forest Research Centre.

The only other North American activity in the 1970s was research at the University of Maine to release an introduced chalcidoid parasitoid of gypsy moth, *Lymantria dispar* (L.), into spruce budworm populations (Minot & Leonard 1976). This research is not discussed further.

Releases and Recoveries

Cephaloglypta laricis Momoi (Hymenoptera: Ichneumonidae)

This ichneumonid parasitoid is specific to todo fir budworm and one other related host in Japan, but may be conspecific with *C. murinanae* Bauer in Europe (H. Zwölfer 1972, personal communication). The parasitoids were air-freighted in several years (Table 74), always as pupae in cocoons lightly packed in insulated containers. The flight was in two stages, first to Belleville, Ontario, for screening of hyperparasitoids and extraneous species then as adults to Fredericton, New Brunswick, requiring about 2 weeks in all. In 1973 shipment was direct to Fredericton so that adult emergence and mating could be studied. In-transit mortality varied from 18 to 58%, perhaps due to overheating at July temperatures.

Females mated within 2 days of emergence, and lived up to 1 month on a diet of sugar, honey, and water. Oviposition on spun-up second-instar budworm hosts began soon after mating and continued vigorously throughout the life of the female. In the four experimental years, some 90 000 spruce budworm larvae were exposed to attack, with ovipositional success varying from 17 to 60% as indicated by dissection of samples. Substantial host larval mortality during early diapause was attributed partly to ovipositional wounding. A problem was to optimize survival of parasitized hosts by exposing candidate larvae long enough to achieve a high percentage of attacked hosts, but short enough to avoid a high percentage of superparasitism and fatal wounding. The exposure patterns usually resulted in a single egg per host; the egg hatched within a few weeks and the parasitoid overwintered as a first-instar larva.

In rearing thousands of prospectively parasitized spruce budworm larvae, only 70 *C. laricis* pre-pupae emerged from their hosts. Of these, 27 successfully pupated in a cocoon, a few adults emerged, and only one female oviposited. Efforts to rear its progeny were unsuccessful. Between autumn and late spring in all years, the percentage parasitism of sampled host stock declined sharply, almost to zero. Hypotheses to account for the decline included ovipositional injury to hosts, lack of cold hardiness in the parasitoid, host diet unfavourable to the parasitoid, encapsulation reaction, and other biochemical defence mechanisms.

Mortality of hosts in hibernacula was 20−62%, higher than the usual 9-22% recorded from populations in their natural environment. Some of the additional mortality was attributed to ovipositional wounding, as *C. laricis* has a large ovipositor compared with its native homologue, *Glypta fumiferanae* (Vier.). However, mortality of hosts was not an important factor in reduction of percentage parasitism. It is unlikely that low winter temperature discriminated against parasitoids within their spruce budworm hosts. In a test for cold hardiness, overwintering larvae of *C. fumiferana*, *G. fumiferanae*, and *C. laricis* were rapidly supercooled down to −60°C. All three species showed ice-crystal formation in the −40 to −43°C range, which is below the coldest natural environment in New Brunswick. Thus the parasitoid stock of Hokkaido provenance appeared to be as hardy as the native stocks of parasitoids and hosts, just as one would expect from a comparison of the monthly mean temperatures of the two regions.

When a stock of second-instar spruce budworm larvae was held in storage from October to May at a constant temperature of 7°C, the rate of decline of parasitism was unusually rapid compared with stock held out-of-doors at much lower temperatures. Evidently the "cold room" was a selectively hostile environment to the parasitoid, but not to the host. This suggested that the decline in parasitism was temperature-related rather than time-related.

Most of the loss of parasitism occurred in post-diapause stocks; however, no evidence implicating host diet appeared, as rearings with either artificial diet or with natural foliage produced healthy hosts and moribund or vanished parasitoids. Dissections revealed very few encapsulated parasitoid larvae, not enough to indicate a mechanism for decline in percentage parasitism. The expected parasitoids seemed to disappear without trace, but occasionally parasitoid head capsules, faintly yellowed and very membranous, were discovered within the body cavity of healthy maturing hosts. Thus the most plausible hypothesis is that the spruce budworm is able to kill and absorb larvae of *C. laricis*, leaving the host unimpaired.

Cephaloglypta murinanae Bauer (Hymenoptera: Ichneumonidae)

A small shipment (Table 74) of pupae was air-freighted from Czechoslovakia to New Brunswick in July 1972. Some 1300 spruce budworm larvae in hibernacula were exposed, and females oviposited vigorously with behaviour similar to *C. laricis*. A January sample of hosts wintered out-of-doors indicated 25% parasitism of survivors and 18% host mortality. Overwintered hosts were successfully reared in May on artificial diet, or on diet followed by foliage, but only four parasitoid pre-pupae matured, two dying within pupal hosts and two emerging from larval hosts but failing to pupate. The disappearance of the expected population of parasitoids is unexplained, but circumstances were similar to those of *C. laricis*.

***Lissonota* sp. (Hymenoptera: Ichneumonidae)**

The biology of this ichneumonid species from *C. diversana* hosts on todo fir in Japan has been reported by Zwölfer (1972, personal communication). Shipments were air-freighted from Hokkaido to New Brunswick in 1972, 1973, and 1975 (Tables 73 and 74). Females mated soon after emergence from the cocoon, and were maintained on a diet of sugar, honey, and water for up to a month in insectary cages. No responses were observed to the presence of free-moving first-instar or spun-up second-instar spruce budworm. However, large samples dissected in winter revealed parasitism rates of 7% in 1972 and 1% in 1973. No more than one egg shell was dissected from any one host, although heavy superparasitism is the rule in *C. diversana* hosts. Although many thousands of spruce budworm hosts were exposed, only one parasitoid emerged as a pre-pupa, and it failed to pupate. Three more mature larvae were dissected from dead hosts, but no evidence of encapsulation or of tissue absorption of the parasitoids was found. Thus *Lissonota* sp. appears to be incompatible with spruce budworm because of low ovipositional stimulus to the parasitoid and some innate resistance by the host.

***Trichogramma minutum* Riley (Hymenoptera: Chalcidoidea)**

The object of this study in Québec in 1970–75 (F.W. Quednau 1974, personal communication, unpublished work reported to the Annual Review of Spruce Budworm Research, Canadian Forestry Service) was to explore the potential of this egg parasitoid for inundative release against rising spruce budworm populations in local epicentres. This native chalcidoid wasp is already present in all eastern North American forests, attacking eggs of scores of species, and typically contributing 5–25% parasitism of spruce budworm eggs. Apparently, scarcity of alternate hosts prevents the numerical response that *T. minutum* might otherwise make to rising budworm density. Quednau believed that about 2.5×10^6 female parasitoids per hectare would be needed for effective release in epicentres.

The research problems facing the Laurentian Forest Research Centre were: (1)the maintenance of year-round research cultures of the parasitoids and suitable hosts; (2)comparison of commercial strains and arboreal ecotypes, and the selection of a type most responsive to laboratory rearing and the spruce budworm forest habitat; (3)determination of the parasitoid female's functional response to host density; (4)refinement of a mass-culture method, and (5)development of techniques for inundative release and assessment of efficacy.

From 1970 to 1975 considerable biological knowledge and research experience was gained in the first three problems. Methods of air-freighting stock from the United States and Mexico were established, and techniques for small-scale rearing and storage of parasitoids and host eggs were developed (Table 74). However, the application of American commercial techniques of mass-rearing was thwarted by quarantine regulations, which prevented the utilization of the angoumois grain moth, *Sitotroga cerealella* (Olivier), the most suitable host for *Trichogramma* cultures. Moreover, there were worker health problems (allergies) associated with mass-rearing.

Progress was made in the selection of a Canadian ecotype well adapted to the forest environment and with superior host-finding potential. However, *T. minutum* was shown to have only a weak functional response to density of spruce budworm egg-masses; that is, in caged-tree experiments with varying parasitoid densities and host densities, the number of attacked host eggs per individual parasitoid was greater when egg-mass density was increased. However, increased density of parasitoids tended to lower success in finding hosts, perhaps because of mutual interference and increased superparasitism. Parasitism greater than 50% was not attained in cage experiments. Thus the prospects for useful impact on spruce budworm dynamics appear low, because only a drastic change in egg parasitism could have any significant effect on generation survival in spruce budworm outbreak densities (Morris 1963).

The research was terminated in 1975 without definitive assessment of the feasibility of application to spruce budworm control.

Agria housei Shewell (Diptera: Sarcophagidae)

The native sarcophagid fly, *Agria housei*, has long been recorded as a pre-pupal and pupal parasitoid of spruce budworm and western spruce budworm, usually at low percentage parasitism. However, it merits consideration for inundative release because it is one of the few budworm parasitoids that can be laboratory-produced in large quantities at acceptable costs; it can be reared directly on pork liver without an intermediate insect host. It has other characteristics that enhance its prospective application to the spruce budworm problem. However, the laboratory stock available for experimentation had been reared for many generations under artificial conditions, causing loss of cold hardiness and diapause induction (House 1967). Therefore there were doubts about the responsiveness of this stock to spruce budworm hosts and about its capacity to revert to univoltine reproduction.

Field and laboratory tests were conducted in 1971 to gain experience in handling the species and to test its response to larval and pupal spruce budworm hosts (Tables 73 and 74). The fly stock was provided by the Agriculture Canada Research Institute, Belleville, Ontario. Laboratory cage experiments designed to test host/parasitoid densities failed to elicit any apparent interaction, and no parasitoid progeny issued. However, some 2 800 adult flies (34% female) were freely released into a balsam fir, *Abies balsamea* (L.) Mill., stand highly infested with spruce budworm pupae, at Fundy National Park. From a sample of 1 114 pupae collected within 10 days of release and within 75 m of release point, 14 *A. housei* puparia were recovered. Based on estimates of the density of available hosts, this rate of recovery indicated an excellent response by the introduced stock. No further specimens were recovered from samples of host pupae in 1972 and 1973, but given the poor cold hardiness inheritance of the stock and the dispersal characteristics of the species, detectable persistence at the release point was not expected. The potential for applied biological control under appropriate pre-outbreak conditions remains unassessed.

Native parasitoids of spruce budworm larvae

The Canadian Forestry Service has maintained a sporadic surveillance of the effectiveness of native parasitoids attacking spruce budworm, especially *G. fumiferanae* and *Apanteles fumiferanae* Vier., which parasitize overwintering larvae. A study at the Maritimes Forest Research Centre was conducted from 1977 to 1980 on the premise that an understanding of the factors limiting parasitoid effectiveness might lead to methods of managing the forest habitat to stimulate parasitoid regulation of spruce budworm. The study was directed to the behavioural biology of adults of these two species (feeding habits, flight periodicity, distribution in the canopy), but did not reveal any clear avenue for practical application (R.S. Forbes 1980, unpublished work, Canadian Forestry Service).

Recommendations

It is recommended that the Canadian Forestry Service (a) maintain contacts with CIBC concerning collection opportunities abroad; (b) investigate the compatibilities of a range of non-indigenous parasitoids with spruce budworm hosts; (c) develop rearing, release, and assessment skills and facilities; and (d) assess opportunities for incorporating parasitoid manipulation in integrated pest management programmes based on spruce budworm population dynamics.

Literature Cited

House, H.L. (1967) The decreasing occurrence of diapause in the fly *Pseudosarcophaga affinis* through laboratory-reared generations. *Canadian Journal of Zoology* 45,149–153.

McGugan, B.M.; Coppel, H.C. (1962) Biological control of forest insects 1910–1958. In: A review of the biological control attempts against insects and weeds in Canada. *Commonwealth Institute of Biological Control Technical Communication* 2,35–127.

Miller, C.A.; Angus, T.A. (1971) *Choristoneura fumiferana* (Clemens), spruce budworm (Lepidoptera: Tortricidae). In: Biological control programmes against insects and weeds in Canada 1959–1968. *Commonwealth Institute of Biological Control Technical Communication* 4,127–130.

Minot, M.C.; Leonard, D.E. (1976) Host preference and development of the parasitoid *Brachymeria intermedia* in *Lymantria dispar* L., *Galleria mellonella* and *Choristoneura fumiferana*. *Environmental Entomology* 5,527–532.

Morris, R.F. (1963) The analysis of generation survival in relation to age-interval survivals in the unsprayed areas. In: Morris, R.F. (*Ed.*) The dynamics of epidemic spruce budworm populations. *Memoirs of the Entomological Society of Canada* 31,32–37.

E. Review of Biological Control Opportunities against Spruce Budworm, *Choristoneura fumiferana* (Clem.)

I.W. VARTY

Any biological control method has to prove its worth on the basis of three criteria — technical feasibility, cost-effectiveness, and environmental compatibility. The method must work effectively over a sufficient area and within a sufficient time frame; it must cost less than the value of the resource saved; and it must respect the integrity of the ecosystem and the sensitivities of the public. This section summarizes the efforts made, the progress achieved, and the opportunities that arose during the period 1969–80, and updates the review by Miller & Varty (1975).

It is generally assumed that the spruce–fir (*Picea* spp. and *Abies* spp.) ecosystem is a favourable setting for effective long-term applied biological control. The forest has stability in space and time, complexity and diversity, and large-scale contiguity for dispersion of organisms. The spruce budworm, however, is a native pest with many natural enemies (predators, parasitoids, pathogens) already in place; theory and practice suggest that it is more difficult to establish an effective biological control agent on a native pest in its evolved habitat than on an introduced pest in an adopted habitat. The search for a self-sustaining, self-dispersing organism that can react favourably to local or regional abundance of spruce budworm is uncertain of results. The alternative is to concentrate research efforts on development of biological insecticides, which require organisms to be artificially mass-propagated and inundatively released over a large area, producing a powerful in-season effect but a limited persistence. The experience of the 1970s is that neither method can ensure success. Nonetheless, the effort is justified for a pest so important as the spruce budworm because the spruce–fir inventory is so valuable.

Progress and Prospects, 1969–80

The goals of biological control of spruce budworm are to decrease defoliation damage to below the unacceptable threshold by remedial treatment of epidemic populations and to prevent spread of damage by early treatment of populations while the infestation is still localized. Much of the scientific investigation conducted from 1969 to 1980 was of an exploratory nature, laying a foundation for further development. Most of the research has been conducted by the Canadian Forestry Service (C.F.S.), particularly within its Forest Pest Management Institute at Sault Ste. Marie, Ontario. There has been little input by other agencies, such as provincial governments, universities, and industry. The federal effort has been sustained and substantial; for example, 21 person years were allocated in 1980, or about 9% of its forest protection research budget. Funding was distributed among studies of virus diseases, bacterial pathogens, fungal diseases, protozoan parasites, genetic manipulation, and insect parasitoids and predators, in that order of precedence. In Canadian experience, most attempts at permanent control of crop pest insects by the introduction of exotic agents have failed, but taken as a whole, biological control has been cost-effective (Beirne 1975).

Viruses

These pathogens have been the most successfully manipulated insect-pest control agents in world agriculture (Tinsley 1979). In North American forests, application has been especially effective against sawflies and tussock moths. Viruses have mostly been used as experimental biological insecticides with weak dispersion and self-sustenance in subsequent years. The Canadian Forestry Service (CFS) has undertaken fundamental and innovative research for more than 30 years, yet a cost-effective operational treatment for spruce budworm has not been achieved so far. In the 1970s, the emphasis was on isolation, mass-propagation, formulation, and spray-testing of baculoviruses. The economic problem was the high cost of virus propagation on diet-reared spruce budworm larvae, currently about $250 per sprayed hectare, or 30–40 times the cost of chemical treatment. The technical problem was the low rate of foliage protection, even when high larval mortality was attained.

The probability of large-scale operational application in the 1980s is low. The future of virus sprays may depend on development of a cheap culture method. Prospects for a self-spreading low-density inoculation method are poor. As no natural epizootic has been recorded, it appears that spruce budworm viruses have low natural infectiveness. Spray efficacy will probably demand high dosage and high density coverage of the target forest.

Bacteria

The commercial production of *Bacillus thuringiensis* Berliner (*B.t.*) and its Canadian registration for forest use have already been achieved. *B.t.* is widely used and effective for North American field crops, but has been generally less successful against forest pests in experimental trials. Its toxicology, formulation, aerial delivery techniques, and field assessment methods have been developed for use against spruce budworm. *B.t.* is essentially a biological insecticide, and is not designed to meet the self-regulating, self-distributing, density-dependent criteria of classicial biological control. Canadian laboratory studies in the 1970s addressed the mode of action of the crystal, its structure, and assays of potency of field preparations; field research was designed to develop formulations, test aerial delivery techniques, and assess efficacy for population reduction and foliage protection. Even so, *B.t.* application is 4–5 times more costly and somewhat less reliable than corresponding treatments with the chemical insecticide fenitrothion. *B.t.* technology is still short of its full potential: potential future developments include selection of more potent strains, preparation of synthetic mimics of the crystal toxin, improvements in formulation, and more effective aerial delivery systems.

Fungi

The study of mycoses caused by entomogenous fungi holds a modest position in research and development of biological methods for agricultural pests (Ferron 1978). International interest has focused mainly on the role of fungal epizootics in pest population regulation, but in practice few commercial products exist and experimentation in forestry, for example in the USSR, has been limited. The object of field application is to initiate an epizootic by artificial dissemination of inoculum, and research has been on the nature of infectiveness, methods of mass propagation, and methods of field distribution.

Canadian research in this field, conducted mostly at the Forest Pest Management Institute and at Memorial University of Newfoundland, was exploratory in the 1970s and was on methods of culture of inoculum and the mode of spore formation and germination. Until 1980 no field spray experimentation for spruce budworm control had been attempted. Fungi are valuable agents for biological control because of their high specificity and high infectiveness in the laboratory. However, their natural incidence in spruce budworm populations is low, so the prospects for stimulating large-scale epizootics are correspondingly weak. Development of an operational cost-competitive treatment apears to be remote.

Protozoa

Protozoan parasitism of agricultural pests is commonplace and is reported to reduce population vigour. Many species occur in forest defoliator insects, including two species of microsporidia in spruce budworm. However, in pest control history the applied use of protozoa has been rare, because they are costly to propagate in quantity, and exhibit low infectiveness after artificial distribution.

Canadian research on microsporidian parasites of spruce budworm was restricted to the Forest Pest Management Institute in the 1970s, which documented their natural incidence and parasitic role. Spray experiments on single trees demonstrated that parasitism rates can be artificially raised, but left the problem of mass production unsolved. Operational application is not yet practicable.

Parasitoids

Canada has enjoyed some success in the importation of exotic parasitoids to control introduced pest insects in both agriculture and forestry, but the record of releases to control native forest pests has been discouraging (Munroe 1971). However, it is not known whether the low success rate in the control of native hosts is associated with the separate evolutions of host and parasitoid, or stems from imperfect testing techniques. In agriculture, inundative release of a few species of natural enemies is routine practice, but experimentation against forest pests has been fragmentary and unsustained (Pschorn-Walcher 1977).

Canadian research on manipulation of spruce budworm parasitism was scant in the 1970s, but was on a sounder base than the hit-and-miss releases of the 1940s and 1950s. Research was on field-cage studies of host and site compatibility with exotic parasitoids and studies of the numerical response of native parasitoids, with the inundative release method in mind.

The case for increasing the Canadian effort with respect to spruce budworm parasitism is not strong if the probability of establishing additional natural enemies decreases with increasing diversity of the natural enemy complement. There is a need for better theoretical guidelines in order to achieve maximum success. The empirical approach — to collect homologous parasitoids abroad, then release and monitor their distribution and persistence — is unpredictable and does not increase understanding of the processes unless accompanied by studies of host-parasitoid compatibility.

Predators

In world agriculture, some striking successes have been achieved by inoculative release of exotic predaceous arthropods, but in Canadian forestry few such releases have been attempted (Munroe 1971). In general, the release of an exotic broad-spectrum predator gives rise to more concern about ecological integrity than the introduction of host-specific parasitoids. In Canada, the only releases of non-native arthropod predators of forest pests in 1969–80 were the successful introductions of two red wood ant species from Italy and Manitoba to Québec (Finnegan 1978); the experiments show encouraging results in the control of spruce budworm populations in young stands. However, in general, the technical and economic feasibilities of predator releases against spruce budworm have not been assessed.

Genetic approaches

Pest insect genetics is an important means of biological control, because it leads to autocidal methods for manipulating populations. The technique of releasing sterile males to interfere with the transmission of genes is best exemplified by the successful screwworm and fruit fly control operations by the United States Department of Agriculture. During the 1970s many agencies around the world attempted pest control based on this principle (Whitten & Foster 1975).

Exploratory research on spruce budworm genetics was conducted by CFS in the late 1970s, the only research agency with a declared interest. Such work is essentially long-term, demanding of innovation and risky as an investment; it includes investigations of induced sterility, gene-damaging methods, and procedures for attaining population control. It may offer opportunities for diminishing the budworm's reproductive success, but has potential mainly for small areas in the early stages of population increase. Prospects for early operational application are dim.

Further research

Clearly none of the biological methods of control is ready to displace chemical spraying in a significant percentage of the area needing protection from spruce budworm. However, *B.t.* is efficacious and acceptable financially for small areas, especially in high value stands and in zones where chemicals are prohibited. The technical research, as recommended by Smirnoff & Morris in this chapter, should be strongly supported, because *B.t.* is close to being cost-competitive. It is the only non-chemical method able to replace the registered chemicals in the event of further restrictions, and it has considerable international research interest.

For other entomopathogens, the main barrier to progress is the lack of methods for mass-propagation of agents at acceptable cost. Research agencies should re-examine the feasibility of hastening such capability, consulting with the many international agencies interested in similar, but problem-specific, developmental research. Because the search for new species or virulent strains of known entomopathogens has shown little promise, the prospects for induced genetic alteration should be monitored by close contact with scientific advances in agriculture. Meanwhile it is desirable to maintain the current steady progress reported already in this chapter.

Research on parasitoids and predators is at low ebb. More basic research on host-parasitoid compatibility and on release and dispersion methods should be conducted. Research on genetic methods of autocidal control have been exploratory, but results have been sufficiently encouraging to justify further work.

The contributors to each section of this chapter have recommended further long-term research. The potential for effective alternatives to chemical control is real, even if slight in most cases. Except for *B.t.*, none of these alternatives is likely to have a significant role in forest protection operations in the 1980s. However, because the spruce budworm is a long-term problem, and spruce–fir forest is a long-term resource, the need for an array of protection techniques for both remedial and preventive intervention is likely to persist.

Literature Cited

Beirne, B.P. (1975) Biological control attempts by introductions against pest insects in the field in Canada. *Canadian Entomologist* 107,225–236.

Ferron, P. (1978) Biological control of insect pests by entomogenous fungi. *Annual Review of Entomology* 23,409–442.

Finnegan, R.J. (1978) Predation by *Formica lugubris* (Hymenoptera: Formicidae) on *Choristoneura fumiferana* (Lepidoptera: Tortricidae). *Canadian Forestry Service Bi-monthly Research Notes*. 34(1),3–4.

Miller, C.A.; Varty, I.W. (1975) Biological methods of spruce budworm control. *Forestry Chronicle* 51,16–19.

Munroe, E.G. (1971) Status and potential of biological control in Canada. In: Biological control programmes against insects and weeds in Canada, 1959-68. *Commonwealth Institute of Biological Control Technical Communication* 4,213–255.

Pschorn-Walcher, H. (1977) Biological control of forest insects. *Annual Review of Entomology* 22,1–22.

Tinsley, T.W. (1979) The potential of insect pathogenic viruses as pesticidal agents. *Annual Review of Entomology* 24,63–87.

Whitten, M.J.; Foster, G.G. (1975) Genetic methods of pest control. *Annual Review of Entomology* 20,461–476.

Choristoneura occidentalis Freeman, Western Spruce Budworm (Lepidoptera: Tortricidae)

R.F. SHEPHERD and J.C. CUNNINGHAM

Pest Status

Two outbreaks of the western spruce budworm, *Choristoneura occidentalis* Freeman, on Douglas fir, *Pseudotsuga menziesii* (Mirb.) Franco, in British Columbia have been well documented and evidence of three more outbreaks during this century has been found by examination of tree radial sections. The current outbreak began in 1970 and still persists in 1980. It is more extensive and has continued longer than the previous outbreak, which lasted from 1954 to 1958.

These outbreaks all occurred in the Frazer and Lillooet River valleys, which form an inter-zone between the moist coastal and dry interior zones of British Columbia. In the western and southern portions of these valleys, defoliation occurred in a mid-slope band but not in the valley bottoms. With time, populations have decreased in the western extremities of the infestation and increased toward the northeast, but during the current outbreak populations have decreased and then increased regionally more than once. Typical damage to trees includes reduction in height and radial growth, dieback, and deformity. Complete tree mortality has occurred in a few small patches but more commonly occurs in the understorey below larger heavily defoliated trees.

Background

The same viruses that infect *C. occidentalis* also infect spruce budworm, *C. fumiferana* (Clem.), and laboratory tests indicate that the former is more susceptible to a nuclear polyhedrosis virus (NPV) and a granulosis virus (GV) (Cunningham unpublished). In addition, populations of western spruce budworm are virtually free of the microsporidian parasites that are found at very high levels in older infestations of spruce budworm; it is possible that infection with microsporidia may be antagonistic to subsequent infection with viruses but this has to be verified.

Field Trials

The first aerial spray trial using NPV on western spruce budworm was conducted in 1976 on a 20.5-ha plot, using a fixed-wing aircraft fitted with boom and nozzle spray equipment. The dosage was 250 x 10^9 polyhedral inclusion bodies (PIB) per hectare; the aqueous formulation contained 25% molasses, 6% IMC 90-001 (International Minerals Corp.) UV protectant, and 1% Chevron® sticker; the emission rate was 9.4 l/ha and larvae were in the fourth, fifth, and sixth instars. Population reduction as a result of treatment, calculated by Abbott's formula (Abbott 1925), was 36%; the incidence of NPV in larvae, determined microscopically, was 6.8% in the treated plot and 1.6% in the check plot respectively. In 1977 the incidence of virus in the treated plot fell to 1.3% (Shepherd & Cunningham unpublished). This poor result was probably due to too low a dosage of virus, applied too late.

In 1978 the same equipment and emission rate were used to treat three 20-ha plots with NPV at a dosage of 750 x 10^9 PIB/ha just after budflush when larvae were at the peak of the fifth instar, with third, fourth, and sixth instars also present. The aqueous formulation contained 25% molasses, 3% Shade® (Sandoz Inc.) (the same product as IMC 90-001), and 0.05% Triton® B-1956 (Rohm and Haas Co. of Canada Ltd.) sticker. Fifteen days after spraying, NPV infection levels, detemined microscopically, were 55, 87, and 25%; and population reductions due to treatment were 0, 26, and 48% respectively. Successful adult emergence was lower in treated than in check plots,and the male:female ratio was 2:1 in the treated plots compared to 1:1 in the check plots (Hodgkinson *et al.* 1979).

In 1979 and 1980, follow-up studies were conducted on two of the three virus-treated plots, the third being abandoned because of a local population decline. In these two plots, NPV infection rates were recorded as 18 and 54% in 1979, and 7 and 17% in 1980. In 1979, population declines of 51 and 62% in the virus treated plots were recorded compared to increases of 46 and 201% in the corresponding check plots. However, this trend was reversed in 1980; populations more than doubled in the virus-treated plots compared to 1979 levels and substantial population declines were noted in the check plots (Shepherd *et al.* 1982).

Bacillus thuringiensis Berliner

In 1978, four 40-ha plots were treated with Thuricide® 16B (Sandoz Inc.), a commercial preparation of *Bacillus thuringiensis* Berliner (*B.t.*). The emission rate was 9.4 l/ha and the dosage 20 x 10^9 I.U./ha. Fifteen days after spraying, populations were reduced by 32, 66, 79, and 91% (Hodgkinson *et al.* 1979).

In the plots treated with *B.t.*, western spruce budworm populations increased in 1979 in two of the four plots; in the third plot the population increased at a lesser rate and in the fourth it showed a small, but significant, decrease compared to corresponding check plots. In 1980 this trend was reversed, with untreated populations decreasing and *B.t.*-treated populations increasing dramatically (Shepherd *et al.* 1982).

Evaluation of Control Attempts

Applications of NPV and *B.t.,* in 1978 did not provide adequate control in that year, but NPV and possibly *B.t.* did provide some carry-over effect and continued to cause a population regulating effect a year later. This carry-over effect might have continued longer, but the plots treated in these trials were small and results were probably masked by immigration of moths. If either NPV or *B.t.* can regulate populations for a number of years after a single application, their use may be justified economically.

The same constraints that apply to the viruses on *C. fumiferana* also apply to *C. occidentalis*. However, the western species is virtually free from microsporidian parasites; preliminary laboratory tests have also shown that it is more susceptible to both NPV and GV than microsporidia-free, laboratory-reared spruce budworm larvae. It therefore appears that western spruce budworm is a better candidate for population regulation with viruses.

Recommendations

More field trials should be conducted to compare the efficacy of NPV and GV.

A test involving an isolated population, where moth immigration would be negligible, should be conducted in order to evaluate the long-term effects of these biological control agents.

Literature Cited

Abbott, W.S. (1925) A method of computing the effectiveness of an insecticide. *Journal of Economic Entomology* 18,265–267.

Hodgkinson, R.S.; Finnis, M.; Shepherd, R.F.; Cunningham, J.C. (1979) Aerial applications of nuclear polyhedrosis virus and *Bacillus thuringiensis* against western spruce budworm. *British Columbia Ministry of Forests/Canadian Forestry Service Joint Report* 10, 19 pp.

Shepherd, R.F.; Gray, T.G.; Cunningham, J.C. (1982) Effects of nuclear polyhedrosis virus and *Bacillus thuringiensis* on western spruce budworm one and two years after aerial application. *Canadian Entomologist* 114,281–282.

Coleophora laricella (Hübner), Larch Casebearer (Lepidoptera: Coleophoridae)

I.S. OTVOS and F.W. QUEDNAU

Pest Status

The larch casebearer, *Coleophora laricella* (Hübner), was accidentally introduced from Europe, probably on nursery stock. It was first recorded in North America in 1886 at Northampton, Massachusetts (Herrick 1912), and in Ottawa in 1905 (Fletcher 1906). The casebearer spread rapidly and by 1947 infested most tamarack, *Larix laricina* (Du Roi) K. Koch, in Newfoundland, the Maritimes, and Ontario, spreading west as far as Thunder Bay (McGugan & Coppel 1962). By 1970 it was present in southeastern Manitoba. It was first found on western larch, *Larix occidentalis* Nutt., around St. Maries, Idaho, in 1957 (Denton 1958), and in British Columbia in 1966 (Molnar *et al.* 1966). The larch casebearer moved rapidly along major larch stands in the valleys, and became established throughout the range of western larch in southeastern British Columbia in varying degrees of infestation. Successive years of severe defoliation cause growth reduction, crown dieback, and subsequent tree mortality (McGugan & Coppel 1962, Turnock *et al.* 1969).

Background

Successful control of the larch casebearer in eastern and central Canada (Webb & Quednau 1971) by the introduction of parasitoids and by similar biological control work in the northwestern United States (Denton 1972, Ryan & Denton 1973) led to a resumption of the use of parasitoids in 1974 to combat the larch casebearer in British Columbia. An intensive survey of larch casebearer parasitoids conducted in British Columbia in 1973 yielded 32 species of Hymenoptera (Miller & Finlayson 1974). In Québec, studies on *Diadegma laricinellum* (Strobl) were continued after 1968, but the project was terminated in 1974 because of low priority.

Releases and Recoveries

Four species of parasitoids, *Agathis pumila* (Ratz.), *Chrysocharis laricinellae* (Ratz.), *Diadegma laricinellum* (Strobl), and *Dicladocerus japonicus* Yshm. have been released in southeastern British Columbia, and a few *D. laricinellum* were also liberated in Québec. The details of these releases are summarized in Table 75.

Agathis pumila (Ratz.) (Hymenoptera: Braconidae)

This species was first introduced into British Columbia in 1969 when parasitized larch casebearers, collected in Montana (USA) were placed in larch casebearer infested stands at Arrow Creek and Fruitvale. The species was recovered in 1972 at both sites. Between 1974 and 1977 more adults of *A. pumila*, this time from Europe, were released at other locations in the southeastern part of the province. In addition, larch casebearer larvae parasitized by *A. pumila* in the laboratory were placed in infested larch stands, and parasitoids from an old larch casebearer infestation in British Columbia were also relocated.

The establishment of the parasitoids was monitored by rearing last-instar larch casebearer larvae and pupae collected at the release sites. These rearings showed that *A. pumila* became established, but its numbers seem to be declining as another parasitoid, *C. laricinellae*, is increasing over the years. Quednau (1970) has shown that the multivoltine *C. laricinellae* increased at the expense of *A. pumila* in Québec.

Table 75 Summary of parasitoid releases and relocations in British Columbia and Québec between 1969 and 1980

Species and locality	Lat.	Long.	Origin	No. and sex of parasitoids released	Year
Agathis pumila (Ratz.)					
Fruitvale	49.06°N	117.33°W	Montana	100 M & F	1969
Arrow Creek	49.07°N	116.26°W	Montana	1000 M & F	1969
Nelway	49.02°N	117.17°W	Austria	87 M + 89 F	1974
Blewett	49.28°N	117.25°W	France	15 M + 15 F	1974
Creston	49.06°N	116.31°W	Italy	10 M + 24 F	1975
East Arrow	49.09°N	116.26°W	Italy	20 M + 32 F	1975
Thrums	49.23°N	117.33°W	Italy	9 M + 32 F	1975
Pass Creek	49.25°N	117.36°W	Italy	17 M + 31 F	1975
Blewett	49.28°N	117.25°W	Italy	14 M + 31 F	1975
North Salmo	49.13°N	117.14°W		400 (parasitized larvae)	1976
Christina Lake	49.02°N	118.10°W	Austria	25 M + 30 F	1976
Ross Spur	49.11°N	117.28°W	Austria	24 M + 28 F	1976
Rossland	49.04°N	117.47°W	Italy & Austria	16 M + 36 F	1977
Cherryville	50.13°N	118.34°W	(relocation from Arrow Creek)	100 M & F	1978
Chrysocharis laricinellae (Ratz.)					
Thrums	49.23°N	117.33°W	Switzerland	6 M + 12 F	1974
Shuttleworth Creek	49.19°N	119.32°W	(relocation from B.C.)	512 M & F	1980
Diadegma laricinellum (Strobl)					
South Slocan	49.23°N	117.33°W	Austria	10 M + 5 F	1974
Thrums	49.23°N	117.33°W	Austria	7 M + 5 F	1974
East Arrow	49.09°N	116.26°W	Italy	9 M + 8 F	1975
Thrums	49.23°N	117.33°W	Italy	9 M + 8 F	1975
North Creston	49.12°N	116.34°W	Italy	12 M + 28 F + 25 parasitized larvae	1976
Joliette*	46.02°N	73.27°W	own rearings	12 F (mated)	1972
Dicladocerus japonicus Yshm.					
Blewett	49.28°N	117.25°W	Japan	52 M + 215 F	1974
Thrums	49.23°N	117.33°W	Japan	5 M + 20 F	1974

* Located in Québec; all others are in British Columbia.

Chrysocharis laricinellae (Ratz.) (Hymenoptera: Eulophidae)

C. laricinellae was apparently introduced in the Pacific Northwest with its host or it was released inadvertently with the early introductions of *A. pumila* (Ryan *et al.* 1974). In 1974 a small number of imported *C. laricinellae* were released in British Columbia; in 1980 greater numbers of the parasitoid were relocated from older infestations in the province.

***Diadegma laricinellum* (Strobl), formerly known as *D. nana* (Grav.), (Hymenoptera: Ichneumonidae)**

Adults of this parasitoid were liberated in small numbers in British Columbia from 1974 to 1976, but none of the species has been recovered. In Québec, 12 mated females from laboratory rearings were released in 1972, but recoveries could not be made because the project was terminated. Québec received small quantities of *D. laricinellum* from Austria between 1968 and 1973. Because the biology of *D. laricinellum* was incompletely known, laboratory studies on competition between it and *A. pumila* were conducted before it was released in the field. These studies have shown that host larvae attacked by *A. pumila* are normally avoided by *D. laricinellum* except at high densities of *A. pumila*, which are rare in nature. *D. laricinellum* proved to be intrinsically inferior and died from the presence of *A. pumila* when both parasitoid species occurred in the same host larva (Quednau, unpublished). Synchronous attacks on larch casebearer larvae by both parasitoid species are unlikely to occur in nature, because *A. pumila* attacks smaller needle-mining stages than *D. laricinellum*. Factors adversely affecting the sex ratio of the progeny of this parasitoid were discussed by Ryan (1980). These factors comprise continuous supply with hosts, overmating, too-early contact with hosts, superparasitism and lack of sufficient time between attacks to encourage the fertilization of parasitoid eggs. Because some of these factors were not used to provide optimum conditions for the proper sex ratio during the laboratory rearings of *D. laricinellum* in Québec, too few female parasitoids were obtained for field liberations (Quednau unpublished).

***Dicladocerus japonicus* Yshm. (Hymenoptera: Eulophidae)**

This species was released in 1974 at Blewett, British Columbia, in relatively small numbers and has not been recovered to date.

Evaluation of Control Attempts

It is too early to evaluate the effect of the parasitoids on the larch casebearer in British Columbia. Data obtained so far indicate that both *A. pumila* and *C. laricinellae* have become established. The latter is now fairly common in this province (Otvos unpublished) and may have contributed to the reduction of the larch casebearer and hence of tree mortality.

Recommendations

Further searches should be made for *D. laricinellum*; because of the failure to recover it, the role of this parasitoid in Canadian larch casebearer populations remains to be evaluated. Its presence in the ecosystem can be only beneficial because it also may serve as a host for *C. laricinellae*. The introduction of *A. pumila* has probably facilitated the distribution of *C. laricinellae* in British Columbia. Although no studies have been made on this subject, winter mortality of the parasitoids should be considerably less in British Columbia than in the eastern part of Canada, and the impact of *C. laricinellae* on larch casebearer populations should be stronger because of the more favourable climate in the west. Life-table studies, although time consuming, are essential to the understanding of host-parasitoid population regulation mechanisms.

Literature Cited

Denton, R.E. (1958) The larch casebearer in Idaho – a new defoliator for western forest. *US Department of Agriculture Forest Service Research Note* 51, *Intermountain Forest and Range Experiment Station, Ogden, Utah*, 6 pp.

Denton, R.E. (1972) Establishment of *Agathis pumila* (Ratz.) for control of larch casebearer, and notes on native parasitism and predation in Idaho. *US Department of Agriculture Forest Service Research Note* INT-164, 6 pp.

Fletcher, J. (1906) Insects injurous to Ontario crops in 1905. Forest and shade trees. *Entomological Society of Ontario Report* 36,89–90.

Herrick, G.W. (1912) The larch casebearer. *Bulletin of the Cornell Agricultural Experiment Station* No. 322.

McGugan, B.M.; Coppel, H.C. (1962) Biological control of forest insects, 1910-1958. In: A review of biological control attempts against insects and weeds in Canada. *Commonwealth Institute of Biological Control Technical Communication* 2,35–216.

Miller, G.E.; Finlayson, T. (1974) Native parasitoids of the larch casebearer, *Coleophora laricella* (Lepidoptera: Coleophoridae), in the west Kootenay area of British Columbia. *Journal of the Entomological Society of British Columbia* 74,14–21.

Molnar, A.C.; Harris, J.W.E.; Ross, D.A. (1966) British Columbia Region. *Annual Report of the Forest Insect and Disease Survey, Ottawa 1967*, pp. 108–123.

Quednau, F.W. (1970) Competition and cooperation between *Chrysocharis laricinellae* and *Agathis pumila* on larch casebearer in Québec. *Canadian Entomologist* 102,602–612.

Ryan, R.B. (1980) Rearing methods and biological notes for seven species of European and Japanese parasitoids of the larch casebearer (Lepidoptera: Coleophoridae). *Canadian Entomologist* 112,1239–1248.

Ryan, R.B.; Denton, R.E. (1973) Initial releases of *Chrysocharis laricinella* and *Dicladocerus westwoodii* for biological control of the larch casebearer in the western United States. *US Department of Agriculture Forest Service Research Note* PNW-200, *Pacific Northwest Forest and Range Experiment Station, Portland, Oregon*, 4 pp.

Ryan, R.B.; Bousfield, W.E.; Miller, G.E.; Finlayson, T. (1974) Presence of *Chrysocharis laricinellae*, a parasite of the larch casebearer, in the Pacific Northwest. *Journal of Economic Entomology* 67,805.

Tunnock, S.; Denton, R.E.; Carlson, C.E.; Janssen, W.W. (1969) Larch casebearer and other factors involved with deterioration of western larch stands in northern Idaho. *US Department of Agriculture Forest Service Research Paper*, INT-68. *Intermountain Forest and Range Experiment Sation, Ogden, Utah*, 10 pp.

Webb, F.E.; Quednau, F.W. (1971) *Coleophora laricella* (Hübner), larch casebearer (Lepidoptera: Coleophoridae) In: Biological control programme against insects and weeds in Canada, 1959-1968. *Commonwealth Institute of Biological Control Technical Communication* 4,131–143.

Coleophora serratella (L.), Birch Casebearer (Lepidoptera: Coleophoridae)

A.G. RASKE

Pest Status

The birch casebearer, *Coleophora serratella* (L.)[1], is native to Europe; Salman (1929) postulated that the insect was accidentally introduced into northeastern North America about 1920. There it has become the most important defoliator of white birch, *Betula papyrifera* Marsh. It was discovered in Maine in 1927 (Salman 1929) and had spread to New Brunswick by 1933, to Nova Scotia by 1937 (Reeks 1951), to Ontario by 1944 (Raizenne 1952), to Prince Edward Island by 1951, to Québec by 1956, and to Manitoba by 1969 (Cochran 1974). It was discovered in western Newfoundland in 1953 and had spread throughout the island by 1971 (Raske 1974*a*). It was discovered in Victoria, British Columbia, in 1962 (probably from a separate introduction) and was causing severe defoliation of ornamental trees in about 100 ha of the city by 1980 (van Sickle personal communication). The pest has also been collected in Alberta (B. Wright 1981, personal communication). This casebearer thus appears to occur throughout Canada (Fig. 11) but the exact distribution in central and western Canada is largely unknown. Identification records for Ontario were recently revised. Thus *C. serratella* has never been positively identified in Ontario but a thorough search has not been undertaken.

The most common form of damage to trees is the browning and reduction of foliage caused by the feeding of late instar larvae in early summer (Bryant & Raske 1975). The larva feeds like a leafminer. It attaches its case to a leaf and mines as far as it can reach without leaving the case, creating somewhat rectangular mined patches. After a larva has eaten all the food within reach, it moves to another part of the leaf and repeats the process. The mined portion turns brown and this causes the scorched appearance of severely damaged trees. The damage tends to become masked later in the summer by the continuous production of new foliage.

The more severe form of damage occurs in early spring when large numbers of early-instar larvae destroy the flushing buds, causing twig, branch, and sometimes tree mortality (Bryant & Raske 1975); many insects starve.

Outbreaks of the birch casebearer have been reported from Victoria, British Columbia, and from Québec to Newfoundland. The insect probably reached St. John's, Newfoundland, in the late 1960s. It was extremely rare in Newfoundland in 1971 (Raske 1974*a*) but had reached outbreak proportions by 1980. In a newly invaded area, populations tend to increase to a super-saturation level. After 1 or 2 years populations decline but continue to cause a high level of chronic defoliation of 20–50%. However, the post-outbreak population levels vary in both location and time.

The life history and habits of the birch casebearer in the northeastern United States were reported by Salman (1929) and by Gillespie (1932). Guèvremont & Juillet (1974) and Raske (1976) described aspects of the life history in Canada. Coshan (1974) and Gepp (1975*a*) have summarized life history studies in Europe. The life cycle of the insect spans two calendar years. In Canada, eggs are laid in July on the underside of leaves. Larvae hatch in August and mine the leaf until early September. After moulting they cut a case from the leaf epidermis, crawl to a different area of the leaf, and feed until fall.

[1] J.C. Bradley (In: Kloet, G.S.; Hinks, W.D. *(Eds.)* 1972 — Handbook for the identification of British insects, Vol. 2, Part 2. Lepidoptera. Royal Entomological Society, London) cites *C. fuscedinella* Zell. as a synonym of *C. serratella* (L.).

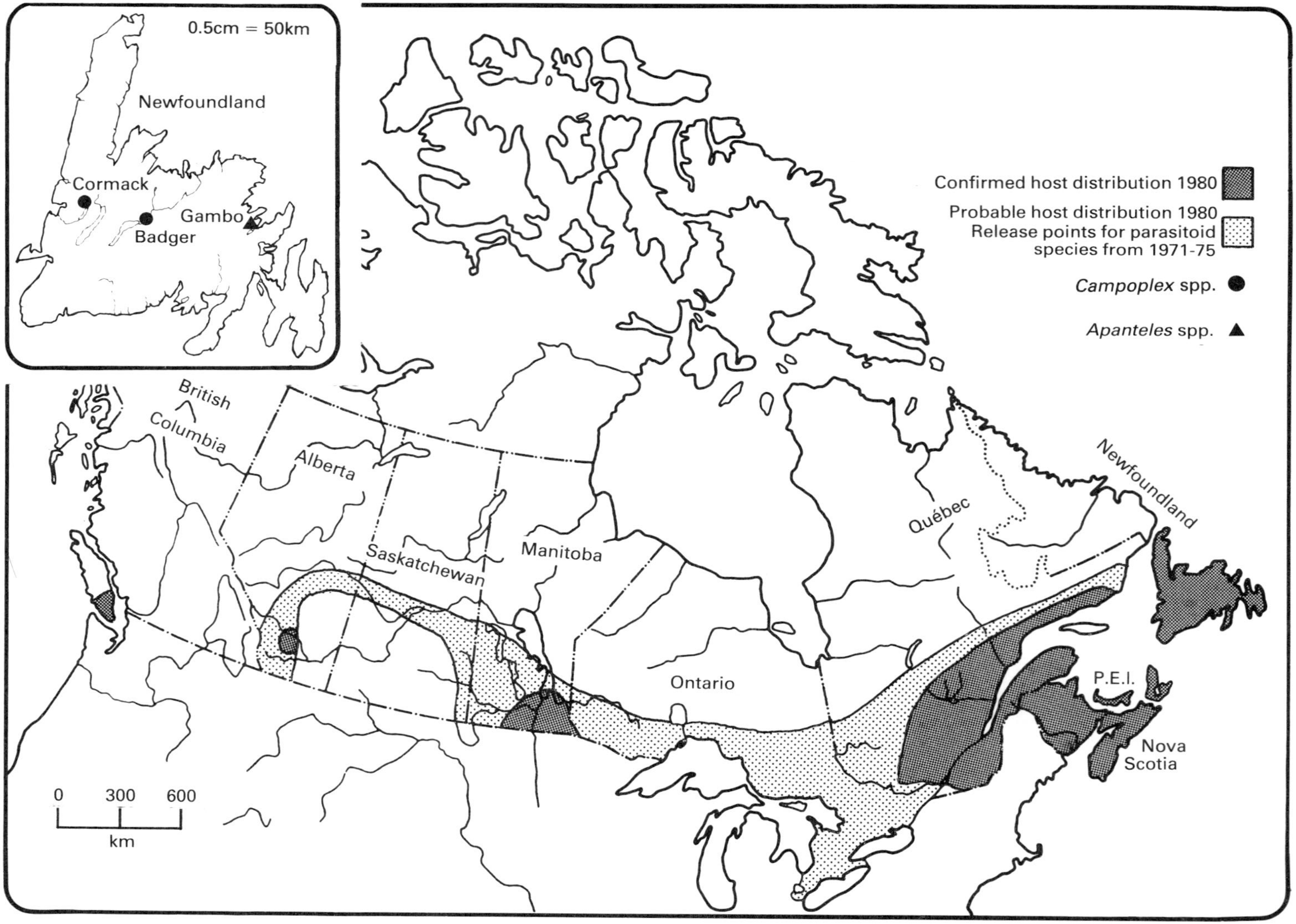

Fig. 11. Distribution of *Coleophora serratella* (L.) in 1980 and release points for parasitoid species, 1971-75

Just before leaf drop they crawl to branch crotches or to the base of leaf buds to overwinter. In the following spring, after moulting, they feed on developing leaves. When fully grown, by early summer, they pupate and adults emerge in July.

Species of alder (*Alnus* spp.) are the main hosts in continental Europe (Gepp 1975*a*), but in the British Isles several species of birch are the most severely attacked trees (Coshan 1974). In North America (Guèvremont & Juillet 1974, Bryant & Raske 1975) white birch is the most severely damaged species, although grey birch, *Betula populifolia* Marsh., may also sustain continuous severe damage. The casebearer may complete its life cycle on several species of hardwood trees and shrubs, but in North America it is known to build up high population levels only on white and grey birch (Guèvremont & Juillet 1974, Bryant & Raske 1975).

Background

In North America, the extent of mortality of various life stages has been determined in Québec (Guèvremont & Juillet 1974) and in Newfoundland (Raske unpublished). Life tables for the two regions differ substantially. In Québec, mortality of first- and second-instar larvae and pupal parasitism seem to be the most important mortality factors. In Newfoundland, highly variable egg mortality, from unknown causes, seems to regulate the population at a chronic high level. This egg mortality (Raske 1974*b*) has not been reported on mainland Canada. In fall, early instar larval mortality in Newfoundland is about 20%. Surviving larvae that overwinter suffer mortality of about 60% annually. In spring, larval mortality among those remaining is about 80% and pupal mortality is 25%. Total losses from parasitism of all stages is about 10%.

The parasitoid complex of this casebearer has been studied both in Europe and in North America where biological control of this pest has been attempted. In its native Europe, parasitoids are more important in regulating the population of the birch casebearer (Pschorn-Walcher 1969 to 1975, Gepp 1975*b*, 1975*c*, Coshan 1974) than in North America (Guèvremont & Juillet 1975, Raske 1978). However, the parasitoid complexes on the two continents resemble one another. Many genera are native to both regions: *Scambus, Itoplectis, Gelis, Orgilus, Agathis, Habrocytus, Cirrospilus,* and *Chrysocharis*. The species on each continent differ, except for *Habrocytus semotus* (Walker). *Apanteles* spp. and *Campoplex* spp. are dominant parasitoids of this casebearer in Europe, but are virtually absent on it in North America. The main difference between Québec and Newfoundland is that *Agathis cincta* was common in Québec and absent in Newfoundland.

Most parasitoid species in North America are parasitic on many lepidopterous insects and are best considered incidental parasitoids that have adapted to this host.

Releases and Recoveries

In 1968 the Canadian Forestry Service, in co-operation with the CIBC, initiated a biological control programme against the birch casebearer in Newfoundland; introductions of European parasitoids began in 1971 and terminated in 1975 (Raske 1977). Two species complexes of parasitoids were released: *Campoplex* (= *Porizon*) spp. (Hymenoptera: Ichneumonidae) and *Apanteles* spp. (Hymenoptera: Braconidae) (Table 76). The *Campoplex* spp. consisted of *C. borealis* (Zett.) and an undescribed species. The *Apanteles* complex included three species: predominantly *A. coleophorae* (Wilk.) and a few each of *A. mesoxanthus* Ruschka and *A. corvinus* Reinh. Living individuals of neither complex

Table 76 Number of parasitoids released against the birch casebearer, *Coleophora serratella* (L.) in Newfoundland from 1971 to 1975 and parasitoid recovery to 1980

Locality and year	No. parasitoids released		Country of origin	Number of hosts reared for parasitoid recovery				Number of parasitoids recovered			
	Male	Female		1973	1974	1975	1980	1973	1974	1975	1980
				Campoplex spp.							
Cormack 1971	215	179	Germany and Austria	5278	2344	5329	1760	2	1	10	0
Badger											
1972	31	26	Germany								
1973	104	98	Germany								
1974	291	230	Germany and Switzerland		961	2053	3106		0	2	2
1975	154	144	Germany								
				Apanteles spp.							
Gambo											
1974	147	114	Germany and Switzerland			202	5527			8*	0
1975	90	59	Germany								

* All progeny from caged trees.

could be identified to species; therefore exact numbers of each species released are not known. *Campoplex* spp. were released in western Newfoundland (Cormack) in 1971 and in central Newfoundland (Badger) from 1972 to 1975. *Apanteles* spp. were released in eastern Newfoundland (Gambo) in 1974 and 1975 (Fig. 11 and Table 76).

In 1971 at Cormack and in 1974 at Gambo, all parasitoids were released on caged white birch trees with three to five host larvae per leaf. At other times parasitoids were released beneath uncaged birch trees infested with the birch casebearer. All caged trees were checked the year after release; when adult parasitoids were seen in the cages, the cages were opened and the parasitoids released. Thus successful oviposition and overwintering in Newfoundland's climate were demonstrated.

Evaluation of Control Attempts

Very few parasitoid progeny have been recovered to date, even though all host insects have been collected at or very near the release site. Successful establishment of *Apanteles* is unlikely and even that of *Campoplex* at Badger is doubtful. Members of these parasitoid genera are strong fliers and disperse well. Therefore, low populations may spread over large areas before their numbers increase significantly at any locality, and it may be 10 years before establishment can be ascertained and even longer before the effectiveness of the parasitoids is determined.

Populations of the birch casebearer are continuing at levels that cause light-to-moderate defoliation of birch in large regions, providing the opportunity for parasitoid populations to increase. Furthermore, the parasitoid is able to survive Newfoundland climatic conditons, and so presumably its population increase is not limited by climate. The cause of the failure of parasitoid increase is unknown, but the extremely low numbers of parasitoids, too low for an adequate number of mates to find each other, is more likely to be the cause than any other factor.

Recommendations

The birch casebearer remains a very promising candidate for successful biological control; it is an introduced insect, and it has an effective parasitoid in its native home. A sizeable mortality factor of late-instar larvae would effectively reduce populations.

The problem of the past attempt to introduce parasitoids was the low ratio of parasitoids emerging in Canada from the birch casebearer pupae sent from Europe. Samples of the pupae, reared in Europe, yielded a ratio of 10 times more parasitoids than the bulk of the collection shipped to Canada. The reason for the low emergence in Canada is not known. The probability of future successful control by the introduction of parasitoids is theoretically high, but larger numbers of parasitoids should be released to increase the chance of successful establishment. Future introduction of parasitoids is justified and recommended. The same *Campoplex* spp. and *Apanteles* spp. are still the most promising candidates. *A. cincta*, a prominent parasitoid of Québec populations of *C. serratella*, should also be introduced into Newfoundland.

Literature Cited

Bryant, D.G.; Raske, A.G. (1975) Defoliation of white birch by the birch casebearer, *Coleophora fuscedinella* (Lepidoptera: Coleophoridae). *Canadian Entomologist* 107,217–223.

Cochran, S.G. (1974) The birch casebearer in Newfoundland. Master's Thesis, McGill University, Montréal, Quèbec, 126 pp.

Coshan, P.F. (1974) The biology of *Coleophora serratella* (L.) (Lepidoptera: Coleophoridae). *Transactions of the Royal Entomological Society of London* 126,169–188.

Gepp, von J. (1975*a*) Zur Biologie von *Coleophora fuscedinella* Zeller (Lep., Coleophoridae). *Zeitschrift für angewandte Entomologie* 78,225–236.

Gepp, von J. (1975*b*) Zum Parasitenkomplex von *Coleophora fuscedinella* (Lep., Coleophoridae) in Österreich. *Zeitschrift für angewandte Entomologie* 79,76–96.

Gepp, von J. (1975*c*) Zür Faunistic von *Coleophora fuscedinella* Z. und ihrer Parasiten in der Steiermark. *Mitteilungen der Abteilung für Zoologie im Landesmuseum Joanneum* 4,169–175.

Gillespie, H. (1932) The birch casebearer in Maine. *Maine Forestry Service Bulletin* 7, 23 pp.

Guèvremont, A.M.; Juillet, J. (1974) Recherches sur la dynamique des populations naturelles de *Coleophora fuscedinella* Zeller (Lépidoptères: Coleophoridae) dans la région de Sherbrooke, Québec. *Phytoprotection* 55,121–134.

Guèvremont, H.; Juillet, J. (1975) Parasites du porte-case du bouleau, *Coleophora fuscedinella* Zeller (Lépidoptères: Coleophoridae) dans la région de Sherbrooke, Québec. *Phytoprotection* 56,1–17.

Pschorn-Walcher, H. (1969 to 1975) Birch casebearer (*Coleophora fuscedinella*): work done in Europe. *Commonwealth Institute of Biological Control Annual Project Statement*, Delémont, Switzerland. (Unpublished reports; one report for each year).

Raizenne, H. (1952) Forest Lepidoptera of southern Ontario and their parasites. *Canadian Department of Agriculture Science Service, Division of Forest Biology*, 277 pp.

Raske, A.G. (1974*a*) Introduction of the parasites *Campoplex* spp. into Newfoundland for the biological control of the birch casebearer. *Canadian Forestry Service Newfoundland Forest Research Centre Information Report* N-X-108, 22 pp.

Raske, A.G. (1974*b*) Mortality of birch casebearer eggs. *Canadian Forestry Service Bi-monthly Research Notes* 30,1–2.

Raske, A.G. (1976) Complexities in the number of larval instars of the birch casebearer in Newfoundland (Lepidoptera: Coleophoridae). *Canadian Entomologist* 108,401–405.

Raske, A.G. (1977) Introduction of the birch casebearer parasites *Campoplex* and *Apanteles* into Newfoundland. *Canadian Forestry Service Bi-monthly Research Notes* 33,19–20.

Raske, A.G. (1978) Parasites of birch casebearer larvae in Newfoundland (Lep.: Coleophoridae). *Entomophaga* 23,103–108.

Reeks, W.A. (1951) The birch casebearer, *Coleophora salmani* Heinr. *Canadian Department of Agriculture, Forest Biology Division Bi-monthly Progress Report* 7,1.

Salman, K.A. (1929) Notes on the immature stages and biology of a birch casebearer. *Annals of the Entomological Society of America* 22,480–488.

Fenusa pusilla (Lepeletier), Birch Leaf-miner (Hymenoptera: Tenthredinidae)

F.W. QUEDNAU

Pest Status

Fenusa pusilla (Lepeletier) is an important endemic pest on ornamental birches, *Betula* spp., in Canada. This insect was accidentally introduced from Europe into the State of Connecticut, USA, in 1923 (Friend 1933). The tree species attacked are *Betula papyrifera* Marsh., *B. populifolia* Marsh., *B. lenta* L., and the European *B. verrucosa* Ehrh. and its varieties. The present distribution of the sawfly in Canada is mainly from Newfoundland to Thunder Bay in northwest Ontario, but it has recently been recorded from the Prairie Provinces (Fig 12). The infestation level of *F. pusilla* in Canada is usually high and remains relatively constant from one year to another. The insect, by its oviposition and the ensuing feeding of larvae in leaf mines, causes discolouration of the foliage by the destruction of parenchyma cells. The affected trees have an unsightly brownish discolouration of the leaves, which drop prematurely. Repeated infestations of the sawfly weaken a tree and make it prone to attack by other insects or diseases. There are about three overlapping generations of *F. pusilla* in a year. The biology of this insect in Québec has been described by Cheng & LeRoux (1965) and Guèvremont (1975).

Background

A preliminary study of biological control agents affecting *F. pusilla* in Europe was undertaken by the Commonwealth Institute of Biological Control (CIBC) at Delémont, Switzerland (Pschorn-Walcher & Eichhorn 1968). The observation that *F. pusilla* is much less abundant in Central Europe than in Canada led to the decision by the Canadian Forestry Service to introduce parasitoids. In 1972 the Newfoundland Forest Research Centre initiated a programme at Pasadena, Newfoundland. In 1974 the project was transferred for technical reasons to the Laurentian Forest Research Centre at Ste-Foy, Québec. The aim of this biological control programme was the implantation of the two hymenopterous parasitoid species, *Lathrolestes nigricollis* Thoms. and *Grypocentrus albipes* Ruthe. For information on the techniques employed for the releases and recoveries the reader is referred to reports by Guèvremont & Quednau (1976, 1978).

Releases and Recoveries

The numbers of parasitoids released in Canada are listed in Table 77.

Lathrolestes nigricollis Thoms. (syn. *Priopoda nigricollis* (Thoms.)) (Hymenoptera: Ichneumonidae)

This is a multivoltine solitary endoparasitoid of semi-mature and mature larvae of the leafminer. Its biology in Europe was described by Eichhorn & Pschorn-Walcher (1973). Additional observations on its mating and oviposition behaviour were made by Quednau & Guèvremont (1975). It is the most important natural enemy of *F. pusilla* in Central Europe. It is probably monophagous and well synchronized with successively overlapping stages of the host in the field. In captivity the female parasitoids are easily mated. Partial encapsulation of parasitoid eggs in the host larva, the incidence of superparasitism, multiple parasitism by other parasitoid species, and entry in diapause

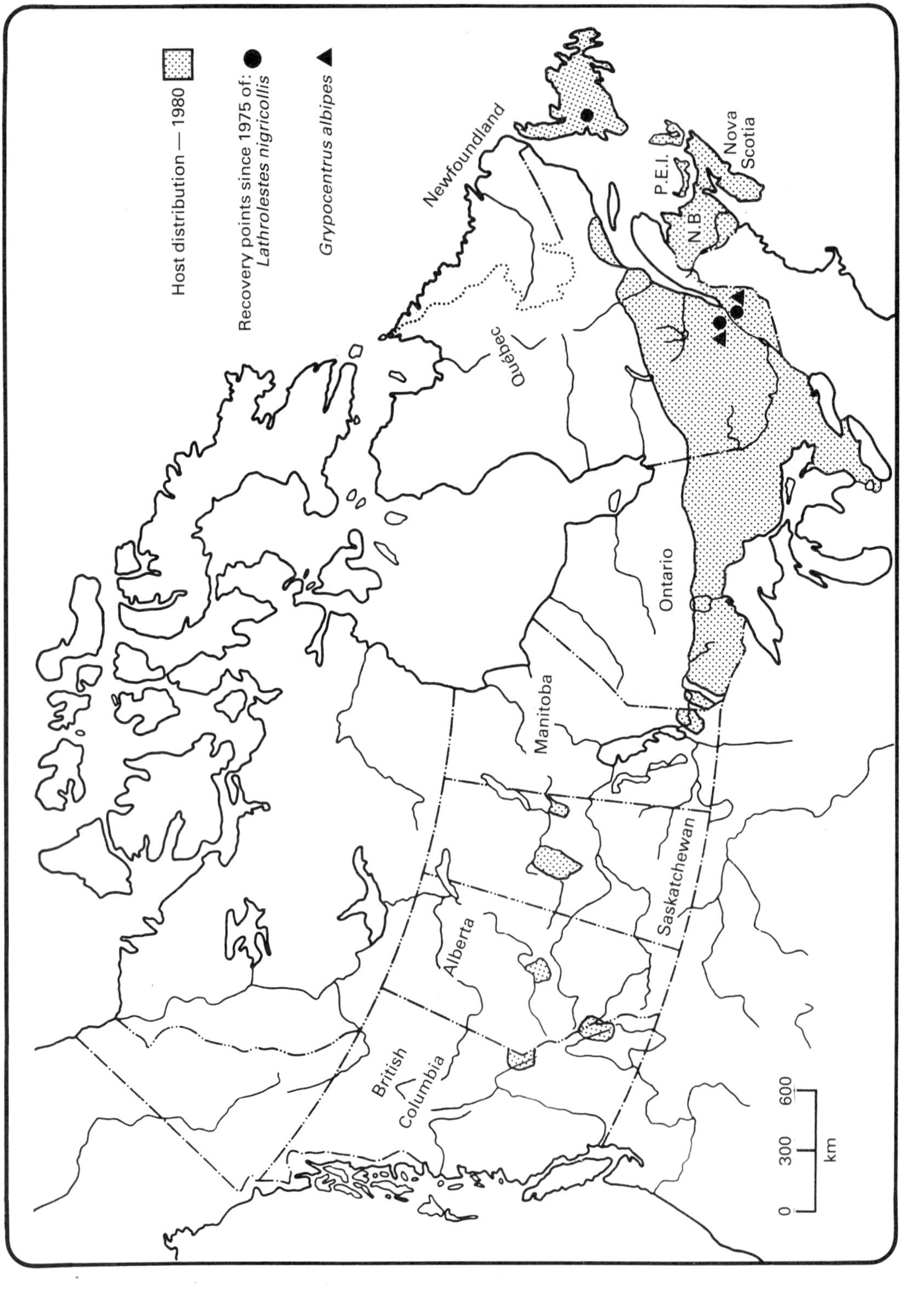

Fig. 12. Distribution of *Fenusa pusilla* (Lep.) in 1980 and recovery points for released parasitoids since 1975

Table 77 Parasitoids utilized for releases against *Fenusa pusilla* (Lepeletier) in Newfoundland and Québec between 1972 and 1978

Species	Year	Origin	No. and sex of parasitoids received	
			Male	Female
Lathrolestes nigricollis Thoms.	1972–73*	Austria	111	127
	1974	Austria	198	352
	1975	Austria	433	578
	1976	Austria	707	698
	1977	Austria	692	719
	1978	Austria	206	173
Grypocentrus albipes Ruthe	1972–73*	Austria	12	18
	1974	Austria	339	287
	1975	Austria	159	134
	1976	Austria	239	146
	1977	Austria	206	104
	1978	Austria	67	42

*Released in Newfoundland, all others in Québec.

of a greater part of the parasitoid's first generation all contribute to a reduction in biological control potential. Encapsulation of the eggs of *L. nigricollis*, however, is not always fatal because the parasitoid larva frequently manages to hatch from an encapsulated egg.

Grypocentrus albipes Ruthe (Hymenoptera: Ichneumonidae)

This species is the second most important parasitoid of *F. pusilla*. It is a multivoltine, solitary, host-specific ectoparasitoid of medium-sized to mature sawfly larvae. Details of the biology were published by Eichhorn & Pschorn-Walcher (1973) and by Guèvremont & Quednau (1977*a*). Mating this species in captivity is more difficult than with the preceding species. The searching capacity at low host densities seems to be greater than that of *L. nigricollis*. The rate of superparasitism is high, and when multiple parasitism with *L. nigricollis* occurs, *G. albipes* proves to be intrinsically superior. A greater part of the parasitoid population enters a diapause rather early in the season.

Evaluation of Control Attempts

It appears that *L. nigricollis* has become established in Newfoundland near Pasadena (lat. 49°02′N, long. 57°36′W), but there is uncertainty about *G. albipes* (Raske & Jones 1975).

In Québec both *L. nigricollis* and *G. albipes* were liberated in 1974 and 1975 at Base de Plein Air, Ste-Foy (lat. 46°47′, long. 71°19′), St-Etienne-de-Lauzon (lat. 46°39′, long. 71°18′), and only *L. nigricollis* at Villeroy (lat. 46°22′, long. 71°51′) and St-Louis-de-Blandford (lat. 46°15′, long. 72°01′). Infestation levels by the sawfly were severe to moderate. Three generations of the parasitoids, the first of which overwintered, were recovered at the end of the summer in 1975 in Ste-Foy and St-Etienne (Guèvremont & Quednau 1977*b*). In 1976–77 releases were continued and new release sites were chosen, such as Neuville (lat. 46°45′, long. 71°35′), Bourg Louis (Laval University grounds, lat. 46°50′, long. 72°53′), Base de Plein Air, Ste-Foy, Cap Rouge (lat. 46°46′, long. 71°21′), and Rue St-Amable in the centre of Québec City (lat. 46°48′, long. 71°13′). In these release sites, recoveries were made in 1976 of only *L. nigricollis*, whereas *G. albipes* apparently did not become established that year. *L. nigricollis* was also released in 1976 at Mont-St- Hilaire (lat. 45°35′, long. 73°13′), Pidgeon (lat. 46°10′, long. 71°38′), Ste-Anastasie (lat. 46°23′, long. 71°34′), Arthabaska (lat. 46°02′, long. 71°54′), Drummondville-Sud (lat. 45°51′, long. 72°28′), and Sherbrooke (lat. 45°22′, long. 71°54′). However, no attempts were made to recover material from these places. In 1978 both *L. nigricollis* and *G. albipes* were recovered in four release sites near Québec City (Neuville, Bourg Louis, St-Etienne, and Ste-Foy). The fact that the parasitoids could be collected after two winters, of which December 1975 and January 1976 were particularly severe, indicates that the insects have the capacity to tolerate the Québec climate. The existence of a second summer generation in Québec of both *L. nigricollis* and *G. albipes* was confirmed in spite of the cold and rainy summer of 1976 (Guèvremont & Quednau 1976). However, the average percentage of parasitism in these evaluations was low, not more than 15%.

In 1978 the biological control programme against *F. pusilla* was terminated because it was of low priority and because implantation of the two parasitoid species in Québec had been successful. No quantitative analysis of the impact of the parasitoid on the birch leafminer has been carried out.

Recommendations

Further work with the two parasitoid species should be oriented toward more basic research to discover possible shortcomings of the biological control agents. Detailed life-table studies would probably reveal weaknesses related to climate or other environmental barriers. Another effort should be made to develop adequate mass-rearing techniques so that enough parasitoid material becomes available for large-scale initial implantations, preferably carried out in cages and not as open releases.

Literature Cited

Cheng, H.H.; LeRoux, E.J. (1965) Life history and habits of the birch leaf-miner, *Fenusa pusilla* (Lepeletier) (Hymenoptera: Tenthredinidae) on blue birch, *Betula caerulea grandis* Blanchard, Morgan Arboretum, Québec, 1964. *Annals of the Entomological Society of Québec* 10,173–188.

Eichhorn, O.; Pschorn-Walcher, H. (1973) The parasites of the birch leaf-mining sawfly (*Fenusa pusilla* Lep.) (Hym.: Tenthredinidae) in Central Europe. *Commonwealth Institute of Biological Control Technical Bulletin* 16,79–104.

Friend, R.B. (1933) The birch leaf-mining sawfly, *Fenusa pumila* Klug. *Bulletin of the Connecticut Agricultural Experiment Station* 348.

Guèvremont, H. (1975) La petite mineuse du bouleau. *Centre de Recherches Forestiale Laurentides, Ste-Foy, Québec, Feuillet d'Information* 13, 7 pp.

Guèvremont, H.C.; Quednau, F.W. (1976) Compte rendu des libérations de deux parasitoides contre *Fenusa pusilla* (Lep.) au Québec, pour l'année 1976. *Centre de Recherches Forestiales Laurentides, Ste-Foy, Québec*, 16 pp.

Guèvremont, H.C.; Quednau, F.W. (1977*a*) Morphologie et biologie de *Grypocentrus albipes* (Hymenoptera: Ichneumonidae), parasite de la petite mineuse du bouleau, *Fenusa pusilla* (Hymenoptera: Tenthredinidae). *Canadian Entomologist* 109,1417–1424.

Guèvremont, H.C.; Quednau, F.W. (1977*b*) Introduction de parasites ichneumonides pour la lutte biologique contre *Fenusa pusilla* (Hymenoptera: Tenthredinidae) au Québec. *Canadian Entomologist* 109,1545–1548.

Guèvremont, H.C.; Quednau, F.W. (1978) Implantation de deux parasitoïdes pour la lutte biologique contre *Fenusa pusilla* (Lep.) au Québec, en 1974 et 1975. *Centre de Recherches Forestiale Laurentides, Ste-Foy, Québec, Rapport d'Information* LAU-X-22, 20 pp.

Pschorn-Walcher, H.; Eichhorn, O. (1968) Birch leaf-mining sawfly (*Fenusa pusilla*): work in Europe in 1968. *Commonwealth Institute of Biological Control, Switzerland, Annual Project Statement*, 11 pp.

Quednau, F.W.; Guèvremont, H. (1975) Observations on mating and oviposition behaviour of *Priopoda nigricollis* (Hymenoptera: Ichneumonidae) a parasite of the birch leaf-miner, *Fenusa pusilla* (Hymenoptera: Tenthredinidae). *Canadian Entomologist* 107,1199–1204.

Raske, A.G.; Jones, R.N. (1975) Introduction of parasitoids of the birch leaf-mining sawfly into Newfoundland. *Canadian Department of the Environment Bi-monthly Research Notes* 31(2),20–21.

Gilpinia hercyniae (Hartig), European Spruce Sawfly (Hymenoptera: Diprionidae)

L.P. MAGASI and P.D. SYME

Pest Status

The European spruce sawfly, *Gilpinia hercyniae* (Hartig), a major pest of spruce (*Picea* spp.) in eastern Canada during the 1930s and early 1940s, has lost its economic importance with the collapse of outbreaks in the Gaspé Peninsula and in northern New Brunswick (Neilson *et al.* 1971). Populations have remained generally low since the early 1940s. The insect, and specifically its biological control by parasitoids and pathogens, has thus received very little attention since the last review in 1969. Consequently, the information presented here is meagre. It deals mainly with major population changes and changes in distribution of the insect during the past decade. No research has been conducted on the biological control of the spruce sawfly in Canada since 1969 and few references are available. Distribution of the pest is summarized in Fig. 13.

Newfoundland

Populations were generally low, with occasional minor increases. Larval numbers in western Newfoundland increased from 1.4/tree in 1970 to 18.0/tree in 1971 and to 24.0/tree in 1972. However, by 1974 the population decreased to 1.0 larva/tree throughout the province and has remained at that level.

The nuclear polyhedrosis virus (NPV) *Borrelinavirus hercyniae*, introduced in 1943, is assumed to be a principal factor in controlling outbreaks but the introduction of invertebrate parasitoids and perhaps even the masked shrew, *Sorex cinereus cinereus* Kerr, have doubtless improved the biological control of this insect.

Maritimes

Population levels were generally low in the region throughout the decade. The highest number recorded was in Pictou County, Nova Scotia, where 17 larvae were found on four white spruce trees, *Picea glauca* (Moench) Voss, in 1978. The NPV *B. hercyniae* was last found in larvae in 1975. The dipterous parasitoid *Drino bohemica* (Mesn.) and the hymenopterous parasitoid *Exenterus vellicatus* Cush. were last reared in 1977, but it should be added that, apart from these rearings, no efforts have been made to identify these control agents.

Neilson *et al.* (1971) in an earlier edition of this book described an increase in the sawfly population after the cessation of chemical insecticide spraying to control spruce budworm, *Choristoneura fumiferana* (Clem.). This situation occurred again in the early 1970s when the monitoring plot was sprayed for 2 years. Sawfly levels increased immediately afterwards and remained high for a number of generations until, presumably, the host and its control agents re-established their balance.

Québec

The insect was present in low numbers throughout its known range, with minor fluctuations. Laboratory rearings showed the presence of the NPV *B. hercyniae* in 1971, 1973, 1974, and 1975, and the virus is considered a major factor in controlling the population.

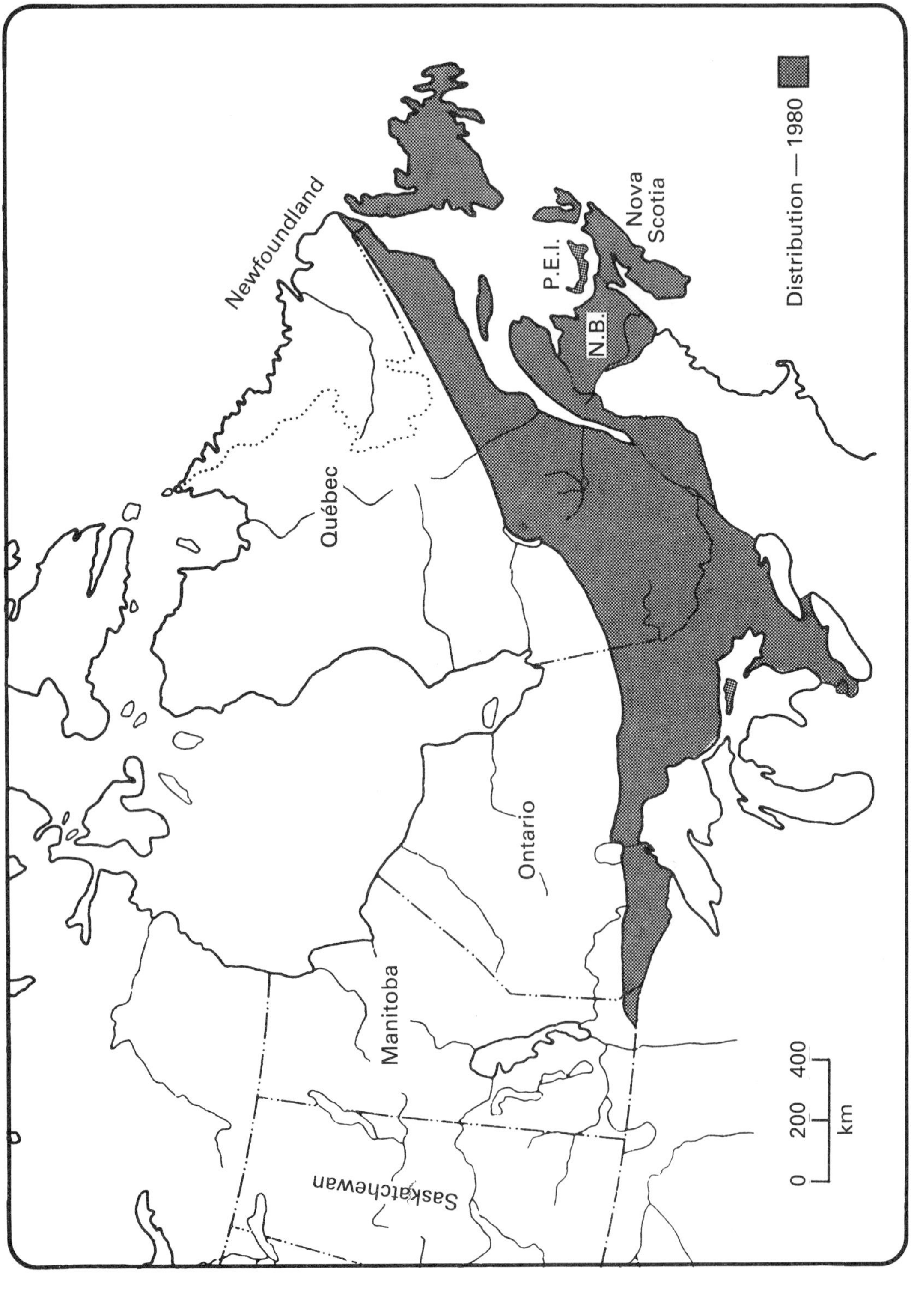

Fig. 13. Distribution of *Gilpinia hercyniae* (Htg.) in 1980

Ontario

The insect has been sporadically reported, and occurs throughout the province within its known range but usually in very low numbers. In 1977, light defoliation of white spruce occurred in parts of southwestern Ontario.

The Forest Insect and Disease Survey participated in a parasitoid identification project with CIBC in 1973. Of 160 sawfly cocoons, 7 adults of *E. vellicatus* emerged (4.4% parasitism); the parasitoids were mostly male.

Manitoba

The insect was first reported in Manitoba in 1969, when it was found in low populations on 5200 km^2 adjacent to the Ontario border. In 1970, populations were reported on a further 2600 km^2, as far north as Otter Falls and Meditation Lake in the Whiteshell Provincial Park, and as far west as the Agassiz and the Sandilands Provincial Forests. It has not extended its range since then and population levels have remained low.

Evaluation of Control Attempts

The European spruce sawfly appears to be controlled, largely by biological agents. However, there may be differences in areas that are regularly subjected to chemical sprays. Neilson *et al.* (1971) stated that when spraying stopped, there was a sharp increase in sawfly populations — at least until the balance between host and control agent was re-established. This phenomenon was observed again in the 1970s. In each case, the spray period lasted only 2–4 years. It is not known what is happening in areas of sustained chemical insecticide application: the European spruce sawfly, obviously, is kept at low populations but possibly not by biological control agents. Nor is it known what the status of these agents is or if they would rapidly re-establish the balance necessary for control once chemical insecticide spraying ceased. High-value spruce plantations, which were practically non-existent in the 1930s and 1940s, are now plentiful and provide a bountiful food supply for the sawfly.

Literature Cited

Neilson, M.M.; Martinez, R.; Rose, A.H. (1971) *Diprion hercyniae* (Hartig), European spruce sawfly. In: Biological control programmes against insects and weeds in Canada, 1959–1968. *Commonwealth Institute of Biological Control Technical Communication* 4,136–143.

Leucoma salicis (L.), Satin Moth (Lepidoptera: Lymantriidae)

L.P. MAGASI and G.A. VAN SICKLE

Pest Status

Satin moth, *Leucoma salicis* (L.), has been a pest throughout Canada, mostly of ornamental and shade trees of the genera *Populus* and *Salix*, with occasional forays into woodland areas in New Brunswick and British Columbia. Biological control of the insect did not receive close attention during the review period from 1959 to 1968, and work was concerned mainly with detecting and monitoring infestations and with laboratory rearing by the Forest Insect and Disease Survey (Forbes & Ross 1971). Since 1969, because of manpower and monetary restraints within the Canadian Forestry Service, minor pests such as the satin moth have received only incidental attention. General surveys by the Forest Insect and Disease Survey were not thorough or consistent, and rearing programmes and assessments of biological control factors hardly existed. Nevertheless, because defoliation by this insect is usually conspicuous and usually damages shade trees, we suspect that few, if any, major outbreaks went unrecorded; all information presented here is based on records of Forest Insect and Disease Survey units of the Canadian Forestry Service and is summarized in Fig. 14. Affected tree species and their regional distribution during 1969–80 are shown in Table 78.

Table 78 Tree species affected by *Leucoma salicis* (L.) in Canada

Host		Nfld.	Maritimes	Qué.	Ont.	B.C.
Trembling aspen	*Populus tremuloides* Michx.	x	x		x	x
Silver poplar	*Populus alba* L.	x	x	x		x
Balsam poplar	*Populus balsamifera* L	x		x		
Willow	*Salix* L.	x	x	x		
Poplar	*Populus* L.	x		x		
Lombardy poplar	*Populus nigra* var. *italica* Moench	x		x	x	
Eastern cottonwood	*Populus deltoides* Bartr.			x		
Black cottonwood	*Populus trichocarpa* Torr. & Gray					x
Largetooth aspen	*Populus grandidentata* Michx.		x			
Carolina poplar	*Populus* x *canadensis* Moench		x			

Newfoundland

Various levels of defoliation occurred in several locations in the province, mostly on the Avalon Peninsula and on the west coast of the island. Tree mortality was reported from Badger, Clarenville, and St. John's in 1974. In Badger, *Apanteles solitarius* (Ratz.), a parasitoid first introduced into Newfoundland in 1936, was recovered in high numbers in 1977; in Corner Brook, a 3-year-old infestation of the satin moth was terminated in 1978 by this same parasitoid, which was first introduced to western Newfoundland in 1942.

A severe infestation has persisted on the Avalon Peninsula since 1977. Complete defoliation of poplars (*Populus* spp.) and willows (*Salix* spp.) was common throughout St. John's and vicinity and in towns in the Conception Bay area. In western Newfoundland, in 1979, severe defoliation of balsam poplar, *Populus balsamifera* L.,

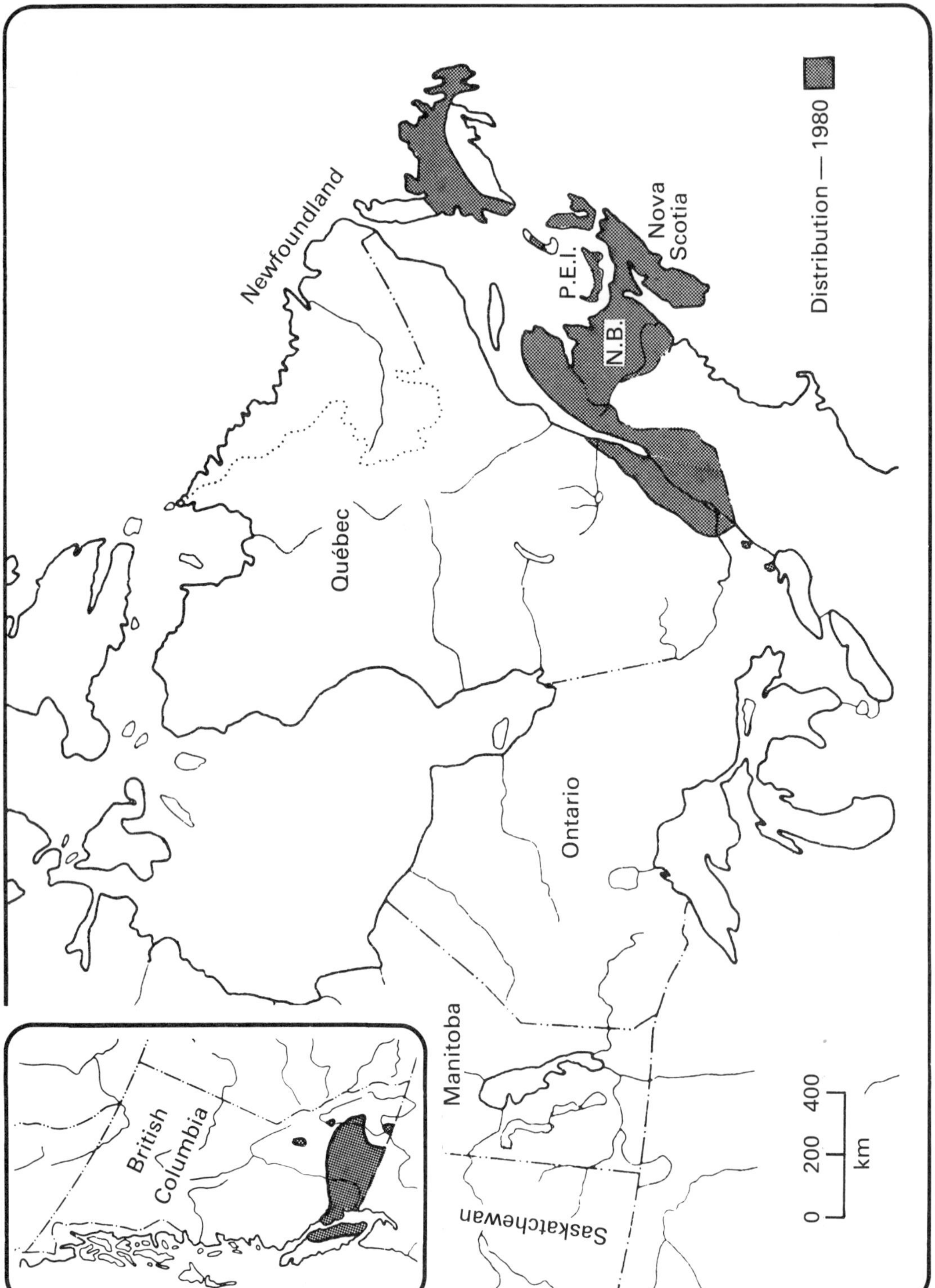

Fig. 14. Distribution of *Leucoma salicis* (L.) in 1980

was recorded at Upper Ferry and light damage occurred on willow in the Stephenville area. In 1980 the outbreak spread from St. Andrews to Fischell's River, mostly along the major rivers where natural stands of balsam poplar occur.

Maritimes

In 1969, over 3200 ha of woodland trees, trembling aspen, *Populus tremuloides* Michx., and largetooth aspen, *Populus grandidentata* Michx., sustained moderate to severe defoliation in southeastern New Brunswick. The outbreak began in 1967, persisted for 3 years, and collapsed in 1970. The collapse was partly due to biological control agents. At Elgin, New Brunswick, cocoons of *A. solitarius* were numerous on tree trunks in the winter of 1969–70 and counts of up to 732 cocoons/m^2 of bark surface were made. A cytoplasmic polyhedrosis virus, known in the maritimes since 1954, infected up to 67% of field-collected laboratory-reared larvae in 1968 near Anagance, within the same area as the *L. salicis* infestation.

Small outbreaks occurred in natural stands in 1974 in south-central New Brunswick, in 1976 in the central and in 1979 in the east-central parts of the province. All these outbreaks affected trembling aspen, but in 1974 largetooth aspen was also affected. None of the outbreaks lasted most than a single season.

The insect was found on ornamental silver and Carolina poplars *Populus alba* L. and *P. canadensis* Moench, and willow, during 1969–80 at locations scattered throughout the region, resulting in varying degrees of defoliation.

The parasitoids reared from satin moth in 1969–70 and their frequency are listed as follows:

Apanteles solitarius (Ratz.)	12 locations; mean parasitism 51% (range 16–100%)
Compsilura concinnata (Mg.)	6 locations; mean parasitism 15% (range 5–55%)
Exorista sp.	1 location; percentage parasitism unknown

The mean parasitism from all rearings was 28% for *A. solitarius* and 4% for *C. concinnata*. No rearings or insect analyses have been done since 1970.

Québec

The general distribution of satin moth has increased considerably during the past decade and now includes all eastern Québec and both sides of the St. Lawrence River. Infestations, causing defoliation that varied from trace to severe, were localized and were reported every year with the exception of 1970. No examinations of parasitism were made from 1969–80.

Ontario

The insect was first reported in Ontario in 1972 from Cornwall and Lancaster Township, where it caused severe defoliation of individual silver poplar trees. By 1976 it had spread to near Ottawa. By 1980 it had reached west of Brockville and the severity of attack was reported on the increase. There are no records of parasitoids from the province, but again no check was made.

British Columbia

The general distribution has not changed appreciably during the reporting period but satin moth infestations were recorded in three new areas; at Revelstoke in 1972, at Avola in the North Thompson Valley in 1975, and at Rossland in 1976. Whether these represent a true range extension or are only outlying areas cannot be stated with certainty because of the discontinuous nature of the tree-host and insufficient collections of insects. Infestations usually occurred on small groups of a few shade trees. However, starting in 1973 a few severe outbreaks on trembling aspen and black cottonwood, *Populus trichocarpa* Torr. & Gray, were observed near Merritt, and by 1976 almost 1500 ha of natural stands were severely defoliated between Aspen Grove and Kamloops. Another outbreak, near Rossland, occurred in 1977 on over 140 ha of trembling aspen and small patches of trees were again defoliated in both these areas in 1978.

In contrast to 67 rearing records during 1959–68, there were only two during 1969–80: in 1972, 3 larvae reared from Revelstoke yielded one native parasitoid, *Exorista mella* (Walker), and in 1976, 24 larvae, collected near Victoria, were reared: 6 were parasitized and 11 *Apanteles melanoscellus* (Ratz.) were recovered.

Field observations recorded, but apparently unsubstantiated by collections, include: (1)in 1969, larvae were heavily parasitized in Lewis Park, Courtenay, where some trembling aspen was 80% defoliated; (2)in 1969, cocoons of *Apanteles* spp. were numerous on the trunks of severely defoliated black cottonwood at Slocan City; (3)in 1975, *A. solitarius* was present in the second and last year of defoliation of black cottonwood over 1 ha along Birkenhead Lake.

In a collection of more than 50 larvae from Lillooet in 1975, 6 were infected with a *Beauveria* sp. of fungus.

Evaluation of Control Attempts

The information presented here deals mainly with major population changes and changes in distribution of the host insect during the past decade; there are few instances where reference to biological control is made. No research has been conducted on the biological control of satin moth in Canada during the review period and none is recommended, given the minor economic importance of the pest. We have no information to alter the assessment given in the previous review, that parasitoids must recieve major credit for helping to shorten periods of pest outbreak and thus for reducing damage (Forbes & Ross 1971).

Literature Cited

Forbes, R.S.; Ross, D.A. (1971) *Stilnoptia salicis* (L.), satin moth. In: Biological control programmes against insects and weeds in Canada, 1959–1968. *Commonwealth Institute of Biological Control Technical Communication* 4,205–212.

Lymantria dispar (L.), Gypsy Moth (Lepidoptera: Lymantriidae)

K.J. GRIFFITHS and F.W. QUEDNAU

Pest Status

The natural distribution of the gypsy moth, *Lymantria dispar* (L.), extends across the whole of Europe, including Norway and Sweden up to about 58°N and all the temperate areas of North Africa and Asia, including Japan. Outbreaks occur periodically throughout most of this range (Leonard 1974). The gypsy moth was accidentally introduced into North America at Medford, Massachusetts, in 1869, and within 20 years had become a serious local pest that required control (Burgess 1914). This pest has continued to cause serious periodic damage in spite of almost continuous work to control it. It is now found in the United States throughout the New England States, New York, New Jersey, and Pennsylvania, and has also been recorded in Delaware, Maryland, Virginia, and West Virginia. Isolated infestations have been reported in Washington, Oregon, California, Michigan, Wisconsin, Illinois, and Ohio (Forest Pest Survey Report, Virginia Division of Forestry, Sept.–Dec., 1980, unpublished report).

The gypsy moth was first recorded in Canada in 1924 in Stanstead and St-Jean Counties, Québec (McLaine 1925). A second invasion occurred in New Brunswick in 1936 (McLaine 1938). Both these invasions apparently died out (Brown 1968) but a third, into Québec south of Montréal in 1959 (Brown 1968), did not disappear and the gypsy moth has continued to spread in all directions from that area. The first invasion into Ontario occurred on Wolfe Island, just south of Kingston in 1969 (Sippell *et al.* 1970), and it was reported on nearby Howe Island and the mainland near Kingston in 1970 (Sippell *et al.* 1971). The Plant Quarantine Division of Agriculture Canada reports that the 1979 distribution of gypsy moth extends from approximately 60 km west of Québec City to Morrisburg, Ontario, and north to Ottawa, and from near Brockville, Ontario, to 25 km west of Belleville, with a northward extension of approximately 70 km. There is also a population in Greater Toronto and in Oakville, Ontario (Fig. 15). A small infestation was detected over a two-block area in Vancouver, British Columbia, in the summer of 1978, and a control operation was undertaken in 1979. There were no male adult recoveries in pheromone traps set out in the area in the fall of 1979 and the infestation is believed to have been averted. On the east coast there seems to be an annual influx of male moths from Maine. In Nova Scotia and New Brunswick males have been recovered in pheromone traps every year since 1971, with a marked increase in both provinces in the past 4 years. There were no male recoveries on Prince Edward Island in the first 6 years of trapping, but in 1979 nearly 47% and in 1980, 19% of the traps produced males. However, egg masses have never been recovered in the maritimes in this period (Magasi 1981).

Background

Since its arrival in North America, the gypsy moth has acquired a large complex of native insect parasites, and predators. Griffiths (1976) lists 22 species of insect parasites and Sabrosky & Reardon (1976) added 4 more to this list. In addition there are 17 species of insect predators, 4 species of mammalian predators (Griffiths 1976), and 46 species of

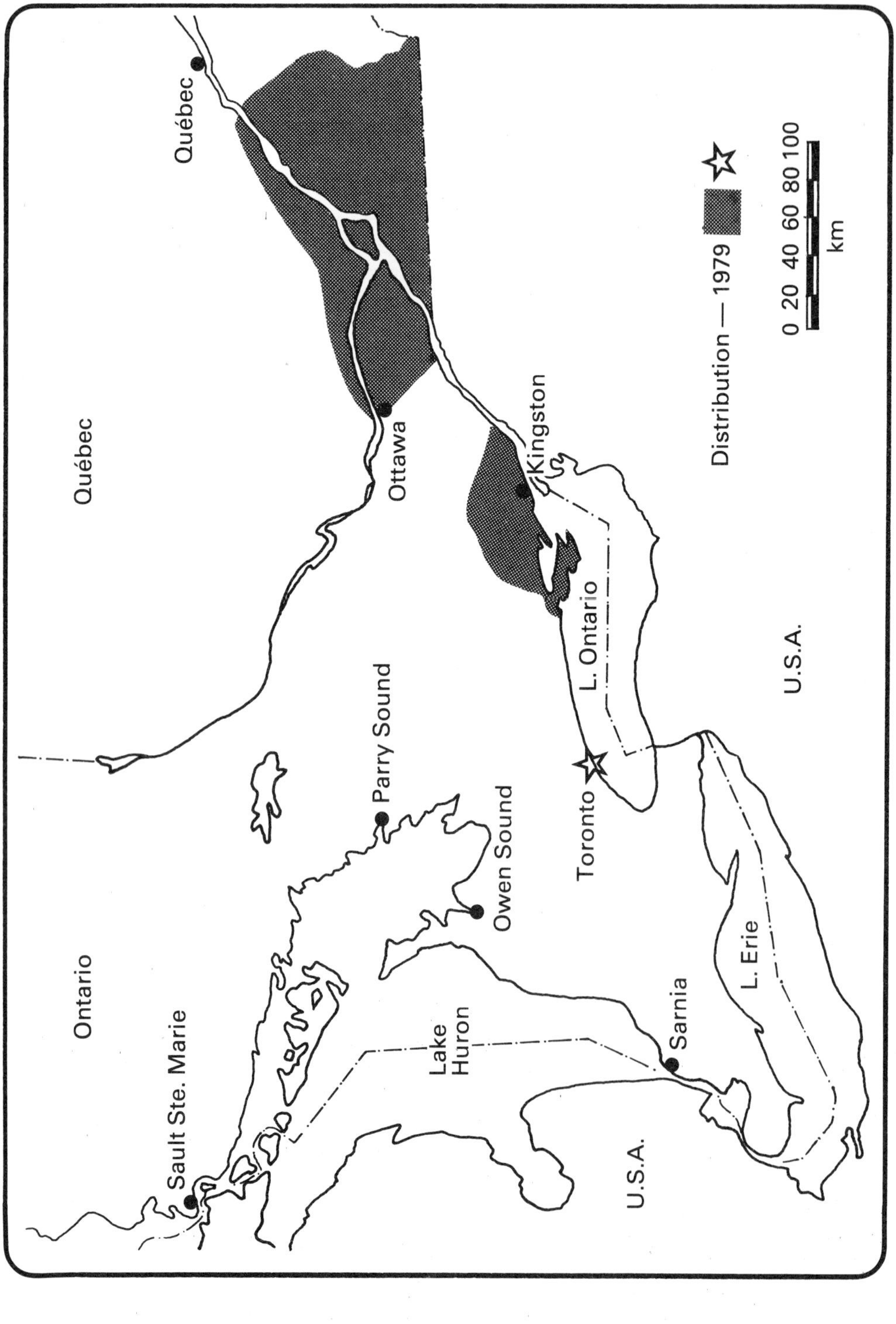

Fig. 15. Distribution of *Lymantria dispar* (L.) in 1980

avian predators (McAtee 1911) recorded from the United States. A major programme of importation and liberation of exotic insect parasites and predators was carried on in the United States from 1905 to 1933 (Dowden 1962) in which 44 species of insect parasites and 9 species of predators were released. Thirteen insect parasites and one predator became established. As pointed out by Ticehurst *et al.* (1978) these natural enemies have not prevented the dispersal of the gypsy moth in North America, but their influence may be responsible for initiating population collapses and for maintaining relatively stable infestations in many areas following a collapse. The introduction programme was revived in the early 1970s and is continuing. Eight species of insect parasite and one species of nematode have been released. Four of the insect species were released unsuccessfully in earlier work (Griffiths 1976). None of these nine species is known to be established.

Fourteen of the 26 species of native parasitoids that attack the gypsy moth in the United States have been recovered from other hosts in areas of Ontario and Québec adjacent to the American infestation. Also 11 of the 17 insect predators, all of the four mammalian predators and 43 of the 46 species of avian predators are known to be in this area of Canada (Griffiths 1976). In addition, four exotic parasitoids established in the United States were present in southern Ontario and Québec before the gypsy moth was recorded. Three of these species, *Compsilura concinnata* (Meigen), *Apanteles lacteicolor* Vier., and *Meteorus versicolor* (Wesm.) had been introduced against the browntail moth, *Euproctis chrysorrhoea* L., and the satin moth, *Leucoma salicis* L., and one, *Exorista larvarum* (L.), apparently dispersed naturally from the United States.

Several studies have been carried out recently to determine which parasitoid species are attacking the gypsy moth in Canada. The first of these investigations, in 1974 and 1975, involved the collection and rearing of gypsy moth larvae from four woodlots near Kingston and three woodlots near Cornwall, Ontario. From these rearings, one native and three introduced parasitoid species were recovered (Table 79) (Griffiths 1977). The second study was carried out in 1977 and 1978 near Havelock, south of Montréal, and Mt-St-Hilaire, east of Montréal, Québec. Larvae and pupae were collected and reared, and one more native and four more exotic species recovered (Table 79) (Madrid & Stewart 1980). A third study, carried out in 1978–80, used Malaise traps to capture adults at Mt-St-Hilaire and Mt-St-Bruno, east of Montréal. In this survey five additional native species were found (Table 79). The final study, carried out in 1979 and 1980, was a repetition of the work done by Griffiths (1977); two new native species and one exotic species, *Ooencyrtus kuvanae* (How.), were recovered (Table 79).

Howard & Fiske (1911) noted that naturally occurring pathogens were playing a role in gypsy moth population regulation in the United States shortly after the introduction of the insect there. Much work has been done on pathogens since then. Bacterial and fungal pathogens are a minor influence (Doane 1971, Podgwaite & Campbell 1972, Majchrowicz & Yendol 1973), but viral pathogens, especially the nuclear polyhedrosis virus (NPV) *Borrelinavirus reprimens* Holmes, are known to be a major factor in reducing outbreaks (Doane 1970, Campbell & Podgwaite 1971). The gypsy moth NPV was registered for use in the United States in 1978, and quantities of it have since been produced and marketed as Gypchek® (US Department of Agriculture) (Doane & McManus 1981).

An NPV was recorded from gypsy moth larvae in southeastern Ontario during studies of gypsy moth parasites carried out in 1974, 1975, 1979, and 1980 (Griffiths 1977, Griffiths & Wallace in press). It was also recorded in southwestern Québec by Cardinal & Smirnoff (1973). Smirnoff has pointed out that for the last 3 years NPV has lowered gypsy moth numbers in Québec (W.A. Smirnoff 1981 personal communication).

Table 79 Parasitoids recovered from the gypsy moth, *Lymantria dispar* (L.), in Canada 1974–80

	Griffiths 1977	Madrid & Stewart 1980	Quednau 1978–80 (unpublished)	Griffiths 1979–80 (unpublished)
Diptera				
Sarcophagidae				
Sarcophaga aldrichi (Park.)		xx	xx	
Tachinidae				
Blepharipa pratensis (Mg.)*		xx	xx	xx
Compsilura concinnata (Mg.)*	xx	xx	xx	xx
Exorista larvarum (L.)*		xx		xx
Exorista mella (Wlk.)			xx	
Parasetigena silvestris (R-D)*	xx	xx	xx	xx
Hymenoptera				
Braconidae				
Cotesia melanoscelus (Ratz.)*	xx	xx	xx	xx
Chalcididae				
Brachymeria intermedia (Nees)*		xx	xx	
Encyrtidae				
Ooencyrtus kuvanae (How.)*				xx
Eupelmidae				
Anastatus disparis Ruschka*			xx	
Ichneumonidae				
Exochus sp.				xx
Itoplectis conquistitor (Say)			xx	xx
Phobocampe disparis (Vier.)*		xx	xx	xx
Pimpla pedalis (Cress.)	xx	xx	xx	xx
Theronia atalantae fulvescens (Cress.)			xx	
Theronia hilaris (Say)			xx	
Scelionidae				
Telonomus sp.				xx

* Exotic parasitoids established in the United States.

Releases and Recoveries

Anastatus disparis Ruschka (Hymenoptera: Eupelmidae)

This parasitoid is a European species, successfully established in the United States by Howard & Fiske (1911). Natural dispersal is slow because female adults cannot fly and the species was subsequently distributed by further releases throughout the gypsy moth's American range (Dowden 1962).

Adults of this species emerge from the previous year's gypsy moth eggs at about the time female moths are ovipositing. Female parasitoids attack the eggs and the parasitic larvae can develop in all embryonic stages of the gypsy moth. However, the proportion of gypsy moth females produced and the fecundity and longevity of the ovipositing female is reduced only if older eggs are attacked. There is usually only one generation per year and the parasitoid overwinters as a mature larva within the host egg. Preliminary work on the overwintering ability of mature larvae of the parasitoid indicated that *A. disparis* apparently could survive over the same geographic area in Canada as the gypsy moth (Sullivan *et al.* 1977). Because parasitism up to 28% had been recorded by Burgess & Crossman (1929), introduction of this species into Canada was undertaken following preliminary field cage and insectary studies (Table 80). The first releases were made near St-Mathias, Rouville County, Québec, in 1979, with further releases near Acton Vale and Drummondville, Bagot and Drummond Counties respectively, in 1980 (Table 81). Establishment of *A. disparis* occurred in all areas where releases were made. The initial occurrence of parasitism was 20–30% when batches of 25 female parasitoids were implanted in small ventilated cages, but only 2% when open releases of batches of 200–400 females were made. Successful hibernation of *A. disparis* at St-Mathias was observed in 1980. A third release of this parasitoid was made in August 1980 near Kilburnie, Ontario, but there have been no recoveries from this release as yet (Table 81).

Table 80 Laboratory and field cage studies of parasitoids against the gypsy moth, *Lymantria dispar* (L.) in Québec

Species	Year	Origin	Number
Anastatus disparis Ruschka	1979	Hungary Austria Romania	845*
		Hungary Austria Romania	2505**
	1980	Hungary	234**

* Field cage studies.
** Laboratory and insectary studies.

Table 81 Open releases and recoveries of parasitoids against the gypsy moth, *Lymantria dispar* (L.)

Species and province	Year	Origin	Number	Year of recovery
Anastatus disparis Ruschka				
Québec	1980	USA Hungary	5 600	1980
Ontario	1980	Hungary	1 020	
Ooencyrtus kuvanae (How.)				
Ontario	1976	USA	25 000	1978

Ooencyrtus kuvanae (How.) (Hymenoptera: Encyrtidae)

This Japanese egg parasitoid was introduced into New England in 1909. Its establishment was immediate and its natural dispersal was augmented by additional large releases throughout the infested area (Burgess & Crossman 1929).

In the United States *O. kuvanae* can have three generations per year on gypsy moth. The first generation cycle takes approximately 6 weeks but the subsequent ones only 3 weeks. Early studies showed that attack by this species was weak. Britton (1935) recorded that in southeastern Massachusetts and in Rhode Island egg parasitism was not more than 10% per year. However, higher attack levels have been recorded in more recent work. Dowden (1962) found attacks of 40–45% in Massachusetts and Connecticut. Weseloh (1972) gives a total summer attack of 29–42% in Connecticut. Thus its introduction into Canada was justified and preliminary investigations of the ability of the overwintering mated female adult to withstand low temperatures were undertaken (Griffiths & Sullivan 1978). It was found that adults were unable to survive continuous exposure to 0°C for 30 days, and it was concluded that this species would be incapable of surviving in the current range of the gypsy moth in Canada. To test this hypothesis, approximately 25 000 adults supplied by the New Jersey Department of Agriculture were released on Wolfe Island, Ontario, on 24 September 1976 (Table 81). Gypsy moth egg masses have been collected in the release area in late August or early September each year since the release. There were no recoveries of *O. kuvanae* until 1978, when 28 adults were recovered from two egg masses. A single adult was obtained from the release area in 1980. In addition, 45 adults of *O. kuvanae* were recovered from two egg masses collected near Glen Norman, Glengarry County, Ontario, in September 1980. Because this recovery site is approximately 180 km from the Wolfe Island release site, it is unlikely that there is any relation between the two; the parasitoids must have dispersed there naturally.

***Borrelinavirus reprimens* Holmes**

Two experimental techniques for the application of NPV were tested by Cardinal & Smirnoff (1973) in Missisquoi and St-Jean Counties, Québec, in 1972. They hand-applied an aqueous solution of NPV at 40 x 10^6 polyhedral inclusion bodies (PIB) per millilitre to egg masses in early May and recorded nearly 100% mortality in the larvae emerging from them. They also sprayed foliage by mist blower with an aqueous solution at 4 x 10^6 PIB/ml when larvae were in the second or third instar, resulting in higher mortality than was seen in larvae on unsprayed foliage.

Bacillus thuringiensis* Berliner *(B.t.)

Jobin & Caron (1982) made aerial applications of *B.t.,* using the commercial product Thuricide® 32B (Sandoz Inc.) in Chambly and Rouville Counties, Québec, in 1979. There were two applications in each area at dosages of 19.6 and 39.52 x 10^9 I.U./ha. All applications included a sticker and chitinase and the *B.t.* was applied at a rate of 9.36 l/ha. Spraying was carried out in late May when 60–70% of larvae were in the second instar. Mortality of larvae in the treated areas 20 days after treatment was 100%, compared to a decrease of 70–76% in control areas over the same time period.

Evaluation of Control Attempts

The establishment of exotic insect parasites on the gypsy moth in Canada is proceeding well, largely through natural dispersal. We now know that 8 of the 13 established exotic species are attacking gypsy moth in the infested zone in Canada and we are confident that a 9th species, *Anastatus disparis* Ruschka, will become established. Two of the remaining four species are already present in Canada on other hosts. They are *Apanteles lacteicolor* Viereck, essentially a parasitoid of the browntail moth, which was established against that species in the maritimes and Québec in 1915 (Tothill 1916); and *M. versicolor,* introduced against browntail moth and satin moth in Canada (Tothill 1916) and recovered in Ontario from *Rheumaptera hastata* L. by Forest Insect and Disease Survey staff. *M. versicolor* is infrequently recovered from the gypsy moth in the United States. The other two exotic species are *Eupteromalus hemipterus* (Walker) and *Monodontomerus aereus* Walker. The former species is a common parasitoid of the satin moth in the United States but is rarely obtained from the gypsy moth there (Burgess & Crossman

1929, Proper 1931). The latter species is a secondary parasitoid as frequently as it is a primary one (Howard & Fiske 1911).

It is too soon to state that the release of egg parasitoids in Canada has resulted in the permanent establishment of these insects. We do know that seven other species of exotic parasitoids previously established in the United States are now attacking gypsy moth in Canada, but, in the absence of careful population studies of the gypsy moth we cannot say what impact any of them is having.

A combination of applications of carbaryl on some areas and of "insecticide soap" on other areas in 1979 in Vancouver apparently resulted in elimination of the population there. It should be pointed out, however, that very few of the isolated infestations recorded in the history of the gypsy moth in the United States have ever been completely eradicated. Spraying with either an NPV or *B.t.* has been effective in small scale trials but the cost of application is many times higher than the cost of applying chemicals.

Recommendations

It is evident that, aside from continuing to release and sample for the two egg parasitoids, *O. kuvanae* and *A. disparis*, there is little more to be done in the introduction of biological agents because there are no more known suitable candidates. The one successfully established exotic predator, *Calosoma sycophanta* L., was introduced unsuccessfully into Canada at a number of locations (McGugan & Coppel 1962) and it is questionable whether further work on this species is justified.

Cardinal & Smirnoff (1973) were able to produce gypsy moth NPV by rearing contaminated larvae in cages in the field. Smirnoff (personal communication) recommends that the NPV so produced be used to inoculate incipient outbreaks by application on either egg masses or foliage to prevent buildup of gypsy moth to outbreak levels.

In view of the above and of the relatively innocuous levels of gypsy moth populations in Canada in 1980, we feel there is no urgent need for further classical biological control work on this species. However, we strongly recommend that careful surveillance of the dispersal of the gypsy moth be continued by staff of the Plant Protection Division of Agriculture Canada. Continued monitoring in the Vancouver area is especially important.

Literature Cited

Britton, W.E. (1935) The gypsy moth. *Connecticut Agricultural Experiment Station Bulletin* 375,623–647.

Brown, G.S. (1968) The gypsy moth, *Porthetria dispar* L., a threat to Ontario horticulture and forestry. *Proceedings of the Entomological Society of Ontario* 98,12–15.

Burgess, A.F. (1914) The gypsy moth and the brown-tail moth, with suggestions for their control. *US Department of Agriculture Farm Bulletin* 564, 24 pp.

Burgess, A.F.; Crossman, S.S. (1929) Imported insect enemies of the gypsy moth and the brown-tail moth. *US Department of Agriculture Technical Bulletin* 86, 147 pp.

Campbell, R.W.; Podgwaite, J.D. (1971) The disease complex of the gypsy moth. I. Major components. *Journal of Invertebrate Pathology* 18(1),101–107.

Cardinal, J.A.; Smirnoff, W.A. (1973) Introduction expérimentale de la polyédrie nucléaire de *Porthetria dispar* L. (Lépidoptères: Lymantriidae) en forêt. *Phytoprotection* 54(1),48–50.

Doane, C.C. (1970) Primary pathogens and their role in the development of an epizootic in the gypsy moth. *Journal of Invertebrate Pathology* 15(1),21–23.

Doane, C.C. (1971) Field application of a *Streptococcus* causing brachyosis in larvae of *Porthetria dispar*. *Journal of Invertebrate Pathology* 17(3),303–307.

Doane, C.C.; McManus, M.L. (Eds.) (1981) The gypsy moth: research toward integrated pest management. *US Department of Agriculture Science Education Agency APHIS Technical Bulletin* 1584, 757 pp.

Dowden, P.B. (1962) Parasites and predators of forest insects liberated in the United States through 1960. *US Department of Agriculture Forest Service Handbook* 226, 70 pp.

Griffiths, K.J. (1976) The parasites and predators of the gypsy moth: a review of the world literature with special application to Canada. *Canadian Forestry Service Sault Ste. Marie, Ontario Report* 0-X-243, 92 pp.

Griffiths, K.J. (1977) A preliminary report on the gypsy moth and its parasites in southeastern Ontario. *Proceedings of the Entomological Society of Ontario* 107,79–84.

Griffiths, K.J.; Sullivan, C.R. (1978) The potential for establishment of the egg parasite *Ooencyrtus kuvanae* (How.) in Ontario populations of the gypsy moth. *Canadian Entomologist* 110,633–638.

Griffiths, K.J.; Wallace, D.R. (in press) The parasites of the gypsy moth in southeastern Ontario. *Canadian Entomologist.*

Howard, L.O.; Fiske, W.F. (1911) The importation into the United States of the parasites of the gypsy moth and the brown-tail moth. *US Bureau of Entomology Bulletin* 91, 344 pp.

Jobin, L.; Caron, A. (1982) Results of aerial treatments with Dimilin and *Bacillus thuringiensis* for gypsy moth *(Lymantria dispar* L.) control in Québec. *Canadian Forestry Service Research Notes* 2(3),18–20.

Leonard, D.E. (1974) Recent developments in ecology and control of the gypsy moth. *Annual Review of Entomology* 19,197–229.

Madrid, F.J.; Stewart, R.K. (1980) Parasitoids and hyperparasitoids of the gypsy moth *Lymantria dispar* (Linnaeus) (Lepidoptera: Lymantriidae) in Québec. *Notes of the Lyman Entomological Museum Research Laboratory* 6, 28 pp.

Magasi, L.P. (1981) Forest pest conditions in the Maritimes in 1980. *Canadian Forestry Service, Fredericton, New Brunswick, Report* M-X-118, 35 pp.

Majchrowicz, I.; Yendol, W.G. (1973) Fungi isolated from the gypsy moth. *Journal of Economic Entomology* 66(3),823–824.

McAtee, W.L. (1911) Economic ornithology in recent entomological publications. *The Auk* 28,282–287.

McGugan, B.M.; Coppel, H.C. (1962) A review of the biological control attempts against insects and weeds in Canada. II. Biological control of forest insects, 1910–1958. *Commonwealth Institute of Biological Control Technical Communication* 2,35–216.

McLaine, L.S. (1925) The outbreak of the gypsy moth in Québec. *Annual Report of the Entomological Society of Ontario* 55,60–62.

McLaine, L.S. (1938) Some notes on the gypsy moth eradication campaign in New Brunswick and the Japanese beetle preventive work. *Annual Report of the Entomological Society of Ontario* 69,43–45.

Podgwaite, J.D.; Campbell, R.W. (1972) Disease complex of gypsy moth. II. Aerobic bacterial pathogens. *Journal of Invertebrate Pathology* 20(3),303–308.

Proper, A.B. (1931) *Eupteromalus nidulans*, a parasite of the brown-tail and satin moths. *Journal of Agricultural Research* 43(1),37–56.

Sabrosky, C.W.; Reardon, R.C. (1976) Tachinid parasites of the gypsy moth, *Lymantria dispar,* with keys to adults and puparia. *Entomological Society of America Miscellaneous Publication* 10(2), 126 pp.

Sippell, W.L.; Rose, A.H.; Gross, H.L. (1970) Ontario region. *Canadian Forestry Service Annual Report of the Forest Insect and Disease Survey 1969,* Ottawa, Ontario, pp. 52–71.

Sippell, W.L.; Rose, A.H.; Gross, H.L. (1971) Ontario region. *Canadian Forestry Service Annual Report of the Forest Insect and Disease Survey, 1970,* Ottawa, Ontario, pp. 49–66.

Sullivan, C.R.; Griffiths, K.J.; Wallace, D.R. (1977) Low winter temperatures and the potential for establishment of the egg parasite *Anastatus disparis* (Hymenoptera: Eupelmidae) in Ontario populations of the gypsy moth. *Canadian Entomologist* 109,215–220.

Ticehurst, M.; Fusco, R.A.; Kling, R.P.; Unger, J. (1978) Observations on parasites of gypsy moth in first cycle infestations in Pennsylvania from 1974–1977. *Environmental Entomologist* 7(3),355–358.

Tothill, J.D. (1916) The introduction and establishment in Canada of the natural enemies of the brown-tail and gypsy moths. *Agricultural Gazette of Canada* 3(2),111–116.

Weseloh, R.M. (1972) Influence of gypsy moth egg mass dimensions and micro-habitat distribution on parasitization by *Ooencyrtus kuvanae. Annals of the Entomological Society of America* 65(1),64–69.

Malacosoma disstria Hübner, Forest Tent Caterpillar (Lepidoptera: Lasiocampidae)

W.G.H. IVES

Pest Status

The forest tent caterpillar, *Malacosoma disstria* Hübner, has had a long history of periodic outbreaks in Canada (Baird 1917, Hodson 1941, Hildahl & Reeks 1960, Sippell 1962, Witter *et al.* 1975). The insect attacks a wide variety of tree species (Prentice 1963), although the preferred host in most areas is trembling aspen, *Populus tremuloides* Michx. Host trees are often completely stripped of foliage during outbreaks, but because the outbreaks usually last for only 3–6 years at any one location (Sippell 1962, Witter *et al.* 1975), and even severely defoliated trees usually refoliate within a few weeks, there is usually very little tree mortality directly attributable to defoliation (Kulman 1971). However, severe mortality has occasionally been reported (Hodson 1941, G.N. Still 1980 personal communication). Observations by the author suggest that such mortality is attributable primarily to the effects of severe drought following several years of moderate to severe defoliation.

Between 1969 and 1980 records compiled by the Forest Insect and Disease Survey (Anon. 1970–79) show that outbreaks of the forest tent caterpillar have occurred in all provinces except Newfoundland (Fig. 16). In New Brunswick, about 9 000 ha of trembling aspen were severely defoliated in 1969, and the infestation was expected to enlarge in 1970. However, a combination of unseasonably warm weather in early May 1970, followed by several frosts, apparently resulted in the starvation of many young larvae. The area of infestation therefore decreased, and the expected outbreak did not materialize. In Nova Scotia, about 50 km² of aspen were defoliated in 1972, and in Prince Edward Island about 28 000 ha in 1974. In Québec, several small areas totalling less than 100 km² were defoliated between 1973 and 1976. In Ontario, remnants of earlier outbreaks persisted into the survey period and new outbreaks began developing soon afterwards. A total of over 100 000 km² east of the Lakehead was defoliated between 1975 and 1977, while the outbreak in northwestern Ontario alone covered about 160 000 km² in 1978. In Manitoba, an incipient outbreak was detected in 1971 and by 1977 most of the southern part of the province was affected. The outbreak collapsed in 1978, except for an area west of Lake Winnipegosis where it collapsed in 1979. In Saskatchewan, the outbreak was slower in developing and did not reach its peak until 1980, when about 128 000 km² were subjected to moderate or severe defoliation. In Alberta, an outbreak in the Wabamun Lake area has persisted throughout the survey period, but the greatest damage occurred in 1980, when about 75 000 km² were defoliated. In British Columbia, over 70 000 ha were defoliated in the central part of the province in 1973, but the outbreak collapsed in 1974.

Background

Populations of forest tent caterpillar are influenced by a variety of environmental factors including weather, competition, predation, parasitism, and diseases. Ives (1973) believed that unusually favourable spring weather 2–4 years before any noticeable defoliation was responsible for triggering outbreaks, and a number of workers report that unfavourable spring weather appears to be a major factor in population collapse (Tothill 1918, Sweetman

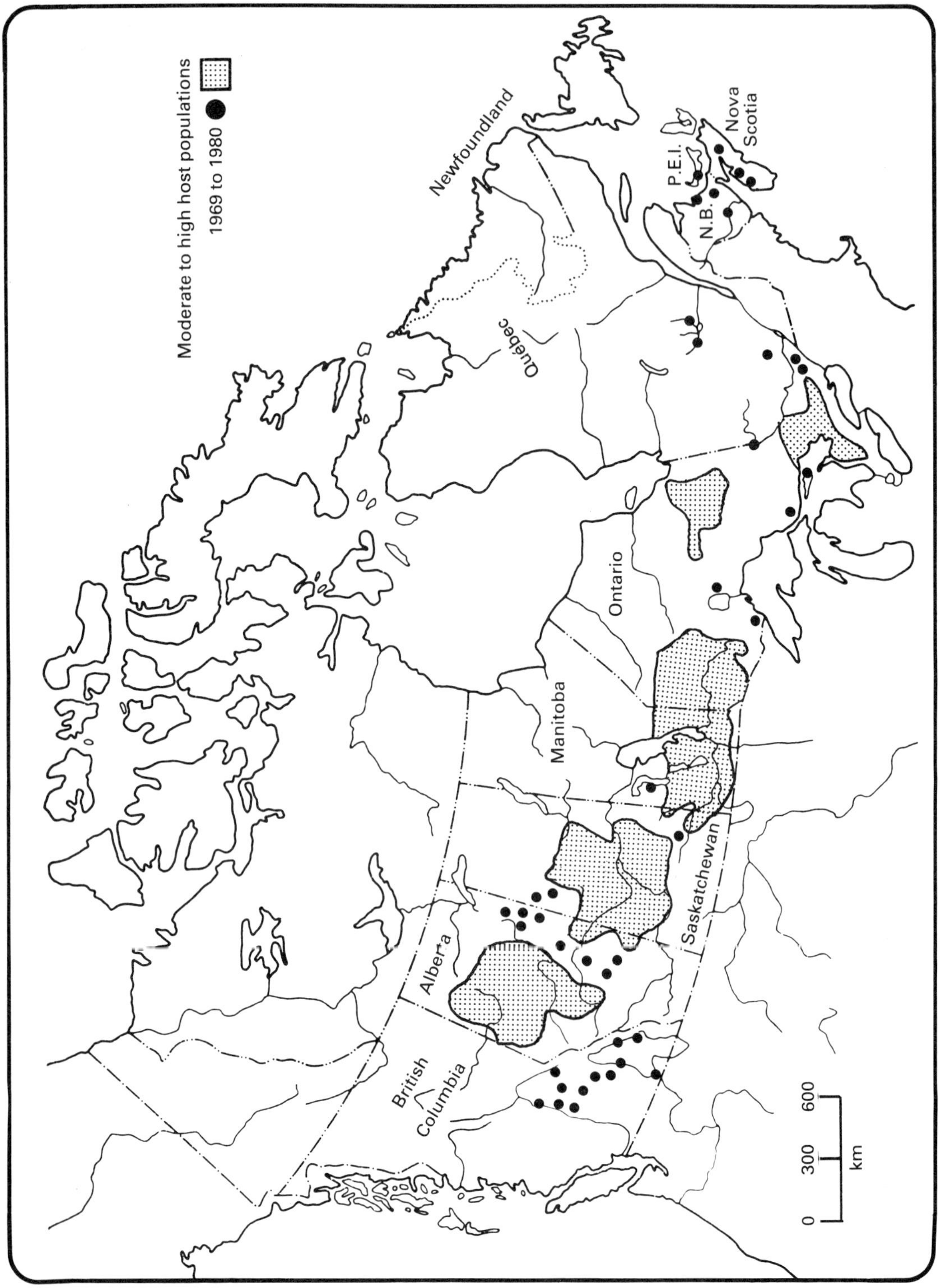

Fig. 16. Distribution of *Malacosoma disstria* Hbn. 1969-1980

1940, Blais *et al.* 1955, Witter *et al.* 1972, Hodson 1977). Unusually cold winter weather is also sometimes responsible for population decline (Witter *et al.* 1975). Hodson (1941, 1977) found that starvation, often coupled with high pupal parasitism by *Sarcophaga aldrichi* Parker, was a major contributor to population collapse. A large variety of insect parasites attacks the forest tent caterpillar (Witter & Kulman 1972), but the most prevalent is usually *S. aldrichi*, which often attacks over 90% of the pupae in the declining phases of an outbreak (Hodson 1941, 1977). Predation by various birds (McAtee 1926, Hodson 1941, Witter & Kulman 1972) and invertebrate predators such as ants (Tothill 1918, Green & Sullivan 1950, Ayre & Hitchon 1968) has been observed on a number of occasions, but it is usually unimportant.

Bacteria, fungi, microsporidia, and viruses all cause diseases in forest tent caterpillar larvae and pupae (Bird 1971). The bacteria isolated from *M. disstria* larvae include *Bacillus cereus* Frankland & Frankland, *Clostridium brevifaciens* Butcher, a *Pseudomonas* species, and *Serratia marcescens* Bizio, but all were rare (Smirnoff 1968). The forest tent caterpillar is also susceptible to *Bacillus thuringiensis* Berliner *(B.t.)*, although this organism is not usually found under natural conditions. Several fungi have been isolated from forest tent caterpillar larvae and pupae, but according to Stairs (1972) only *Beauveria bassiana* (Balsamo) Vuillemin and *Entomophthora* sp. were of any importance, and then only under humid conditions. A microsporidium, *Nosema disstriae* Thompson, has killed up to 90% of early-instar larvae in some areas (Smirnoff 1968). It also causes a reduction in the size of the insects and may affect their vigour, so that the full significance of infection is hard to assess (Bird 1971).

Infection with a nuclear polyhedrosis virus (NPV) is often thought to be one of the most important factors in terminating forest tent caterpillar outbreaks. For example, Stairs (1966) stated that "Epizootics of nuclear polyhedrosis virus disease are known to be associated with the rapid decline of tent caterpillar populations, *Malacosoma* spp." Although this statement may be true for the tent-forming species, the evidence is equivocal for the forest tent caterpillar. Of the references cited by Stairs (1966), the following three deal with *M. disstria*. Bergold (1951) obtained high mortality among larvae in cages when they were sprayed with the virus under experimental conditions. Sippell (1952) demonstrated that a large percentage of egg bands collected in late fall were infected with virus. Neither experiment attempted to show that the virus had an effect on population trends. The third paper (Chapman & Glaser 1915) states: "Though wilt disease was also reported to have occurred in many places in the forest tent caterpillar, neither of us have seen more than a few typical cases from the field. Many caterpillars were sent in by field men but only a few of them proved to have typical wilt." Smirnoff (1968) conducted a long-term study of *M. disstria* diseases in Québec, and found NPV to be unimportant. Similarly, population studies by Hodson (1941) showed that the virus was responsible for less than 1% mortality in Itasca State Park in 1936, and Witter *et al.* (1972), on the basis of intensive sampling, considered the virus to be unimportant in northern Minnesota during 1967–69.

Although NPV appears to be unimportant as a control agent under natural conditions, it is possible that artificial dissemination of the virus into field populations might increase the amount of virus infection, thus increasing its effectiveness. Stairs (1964) used a small backpack mist blower to apply various dosages of virus to different larval instars of the forest tent caterpillar in 1963. A concentration of 10^7 polyhedral inclusion bodies (PIB) per millilitre applied to second- and third-instar larvae caused 92% mortality, most of the larvae dying within 10 days of the application. Stairs (1965) found virus-infected larvae in 1964 in both of the areas that had been sprayed in 1963, but not in an unsprayed area. The virus also occurred at considerable distances from the sprayed areas, and Stairs believed that he had been successful in artificially initiating an epizootic.

Releases and Recoveries

Biological control attempts against the forest tent caterpillar during the past decade have involved the bacterium *B. thuringiensis*, the microsporidium *N. disstriae*, and an NPV.

Bacillus thuringiensis Berliner

Because some formulations of *B. thuringiensis* have been registered for control of the forest tent caterpillar, and have therefore been used commercially it is difficult to provide an accurate record of either the amounts used or the effectiveness of the applications. The following is a summary of the more readily available information.

The most comprehensive trials of *B.t.* against the forest tent caterpillar were conducted in Ontario in 1975, 1977, and 1978 (G.M. Howse 1981 personal communication). In 1975, small-scale aerial spray trials with Thuricide® 16B (Sandoz Inc.) on trembling aspen stands in southern Ontario showed that the preparation had promise. One application at a rate of 10 x 10^9 I.U./ha reduced larval populations by 71%; two applications at the same rate caused 92% mortality. The treatment was applied too late in the season to save any foliage. In 1977, evaluations were made of the effectiveness of two applications of Thuricide® 16B (10 x 10^9 I.U./ha at each application) in controlling forest tent caterpillar populations over a total of about 850 ha of trembling aspen and sugar maple, *Acer saccharum* Marsh., in three provincial parks in southern Ontario. Resultant population reduction on trembling aspen ranged from 30 to 79%, and defoliation in the treated areas ranged from 5 to 46% compared to 82 to 100% in untreated areas. Similar results were obtained for infestations on sugar maple, where population reductions due to treatment ranged from 38 to 69%, and defoliation in sprayed areas was about 15%, compared to 100% in an unsprayed area. A further 940 ha were sprayed with Thuricide® 16B in 1977, but no assessment of its effectiveness was made. In 1978, either one or two applications of Thuricide® 16B, mostly at the rate of 7.5 x 10^9 I.U./ha for each application, were made on a total of about 515 ha of forest tent caterpillar infestations on trembling aspen, sugar maple, and red oak, *Quercus rubra* L., in southern Ontario. Insect populations were lower than in 1977, so the comparison of defoliation was not a practical way of showing effectiveness. However, population reduction due to treatment ranged from 79 to 100% on trembling aspen, from 71 to 100% on red oak, and was 100% in both treated infestations on sugar maple. In addition, 250 ha of sugar maple were sprayed with one application of Dipel® WP (Abbott Ltd.) (6.0 x 10^9 I.U./ha) in late May. Population reduction due to treatments was 98%.

Data for western Canada are more fragmentary. In Alberta, a hydraulic sprayer was used in 1974 to treat two 0.2-ha plots of trembling aspen with Dipel® WP (9 x 10^9 I.U./ha) and a further two 0.2-ha plots with Thuricide® HPC (Sandoz Inc.) (5 x 10^9 I.U./ha) (Drouin & Kusch 1975). Control was estimated at 75 and 70% respectively. In 1976, two 0.4-ha plots of trembling aspen were treated with Dipel® WP (9 x 10^9 I.U./ha). Control was estimated at 75 and 85%, based on the reduction in amount of defoliation (Drouin & Kusch 1977). In addition, several applications were made by private operators under contract to various agencies.

In Saskatchewan, Dipel® WP was used to control forest tent caterpillar outbreaks in Prince Albert and Saskatoon. In Prince Albert, about 325 ha within the city limits were treated in 1980 when the larvae were in the third and fourth instars, with good results (P. Kabatoff 1981 personal communication). The chemical insecticide malathion had been used earlier, when the larvae were in the first and second instars, because it was felt that there was insufficient foliage at that time for Dipel® WP to be effective. In Saskatoon, about 800 ha of city parks and about 40 000 avenue trees were treated with Dipel® WP in 1980 (D. Scott 1981 personal communication). About 340 kg of Dipel® WP were used, and results were generally satisfactory.

In Manitoba, Thuricide® 16B (17.8 x 10^{9} I.U./ha) was applied to street plantings of Manitoba maple, *Acer negundo* L., and white elm, *Ulmus americana* L., in 1978 (R.A. Ellis 1981 personal communication). A total of 129 trees on two sites was treated on 11 May with a Meyers mist blower. Population reductions 1 day after treatment were 75 and 100%.

Nosema disstriae Thompson

Preliminary laboratory tests with *N. disstriae* (Wilson 1977) showed that this microsporidium eventually caused both larval and pupal mortality, but feeding continued for a prolonged period. Dosages of about 5 x 10^{5} and 5 x 10^{7} spores were fed in plastic cups to groups of 10 second-instar larvae on a synthetic diet and caused 38 and 83% mortality after a period of 29 and 31 days respectively. Corresponding pupal mortalities were 69 and 100%. Similar experiments (Wilson 1979), using dosages of 3 x 10^{4}, 3 x 10^{6}, and 3 x 10^{8} spores per cup caused 69.4, 100 and 100% larval mortality, respectively, when applied to second-instar larvae, and 26.7, 48.9, and 98.1% mortality when fed to fourth-instar larvae. There was also some evidence that female fecundity was reduced.

A combination of the microsporidian parasites *N. disstriae* and *Pleistophora schubergi* Zwölfer was tested against the forest tent caterpillar in 1977 (Wilson & Kaupp 1977). A backpack mist blower was used to apply 1.8 x 10^{11} spores per individual trembling aspen tree. In one case the mixture was 50% *N. disstriae* and 50% *P. schubergi*; the other contained 4% *N. disstriae* and 96% *P. schubergi*. The 50:50 mixture resulted in an average level of infection of 76% for *P. schubergi* and 20% for *N. disstriae* with 0 and 23% respectively in the check area. Spraying with the 4:96 mixture gave infection levels of 85% for *P. schubergi* and 25% for *N. disstriae*. In all cases the levels of infection by *P. schubergi* were highly significant, but this was not the case for *N. disstriae*.

Nuclear polyhedrosis virus

The preliminary results of field trials with the NPV on the forest tent caterpillar (Stairs 1964, 1965) encouraged other workers to conduct additional studies during the past decade. Concentrations of 1 x 10^{4}, 2 x 10^{4}, and 37 x 10^{11} PIB/ml caused 83.3, 92.9, and 100% mortality, respectively, when applied to egg bands in the laboratory (Elgee 1971), demonstrating that spumaline contaminated with virus was capable of infecting larvae as they hatched. Other small-scale spray tests on late-instar larvae in New Brunswick showed that a small percentage of larvae in the next generation could be infected under field conditions by transmission through the eggs (Forbes *et al.* 1972).

Ives & Muldrew (1978) conducted preliminary tests on the effectiveness of egg-band contamination in the transmission of NPV under field conditions. Two methods of application were tried in 1976. In the first, twigs bearing egg bands were dipped in a suspension containing 10^{8} PIB/ml and were then tied near an unsprayed egg band on samplings in the field; all other egg masses were removed from each sampling. In the second, a hand sprayer was used on 27 April to treat, with the same concentration of virus, about 200 egg bands on young trembling aspen. The larvae in the area began hatching 5 days later. After 20 days over 60% of the larvae from the paired egg bands were dead. After 23 days, over 90% of the larvae from the sprayed egg bands were also dead. Nearly all larvae from both treatments were dead by 29 days, compared to a corresponding mortality of about 13% in a nearby unsprayed area. Additional tests on virus transmission through contaminated spumaline were conducted in 1977, using laboratory-produced virus (J.C. Cunningham 1977 personal communication). On 14 April two isolated strips of heavily infested young trembling aspen were sprayed to drip-point with a backpack mist blower, using various concentrations of virus protected from ultraviolet radiation by 25 g/l Shade® (Sandoz Inc.). Concentrations of 5 x 10^{5}, 10^{6}, 10^{7}, and 10^8 PIB/ml caused subsequent larval mortalities of approximately 69, 81, 95, and 100% respectively, compared to mortalities of about 48 and 58% in nearby unsprayed areas.

The effect of virus on larvae was also tested under field conditions in 1976 (Ives & Muldrew 1978). A hydraulic sprayer was used on 6 May to treat two narrow 0.2-ha plots with an aqueous suspension containing about 10^7 PIB/ml. Total larval mortality approximated 95 and 100%, compared to about 36% in an unsprayed area. However, late-instar larvae from the adjacent unsprayed area then moved into the plot and very little foliage was saved. Nevertheless, there was apparently a reasonable amount of virus carry-over into subsequent generations, although it was impossible to demonstrate any spread of the virus. Follow-up sampling in the original plot in 1977, 1978, and 1979 provided corresponding estimates of total larval mortality of 75, 84, and 31% compared with 12, 40, and 12% in an unsprayed area 1 km north of the plot.

The effectiveness of aerial application of NPV in controlling forest tent caterpillar populations was tested experimentally in Alberta in 1978, 1979, and 1980 (Ives *et al.* 1982). A Bell 47 G3 B1 helicopter fitted with saddle tanks and a 12-m boom equipped with 26 Diaphragm Teejet nozzles operating at a pressure of 137 kPa was used for all applications. Although virus concentrations varied, all preparations contained 25% v/v animal feed grade molasses, 50 g/l Shade®, and 6 ml/l Atplus® spreader-sticker (Atlas Chemical Industries, Inc.) for improved adhesion. The virus used in 1978 was prepared in the laboratory (J.C. Cunningham 1978 personal communication). The remaining virus was obtained from field-collected larvae, infected by spraying small trembling aspen trees with virus when most larvae were in the fourth instar. The larvae were harvested just before pupation, macerated, and allowed to rot. The virus was then concentrated by sedimentation and several washings in distilled water. The plots varied in area from approximately 2 to 8 ha, but most were about 4 ha. Larval mortality was assessed by placing samples of about 300 larvae in heavy paper bags and counting living and dead larvae after 3–5 days. Dead larvae were not checked microscopically for virus, so some mortality may be due to other causes. However, most dead larvae showed typical symptoms of virus infection. The results are summarized in Table 82.

In 1978, total larval mortality, mostly due to virus, in two unsprayed plots was 68 and 80%, and this hampered comparisons. Nevertheless, the heavier concentrations of virus increased both the daily larval mortality rate and the total larval mortality. In 1979, one or two applications of virus before the eggs hatched doubled both the daily larval mortality rate and the total larval mortality. When applied to first-instar larvae, both dosages increased early and mid-season daily rates of larval mortality (and hence total larval mortality), but had no effect on late-season daily larval mortality. However, the effects were minimal when third- and fourth-instar larvae were treated. In 1980, the effects of the massive doses were statistically significant but were otherwise disappointing. The poor results may have been due, at least in part, to the warm dry weather in the spring of 1980, which would tend to reduce secondary infection.

No increase in the amount of virus carry-over could be demonstrated in the areas treated in 1978 or 1979, nor was there any noticeable decrease in forest tent caterpillar populations that could be attributed to virus infection. For example, we used two of the areas sprayed in 1978 for our 1980 tests. In spite of the high incidence of virus in 1978 there were still enough larvae present in 1980 to cause complete defoliation in that year. Also, populations were still high in 1980 in stand 3 (Table 82) which had a total larval mortality of 84 and 97% in the two treated plots in 1978. In 1978, virus-killed larvae literally coated the tops of some of the smaller trees, yet one of these trees, about 8 cm in diameter at breast height, had over 60 egg bands on it in the fall of 1980. Hodson (1941) considered that only 9 egg bands were required to produce enough larvae to cause complete defoliation on a tree of this size.

Table 82 Daily mortality and crude estimates of total larval mortality for forest tent caterpillar, *Malacosoma disstria* (Hübner), larvae collected from aspen, *Populus* spp., stands subjected to aerial applications of virus in 1978, 1979, and 1980.

Year	Stage when treated	Stand	PIB/ha	Daily larval mortality (%) Early season	Mid season	Late season	Approximate total larval mortality (%)
1978	II instar	1	4.4×10^9	5.2	2.3	0.8	64
			3.1×10^{10}	4.2	2.8	0.6	63
		2	3.1×10^{10}	3.8	4.7	2.4	78
			6.1×10^{10}	6.2	4.5	3.3	81
			4.6×10^{11}	6.8	6.4	2.4	87
		3	4.4×10^9	7.4	5.2	2.5	84
			3.6×10^{11}	8.8	10.5	2.7	97
		4	None	1.6	4.3	1.8	68
			3.6×10^{11}	4.8	7.6	1.4	92
		5	None	3.0	6.3	2.2	80
		6	6.1×10^{10}	5.4	8.9	3.2	92
1979	Eggs	7	None	1.0	0.8	0.3	25
			1.5×10^{11}	1.9	2.0	1.2	50
			3.0×10^{11}	2.6	1.6	0.8	49
	I instar	8	None	1.3	0.6	3.2	48
			1.5×10^{11}	2.2	1.8	3.2	64
			3.0×10^{11}	4.5	2.5	3.2	76
	III & IV instars	9	None	0.7	0.1	1.6	36
			1.5×10^{11}	0.4	0.6	1.8	36
			3.0×10^{11}	0.9	0.9	2.0	54
1980	I & II instars	4	None	0.3	0.2	1.6	30
			9.0×10^{12}	1.2	0.9	2.1	44
		1	1.8×10^{13}	0.6	1.6	2.6	55

Evaluation of Control Attempts

The results obtained with *B. thuringiensis* clearly indicate that this pathogen is the best prospect among those tested for biological control of the forest tent caterpillar. Commercial preparations are available at prices that are competitive with chemical insecticides, and the preparations are safe. On the basis of the Ontario tests a double application seems best, about a week apart, of $8-10 \times 10^9$ I.U./ha. The first application should be made when most of the larvae are in the second instar. Trials with NPV show no evidence that the virus is causing any significant population reduction, even though infection rates were high. No final conclusions on the possible use of *N. disstriae* in biological control can be made until it is tested free of contamination by other microsporidia. Wilson (1979) still considers that *N. disstriae* is a potentially useful agent for control of forest tent caterpillar.

Recommendations

B. thuringiensis may be subject to ultraviolet degradation, and this should be determined. Additional field testing to demonstrate the effectiveness of various dosages, formulations, and timing, under a variety of environmental conditions, are probably warranted before large-scale use can be recommended. The timing of application may be fairly critical if foliage protection over large areas is required, as larvae usually remain in the second instar for only a few days (Hodson 1941).

Further small-scale field testing of *N. disstriae* should be conducted to determine how infection rates are affected by environmental conditions. No further testing of NPV appears warranted, unless a more virulent strain is discovered.

Literature Cited

Anonymous. (1970–79) Annual reports of the Forest Insect and Disease Survey, 1969–1976. Canadian Forestry Service, Ottawa, Ontario.

Ayre, G.L.; Hitchon, D.E. (1968) The predation of tent caterpillars, *Malacosoma americana* (Lepidoptera: Lasiocampidae) by ants (Hymenoptera: Formicidae). *Canadian Entomologist* 100,823–826.

Baird, A.B. (1917) An historical account of the forest tent caterpillar and the fall webworm in North America. *Annual Report of the Entomological Society of Ontario 1916*,47,73-87.

Bergold, G.H. (1951) Notes on polyhedral disease of forest tent caterpillar, *Malacosoma disstria. Canadian Department of Agriculture, Scientific Service, Division of Forestry Biology Bi-monthly Progress Report,* 7(4),1–2.

Bird, F.T. (1971) *Malacosoma disstria* Hübner, forest tent caterpillar (Lepidoptera: Lasiocampidae). *In*: Biological control programmes against insects and weeds in Canada 1959–1968. *Commonwealth Institute of Biological Control Technical Communication* 4,144–147.

Blais, J.R.; Prentice, R.M.; Sippell, W.L.; Wallace, D.R. (1955) Effects of weather on the forest tent caterpillar *Malacosoma disstria* Hbn., in central Canada in the spring of 1953. *Canadian Entomologist* 87,1–8.

Chapman, J.W.; Glaser, R.W. (1915) A preliminary list of insects which have wilt, with a comparative study of their polyhedra. *Journal of Economic Entomology* 8,140–149.

Drouin, J.A.; Kusch, D.S. (1975) Pesticide field trials on shade and shelterbelt trees in Alberta and Saskatchewan, 1974. *Canadian Forestry Service, Northern Forest Research Centre, Edmonton, Information Report* NOR-X-131.

Drouin, J.A.; Kusch, D.S. (1977) Pesticide field trials of shade and shelterbelt trees in Alberta, 1976. *Canadian Forestry Service Northern Forest Research Centre, Edmonton, Information Report* NOR-X-184.

Elgee, D.E. (1971) Disease in forest tent caterpillar larvae hatching from egg bands contaminated with virus. *Canadian Forestry Service Bi-monthly Research Notes* 27(6),41.

Forbes, R.S.; Underwood, G.R.; van Sickle, G.A. (1972) Maritimes Region. *Annual Report of the Forest Insect and Disease Survey, 1971.* Canadian Forestry Service, Ottawa, Ontario, pp. 19–33.

Green, G.W.; Sullivan, C.R. (1950) Ants attacking larvae of the forest tent caterpillar, *Malacosoma disstria* Hbn. (Lepidoptera: Lasiocampidae). *Canadian Entomologist* 82,194–195.

Hildahl, U.; Reeks, W.A. (1960) Outbreaks of the forest tent caterpillar, *Malacosoma disstria* Hbn., and their effects on stands of trembling aspen in Manitoba and Saskatchewan. *Canadian Entomologist* 92,199–209.

Hodson, A.C. (1941) An ecological study of the forest tent caterpillar, *Malacosoma disstria* Hbn., in northern Minnesota. *University of Minnesota Agricultural Experiment Station Technical Bulletin* 148.

Hodson, A.C. (1977) Some aspects of forest tent caterpillar population dynamics. *In:* Insect ecology papers presented in the A.C. Hodson ecology lectures. *University of Minnesota Agricultural Experiment Station Technical Bulletin* 310,5–16.

Ives, W.G.H. (1973) Heat units and outbreaks of the forest tent caterpillar, *Malacosoma disstria* (Lepidoptera: Lasiocampidae). *Canadian Entomologist* 105,529–543.

Ives, W.G.H.; Muldrew, J.A. (1978) Preliminary evaluations of the effectiveness of nucleopolyhedrosis virus sprays to control the forest tent caterpillar in Alberta. *Canadian Forestry Service, Northern Forest Research Centre, Edmonton, Information Report* NOR-X-204.

Ives, W.G.H.; Muldrew, J.A.; Smith, R.M. (1982) Experimental aerial application of forest tent caterpillar baculovirus. *Canadian Forestry Service, Northern Forest Research Centre, Edmonton, Information Report* NOR-X-240.

Kulman, H.M. (1971) Effects of insect defoliation on growth and mortality of trees. *Annual Review of Entomology* 16,289–324.

McAtee, W.L. (1926) The relation of birds to woodlots in New York State. *Roosevelt Wild Life Bulletin* 4(1),7–152.

Prentice, R.M. (1963) Forest Lepidoptera of Canada recorded by the Forest Insect Survey. *Canadian Department of Forestry, Forest Entomology and Pathology Branch, Ottawa, Publication* 1013.

Sippell, W.L. (1952) Winter rearing of the forest tent caterpillar, *Malacosoma disstria* Hbn. *Canadian Department of Agriculture, Scientific Service, Division of Forest Biology Bi-monthly Progress Report* 8(4),1–2.

Sippell, W.L. (1962) Outbreaks of the forest tent caterpillar, *Malacosoma disstria* Hbn., a periodic defoliator of broad-leaved trees in Ontario. *Canadian Entomologist* 94,408–416.

Smirnoff, W.A. (1968) Microorganisms isolated from *Malacosoma americanum* and *Malacosoma disstria* in the Province of Québec. *Department of Forestry and Rural Development, Forestry Branch Bi-monthly Research Notes* 24(1),4.

Stairs, G.R. (1964) Dissemination of nuclear polyhedrosis virus against the forest tent caterpillar, *Malacosoma disstria* (Hübner) (Lepidoptera: Lasiocampidae). *Canadian Entomologist* 96,1017–1020.

Stairs, G.R. (1965) Artificial initiation of virus epizootics in forest tent caterpillar populations. *Canadian Entomologist* 97,1059–1062.

Stairs, G.R. (1966) Transmission of virus in tent caterpillar populations. *Canadian Entomologist* 98,1100–1104.

Stairs, G.R. (1972) Pathogenic microorganisms in the regulation of forest insect populations. *Annual Review of Entomology* 17,355–372.

Sweetman, H.L. (1940) The value of the hand control for the tent caterpillars, *Malacosoma americana* Fabr. and *Malacosoma disstria* Hbn. (Lepidoptera: Lasiocampidae). *Canadian Entomologist* 72,245–250.

Tothill, J.D. (1918) The meaning of natural control. *Proceedings of the Entomological Society* 4,10–14.

Wilson, G.G. (1977) Effects of the microsporidia *Nosema disstriae* and *Pleistophora schubergi* on the survival of the forest tent caterpillar, *Malacosoma disstria* (Lepidoptera: Lasiocampidae). *Canadian Entomologist* 109,1021–1022.

Wilson, G.G. (1979) Effects of *Nosema disstriae* (Microsporidia) on the forest tent caterpillar, *Malacosoma disstria* (Lepidoptera: Lasiocampidae). *Proceedings of the Entomological Society of Ontario* 110,97–99.

Wilson, G.G.; Kaupp, W.J. (1977) Application of *Nosema disstriae* and *Pleistophora schubergi* (Microsporida) against the forest tent caterpillar in Ontario, 1977. *Canadian Forestry Service, Sault Ste. Marie, Information Report* FPM-X-4.

Witter, J.A.; Kulman, H.M. (1972) A review of the parasites and predators of tent caterpillars (*Malacosoma* spp.) in North America. *University of Minnesota Agricultural Experiment Station Technical Bulletin* 289.

Witter, J.A.; Kulman, H.M.; Hodson, A.C. (1972) Life tables for the forest tent caterpillar. *Annals of the Entomological Society of America* 65,25–31.

Witter, J.A.; Mattson, W.J., Kulman, H.M. (1975) Numerical analysis of a forest tent caterpillar (Lepidoptera: Lasiocampidae) outbreak in northern Minnesota. *Canadian Entomologist* 107,837–854.

Neodiprion abietis (Harris), Balsam Fir Sawfly (Hymenoptera: Diprionidae)

J.C. CUNNINGHAM

Pest Status

The balsam fir sawfly, *Neodiprion abietis* (Harr.) complex, occurs from New England to the Great Lakes and south to Missouri in the USA, from coast to coast in southern Canada (Baker 1972), and also in Newfoundland. In Canada, periodic outbreaks, usually of short duration, are common on balsam fir, *Abies balsamea* (L.) Mill., and spruce, *Picea* spp., from Saskatchewan eastwards. Tree mortality is rare and is usually reported in conjunction with other pests such as spruce budworm, *Choristoneura fumiferana* (Clem.), eastern blackheaded budworm, *Acleris variana* (Fern.), and balsam woolly adelgid, *Adelges piceae* (Ratz.) (Rose & Lindquist 1977).

Balsam fir sawfly overwinters as eggs in diapause and spring hatching is fairly uniform. Larvae are gregarious and feed on old needles, but eat only parts of them, thus damaging many more needles than they consume. In the later instars, the colonies disperse and the larvae become solitary. There are five larval instars, but males do not feed during their last instar (D.R. Wallace & C.R. Sullivan 1980 personal communication). In 1980, traces of defoliation caused by this insect were found in western and central Newfoundland and light to moderate defoliation was observed in the northern, eastern, and Algonquin regions of Ontario (Sterner & Davidson 1981).

Background

A nuclear polyhedrosis virus (NPV) was first reported in balsam fir sawfly by Steinhaus (1949). This disease has subsequently been reported in balsam fir sawfly populations in several Canadian provinces including Saskatchewan (Brown 1951), Alberta and Manitoba (Cumming 1954), and Québec (Martineau & Lavallée 1971).

Field Trial

Ground spray trials on single trees were conducted in Ontario in 1972 to test NPV on balsam fir sawfly (Olofsson 1973). Aqueous suspensions of 10^7, 10^6, 10^5, 10^4, and 10^3 polyhedral inclusion bodies (PIB) per millilitre plus 2.5% IMC 90-001 (International Minerals Corporation) UV protectant were tested at two application times, 8 days apart, on first- and third-instar larvae. A backpack mist blower was used and a 360-ml suspension was applied to each 2.5-m tall balsam fir tree, five trees being treated with each dosage on each date. There were 5–10 sawfly colonies per tree.

The percentage of infected larvae following the applications is shown in Table 83. Infection is directly related to mortality with this NPV.

Table 83 Infection and defoliation following application of NPV on first- and third-instar balsam fir sawfly, *Neodiprion abietis* (Harr.), larvae in a ground spray trial (from Olofsson 1973)

Dosage PIB/ml	Percentage virus infection		Defoliation*	
	I instar treated	III instar treated	I instar treated	III instar treated
10^7	100	100	1	3
10^6	100	88	2	4
10^5	96	60	3	5
10^4	72	24	4	5
10^3	60	4	4	5
Check	0	0	5	5

* Defoliation was rated on a scale of 1 to 5 where 1 was very light and 5 severe.

Recommendations

The limited ground spray trials conducted in 1972 demonstrated that this NPV is a potentially useful biological control agent for balsam fir sawfly. However, as this species feeds mainly near the tops and mid-crowns of mature trees, an aerial spray application would be required for complete coverage. Further testing has not been conducted and aerial spray trials are required before definitive conclusions can be made and recommendations formulated.

Olofsson (1973) showed that timing of the application is critical and he recommended applying the NPV on late first-instar larvae. By the time they develop to the third instar, feeding damage is severe and the larvae require a much higher dosage of virus. For ground spray applications, Olofsson recommended a dosage of 10^5 PIB/ml and he calculated that one fully grown, NPV-infected larva produced 1 l of spray of this concentration. However, this dosage would have to be increased for an aerial application. If the economic importance of balsam fir sawfly justifies the development of non-chemical pesticide control strategies, this NPV merits further research.

Literature Cited

Baker, W.J. (1972) Eastern forest insects. *US Department of Agriculture Miscellaneous Publication* 1175. US Government Printing Office, 642 pp.

Brown, C.E. (1951) Forest insect survey notes. *Canadian Department of Agriculture Bi-monthly Progress Report* 7(2),2.

Cumming, M. (1954) Diseases of insects. *Canadian Department of Agriculture Bi-monthly Progress Report* 10(3),3.

Martineau, R.; Lavallée, A. (1971) Québec region. *Canadian Forestry Service, Annual Report of the Forest Insect and Disease Survey,* pp. 34–53.

Olofsson, E. (1973) Evaluation of a nuclear polyhedrosis virus as an agent for the control of balsam fir sawfly, *Neodiprion abietis* (Harr.). *Canadian Forestry Service, Sault Ste. Marie, Information Report* IP-X-2, 30 pp.

Rose, A.H.; Lindquist, O.H. (1977) Insects of eastern spruces, fir and hemlock. *Canadian Forestry Service, Forestry Technical Report* 23, 159 pp.

Steinhaus, E.A. (1949) Principles of insect pathology. New York; McGraw-Hill, 757 pp.

Sterner, T.E.; Davidson, A.G. (comp.) (1981) Forest insect and disease conditions in Canada 1980. *Canadian Forestry Service, Forest Insect and Disease Survey,* 43 pp.

Neodiprion lecontei (Fitch), Redheaded Pine Sawfly (Hymenoptera: Diprionidae)

J.C. CUNNINGHAM and P. DE GROOT

Pest Status

The redheaded pine sawfly, *Neodiprion lecontei* (Fitch), is a serious defoliator of young hard pine, *Pinus* spp., plantations in Ontario, Québec, and, to a lesser extent, New Brunswick. It is the most serious pest of red pine, *Pinus resinosa* Ait., in these regions due to the increase in the number of plantations over the last 20 years. Red pine is the principal species attacked, but jack pine, *P. banksiana* Lamb., and Scots pine, *P. sylvestris* L., are also attacked. Damage can be severe, especially when trees are less than 2–3 years old; one sawfly colony can totally defoliate and kill several trees (Benjamin 1955). Heavy defoliation results in reduced height growth, branch mortality, and deformity if the leader is killed (Schaffner 1951, Benjamin 1955, MacAloney & Wilson 1964).

This sawfly has only one generation per year in Canada, whereas in the southern parts of its range it may have as many as five (Benjamin 1955). In Canada, adult sawflies emerge, from mid-June to early July, from overwintered cocoons in the duff layer of the soil. The number of eggs laid by females varies considerably, but it is not unusual to have a complement of 80–140 eggs. Egg development varies with the ambient temperature and, as temperatures increase, the incubation period shortens. An incubation period of 3–4 weeks is typical in southern Ontario and Québec. The sawfly larvae are gregarious and feed on foliage of all ages, although they show a preference for the previous year's foliage. Larvae are fully grown in 25–30 days from time of hatching, and are about 25 mm long. Populations of sawflies are not evenly distributed throughout plantations but occur in "hot spots", often along a hardwood perimeter or on knolls.

In the past, most attempts at control have been by applying chemical insecticides from the ground. In many cases, enough larvae escaped treatment to produce a population the following year large enough to require further treatment. This "insecticide treadmill" requires considerable man-power and may have to be continued for several years until the trees have grown enough to be out of danger or until the insect population has been regulated by natural control factors such as egg parasitoids (Benjamin 1955, Griffiths 1956, Rauf *et al.* 1979).

Background

In 1950, a nuclear polyhedrosis virus (NPV) disease of the redheaded pine sawfly was found in Ontario (Bird 1961). In a series of laboratory and ground spray trials, Bird demonstrated that this virus was highly infectious to sawfly larvae, that virus can be introduced into the population and initiate an epizootic, and that it can be transmitted from one generation to the next (Bird 1961, 1971).

Field Trials

Virus for field trials is produced *in vivo*. Because there are no known synthetic diets for sawflies reared in the laboratory, it is necessary to use fresh foliage, which is labour intensive and expensive. Hence to produce virus, a plantation with a high insect population, which has already been severely damaged in previous years, is selected. When larvae reach the fourth instar, trees are sprayed with a mist blower using a suspension of 10^6 polyhedral inclusion bodies (PIB) per millilitre at an application rate of about 20 l/ha. Seven or 8 days after spraying, the first diseased and dead larvae are found, and colonies of moribund larvae are clipped from the trees and taken to the laboratory where they are picked off the foliage and frozen. Collections are made every day until all colonies have died or pupated, which can take up to 30 days. The frozen larvae are lyophilized, ground to a fine powder, and the number of PIBs per gramme measured. This material is stored at 2°C until required and is then suspended in water. Recent experiments have shown that suspension in an emulsifiable oil may be preferable, but such preparations have not yet been field tested.

Spraying of an entomopathogen must be shown to be efficacious and safe before it is approved for operational use by the Canadian federal government. Extensive safety testing of redheaded pine sawfly NPV has been undertaken at Ontario Veterinary College, University of Guelph, where this virus was found to present no hazard to mammals (Forsberg, Valli & Dwyer unpublished), birds (Valli & Claxton, unpublished), and fish and *Daphnia pulex* (Hicks *et al.* 1981). The impact of the virus on non-target organisms was monitored in conjunction with the 1977 aerial spray trials in Ontario. In addition, beehives were placed in the treated plantations. No deleterious effects were noted (Kingsbury *et al.* 1978).

A petition for the registration of this virus will be submitted to Canadian authorities in March 1981 for approval. If accepted this will be the first virus registered for regulation of insect populations in Canada.

Field trials during the review period started with a small ground spray trial with NPV, conducted in Québec in 1970. In 1976, an intensive research programme was initiated to establish an operational method of applying this virus to control redheaded pine sawfly. Aerial spray trials were conducted every year in Ontario from 1976 to 1980, and in Québec in 1978 and 1979. Ground applications were also made during this period; the number of plantations and areas treated are summarized in Table 84. Details of these trials are given below.

Table 84 Summary of field trials conducted with NPV of redheaded pine sawfly, *Neodiprion lecontei* (Fitch), between 1976 and 1980

	Ontario			Québec		
Year	Area treated (ha)	No. of plantations from air	No. of plantations from ground	Area treated (ha)	No. of plantations from air	No. of plantations from ground
1976	43	3				
1977	52	3				
1978	26	2		700	36	1
1979	34	4		330	42	1
1980	540	8	88	21		12

Ground spray trials in Québec in 1970

Purified NPV at a concentration of 1.3×10^6 PIB/l and lyophilized NPV-infected larvae (with an undetermined PIB content) at 30, 60, 130, and 260 mg/l were sprayed on third-instar larvae using a backpack mist blower. First mortality was observed 14 days after spraying and by 48 days 100% mortality was observed in the area treated with the lyophilized NPV-infected larval material, and 94.9 and 97.4% mortality in two areas treated with the purified virus (Anon. 1970).

Aerial spray trials in Ontario in 1976

Three red pine plantations with a total area of 43.2 ha were sprayed with three different dosages of virus, 1.25 x 10^9, 3.75 x 10^9, and 6.75 x 10^9 PIB/ha, at 9.4 l/ha, using a fixed-wing aircraft with boom and nozzle equipment, when most larvae were in the second instar. The formulation contained 60 g/l IMC 90-001 (International Minerals Corporation) sunlight protectant. Pre-spray colony counts for the three plots were 59, 138, and 173 colonies/50 trees, respectively, and 119 colonies/50 trees in an untreated check area. Seventeen days after spraying, the percentage of larvae infected with virus in the treated plots was 92, 96, and 98%, respectively, and 4% in the check area. When the last colony count was made at 22 days, 7, 46, and 1 colony/50 trees were recorded in the treated plots and 115/50 trees in the check plot. There was a good level of foliage protection in the treated areas (Kaupp & Cunningham 1977).

Aerial spray trials in Ontario in 1977

Two red pine and one jack pine plantation with a combined area of 48 ha were treated at a dosage of 5.5 x 10^9 PIB/ha at 9.4 l/ha. A fixed-wing aircraft with boom and nozzle equipment was used to spray second- and third-instar larvae on red pine, and a fixed-wing aircraft with Micronair (Micronair® (Aerial) Ltd. U.K.) equipment to spray fourth-instar larvae on jack pine. The aqueous formulation contained 25% v/v molasses and 60 g/l IMC 90-001.

In the jack pine plantation, 5 days after spraying the number of colonies/100 trees fell from 132 to 47 while the population in the untreated check plot remained constant. This rapid population decline prompted an investigation of the tank mix and it was found to contain 522μg/ml of the chemical insecticide phosphamidon (Sundaram 1977 personal communication). By 26 days after spraying, the population density had fallen to 2 colonies/100 trees.

In the two red pine plantations, pre-spray counts of colonies/100 trees were 163 and 54, and 217 in a check plot. Counts 26 days after spraying were 7 and 0 in the treated plots and 200 in the check area. The NPV also spread to an untreated plantation adjacent to these plantations and destroyed the redheaded pine sawfly population, but not before severe defoliation had occurred. Defoliation was light in all the treated areas, although it would have been heavy in the jack pine plantation without the unintentional contamination by phosphamidon in the aircraft spray tank (Kaupp *et al.* 1978).

Aerial spray trials in Ontario in 1978

Two red pine plantations with a total area of 26 ha were treated when larvae were mainly in the second instar. The spray was applied with a fixed-wing aircraft equipped with a Micronair® spray delivery system. The dosage was 5 x 10^9 PIB/ha in 9.4 l/ha aqueous formulation containing 12.5% v/v molasses and 32 g/l Shade® (Sandoz Inc.) (formerly IMC 90-001). Over 90% of sawfly colonies were diseased or dead 16 days after spraying and by 23 days no healthy colonies remained. Defoliation was lighter in treated than in check areas (de Groot *et al.* 1979).

Aerial and ground spray trials in Québec in 1978

A large operation was undertaken and 37 red pine plantations covering 700 ha were treated. Two types of aerial spray equipment were used, a fixed-wing aircraft equipped with Micronair® and a helicopter equipped with Beecomist (Beeco Products Co., USA). In addition, one plantation was treated from the ground using backpack mist blowers. Dosage was 5 x 10^9 PIB/ha for aerial applications on second-instar larvae and 10^{10} PIB/ha on third- and fourth-instar larvae for the ground application. Volume applied in all treatments was 9.4 l/ha. Some NPV was formulated in 25% v/v molasses and 30 g/l Shade® and some was applied in water alone. Two application strategies were used, with some plantations getting total coverage and others partial coverage using widely spaced spray swaths. By 29 days after spraying, plantations sprayed by fixed-wing aircraft had an average mortality of 97.7% with total coverage and 72.5% with partial coverage. Plantations treated by helicopter had an average mortality of 99.4% with total coverage and 90.8% with partial coverage. The plantation treated with mist blowers had 98.4% mortality and untreated check plots had 29.5% mortality. The saving of foliage was satisfactory in all treatments except for the ground spray application where the application volume of 9.4 l/ha was too low and insect development was too far advanced (Desaulniers & Cunningham unpublished).

Aerial spray trials in Ontario in 1979

Red pine plantations with a total area of 33.6 ha were treated at a dosage of 5 x 10^9 PIB/ha at 9.4 l/ha when larvae were in the first and second instars. A fixed-wing aircraft with Micronair® equipment was used. A comparison was made between a suspension of lyophilized, NPV-infected larvae with Rhodamine B dye as a tracer and the same virus suspension formulated in 25% v/v molasses and 60 g/l Shade®. There was no difference in the efficacy of the two treatments and it was concluded that it is unnecessary to add adjuvants to this virus. Both treatments gave 100% mortality, none of the larvae developed past the third instar and defoliation was minimal (Cunningham & de Groot unpublished).

Aerial and ground spray trials in Québec in 1979

Forty-three red pine plantations with a total area of 330 ha were treated; 323 ha were sprayed from the air and 7 ha from the ground. Two helicopters were used for the aerial spray, one equipped with Beecomist® units and one with boom and nozzle. From the air, application rate was 9.4 l/ha and dosage was 5 x 10^9 PIB/ha. Backpack mist blowers were used for the ground spray application; volume sprayed was 18.8 l/ha and dosage was 10^{10} PIB/ha. Spraying was carried out when 80–90% of the eggs were hatched. The pre-spray population densities were 63 colonies/50 trees in the area treated with Beecomist®, 110/50 trees in the area treated with boom and nozzle, 110/50 trees in the area treated with mist blowers and 22/50 trees in the check area. At 35 days after spraying, mortality in these areas was 87, 98, 94 and 7%, respectively. Defoliation in the treated plantations was practically nil and only light defoliation was recorded in the check plots because of the low sawfly population density (Bordeleau unpublished).

Aerial and ground spray trials in Ontario in 1980

Red pine plantations with a total area of 539.8 ha were treated. Experimental aerial spray trials were conducted on four plantations with a total area of 33 ha. A fixed-wing aircraft with Micronair® equipment was used and dosage on all plantations was 5 x 10^9 PIB/ha. Highly purified virus at 9.4 l/ha emitted volume was compared to lyophilized, NPV-infected larval material at the same volume. A comparison was also made with reduced emitted volumes of lyophilized larval material at 4.7 and 2.4 l/ha. Applications were made on first- and second-instar larvae. There was no difference in effectiveness between the crude and purified material or between the three emitted volumes. By 31 days after spraying, 100% mortality was achieved in all four treatments.

Semi-operational aerial spray trials were conducted on a further four plantations with a total area of 70 ha. Lyophilized NPV-infected larval material was used at 5 x 10^9 PIB/ha and the emitted volume was 9.4 l/ha. Larvae were mainly in the first and second instars, with a few unhatched egg clusters present, at the time of application. When the final survey was made 28 days after spraying, mortality ranged from 82 to 99%.

Ground spray trials were conducted by Ontario Ministry of Natural Resources staff in five districts, and 88 plantations with a total area of 436.8 ha were treated entirely or partially over a 47-day period. Insect development ranged from first and second instar in the early treatments to fourth and fifth instar in the later ones (Cunningham & de Groot unpublished).

Ground spray trials in Québec in 1980

A mist blower was used to treat 12 red pine plantations when larvae were in the first and second instars with 90% of the eggs hatched. The dosage was 5 x 10^9 PIB/ha and two application volumes were tested, 18.7 and 93.6 l/ha. Better results were obtained with the lower volume and 100% mortality was achieved 20 days after spraying. Thirty-five days after spraying, 97% mortality was reached with the higher volume. Defoliation in the treated plantations was negligible because of comparatively low larval population densities (an average of 21 colonies/25 trees on the high volume application and 8 colonies/100 trees on the low volume application) combined with the effectiveness of the treatments (Bordeleau unpublished).

Surveys of plantations aerially sprayed in Ontario in years following the year of application

In order to assess the durability of the virus treatments, surveys have been conducted annually on all the plantations aerially sprayed experimentally in Ontario, which by 1980 numbered 15 (de Groot *et al.* 1979, Cunningham & de Groot unpublished). A summary of these surveys is shown in Table 85. Although not all these plantations remained

Table 85 Survey of plantations treated with NPV of redheaded pine sawfly, *Neodiprion lecontei* (Fitch), in years following the year of application

Year	Plot	Number of colonies/100 trees					
		Pre-spray estimate	1977	1978	1979	1980	1981
1976	1	118	0	0	0	0	0
	2	176	0	0	0	0	0.75
	3	346	0	0	0	0	0
1977	1	163		0	0	0	0.75
	2	81		0	0	1.75	0
	3	132		0	0	0	0.5
1978	1	255			0.5	0	0
	2	174			2.5	1.25	2.0
1979	1	124				0.25	0
	2	77				0	0.25
	3	5				0	0
1980	1	52					0
	2	173					0
	3	225					0
	4	25					0

completely free from redheaded pine sawfly colonies, population densities remained very low and well below the threshold causing economic damage. The highest population observed was 2.5 colonies/100 trees. When the last survey was made in 1981, sawflies were found in only five of the treated plantations. There seems to be no build-up of the population even when colonies have been noted in a particular plantation for several consecutive years.

Evaluation of Control Attempts

A dosage of 5 x 10^9 PIB/ha applied from the air gives satisfactory control and foliage protection provided the treatment is on first-, second-, or third-instar larvae.

A suspension of lyophilized virus-infected larvae in water is adequate and spray adjuvants do not seem to enhance efficacy. However, bacterial growth occurs in such a suspension unless it is used promptly and research on the feasibility of an emulsifiable oil formulation is currently underway.

A satisfactory level of kill has been achieved in all the trials, but the virus is slow-acting compared to a chemical insecticide. Significant insect mortality is usually noted about 15 days after spraying, but it often takes 24–35 days until complete mortality is achieved. These figures vary depending on such factors as spray deposit, ambient temperature, and insect development.

Virus spreads from virus-infected redheaded pine sawfly colonies to healthy colonies in the year of application; transmission is thought to be by predacious and scavenging insects. Hence, where the sawfly populations are light, total spray coverage of the entire area is not required and virus introductions can be made.

Following aerial application of virus, plantations have been entirely free from sawflies or harboured only very low populations in subsequent years. Retreatment has never been required to date.

In conclusion, the use of this highly effective biological control agent is recommended in preference to chemical insecticides. It appears that one treatment of the virus, correctly applied, protects plantations for several years, by which time trees have grown sufficiently to be much less susceptible to redheaded pine sawfly damage. When produced in the field, the cost of this virus is very economical. It takes approximately 50 virus-infected sawfly larvae to produce the 5 x 10^9 PIB dosage necessary to treat a hectare. On a non-profit basis, this has been estimated to cost about $2.50/ha.

Recommendations

Early detection and treatment of infestations is vital in order to obtain satisfactory results with a virus application. Once larvae reach the fourth instar, treatments will not give complete mortality and little or no foliage will be saved. Education of users in the recognition of eggs and newly hatched larvae is an important aspect of any programme involving the use of this virus.

The methodology for aerial application (timing, dosage, and volume) is well established and documented. However, because of adverse public reaction to any aerial application it is expected that most treatments will be made with ground-based equipment. It is difficult to make firm recommendations because so many different types of equipment and application techniques are used. Methods vary from mist blowing every fourth or fifth row of trees in a plantation to directly spraying individual colonies with a pressure hydraulic sprayer. Obviously, recommended dosage and volume of liquid will be very different for these two strategies.

The problem of bacterial growth in the concentrated suspension of lyophilized, virus-infected larvae in water that is supplied to clients may well be overcome by suspending the material in an emulsifiable oil. This preparation requires field-testing before a firm recommendation can be made.

Literature Cited

Anonymous. (1970) Virus trials to control red-headed pine sawfly in Québec plantations. *Canadian Forestry Service, Ottawa, Information Report* DPC-X-1, 29 pp.

Benjamin, D.M. (1955) The biology and ecology of the red-headed pine sawfly. *US Department of Agriculture, Forestry Service Technical Bulletin* 1118, 57 pp.

Bird, F.T. (1961) Transmission of some insect viruses with particular reference to ovarial transmission and its importance in the development of epizootics. *Journal of Insect Pathology* 3,352–380.

Bird, F.T. (1971) *Neodiprion lecontei* (Fitch), red-headed pine sawfly (Hymenoptera: Diprionidae). *In*: Biological control programmes against insects and weeds in Canada, 1959–1968. *Commonwealth Institute of Biological Control Technical Communication* 4,148–150.

de Groot, P.; Cunningham, J.C.; McPhee, J.R. (1979) Control of redheaded pine sawfly with a baculovirus in Ontario in 1978, and a survey of areas treated in previous years. *Canadian Forestry Service, Sault Ste. Marie, Information Report* FPM-X-20, 14 pp.

Griffiths, K.F. (1956) Parasitism of the eggs of *Neodiprion lecontei*. *Canadian Department of Agriculture Bi-monthly Progress Report* 12(4), 2.

Hicks, B.D.; Geraci, J.R.; Cunningham, J.C.; Arif, B.M. (1981) Effects of red-headed pine sawfly, *Neodiprion lecontei*, nuclear polyhedrosis virus on rainbow trout, *Salmo giardneri*, and *Daphnia pulex*. *Environmental Science and Health* B16,493–509.

Kaupp, W.J.; Cunningham, J.C. (1977) Aerial application of a nuclear polyhedrosis virus against red-headed pine sawfly, *Neodiprion lecontei* (Fitch). *Canadian Forestry Service, Sault Ste. Marie, Information Report* IP-X-14, 26 pp.

Kaupp, W.J.; Cunningham, J.C.; de Groot, P. (1978) Aerial application of baculovirus on red-headed pine sawfly, *Neodiprion lecontei* (Fitch), in 1977. *Canadian Forestry Service, Sault Ste. Marie, Information Report* FPM-X-1, 31 pp.

Kingsbury, P.; McLeod, B.M.; Mortensen, K. (1978) Impact of applications of the nuclear polyhedrosis virus of the red-headed pine sawfly, *Neodiprion lecontei* (Fitch), on non-target organisms in 1977. *Canadian Forestry Service, Sault Ste. Marie, Information Report* FPM-X-11, 27 pp.

MacAloney, J.H.; Wilson, L.F. (1964) The red-headed pine sawfly. *US Department of Agriculture Forest Pest Leaflet* 14.

Rauf, A.; Benjamin, D.M.; Coppel, H.C.; Banash, S.E. (1979) Parasitism of white pine sawfly and redheaded pine sawfly eggs by the egg parasite, *Closterocerus cinctipennis* Ashmead. *Forest Research Notes, University of Wisconsin, Madison,* No. 277, 3 pp.

Schaffner, J.V., Jr. (1951) The red-headed pine sawfly. *US Department of Agriculture Leaflet* 296.

Neodiprion sertifer (Geoffroy), European Pine Sawfly (Hymenoptera: Diprionidae)

K.J. GRIFFITHS, J.C. CUNNINGHAM and I.S. OTVOS

Pest Status

The European pine sawfly, *Neodiprion sertifer* (Geoffr.), was first recorded in North America in New Jersey in 1925 (Schaffner 1939) and in Canada near Windsor, Ontario, in 1939 (Raizenne 1957). In Ontario it dispersed steadily until, by 1968, it was found throughout Ontario south and west of a line from Victoria Harbour to Belleville (Fig. 17). Transport of infested nursery stock resulted in isolated populations of *N. sertifer* well beyond this area — on Manitoulin Island, first recorded in 1966 (Sippell *et al.* 1966) and in Sault Ste. Marie and North Bay, first recorded in 1968 (Sippell *et al.* 1969).

In the decade since 1968 the area occupied by *N. sertifer* has continued to expand, moving slightly to the north near Georgian Bay, but mainly to the east along the St. Lawrence River, until by 1978 the continuous distribution extended almost to the Québec border along the St. Lawrence and as far north as Ottawa (Fig. 17) (Lindquist & Miller 1979).

The isolated infestations on Manitoulin Island and at Sault Ste. Marie continued throughout the period 1968–78, but no *N. sertifer* have been found in North Bay since 1969. *N. sertifer* larvae were collected for the first time at several locations in Ottawa in 1969 and by 1973 they were also found in the surrounding areas. They apparently remained there as Ottawa is now in an area of continuous distribution. An isolated light infestation was found north of Thessalon, Ontario, in 1974. The area was treated from the ground with nuclear polyhedrosis virus (NPV) in 1975 and 1976, and there have been no further recoveries of *N. sertifer*.

In 1970 there were several heavy infestations in southwestern Ontario, where the insect has been present the longest. Infestations causing moderate to severe defoliation continued in this area until 1972. A decrease in numbers was noted in 1973, and this decrease continued until by 1977 generally low populations were recorded throughout the area.

In 1974 *N. sertifer* was recorded for the first time in Canada beyond the borders of Ontario. A single colony was recovered from Scots pine, *Pinus sylvestris* L., in Charlesbourg, Québec, a few kilometres northeast of Québec City (Martineau & Lavallée 1975), and a few colonies were recovered from Scots pine at Windsor Lake near St. John's, Newfoundland (Clark & Singh 1975). No further recoveries were made in Québec, but *N. sertifer* has persisted and spread in Newfoundland. It was reported on ornamental pines *Pinus* spp. and in the few pine plantations within a radius of 15 km of St. John's by Otvos & Griffiths (1979). It was established by rearing some larvae that no parasitoids were present. The first discovery of *N. sertifer* in Nova Scotia was made in 1980, when it was obtained from ornamental pines at Little Harbour, Pictou County, and Truro, Colchester County (Magasi 1981).

Background

Studies of the population dynamics of *N. sertifer* carried out at the Great Lakes Forest Research Centre, Sault Ste. Marie, Ontario, had started in 1960 and were continued until 1972. An overview of this work is given in Lyons *et al.* (1971) and Lyons (1977*a*). In brief, outbreaks in southern Ontario typically occur in young plantations where few natural control agents are present. Density peaks in 4–6 years, then declines to a low

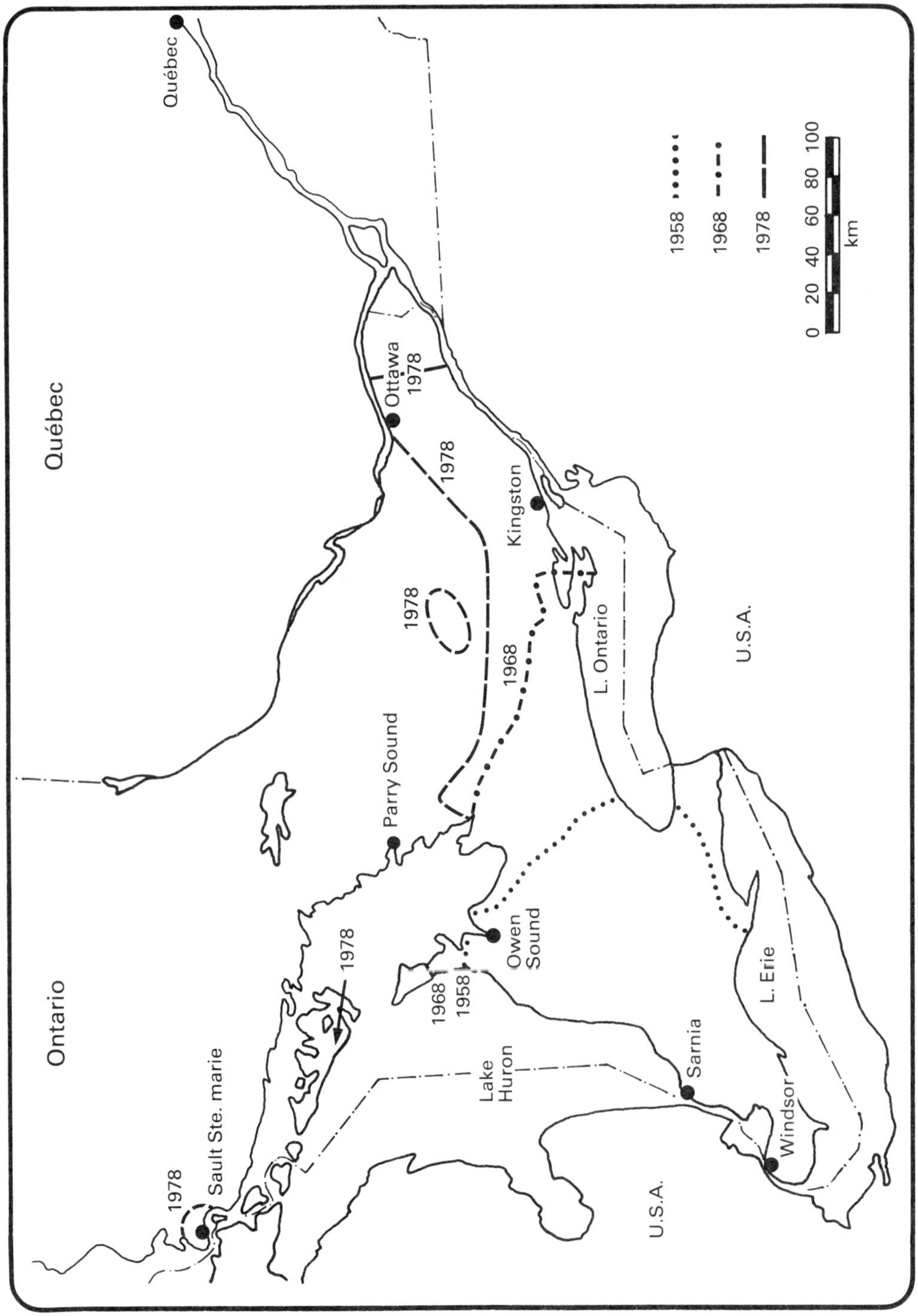

Fig. 17. Distribution of *Neodiprion sertifer* (Geoff.) in 1958, 1968 and 1978

level and does not generally increase to outbreak levels again. Eventual decline of an outbreak was related to starvation of late-feeding larvae, parasitism of pre-spinning larvae by *Exenterus* spp. and predation and parasitism of cocooned larvae. When outbreaks occur in stands of larger trees, they last longer and are associated with varying amounts of prolonged diapause (Lyons 1977*a*).

Until 1958 the recorded insect parasitoid complex of *N. sertifer* consisted of 14 native primary parasitoids, 3 native hyperparasitoids, and 4 established European parasitoids. In the following decade another 14 indigenous parasitoids were recovered; recovery was made of *Drino bohemica* Mesn., another of the European species introduced during the 1930s (McGugan & Coppel 1962); and one new exotic species, *Lophyroplectus luteator* Thunb., was established (Griffiths & Lyons 1968). In the period under review only two species have been added to the parasitoid list. One of these, *Exenterus affinis* Roh., a native ichneumonid, was recovered from *N. sertifer* as early as 1964, but was not separated from the other species of *Exenterus* attacking this host until 1977. It is recovered infrequently (Lyons 1977*b*). The second species is *Exenterus abruptorius* (Thunb.), a European parasitoid of *N. sertifer*, introduced in the 1930s but not recovered since the initial release. Lyons (1977*b*) recovered two specimens in a cocoon collection made in Dufferin County in May, 1972. Since then five more *E. abruptorius* from four other localities have been identified after examination of over 6 000 pinned specimens of *Exenterus* obtained from *N. sertifer* collections throughout southwestern Ontario since 1952. Further discussion of this species is given in the following section. Another recovery of the rare exotic species *D. bohemica* was recorded in 1972 from a collection made at Gore Bay on Manitoulin Island.

Dispersal of *L. luteator* from the releases made in 1962 and 1964 (Griffiths & Lyons 1968) has been followed closely. By 1969, 7 years after the 1962 release in Grey County, recoveries had been made up to 19.3 km east of the release point. However, by 1971, 7 years after the 1964 Norfolk County release, the greatest dispersal was only 3.4 km (Griffiths 1973). Rose (1976) reported that by 1975 *L. luteator* had spread throughout the whole of the *N. sertifer* distribution area east and north of the Grey County release point. It is possible, however, that the northern dispersal into the Bruce Peninsula was the result of a release made near Hepworth, Ontario, in 1970 (Griffiths & Lyons 1980). Rose noted that *L. luteator* had spread 305 km in an easterly direction in the 13 years since its release. There had been little or no movement south or west from the 1962 release point and little further dispersal from the 1964 release in Norfolk County.

Investigations on the biology of the principal parasitoids of *N. sertifer* and the interactions between them have continued during the last decade. It was found that females of *Pleolophus basizonus* (Grav.), which oviposit in cocooned pre-pupae, show a strong rejection of hosts containing larvae or pupae of its own species, although they do not discriminate between unattacked hosts and those containing eggs of its own species (Griffiths 1972). Also, it was found that *P. basizonus* females show a strong avoidance of cocoons in which there are eggs, larvae, or pre-pupae of the larval parasitoid *L. luteator* (Griffiths 1976) and avoid cocoons containing developing *E. abruptorius* and *Dahlbominus fuscipennis* (Zett.). On the other hand, *D. fuscipennis* does not avoid hosts containing *P. basizonus* or *E. abruptorius* and is usually successful in competition with them. *Exenterus abruptorius, E. nigrifrons* (formerly *E. canadensis* Prov.) and *E. amictorius* (Panz.) all superparasitize and none avoids attacking hosts containing eggs of either of the other two species. Neither *E. abruptorius* nor *E. nigrifrons* has an advantage over the other when competing on the host, but *E. amictorius* is usually successful when competing with the other species. This information on *Exenterus* spp., plus data on the three species obtained in insectary experiments over 6 years, was incorporated into a simulation model to investigate the interactions of the three parasitoids and their host. More recently a more sophisticated simulation model of *N. sertifer* and the living and non-living factors that influence it has been developed. It will be used to investigate the complex interactions between all the biological control agents involved in *N. sertifer* population dynamics.

The first trial with the NPV of European pine sawfly was conducted near Strathroy, Ontario, in 1950, with material obtained from Sweden (Bird 1950). Since then it has been used as a biological control agent in more countries than any other insect virus, with applications in Canada, USA, UK, Germany, Finland, Norway, Austria, Italy, Yugoslavia, Poland, and the USSR (Cunningham & Entwistle 1981). Many of these applications were operational sprays not reported in the scientific literature, and it is impossible to assess the number of hectares treated to date.

Releases and Recoveries

Lophyroplectus luteator Thunb. (Hymenoptera: Ichneumonidae)

There have been a number of relocations of this established species in the past decade. In June 1970, 49 male and 31 female adults were released in a jack pine stand, *Pinus banksiana* Lamb., near Hepworth, Ontario, approximately 27 km north of the 1962 release point at Chatsworth, Ontario (Table 86). The previous year the most northerly recovery point was 9.3 km from the 1970 release site (Griffiths 1973). Parasitoids were recovered at Hepworth in the year of release and in both 1971 and 1972, when work was stopped (Griffiths & Lyons 1980).

In 1972, 1150 *L. luteator* adults with an estimated sex ratio of 36.8% females were released into a Scots pine plantation in Gordon Township on Manitoulin Island, Ontario (Table 86). *N. sertifer* was recovered on Manitoulin Island in 1965 and has been recorded there every year since. Adults of this parasitoid were recovered in the release plantation every year until 1976 (when the study stopped). At that time, parasitism was 14.9%. In the first generation after the release, parasitoids were obtained up to 900 m from the release site (Griffiths & Lyons 1980).

A total of 1 113 adults, approximately 36.8% of them females, were released in the city of Sault Ste. Marie, Ontario, in 1972, 4 years after *N. sertifer* had first been recovered there (Table 86). Parasitism in the release area had reached 9.5% by 1975 (Griffiths & Lyons 1980).

A collection of *N. sertifer* was made on Manitoulin Island in 1976 to obtain *L. luteator* for release against *N. sertifer* in Wisconsin in 1977. A total of 408 male and 305 female adults was released in a red pine plantation near La Grange, Wisconsin, in late May, 1977. One male adult was recovered from the 1870 larvae collected there in 1978 (Kraemer *et al.* 1979). There has been no further information on this release.

When *N. sertifer* was found in Newfoundland, arrangements were made to ship parasitoids from Ontario to the Newfoundland Forest Research Centre for release against it. The first shipment of *L. luteator* was made in the spring of 1978. The parasitoids had been obtained from larval collections made the previous summer in southwestern Ontario. In 1978 a total of 178 male and 128 female adults was released at two locations near St. John's. Further releases of similar material were made in 1979, when 28 males and 10 females were released, and in 1980 when 173 males and 294 females were released. Also in 1980, an additional 25 male and 26 female adults from Austria and Finland were released (Table 86). To date there has been no recovery of *L. luteator*.

Dipriocampe diprioni (Ferrière) (Hymenoptera: Tetracampidae)

This European species, previously referred to as *Tetracampe diprioni* Ferrière, attacks the newly laid eggs of *N. sertifer* in September and October. A number of releases have been made in Ontario. McGugan & Coppell (1962) list four, all made in the spring, and Griffiths *et al.* (1971) list a further three poorly timed releases. However, the latter authors record a release of nearly 1600 adults in late September, 1961, in a stand

Table 86 Open releases and recoveries of parasitoids against the European pine sawfly, *Neodiprion sertifer* (Geoffr.)

Species and province	Year	Origin	Number released	Year of recovery
Dipriocampe diprioni Ferrière				
Ontario	1972	Austria	78	
	1974	Germany	183	1975
	1978	Italy	33	
	1979	Italy Switzerland and Austria	272	
Exenterus abruptorius Thnb.				
Ontario	1972	Austria	648	
	1973	Austria	1021	
	1979	Austria	967	
	1979	Italy	1035	
	1980	Finland	235	
	1980	Austria	90	
Newfoundland	1980	Finland	75	
Lophyroplectus luteator (Thnb.)				
Ontario	1970	Ontario	80	1970
	1972	Ontario	2263	1972
Newfoundland	1978	Ontario	306	
	1979	Ontario	38	
	1980	Ontario	467	
	1980	Austria	50	
	1980	Finland	1	
Pleolophus basizonus (Grav.)				
Newfoundland	1977	Ontario	1007	1977
	1978	Ontario	2104	
	1979	Ontario	1007	
	1980	Finland	25	

infested with *N. sertifer* near Williamsford, Ontario. Eggs of *N. sertifer* were collected in the release area and examined in the spring of 1962 and spring egg collections were made annually until 1968 in a population study plot 0.8 km away. No recoveries of the parasitoid were obtained.

Because the egg stage of *N. sertifer* in Canada was essentially an untapped resource for biological control, it was decided that more study on European egg parasitoids was justified in spite of earlier failures. The work was carried out by CIBC European Laboratory scientists. As a result of this work (Pschorn-Walcher & Eichhorn 1971, 1973) it was found that *D. diprioni* had a univoltine strain adapted to *N. sertifer* and a morphologically similar bivoltine strain which occurs rarely in *N. sertifer*. Earlier failures may have been the result of using the wrong strain and further work on introduction, using the univoltine strain, was justified. This work started in 1971 with a preliminary test which indicated that *D. diprioni* had attacked Canadian *N. sertifer* and had overwintered successfully outdoors near Chatsworth, Ontario.

A release of 32 male and 46 female adults was made in Dufferin County Forest, 5 September 1972 (Table 86), but this area was sprayed inadvertently with a chemical insecticide in the summer of 1973 and the *N. sertifer* population there was destroyed. Another release was made, in Grey Country, 6 September 1974, of 29 males and 154 females (Table 86). There was 1.3% parasitism in 87 egg clusters collected at the release site in the spring of 1975. Parasitoids were obtained, but in decreasing numbers in both 1976 and 1977; none was obtained in the following 3 years (Table 87).

Releases of *D. diprioni* have been made at two other localities, one in Simcoe County, consisting of 11 male and 22 female adults, on 16 September 1978, and the other, of 90 male and 182 female adults, in Peterborough County, 28 August and 30 September 1979 (Table 86). There have been no recoveries from either release (Table 87).

Laboratory rearings of *D. diprioni* were started in 1974 using adults obtained from Switzerland and Germany. Work continued until 1977 using the offspring of the original stock augmented with adults from field collections of *N. sertifer* eggs and further shipments from Europe (Table 88) but it was not possible to maintain stock beyond the third generation.

Table 87 Recovery of *Dipriocampe diprioni* Ferrière

Year	Number of clusters collected	Number of eggs examined	Number of egg clusters with attack	Number of eggs attacked	Sex ratio (% females)
Grey County, Ontario					
1975	87	6052	7	80 (1.3%)	53
1976	57	3094	2	34 (1.1%)	50
1977	86	3819	1	7 (0.2%)	71
1978	6	430	0	0	
1979	144	?	0	0	
1980	28	?	0	0	
Simcoe County, Ontario					
1979	52	?	0	0	
1980	48	?	0	0	
Peterborough County, Ontario					
1980	47	?	0	0	

Exenterus abruptorius (Thunb.) (Hymenoptera: Ichneumonidae)

Since the recovery of a small number of adults of this species in southwestern Ontario in the spring of 1972, noted earlier, there has been renewed interest in trying to establish it at higher densities and in determining why it has remained so rare since its introduction in the period from 1936 to 1949 (McGugan & Coppel 1962). A number of well-timed releases have been made in Ontario (Table 86) but because *N. sertifer* populations are generally low, releases could only be made in areas of low host density. At these population levels large cocoon collections would have been necessary to determine if establishment had occurred, and the cost made sampling unpractical.

A release of 43 male and 32 female adults obtained from Finland was made in St. John's, Newfoundland, in July 1980 (Table 86).

Studies on the biology of this species were started with insectary experiments in 1969 and 1970 and continued with laboratory experiments from 1978 to 1980, using *E. abruptorius* adults from various European locations (Table 88). Although analysis of results is not complete it is already clear that no single physical factor has been limiting the numbers of this species in Ontario.

Table 88 Laboratory and field cage studies of parasitoids against the European pine sawfly *Neodiprion sertifer* (Geoffr.)

Species and province	Year	Origin	Number
Dipriocampe diprioni Ferrière			
Ontario	1970	Austria	196
	1971	Austria	228
	1974	Germany	73
	1975	Austria	19
	1976	Switzerland	136
	1977	Switzerland	72
Exenterus abruptorius Thnb.	1969	Austria	579
Ontario	1970	Austria	244
	1978	Austria	242
	1978	Germany	64
	1978	{ Germany Austria	75
	1978	Italy	211
	1979	Finland	139

***Pleolophus basizonus* (Grav.) (Hymenoptera: Ichneumonidae)**

As stated earlier, shipments of parasitoids to Newfoundland were started soon after it was known that *N. sertifer* was established there. The first species to be shipped was *P. basizonus,* a multivoltine parasitoid of cocooned pre-pupae successfully introduced from Europe in the 1930s (McGugan & Coppel 1962). *N. sertifer* cocoons containing developing *P. basizonus* were shipped to Newfoundland in the summer of 1977. They were obtained from a rearing programme conducted in Sault Ste. Marie, Ontario, during the winter preceding shipment, and were reared in St. John's. Emerging parasitoids were released twice weekly between 21 July and 15 August at Windsor Lake, 9.7 km from St. John's. A total of 631 males and 376 females was released in an infested plantation of mixed jack pine and Scots pine (Otvos & Griffiths 1979) (Table 86). Approximately 130 adult *P. basizonus* were recovered from 600 *N. sertifer* cocoons "planted" in the release area in the summer of 1977, indicating that the released adults had attacked successfully in the field (Otvos & Griffiths 1979). Further releases of large numbers of this parasitoid obtained from rearings in Sault Ste. Marie were made in 1978 and 1979. A small number of adults from Finland was also released in 1980 (Table 86). There have been no further recoveries of adults of this species from field-collected cocoons to date. Monitoring of this release is continuing.

Pathogens

In Ontario, European pine sawfly NPV was extensively used in the 1950s and 1960s by Christmas tree growers and provincial government forestry officials.

There are several methods of producing this virus in host insect larvae, the simplest and cheapest being to do so in the field. Plantations with a suitable insect population density are found and fourth-instar larvae are sprayed using a mist blower to disseminate NPV at a concentration of 10^6 PIB/ml. The first dead larvae are found about 7 days after spraying and colonies of dead and dying larvae are collected daily until about 14 days after spraying. Larvae are removed from the foliage, frozen, lyophilized, ground to a fine powder, and the concentration of PIB per gramme determined. About 10^8 PIB are obtained from one dead larva, so, using a dosage of 5×10^9 PIB/ha, about 50 virus-killed larvae are required. A virus treatment is therefore very economical. Between 1970 and 1975 a considerable quantity of this virus was produced by Ontario Ministry of Natural

Resources staff in Simcoe District. The material was processed and standardized at the Forest Pest Management Institute and is available for control operations should European pine sawfly outbreaks occur.

In the 1970s only one operational virus spray treatment was applied. This was at Sandbanks Provincial Park, Quinte Island, in 1975 when 125 ha of Scots pine, jack pine, and red pine, *Pinus resinosa* Ait., were aerially sprayed with NPV. A dosage of 5.1×10^{10} polyhedral inclusion bodies (PIB) per hectare in an aqueous suspension containing 30 g/l IMC 90-001 (International Minerals Corp.) (a UV screening agent) and 0.01% Chevron® sticker (Chevron Chemical Co.) was applied when larvae were mainly in the first instar. Two weeks after the treatment 94% of the colonies were dead and the remaining 6% diseased. After 3 weeks, no healthy sawfly colonies could be found in the park, whereas untreated colonies, removed from the park before virus application, remained healthy. The operation was considered to be an unqualified success and, except for browning of 1-year-old needles near branch tips, there was no defoliation (Cunningham *et al.* 1975). The park remained free from European pine sawfly for 4 years following the treatment, but in 1980 a few colonies were observed.

Evaluation of Control Attempts

An exotic pest such as *N. sertifer* that attacks a host with a wide geographic range presents two problems. The first is prevention of spread beyond the infested area, the second is permanent regulation within the infested area. We have not yet been successful in solving either problem. *N. sertifer* reached Manitoulin Island in 1966 and the cities of Sault Ste. Marie and North Bay, Ontario, in 1968, all as a result of transport of infested nursery stock. In 1969 it was found for the first time in Ottawa, Ontario; in 1974 near Thessalon, Ontario; near Québec City, Québec; and near St. John's, Newfoundland. In all but two of these movements beyond the infested area the pest is still there and in only one of the two areas where it is no longer present can we claim to have brought about the eradication by our efforts, viz using a virus application at Thessalon. The dispersal of *N. sertifer* occurred in spite of a fumigation procedure (Monro & Kirby 1963).

Ontario Forest Insect and Disease Survey records show generally low *N. sertifer* populations in the continuously infested area since several heavy infestations in southwestern Ontario declined in 1973. However, in the absence of intensive population studies we are unable to state what role native or introduced parasitoids have played in this decline. We do know from Lyons (1977*a*, 1977*b*) that parasitism by *Exenterus* spp. (mainly the native species) and predation and parasitism of cocooned pre-pupae (essentially all by the introduced *P. basizonus*) are factors in the decline of outbreaks. We have also noted, by sampling cocoons obtained from mass larval collections made at various sites in Simcoe and Dufferin counties over the past 4 years, that *L. luteator* is common in these areas. Parasitism by this species in seven collections made between 1977 and 1980 ranged from 25 to 54%. These are the only indications we have that parasitoids are contributing to current low *N. sertifer* populations.

The only attempt at introducing new species of parasitoids in the past decade — that with *D. diprioni* — appears to have been unsuccessful. The discovery that *E. abruptorius* had become established following the releases in the 1930s and 1940s has little practical value because the species is present in such low numbers that it has essentially no impact on *N. sertifer* numbers.

The only success we have had in the use of parasitoids in the past decade was in the relocation of established parasitoids. *L. luteator* has been established in Ontario wherever *N. sertifer* is present. *P. basizonus* appears to have been successfully transferred to Newfoundland.

There is no doubt that European pine sawfly NPV is a highly efficacious biological control agent when applied on early instar larvae, and treatments continue to give control in the years following application. The longevity of this control depends on such factors as the initial population density of the host and the size of the trees, but infectious PIBs remain in the environment for at least 3 years following a virus epizootic (Kaupp 1981). Given this respite from defoliation, trees grow considerably and are much less susceptible to damage if further European pine sawfly outbreaks occur.

To date, there are no viruses registered for operational use in Canada. In the USA, a petition has been prepared for registration of European pine sawfly virus under the name "Neochek-S". Extensive animal safety testing has been conducted by the Americans, but it is not known if these data will be acceptable to Canadian authorities. With the limited areas of European pine sawfly infestation currently reported, registration of its NPV in Canada is not of high priority in comparison to viruses of certain other insect pests. Even if European pine sawfly NPV is eventually registered in Canada, there is little commercial incentive for production and marketing owing to its limited and specialized use. Virus production and distribution, therefore, will almost certainly remain in the hands of federal and provincial governments.

Recommendations

In view of the successful establishment of *L. luteator* in five areas of Ontario, all following a single release, the failure to establish this species in Newfoundland following three releases in 3 successive years suggests a fundamental difference between the situations in Ontario and Newfoundland. We recommend that no more releases be made there after the 1981 release until further study reveals success or the reasons for failure.

It was pointed out by Griffiths *et al.* (1971) that there are 62 species of native parasitoids known to attack other pine-feeding diprionids in Ontario that had not been recovered from *N. sertifer*. We have not recovered any of them in the limited sampling in the last decade in southern Ontario. It is possible that some of them are now attacking *N. sertifer* in northern Ontario and we recommend that sampling be done to determine if this is so.

No more releases of *D. diprioni* or *E. abruptorius* should be made. However, sampling to determine whether the former species has become established can be continued with little cost for several years. Collections to determine the presence of *E. abruptorius* should only be undertaken when they can provide adequate numbers of hosts in order to justify the work that the processing of collections requires.

As stated earlier, the operational use of NPV at Sandbanks Provincial Park in Ontario was considered to be an unqualified success. However, after assessing aerial applications of European pine sawfly NPV and evaluating reports of operations conducted outside Canada, two recommendations can be made. Firstly, the dosage of virus applied is about 10 times more than required for satisfactory control. A dosage of 5×10^9 PIB/ha is adequate. Secondly, adjuvants in the tank mix are unnecessary and virus in water alone will suffice.

Although it is not within the discipline of biological control, we are concerned about the dispersal of the European pine sawfly beyond its present boundaries in Canada. Sullivan (1965) pointed out that the overwintering eggs of this species can survive a temperature of −26°C and that, with conditioning and selection, they may be able to withstand temperatures even lower. Thus the possibility of this pest spreading throughout the range of its widely distributed hosts, Scots pine, jack pine, and red pine, is great unless efforts are made to prevent it. We recommend that some attempt be made to fumigate stock before transport to reduce this possibility, using techniques outlined by Monro & Kirby (1963).

Literature Cited

Bird, F.T. (1950) The dissemination and propagation of a virus disease affecting European pine sawfly, *Neodiprion sertifer* (Geoff.). *Canadian Department of Agriculture Bi-monthly Progress Report* 6,2–3.

Clark, R.C.; Singh, P. (1975) Newfoundland region. *Canadian Forestry Service, Annual Report of the Forest Insect and Disease Survey* 1974.

Cunningham, J.C.; Kaupp, W.J.; McPhee, J.R.; Sippell, W.L.; Barnes, C.A. (1975) Aerial application of a nuclear polyhedrosis virus to control European pine sawfly. *Canadian Forestry Service Bi-monthly Research Notes* 31,39–40.

Cunningham, J.C.; Entwistle, P.F. (1981) Control of sawflies by baculovirus. In: Burges, D. (*Ed.*) Microbial control of pests and plant diseases, 1970–1980. London; Academic Press.

Griffiths, K.J. (1972) Discrimination between parasitized and unparasitized hosts by *Pleolophus basizonus* (Hymenoptera: Ichneumonidae). *Proceedings of the Entomological Society of Ontario* 102,83–91.

Griffiths, K.J. (1973) The dispersal of the introduced parasitoid *Lophyroplectus luteator* (Hymenoptera: Ichneumonidae) following its release against *Neodiprion sertifer* (Hymenoptera: Diprionidae) in Ontario. *Canadian Entomologist* 105,833–836.

Griffiths, K.J. (1976) Interactions between *Pleolophus basizonus* and *Lophyroplectus luteator* (Hymenoptera: Ichneumonidae), two parasitoids of *Neodiprion sertifer* (Hymenoptera: Diprionidae). *Entomophaga* 21,13–17.

Griffiths, K.J.; Lyons, L.A. (1968) The establishment of *Lophyroplectus luteator* (Hymenoptera: Ichneumonidae) on *Neodiprion sertifer* (Hymenoptera: Diprionidae) in Ontario. *Canadian Entomologist* 100,1095–1099.

Griffiths, K.J.; Lyons, L.A. (1980) Further releases of *Lophyroplectus luteator* (Hymenoptera: Ichneumonidae), an introduced parasite of *Neodiprion sertifer* (Hymenoptera: Diprionidae). *Canadian Entomologist* 112,421–426

Griffiths, K.J.; Rose, A.H.; Bird, F.T. (1971) *Neodiprion sertifer* (Geoff.), European pine sawfly (Hymenoptera: Diprionidae). *In*: Biological control programmes against insects and weeds in Canada 1959–1968. *Commonwealth Institute of Biological Control Technical Communication* 4,150–162.

Kaupp, W.J. (1981) Studies on the ecology of the nuclear polyhedrosis virus of the European pine sawfly, *Neodiprion sertifer* (Geoff.). Ph.D. Thesis, Oxford University, U.K.

Kraemer, M.E.; Coppel, H.C.; Hall, D.J. (1979) The establishment of *Lophyroplectus luteator*, a larval parasitoid of the European pine sawfly, *Neodiprion sertifer*, in Wisconsin. *University of Wisconsin Forest Research Note* 225, 3 pp.

Lindquist, O.H.; Miller, W.J. (1979) The pine sawfly that came to dinner. *Your Forests* 12(1),10–11.

Lyons, L.A. (1977*a*) On the population dynamics of *Neodiprion* sawflies. *In*: Kulman, H.M.; Chiang, H.C. (Eds.). Insect ecology — papers presented in the A.C. Hodson Lectures. *University of Minnesota Agriculture Experiment Station Technical Bulletin* 310,48–55.

Lyons, L.A. (1977*b*) Parasitism of *Neodiprion sertifer* (Hymenoptera: Diprionidae) by *Exenterus* spp. (Hymenoptera: Ichneumonidae) in Ontario, 1962–1972, with notes on the parasites. *Canadian Entomologist* 109,555–564.

Lyons, L.A.; Sullivan, C.R.; Wallace, D.R.; Griffiths, K.J. (1971) Problems in the management of forest pest populations. *Proceedings of the Tall Timbers Conference on Ecological Animal Control 1971,* pp. 129–140.

Magasi, L.P. (1981) Forest pest conditions in the Maritimes in 1980. *Canadian Forestry Service, Fredericton, New Brunswick, Report* M-X-118, 35 pp.

Martineau, R.; Lavallée, A. (1975) Québec region. *Canadian Forestry Service, Annual Report of the Forest Insect and Disease Survey 1974,* pp. 37–56.

McGugan, D.M.; Coppel, H.C. (1962) Biological control of forest insects, 1910–1958. *In*: A review of the biological control attempts against insects and weeds in Canada. *Commonwealth Institute of Biological Control Technical Communication* 2,35–127.

Monro, H.A.A.; Kirby, C.S. (1963) Fumigation of nursery stock as a possible means of retarding the spread of the European pine sawfly. *Canadian Department of Forestry, Entomology and Pathology Branch Bi-monthly Progress Report* 19(3),2.

Otvos, I.S.; Griffiths, K.J. (1979) Cocoon parasitoid of the European pine sawfly introduced into Newfoundland from Ontario. *Canadian Forestry Service, Bi-monthly Research Notes* 35(4),20–21.

Pschorn-Walcher, H.; Eichhorn, O. (1971) Untersuchungen uber die Eiparasitien (Hym., Chalcidoidea) der rotgelben Kiefern-Buschhorn-blattwespe *Neodiprion sertifer* Geoff. (Hym. Diprionidae). *Anzeiger für Schaedlingskunde, Pflanzenschutz, Umweltschutz* 44(7),97–103.

Pschorn-Walcher, H.; Eichhorn, O. (1973) Studies on the biology and ecology of the egg parasites (Hym.: Chalcidoidea) of the pine sawfly *Neodiprion sertifer* (Geoff.) (Hym.: Diprionidae) in central Europe. *Zeitschrift für angewandte Entomologie* 74,286–318.

Raizenne, H. (1957) Forest sawflies of southern Ontario and their parasitoids. *Canadian Department of Agriculture, Division of Forest Biology Publication* 1009, 45 pp.

Rose, A.H. (1976) The dispersal of *Lophyroplectus luteator* (Hymenoptera: Ichneumonidae) on *Neodiprion sertifer* (Hymenoptera: Diprionidae) in southern Ontario. *Canadian Entomologist* 108,1395–1398.

Schaffner, J.V. (1939) *Neodiprion sertifer* (Geoff.) a pine sawfly accidentally introduced into New Jersey from Europe. *Journal of Economic Entomology* 32,887–888.

Sippell, W.L.; Dance, B.W.; Rose, A.H. (1966) *Canadian Forestry Service, Report of the Insect and Disease Survey 1965*, pp. 44–65.

Sippell, W.L.; Gross, H.L.; Rose, A.H. (1969) *Canadian Forestry Service, Report of the Insect and Disease Survey 1968,* pp. 54–66.

Sullivan, C.F. (1965) Laboratory and field investigations on the ability of eggs of the European pine sawfly, *Neodiprion sertifer* (Geoffroy) to withstand low winter temperatures. *Canadian Entomologist* 97,978–993.

Neodiprion swainei (Middleton), Swaine Jack Pine Sawfly (Hymenoptera: Diprionidae)

R.J. FINNEGAN and W.A. SMIRNOFF

Pest Status

At the time of the last review by McLeod & Smirnoff (1971), the population of *Neodiprion swainei* (Middleton), Swaine jack pine sawfly, was approaching endemic levels over most of its range in eastern Ontario and southwestern Québec, where it has occurred extensively since the early 1950s. This was thought to be due partly to the natural collapse of a prolonged infestation and partly to the application of chemical insecticides in 1965 (McLeod 1968). McLeod & Smirnoff (1971) had expected a recurrence of *N. swainei* in the 1970s, but this did not materialize, probably due to earlier applications of chemical insecticides against both Swaine jack pine sawfly, *N. swainei*, (McNeil *et al.* 1979) and spruce budworm, *Choristoneura fumiferana* (Clem.), (McLeod 1975) in the same general area. The population remained low throughout the 1970s, with the occurrence of only a few localized light infestations in northern Ontario and western Québec.

Background

Parasitoids, predators, and pathogens have all been introduced in attempts to control sawfly populations. McLeod & Smirnoff (1971) reported that "Before 1958, only small numbers of *Pleolophus basizonus* (Grav.) and *Drino bohemica* Mesn. were released against *Neodiprion swainei* in a few localities in Ontario. However, massive releases of these two parasitoids, as well as *Exenterus amictorius* Panzer and *Dahlbominus fuscipennis* (Zett.) were also made in the early 1940s against *Diprion hercyniae* Htg. in both Ontario and Québec. All four species have been recovered from *N. swainei* in a number of localities in Québec since 1958 ... but none in Ontario. *E. amictorius* and *P. basizonus* were most abundant, *D. fuscipennis* was less common, and only one specimen of *D. bohemica* was found". A brief description of the interaction of each parasitoid with *N. swainei* was given. McLeod & Smirnoff also reported on the discovery and successive recoveries of the *Borrelina* viral pathogen. The results of one experimental aerial dispersion of the virus in 1960 and a second in 1964 were reported as "excellent" and "not as good" respectively. The ecology of the virus was briefly considered, stating that certain predators, such as wasps and pentatomids act as vectors of the virus. In conclusion, McLeod & Smirnoff stated that the releases of *E. amictorius* and *P. basizonus* seemed to have been beneficial, and that the use of virus as a control was promising.

Releases and Recoveries

During the past decade, no further release of parasitoids, and few recoveries, were made. Nonetheless, considerable research was completed during this period. In anticipation of an impending resurgence of the *N. swainei* population in Québec, two foreign species of predacious red wood ants, *Formica lugubris* Zett. and *F. obscuripes* Forel, were introduced into susceptible stands of jack pine *Pinus banksiana* Lamb., in the St. Maurice river valley. Observations were reported on other insect predators as well as on birds and small mammals. Studies on the virulence, formulation of sprays, and application of pathogens were also continued.

Parasitoids

McLeod (1971) examined the variation of *N. swainei* cocoon survival and found that only about 10% of the mortality was caused by parasitoids. Later, McLeod (1972) compared host discrimination with density responses during oviposition by parasitoids of *N. swainei*; he observed that the ability of females to discriminate against hosts containing previously deposited parasitoid progeny was variable. Although there was no discrimination at the beginning of the spinning period of *N. swainei*, it was quickly acquired and persisted to the end of the spinning period.

Price (1970*a*, 1970*b*, 1970*c*) showed that "Female parasitic insects in the genera *Pleolophus, Ensasys,* and *Matrus* (Hymenoptera: Ichneumonidae) search the ground cover for hosts and avoid areas they had already inspected" and that "females respond to their own trail odour, and recognition occurs also between conspecific, congeneric and intergeneric individuals". Later Price (1971) discussed the niche breadth and dominance of parasitic insects sharing the same host species. In 1972, he indicated that as a result of recognition of repellent trail odours, females of *P. basizonus* are able to discriminate against parasitized hosts of *N. swainei* between narrow limits "set by the probability of a female finding an unparasitized host", and that they showed "mutual interference in egg laying at high parasitoid density" (Price 1972*a*). He also reported on the immediate and long-term effects of chemical insecticide application on insect parasites in jack pine stands in Québec (Price 1972*b*); on the activity patterns and impact of parasitoids on *N. swainei* (Price 1972*c*); and on the utilization of the same host by different insect parasites (Price 1972*d*).

Predators

Finnegan (1971) reported on preliminary studies on the use of predacious ants to control forest pests in Québec, and concluded that none of the species found was particularly promising. He further discussed (Finnegan 1974) both the harmful and beneficial aspects of red wood ants as insect predators, the reason for their effectiveness, and recommendations for selecting desirable species for propagation. Finnegan (1975, 1977) later introduced two species of red wood ants, *F. lugubris* from central Europe and *F. obscuripes* from southern Manitoba, into a jack pine stand at Lac à la Chienne, about 70 km southwest of La Tuque, Québec, in the St. Maurice river watershed. The two species of ants belong to the *F. rufa* group and are aggressive predators of sawfly larvae, cocoons, and adults, as well as a variety of other defoliating insects (Cotti 1963, Finnegan 1978, McNeil *et al.* 1978). These introductions were made in anticipation of a recurrence of *N. swainei* infestations in the late 1970s. When the initial introductions had become established, additional releases (from Europe and Manitoba) were made in subsequent years. Table 89 shows the dates and approximate numbers of each ant species (workers and sexuals) released in the St. Maurice river watershed. The status of each species in 1981 is also given in Table 89, based on the number, average size, and insect vigour of the nests. Two secondary introductions of *F. obscuripes* (emanating from the first) were made in 1980 in the Gilardo Dam area, about 75 km west of La Tuque, Québec.

Ilnytzky (1974) reported on the species of ants found in a young jack pine stand in Québec, and on their distribution, population level, and predation of *N. swainei*. He studied 10 species of ants, of which *F. fusca* L. was the most abundant, and found that together with several other *Formica* species and *Camponotus herculeanus* (L.) they carried emerging and ovipositing adults, as well as cocoons, into their nests. He concluded that ants may be an important factor regulating *N. swainei* populations.

Table 89 Open releases and recoveries of predacious red wood ants against Swaine jack pine sawfly, *Neodiprion swainei* (Middleton), in the St. Maurice river watershed in Québec

Ant species	Date of release	Approximate number released	Status in 1981: Recovery	Status in 1981: Estimated increase
Formica obscuripes Forel	1972	4 000 000	About 20 nests established (with sexuals present)	15-fold
	1975	1 000 000	About 75 nests established (with sexuals present)	20-fold
F. lugubris Zett.	1973	1 000 000	Ten nests established (with few sexuals)	
	1975	2 000 queens	(added to previous nests)	?
	1976	5 000 queens	(added to previous nests)	?

McLeod (1971) reported, in a preliminary analysis of the variation in cocoon survival of *N. swainei*, that small animal predators accounted for about 47% of the mortality of cocoons, and insect predators for about 16%. McLeod (1974) also found that populations of resident breeding birds in jack pine stands in Québec varied little from 1964 to 1973, but that the mole population showed a steady, statistically significant increase over the 9 years of sawfly infestation.

Tostowaryk (1971*a*) described in detail the life history and behaviour of *Podisus modestus* (Dallas), a predacious pentatomid, which preys on *N. swainei* larvae as well as on other defoliators. He gave details of the habits and abundance of the predator, and also discussed the relationship between parasitism and predation of diprionid sawflies (Tostowaryk 1971*b*). Experimental results of the functional response of *P. modestus* to densities of *N. swainei* and *N. pratti banksianae* Rohwer (Tostowaryk 1971*c*) showed that the response was two-fold: a domed curve when the sawfly larvae were active and relatively large with respect to the size of the predator, and a negatively accelerated curve when the sawfly larvae were freshly killed and relatively small with respect to the predator's size. He also presented a list of coleopterous predators of *N. swainei* (Tostowaryk 1972), and stated that it seemed these predators responded in a density-independent manner to the prey. Tostowaryk (1973) continued his observations on carabids found in jack pine stands. He noted that the three most common species of carabids preyed to a limited extent on cocoons of *N. swainei*.

Pathogens

A highly pathogenic strain of a *Borrelina* species of virus was studied (Smirnoff 1961*a*, McLeod & Smirnoff 1971) and a formulation suitable for aerial dispersion was developed (Smirnoff 1964). Aerial dispersion of the virus provided excellent control of the insect (Smirnoff *et al.* 1962). The dosages used were 4.5 l/ha and 37 l/ha, at a concentration of 2 x 10^6 polyhedra/ml. Cold weather during and after treatment caused a reduction in larval mortality, but because of trans-ovum and trans-ovarial transmission of the disease from parent to progeny, there was a long-term control effect (Smirnoff 1962).

The control efficiency of the transmitted virus varied according to temperatures in July and August. With average and above-average temperatures, larval mortality caused by applied virus was 60–80%; when temperatures were below average, larval mortality was 40–60% (Smirnoff 1967*a*). However, in both cases, larval populations were reduced and maintained at endemic levels.

This phenomenon was exploited to develop a method of spot introduction in an infested stand, using either aerial or ground applications, with low concentrations of virus (200 000 polyhedra/ml of virus at a rate of 4.5–9.3 l/ha) when larvae reached the fifth instar. Another method consisted of feeding sixth-instar larvae (prior to spinning) with virus-contaminated foliage, which produced virus-infected cocoons. The cocoons were stored in soil and placed in wooden boxes with screened tops for 1 or 2 years. When needed, the cocoons were placed in 20 x 10 x 5 cm cages covered with galvanized steel cloth (3 holes/cm), which permits emerging females to pass through the holes. The small cages were dispersed in stands with healthy sawfly populations to provoke centres of viral epizootics (Smirnoff 1972*a*).

For virus production, a carousel method (Smirnoff 1964) was used for rearing larvae in the field; in the laboratory, wooden racks (2.5 x 1 m) with screen bottoms were used (Smirnoff 1964). More recently, a pilot facility for virus production was established. It consists of a multi-drawer container measuring 1.0 x 1.0 x 2.2 m. Each drawer (17 cm high) has a screen top and bottom (screen size varies according to larval size), and a small sliding plate placed under each drawer to collect frass. Good natural ventilation is important to prevent the development of bacteria and fungi. In our pilot facility 23 units, with seven drawers each, provided rearing facilities for 12 million larvae. A method was also developed for preparing virus in the field. This simply consists of spraying an aqueous suspension of virus on fourth-instar larvae and collecting diseased larvae before death.

Several tests revealed that virus infection can also be transmitted in the field by hemipterous and hymenopterous predators and parasitoids of *N. swainei* larvae (Smirnoff 1959, 1961*b*, 1975), most commonly the wasp species *Vespula rufa consobrina* (Saussure) and *Dahlbominus fuscipennis* (Zett.).

Bird (1949) reported the presence of external cores in the abdominal cavity of sawfly larvae and considered them as cancerous tumours. Subsequent research revealed that they resulted from a normal defensive reaction of the organism following viral infection or other injuries. The normal lysis of the midgut epithelium in the abdominal cavity is interrupted during moulting, causing formation of a non-cancerous cyst (Smirnoff 1968, 1969).

During studies of *N. swainei* virus, other microorganisms were found: a species of *Herpetomonas* (a commensal parasitoid) described as *Herpetomonas swainei* Smirnoff & Lipa (Smirnoff & Lipa 1970, Smirnoff 1971*a*); a pathogenic species of microsporidium (*Thelohania* sp.); a few secondary bacteria; and the fungi *Beauveria bassiana* Vuillemin and *Paecilomyces farinosus* (Holm ex. S.F. Grey) Brown & Smith, which were occasionally present in cocoons. *H. swainei* was present with the virus and the microsporidium but no synergism was found. However, these microorganisms can reduce the number of eggs laid by 20–30% (Smirnoff 1974).

Studies were conducted to determine the effect of various wavelengths of visible solar light on the nuclear-polyhedrosis virus of *N. swainei* (Smirnoff 1967*b*, 1972*b*). As a result, evening treatment with virus is recommended to reduce the negative effect of sunlight, and at that time larvae are more thoroughly contaminated by the virus freshly deposited on foliage.

The results of investigations carried out on healthy and virus-infected *N. swainei* populations have shown that virus infection provoked the following strong metabolic disturbances in the host insect: an increase of transaminase activity of up to 2000-fold; increased dehydrogenase and phosphatase activity; an increase in uric acid and sulphur;

and a decrease in free glycerol, potassium, and especially total lipids (energy resources), which were decreased by 50% (Smirnoff 1971*b*).

Results showed that biochemical analysis can be used to determine the pathological state of an *N. swainei* population; to follow, metabolically, the evolution of a virosis in the insect, thus permitting a prognosis of its population; and to evaluate whether the population is expanding or decreasing. The metabolic study done during 1979 and 1980 indicated that the *N. swainei* population of the Saguenay – Lac St-Jean region, Québec, is expanding.

The influence of urea fertilization on the physiology and growth of jack pine and on the behaviour of *N. swainei* (either healthy or virus-infected) was determined. It was found that a treatment with urea in natural forest stands increased the length and weight of jack pine needles proportionally to the amount of nitrogen assimilated. Three years after nitrogen fertilization, the ingestion of foliage by *N. swainei* was reduced by approximately 50%, and the natural mortality of the tenthredinid was increased by 50%. It was found that the sensitivity of larvae to virus infection, as well as the mortality, increased. Also, modifications in the physiological state of jack pine needles provoked changes in insect behaviour. For example, adults seemed to have more difficulty in laying eggs (twisted ovipositor) on needles of fertilized trees than on unfertilized ones (Smirnoff & Bernier 1973, Smirnoff & Valéro 1975).

Evaluation of Control Attempts

Large-scale transplantations of red wood ants over long distances have only been tried a few times on the North American continent, and the costs involved can only be roughly estimated. The transplantations from central Europe were made co-operatively with scientific institutions (University of Pavia in Pavia, Italy, and CIBC in Delémont, Switzerland) and as a consequence had many unrevealed costs. The introduction from southern Manitoba, on the other hand, was funded completely by the Laurentian Forest Research Centre. Because the operations were experimental and of a scientific nature, no charges were made for the ant colonies collected. However, future large-scale operations may have to take this factor into consideration.

The total costs can be divided among four operations: collecting, packaging, transportation, and release. The relative importance of each can vary considerably according to time involved, distances to be covered and accessibility of sites. Table 90 gives estimates of two large transplantations made to the province of Québec: one from Manitoba in 1972, the other from Italy in 1973.

As there has not been any measurable *N. swainei* population in recent years in jack pine stands chosen for the introduction of ants, it has not yet been possible to measure

Table 90 Cost estimates for the transplantation of red wood ant colonies

Date	Collecting (man days)	Transportation	Dispersal (man days)	Packaging	Approximate total costs
1972	25	Southern Manitoba to Québec, with motor vehicle and trailer — $2 000	10	$1 000	$6 500
1973	40	Northern Italy to Québec motor vehicles and by air — $5 000	10	$1 000	$9 500

the effect of predation on the sawfly. However, ants have been observed preying effectively on *C. fumiferana* and other insects in the St. Maurice river watershed and elsewhere in the province of Québec where they have also been introduced (Finnegan 1978, 1979, McNeil *et al.* 1978). The present ant populations are now being reinforced and encouraged, so that a substantial, well-established population will be present over a reasonably large area when the next *N. swainei* infestation occurs.

The strain of *Borrelina* virus that infests *N. swainei* is a valuable biological control agent both on a short-term basis (because of its high degree of pathogenicity) and on a long-term basis (because it is transmitted from progeny to progeny by infected parents). The virus is specific to the Swaine jack pine sawfly, which makes it safe for other living organisms. This is of particular significance where jack pine stands are close to, or intermixed with, commercial blueberry areas, and spraying with chemicals is undesirable (Smirnoff 1971*c*).

Recommendations

N. swainei is an insect that lends itself well to biological control. Often the severity and rate of spread of a typical infestation are sufficiently moderate to permit less dramatic measures than the application of heavy doses of chemical pesticides. It is our opinion that in time of infestations, *Borrelina* virus could be applied as an effective primary control agent, to reduce the sawfly population to an acceptable level. The endemic sawfly population left after applying the virus acts as a breeding-ground for the development and transmission of virus to future sawfly populations, and thus should help maintain stable parasitoid and predator numbers.

The following work is thus recommended. Regulation of sawfly populations by the virus should be tested by releasing the virus in an isolated jack pine stand that has a history of sawfly infestations. The time may be right in 1983 or 1984 to begin such a study, for there are signs of a pending increase in sawfly numbers in various sectors of the province of Québec in 1982.

Predacious red wood ants in large permanent colonies may not only act as effective predators, but may also encourage an increase in the population of small predacious mammals, such as shrews and moles, and various insectivorous nesting birds. Chemical pesticides should not be used in the area tested or in a buffer zone of at least 8 km in width.

Literature Cited

Bird, F.T. (1949) Tumors associated with a virus infection in an insect. *Nature* 163,777–778.

Cotti, G. (1963) Bibliografia ragionata 1930–1961 del grouppo *Formica rufa* in Italiano, Deutsch, English. *Ministerio dell'Agricoltura e delle Foreste, Italia, Collana Verde* 8, 414 pp.

Finnegan, R.J. (1971) An appraisal of indigenous ants as limiting agents of forest pests in Québec. *Canadian Entomologist* 103,1489–1493.

Finnegan, R.J. (1974) Ants as predators of forest pests. *Entomophaga. Mémoire Hors Serie* 7,53–59.

Finnegan, R.J. (1975) Introduction of a predacious red wood ant, *Formica lugubris* (Hymenoptera: Formicidae), from Italy to eastern Canada. *Candian Entomologist* 107,1271–1274.

Finnegan, R.J. (1977) Establishment of a predacious red wood ant, *Formica obscuripes*, (Hymenoptera: Formicidae), from Manitoba to eastern Canada. *Canadian Entomologist* 109,1145–1148.

Finnegan, R.J. (1978) Predation by *Formica lugubris* (Hymenoptera: Formicidae) on *Choristoneura fumiferana* (Lepidoptera: Tortricidae). *Canadian Forestry Service Bi-monthly Research Notes.* 34(1),3–4.

Finnegan, R.J. (1979) Predacious red wood ants in Québec forests. *Bulletin SROP* 1979-11-3,65–75.

Ilnytzky, S. (1974) Formicid predators of the sawfly *Neodiprion swainei* Midd. *Canadian Forestry Service Information Report* LAU-X-8, 14 pp.

McLeod, J.M. (1968) Results of an aerial spraying operation against the Swaine jack pine sawfly, *Neodiprion swainei* Middleton in Québec utilizing the insecticide phosmadon. *Forestry Chronicle* 44,14–20.

McLeod, J.M. (1971) The Swaine jack pine sawfly, *Neodiprion swainei* Midd. life system: a preliminary analysis of variation in cocoon survival. *Canadian Forestry Service Information Report* Q-X-25, 8 pp.

McLeod, J.M. (1972) A comparison of discrimination and of density responses during oviposition by *Exenterus amictorius* and *E. deprionis* (Hymenoptera: Ichneumonidae), parasites of *Neodiprion swainei* (Hymenoptera: Diprionidae), *Canadian Entomologist* 104,1313–1330.

McLeod, J.M. (1974) Bird population studies in the Swaine jack pine sawfly system. *Canadian Forestry Service Information Report* LAU-X-10, 86 pp.

McLeod, J.M. (1975) Possible residual effect of fenitrothion on Swaine jack pine sawfly following aerial applications against the spruce budworm in Québec. *Annals of the Entomological Society of Québec* 20(2),82–85.

McLeod, J.M.; Smirnoff, W.A. (1971) *Neodiprion swainei* Midd., Swaine jack pine sawfly (Hymenoptera: Diprionidae). *In*: Biological control programmes against insects and weeds in Canada, 1959–1968. *Commonwealth Institute of Biological Control Technical Communication* 4,162–167.

McNeil, J.N.; Delisle, J.; Finnegan, R.J. (1978) Seasonal predatory activity of the introduced red wood ant *Formica lugubris* Zett. (Hymenoptera: Formicidae) at Valcartier, Québec in 1976. *Canadian Entomologist* 110,85–91.

McNeil, J.N.; Greenhalgh, R.; McLeod, J.M. (1979) Persistence and accumulation of fenitrothion residues in jack pine foliage and their effect on the Swaine jack pine sawfly *Neodiprion swainei* (Hymenoptera: Diprionidae). *Environmental Entomology* 8(4),752–755.

Price, P.W. (1970*a*) Characteristics permitting co-existence among parasitoids of a sawfly in Québec. *Ecology* (51)3,445–454.

Price, P.W. (1970*b*) Biology of and host exploitation by *Pleolophus indistinctus* (Hymenoptera: Ichneumonidae). *Annals of the Entomological Society of America* 63,1502–1509.

Price, P.W. (1970*c*) Trail odors: recognition by insects parasitic on cocoons. *Science* 170,546–547.

Price, P.W. (1971) Niche breadth and dominance of parasitic insects sharing the same host species (Hymenoptera: Diprionidae) *Neodiprion swainei*. *Ecology* 52(4),584–596.

Price, P.W. (1972*a*) Behaviour of the parasitoid *Pleolophus basizonus* (Hymenoptera: Ichneumonidae) in response to changes in host and parasitoid density. *Canadian Entomologist* 104(1),129–140.

Price, P.W. (1972*b*) Immediate and long-term effects of insecticide application of parasitoids in jack pine stands in Québec. *Canadian Entomologist* 104,263–270.

Price, P.W. (1972*c*) Activity patterns of the Swaine jack pine sawfly *Neodiprion swainei* (Hymenoptera: Diprionidae) and parasitoid impact on the host. *Canadian Entomologist* 104,1003–1016.

Price, P.W. (1972*d*) Parasitoids utilizing the same host: adaptive nature of differences in size and form (Hymenoptera: Diprionidae) *Neodiprion swainei*. *Ecology* 53(1),190–195.

Smirnoff, W.A. (1959) Predators of *Neodiprion swainei* Midd. (Hymenoptera: Tenthredinidae) larval vectors of virus disease. *Canadian Entomologist* 91,246–248.

Smirnoff, W.A. (1961*a*) A virus disease of *Neodiprion swainei* Midd. (Hymenoptera: Tenthredinidae). *Journal of Insect Pathology* 3,29–47.

Smirnoff, W.A. (1961*b*) Predators of larvae of *Neodiprion swainei* Midd. (Hymenoptera: Tenthredinidae). *Canadian Entomologist* 93,272–276.

Smirnoff, W.A. (1962) Trans-ovum transmission of a virus of *Neodiprion swainei* Midd. (Hymenoptera: Tenthredinidae). *Journal of Insect Pathology* 4,192–200.

Smirnoff, W.A. (1964) Preparation and application of viral material in biological control of the jack pine sawfly. *For. Chron.* 40,187–194.

Smirnoff, W.A. (1967*a*) Influence of temperature on the development of a virus disease of *Neodiprion swainei* (Hymenoptera: Tenthredinidae). *Canadian Entomologist* 99,244–249.

Smirnoff, W.A. (1967*b*) Influence of light on the development of a nuclear-polyhedrosis of *Neodiprion swainei*. *Journal of Invertebrate Pathology* 9,269–271.

Smirnoff, W.A. (1968) The nature of cysts found in pupae and adults of *Neodiprion swainei* Midd. *Canadian Entomologist* 100,313–318.

Smirnoff, W.A. (1969) The formation of cysts as a defensive reaction in pupae and adults of *Neodiprion swainei* Midd. (Hymenoptera: Tenthredinidae). *National Cancer Institute Monograph* 31,475–479.

Smirnoff, W.A. (1971*a*) Transmission of *Herpetomonas swainei* sp. n. by means of *Neodiprion swainei* (Hymenoptera: Tenthredinidae) parasitoids. *Canadian Entomologist* 103,630.

Smirnoff, W.A. (1971*b*) Mouvement du potassium et du soufre chez la larve de *Neodiprion swainei* (Hymenoptera: Tenthredinidae) au cours de l'infection virale (polyédrie nucléaire). *Review of Canadian Biology* 30,301–305.

Smirnoff, W.A. (1971*c*) Management of blueberry fields to protect the surrounding jack pine stands against *Neodiprion swainei* Midd. *Phytoprotection* 52,24–27.

Smirnoff, W.A. (1972*a*) Promoting virus epizootics in populations of the Swaine jack pine sawfly by infected adults. *Bioscience* 22,662–663.

Smirnoff, W.A. (1972*b*) The effect of sunlight on the nuclearpolyhedrosis virus of *Neodiprion swainei* with measurement of the solar energy received. *Journal of Invertebrate Pathology* 19,179–188.

Smirnoff, W.A. (1974) Réduction de la viabilité et de la fécondité de *Neodiprion swainei* (Hyménoptères: Tenthredinidae) par le flagellé *Herpetomonas swainei* sp. n. (Protozoaires). *Phytoprotection* 55,64–66.

Smirnoff, W.A. (1975) Parasites and predators as vectors of insect viruses. Baculoviruses for insect pest control: safety considerations. *American Society of Microbiology* pp. 131–133.

Smirnoff, W.A.; Bernier, B. (1973) Increased mortality of the Swaine jack pine sawfly and foliar nitrogen concentrations after urea fertilization. *Canadian Journal of Forestry Research* 3,112–121.

Smirnoff, W.A.; Lipa, Jerzy J. (1970) *Herpetomonas swainei* sp. n. a new flagellate parasite of *Neodiprion swainei* (Hymenoptera: Tenthredinidae). *Journal of Invertebrate Pathology* 16,187–195.

Smirnoff, W.A.; Valéro, J. (1975) Effets à moyen terme de la fertilisation par urée ou par potassium sur *Pinus banksiana* L. et le comportement de ses insectes dévastateurs: tels que *Neodiprion swainei* (Hymenoptera: Tenthredinidae) et *Toumeyella numismaticum* (Homoptera: Coccidae). *Canadian Journal Forest Research* 5,236–244.

Smirnoff, W.A.; Fettes, J.J.; Haliburton, W. (1962) A virus disease of Swaine's jack pine sawfly, *Neodiprion swainei* Midd. sprayed from an aircraft. *Canadian Entomologist* 94,477–486.

Tostowaryk, W. (1971*a*) Life history and behavior of *Podisus modestus* (Hemiptera: Pentatomidae) in boreal forest in Québec. *Canadian Entomologist* 103,662–679.

Tostowaryk, W. (1971*b*) Relationship between parasitism and predation of diprionid sawflies. *Annals of the Entomological Society of America* 64,1424–1427.

Tostowaryk, W. (1971*c*) The effect of prey defense on the functional response of *Podisus modestus* (Hemiptera: Pentatomidae) to densities of the sawflies. *Neodiprion swainei* and *N. pratti banksiana* (Hymenoptera: Neodiprionidae). *Canadian Entomologist* 104,61–69.

Tostowaryk, W. (1972) Coleopterous predators of the Swaine jack pine sawfly *Neodiprion swainei* Middleton (Hymenoptera: Diprionidae). *Canadian Journal of Zoology* 50,1139–1146.

Tostowaryk, W. (1973) Population estimation and feeding behaviour of adult carabids in jack pine stands, Québec. *Canadian Forestry Service Information Report* Q-X-32, 23 pp.

Operophtera bruceata (Hulst), Bruce Spanworm (Lepidoptera: Geometridae)

W.G.H. IVES

Pest Status

The Bruce spanworm, *Operophtera bruceata* (Hulst), attacks a variety of deciduous trees (Prentice 1963), and is very similar in appearance to the winter moth, *Operophtera brumata* (L.). Adults of the two species can be distinguished by differences in genitalia in males and in vestigial wing length in females (Eidt *et al.* 1966). Pupae can be separated by differences in the form of the cremaster, and larvae can be distinguished by the placement of the ocelli (Eidt & Embree 1968). In eastern Canada, sugar maples, *Acer saccharum* Marsh., and beech trees, *Fagus grandifolia* Ehrh., are favoured hosts, but in western Canada, trembling aspen, *Populus tremuloides* Michx., seems to be the principal host, particularly in Alberta. Although the Bruce spanworm is not usually considered a major pest, a number of localized infestations, as well as more extensive outbreaks in which damage has been primarily due to this insect, have been reported by the Forest Insect and Disease Survey (Anon. 1952–79). The following account of infestation history is based on these reports, except where otherwise noted.

In central and eastern Canada, the first reported damage occurred in a small stand of sugar maple near Merrivale, Ontario, in 1951 and 1952. In Newfoundland a small infestation was reported on white birch, *Betula papyrifera* Marsh., in 1956. The first reported damage in the maritimes occurred in 1962, when a patch of sugar maple in Nova Scotia was severely defoliated, and a larger area of defoliation was reported in New Brunswick. Infestations in both Maritime Provinces increased in scope and intensity in 1963, and the outbreak reached its peak in 1964, when over 200 000 ha of severe defoliation occurred in Nova Scotia alone (Harrington 1968). Declines in populations were noted in 1965, and by 1966 the outbreak had collapsed. A short-lived outbreak also occurred in Québec, where the insect reached epidemic proportions in many areas in 1963. However, populations were lower in 1964, and only small pockets of defoliation were reported in 1965. At its peak, this outbreak covered some 39 000 km^2 (Martineau & Monnier 1966). In Ontario, where infestations of Bruce spanworm were first reported, no further damage occurred until 1963, when a small pocket of aspen defoliation was reported near Sudbury. In 1965, three widely separated areas of severe sugar maple defoliation occurred: one was in Algonquin Park; another was on an island in Lake Huron; and the third was on hilltops and ridges near Sault Ste. Marie. The outbreak decreased in intensity in 1966 and collapsed completely in 1967.

In western Canada, damage to trembling aspen attributable to the Bruce spanworm was first reported from the Obed area of Alberta in 1956. Widespread areas of defoliation, much of it severe, were reported in 1957 and again in 1958, when about 130 000 km^2 were moderately or heavily infested (Brown 1962). In 1959 some infestations had subsided, and all outbreaks collapsed in 1960. Severe defoliation of trembling aspen in northern British Columbia was reported in 1958 and 1959, but these infestations also collapsed in 1960.

No further significant damage attributable to the Bruce spanworm was reported until 1968, when extensive areas of moderate to severe defoliation of trembling aspen were again observed in Alberta. Both the intensity and size of this outbreak increased in 1969. Extensive damage also occurred in 1970, but only patches of aspen were severely defoliated in 1971. In Québec, several maple groves were infested in 1970, and these were completely stripped in 1971 and again in 1972. The insect was somewhat less abundant in 1973, and populations continued to decline in 1974 and 1975. In Ontario, pockets of severe defoliation of sugar maple, white birch, and beech were reported along the south boundary of Algonquin Park in 1974, and the outbreak expanded and intensified in 1975, apparently collapsing in 1976. Severe defoliation of trembling aspen also occurred in the Lake Nipigon area in 1976. The outbreak increased in area in 1977, but populations declined in most areas in 1978. In Alberta, severe defoliation by Bruce spanworm was again noted in the Obed area in 1978, and this outbreak continued in 1979 and 1980.

Background

Most of the reported outbreaks of Bruce spanworm have been of short duration. Consequently there was little if any tree mortality, even when defoliation was complete. One of the reasons for the short duration of the outbreaks, at least in eastern Canada, appears to be a nuclear polyhedrosis virus (NPV) which was described by Smirnoff (1964). The virus apparently causes appreciable mortality and thus hastens the collapse of the outbreaks. Although there were several reports of the virus occurring in the maritimes and in Québec, none of the Forest Insect and Disease Survey reports from western Canada mentioned virus, and no experimental assessments of its virulence could be found in the literature. A small-scale field experiment, using NPV from eastern Canada, was therefore conducted in the Obed area of Alberta in 1979.

Field Trials

A small supply of NPV was obtained by infecting larvae reared in a laboratory (J.C. Cunningham 1979 personal communications). Bruce spanworm eggs were obtained by collecting moss from around the base of infested trees near Obed in the fall of 1978. The eggs were incubated during the winter, and the emerging larvae were reared on an artificial diet in plastic cups. A total of 5 000 fifth-instar larvae was infected with virus. The dead larvae were collected and the virus purified by differential centrifugation; 4×10^{11} polyhedral inclusion bodies (PIB) were obtained. The results of the field trial have already been described (Ives & Cunningham 1980) and only a brief description of the experiment need be given here. A strip of moderately infested trembling aspen, mostly under 10 m in height, was selected for the virus trials. Various concentrations of virus in an aqueous formulation containing 50 g/l Shade® (Sandoz Inc.), 1% Chevron® sticker and 25% v/v animal feed-grade molasses were applied until liquid dripped from the foliage (about 350 l/ha) with a backpack mist blower. Virus concentrations of 10^6 and 10^7 PIB/ml were applied on 24 May. The larvae were in the first instar, the temperature was 19°C, relative humidity 42%, and wind speed 0–8 km/h. Leaves on the trees receiving 10^7 PIB/ml were 1.5–2.0 cm in diameter. The remaining trees had not yet flushed. A second application of virus, on a different group of trees, was made on 1 June. Larvae were in the second and third instars, the temperature was 21°C, R.H. 29%, and wind speed 3–11 km/h. The leaves were fully flushed, and three concentrations of virus (10^6, 10^7, and 10^8 PIB/ml) were tested.

Two methods of assessing impact were used: determination of percentage mortality by rearing of individual larvae until death or pupation, and estimation of percentage defoliation by examining large numbers of individual leaves to determine the approximate amount eaten. Neither method was completely satisfactory. Aseptic techniques were not used in the collection of larvae, and all larvae for each sample were placed in the same container, so that contamination after collection probably occurred. The foliage samples were not collected until mid-August, and there was some evidence that the petioles of the more severely damaged leaves had fallen by then. Consequently, the larval rearings may have tended to overestimate rates of infection, while the defoliation may have been underestimated. A further complication was the presence of naturally occurring virus.

Two samples of late-instar larvae were collected about 0.4 km from the experimental area. Microscopic examination of living larvae from those samples revealed that 13 and 21% respectively were infected with NPV (Ives & Cunningham 1980).

The percentages of virus-caused mortality of larvae collected 7 days after the 24 May application were 80 and 98% for larvae receiving 10^6 and 10^7 PIB/ml, compared to a mean of 52% for the two untreated check areas. Corresponding mortalities among larvae collected 10 days after the 1 June application were 94, 100, and 100% for dosages of 10^6, 10^7 and 10^8 PIB/ml respectively, compared to a mean mortality of 68% for larvae collected in the check area. All treatments afforded some degree of foliage protection, but only the 24 May application of 10^7 PIB/ml and the 1 June application of 10^8 PIB/ml provided protection that could be detected visually.

Evaluation of Control Attempt

Bruce spanworm NPV appears to be virulent, on the basis of the limited amount of information available. However, it is expensive to produce, because it must be propagated in a small host. Its use in artificially regulating populations of Bruce spanworm does not currently seem to be practical, partly because of cost and partly because the rate of natural virus infection usually appears to terminate outbreaks before appreciable damage occurs.

Recommendations

Ultimately, this NPV might have potential as an applied biological insecticide, but this will not occur unless more economical methods for propagating insect viruses are developed. No further testing of this virus is recommended in the immediate future.

Literature Cited

Anonymous (1952–1979) Annual reports of the Forest Insect and Disease Survey, 1951–1976. Ottawa, Ontario; Canadian Forestry Service.

Brown, C.E. (1962) The life history and dispersal of the Bruce spanworm, *Operophtera bruceata* (Hulst) (Lepidoptera: Geometridae). *Canadian Entomologist* 94,1103–1107.

Eidt, D.C.; Embree, D.G. (1968) Distinguishing larvae and pupae of the winter moth, *Operophtera brumata*, and the Bruce spanworm, *O. bruceata* (Lepidoptera: Geometridae). *Canadian Entomologist* 100,536–539.

Eidt, D.C.; Embree, D.G.; Smith, C.C. (1966) Distinguishing adults of the winter moth, *Operophtera brumata* (L.), and Bruce spanworm, *O. brumata* (Hulst) (Lepidoptera: Geometridae). *Canadian Entomologist* 98,258–261.

Harrington, W. (1968) Relation of light trap catches of Bruce spanworm to infestations in central Nova Scotia. *Canadian Department of Fisheries and Forestry Bi-monthly Research Notes* 24(5),39–40.

Ives, W.G.H.; Cunningham, J.C. (1980) Application of nuclear polyhedrosis virus to control Bruce spanworm (Lepidoptera: Geometridae). *Canadian Entomologist* 112,741–744.

Martineau, R.; Monnier, C. (1966) Recent outbreak of the Bruce spanworm in Québec. *Canadian Department of Forestry and Rural Development Bi-monthly Research Notes* 22(6),5.

Prentice, R.M. (1963) Forest Lepidoptera of Canada recorded by the Forest Insect Survey. *Canadian Department of Forest Entomology and Pathology Branch, Ottawa, Publication* 1013.

Smirnoff, W.A. (1964) A nucleopolyhedrosis virus of *Operophtera bruceata* (Hulst) (Lepidoptera: Geometridae). *Journal of Insect Pathology* 6,384–386.

Operophtera brumata (L.), Winter Moth (Lepidoptera: Geometridae)

D.G. EMBREE and I.S. OTVOS

Pest Status

The winter moth, *Operophtera brumata* (L.), was accidentally introduced into Nova Scotia, probably in the mid 1930s, and later spread to New Brunswick and Prince Edward Island. Recently it has been found in British Columbia, and in western Oregon in the United States.

Populations of the insect in the maritimes have remained consistently low during the 20-year review period because of the establishment and population build-up of the tachinid parasitoid *Cyzenis albicans* (Fall.) and the ichneumonid parasitoid *Agrypon flaveolatum* (Gravely) in 1961 (Embree 1971*a*). The gradual spread of the insect since it was first detected in 1949 has virtually ceased and it is almost non-existent in forested areas. However, populations have persisted in apple, *Malus* spp., orchards and on shade trees, particularly oaks, *Quercus* spp., and lindens, *Tilia* spp., (Sterner & Davidson 1981). Chemical control is occasionally required in orchards but generally parasitoid numbers have remained sufficiently high to keep the winter moth under control.

In British Columbia, the winter moth is found in and around Victoria on Vancouver Island and in Richmond near Vancouver. As in Nova Scotia (Hawboldt & Cuming 1950), the insect was initially misidentified and was not suspected to be *O. brumata* until 1976, 4 years after it was causing significant damage to Garry oak, *Quercus garryana* Dougl., in Victoria. By 1977 the outbreak had expanded over large areas and an estimated 120 km^2 were severely affected in Victoria on the Saanich Peninsula. The centre of the outbreak appeared to be in the Wilkinson Road area in Victoria where several commercial nurseries are located, suggesting that the insect might have been introduced into British Columbia on nursery stock (Gillespie & Finlayson 1981). In British Columbia the larvae were found to feed on *Acer, Crataegus, Malus, Prunus, Populus, Quercus*, and *Salix* spp., and other deciduous trees and shrubs and to pose a threat to commercial orchards and shade trees in urban areas. Parasitism by native species of parasitoids is less than 5% (Gillespie *et al.* 1978).

In the United States, outbreaks of this pest are confined to Multnomach, Clackamus, and Washington counties in western Oregon, where the insect is found in abandoned orchards and is becoming a pest of filberts, *Corylus* spp.

Background

Biological control of the winter moth in the maritimes by the introduced parasitoids *C. albicans* and *A. flaveolatum* has been successful and remains so because it is helped by the behaviour of the winter moth which results in most of the host larvae starving before suitable food is available. Most of the time, winter moth hatching occurs before budburst of its principal forest host, red oak, *Quercus rubra* L., and other forest species, because less heat is required for hatching than for budburst. Over 98% of the newly hatched larvae may starve because they cannot feed on closed buds and the few survivors are then controlled by the parasitoids (Embree 1965).

On hatching, larvae initially spin downwards and are then carried up to the tree foliage on rising air currents. Such currents are most apt to occur in the early morning when the air begins to warm and rise. By mid-morning, when the land is fully heated, wind patterns develop. Winter moth hatching is synchronized to this pattern of wind movement, with hatching beginning at dawn and virtually ceasing by mid-morning. By this efficient means larvae are transported from hatching sites in crevices and lichens on the lower tree trunks to the tree crowns. Once there, the larvae either survive or die depending on whether the buds have opened. The wind may also transport the larvae several miles, but once established on the foliage, they do not dislodge easily until the late fifth instar in early summer, when they drop to the ground to pupate (Embree 1970). No other means of dispersal exists because the females, which are active in November and December, have vestigial wings and are flightless.

Releases and Recoveries

No releases of parasitoids have been made in the maritimes since 1965 (Embree 1971*b*). Releases of parasitoids in British Columbia were conducted in 1979 and 1980 and are listed in Table 91.

The British Columbia operation was financed and administered by the Crop Protection Branch (formerly Entomology and Plant Pathology Branch) of the British Columbia Ministry of Agriculture and Food in co-operation with the Pacific and Maritimes Forest Research Centres of the Canadian Forestry Service. Collections of last instar larvae were made in apple orchards in Nova Scotia in 1978 and 1979. Collections were also made in Germany by the Commonwealth Institute of Biological Control.

Winter moth pupae were sent to the Biocontrol Unit of the Research Branch of Agriculture Canada in Ottawa, where pupae were cold-treated and then reared under quarantine conditions. Only the adult parasitoids were shipped to British Columbia from the German collection but about two-thirds of the Nova Scotia collection were sent to Victoria for parasitoid rearing. The remainder were reared in Ottawa and the adults of the two parasitoids were sent to Victoria.

Based on studies in Nova Scotia, the control strategy in British Columbia has been to treat the infested area with repeated releases, each consisting of a minimum of 50 mated females of each species.

Agrypon flaveolatum (Grav.) (Hymenoptera: Ichneumonidae)

This parasitoid becomes decreasingly effective as host densities increase, and is therefore most effective at low population density (Embree 1966). The adult parasitoids develop in winter moth pupae and emerge the following spring to parasitize early instars of host larvae.

The first introduction of *A. flaveolatum* was in 1979 when a total of 1 654 males and 1 700 females was released at 33 locations in the greater Victoria area in stands infested with winter moth. Fifty pairs of parasitoids were released in each location after being held in mating cages for 1–3 days. The parasitoids in the cages were provided with twigs of cherry blossoms and bottles of sugared water with feeding wicks. In 1980 an additional 2 213 males and 2 813 females were released at 25 locations; of these a small proportion came from Germany, and the rest from Nova Scotia. This parasitoid has not yet been recovered in British Columbia.

Cyzenis albicans (Fall.) (Diptera: Tachinidae)

This parasitoid exhibits a sigmoid functional response curve and is most effective at high host densities (Embree 1966). The female, which can lay as many as 1300 eggs, oviposits in the vicinity of feeding damage by defoliators. Associated species such as the fall cankerworm, *Alsophila pometaria* (Harr.), which are resistant to parasitism by *C. albicans*, also consume the parasitoid eggs. With increasing winter moth populations *C. albicans* becomes increasingly effective.

Table 91 Number of *Cyzenis albicans* (Fall.) and *Agrypon flaveolatum* (Grav.) released in the greater Victoria region up to 1980 against the winter moth, *Operophtera brumata* (L.)

	1979				1980			
	Cyzenis albicans		*Agrypon flaveolatum*		*Cyzenis albicans*		*Agrypon flaveolatum*	
Location	M	F	M	F	M	F	M	F
1. Gonzales Weather Station (302 Denison Rd.)	150	148	50	50	200	200	64	72
2. S. of Royal Oak Shopping C. (791, 845 Mann Ave.)	150	147	50	50	150	184	54	170
3. Summit Park Reservoir (3 blocks off Hillside & Cook)	150	150	97	95	224	75	54	169
4. Central Park			50	50	215	90	64	73
5. Mt. Tolmie Park	150	150	50	50	505	956	269	461
6. Cattle Point (on Beach Drive)	150	134	50	50	252	365	223	170
7. Pemberton Park	141	150	50	50	150	150	50	50
8. Oakland School & Park	210	157	50	50				
9. 633 Taylor (off William Head)			50	39				
10. Uplands Rd. at Cotswald Rd.	53	65	50	50	150	150	236	120
11. Playfair Park	200	155	50	50	150	150		
12. Christmas Hill (Sacred Heart Church)	150	150	45	45	150	150	46	73
13. 4021 Wilkinson Rd.	159	180	45	46	150	150	100	100
14. Stelly's X-Rd. & W. Saanich Rd.			50	50	150	150	62	65
15. E. Saanich Rd. – Mt. Newton Rd.			50	49	150	150	62	65
16. Colville Rd. — E. of Admiral Rd.			50	50	215	250		
17. Highrock Park	150	150	50	50	215	250	78	115
18. W. Saanich Rd. & Markham Rd.	50	141	50	50	180	180	20	20
19. 5271 Old W. Saanich Rd.	133	138	60	60	306	302	152	318
20. Claremong & Del Monte Ave.			50	50	640	502	50	50
21. Cook & Beach Drive			50	50				
22. W. Burnside & Interurban			50	50	355	445	78	108
23. Yeta Place & Mount View			50	50				
24. Torquay & Kenmore			50	50				
25. Conway & Interurban			50	50			86	95
26. St. Ann & Cranmore			50	50	150	150	50	50
27. Craigdarroch & Joan Cres.			50	50	258	290	48	57
28. Royal Oak & Falaise Dr.			50	50			72	100
29. Blenkinsop & Lohbrunner			50	50	140	130	50	50
30. Blenkinsop & McKenzie			50	50	150	150	50	50
31. Feltham & Gordon Hd. Rd.			50	50				
32. 8th St. & Orchard, Sydney, near Victoria			50	50	150	150	95	112
33. Carey Rd. & Marigold			57	66			100	100
Sub-total	1996	2015	1654	1700	5355	5719	2213	2813
Grand-total	4011		3354		11 074		5026	

In 1979, 159 males and 180 females from Europe and 1 837 males and 1 835 females from Nova Scotia were released at 14 locations in the greater Victoria area. Parasitoids were kept in mating cages (150 pairs per cage) for 1–3 days before release to ensure mating, and about 150 pairs per site were released.

In 1980, 5 355 males and 5 719 females, all from Nova Scotia, were released at 24 locations (including most of the 1979 release sites). To date there has been no positive recovery of this parasitoid.

Pathogens

In 1979, three concentrations of nuclear polyhedrosis virus (NPV), 10^6, 10^7, and 10^8 polyhedral inclusion bodies (PIB) per millilitre, were used as early sprays (just after larval hatch and before budbreak) and late sprays (8 days later after the buds flushed) against the winter moth in British Columbia. They were applied at the rate of 1 l/tree using a backpack mist blower. The best result was obtained with the early spray formulation containing 10^8 PIB/ml. Winter moth mortality was 46% and defoliation was significantly reduced. However, the cost of virus production and application of the spray is high and at present the virus is uneconomical as a biological control agent, especially as the virus application does not have a carry-over effect for the following year (Cunningham *et al.* 1981).

Dipel® (Abbot Laboratories) and Thuricide® (Sandoz Inc.) formulations of *Bacillus thuringiensis* were also tested against winter moth larvae in British Columbia in 1977. Although both of these sprays caused some larval mortality, neither prevented complete defoliation of the treated trees (Tonks *et al.* 1980).

Evaluation of Control Attempts

Close study and monitoring of biological control of the winter moth in the maritimes was suspended in 1970. However, population trends are followed as part of the normal programme of the Forest Insect and Disease Survey of the Maritimes Forest Research Centre. A collection of 20 000 winter moths was made in Nova Scotia during 1981 to obtain the parasitoids *A. flaveolatum* and *C. albicans* for release in western Oregon. Twenty years have elapsed since the initial collapse of the winter moth outbreak in the Maritime Provinces. *C. albicans* and *A. flaveolatum* are still present, and apparently effective, in the residual winter moth population.

In British Columbia the release sites are being monitored by the Pacific Forest Research Centre through mass collections of larvae and additional behavioural and population studies of *O. brumata, C. albicans,* and *A. flaveolatum*. It is still too early to know whether the parasitoids will become established.

Recommendations

Introductions of other species of parasitoids into winter moth populations in orchards might upset the present favourable host parasitoid balance in the Maritime Provinces and therefore should not be attempted.

In British Columbia no further releases are recommended until it has been determined whether *A. flaveolatum* and *C. albicans* have become established. Population studies and accurate population estimates are required to determine whether the parasitoids are effective if and when they become established.

Use of viruses offers potential for control. Questions to be answered include whether an NPV occurs naturally in British Columbia populations and whether it can be introduced into NPV-free populations. However, before operational control attempts are investigated further, work on the role of parasitoids introduced into British Columbia should be completed.

Applications of pathogens and parasitoids could be combined in some cases, particularly on early-leafing tree species where leafing is well synchronised with winter moth egg hatching.

Literature Cited

Cunningham, J.C.; Tonks, N.V.; Kaupp, W.J. (1981) Viruses to control winter moth *Operophtera brumata* (Lepidoptera: Geometridae), *Journal of the Entomological Society of British Columbia* 78,17–24.

Embree, D.G. (1965) The population dynamics of the winter moth in Nova Scotia; 1954–62. *Memoirs of the Entomological Society of Canada* 46, 57 pp.

Embree, D.G. (1966) The role of introduced parasitoids in the control of the winter moth in Nova Scotia. *Canadian Entomologist* 98,1159–1168.

Embree, D.G. (1970) The diurnal and seasonal pattern of hatching winter moth eggs, *Operophtera brumata* (Geometridae: Lepidoptera). *Canadian Entomologist* 102,759–768.

Embree, D.G. (1971*a*) The biological control of the winter moth in eastern Canada by introduced parasites. In: Huffaker, C.B. (*Ed.*) Biological control New York; Plenum Press, pp. 217–226.

Embree, D.G. (1971*b*) *Operophtera brumata* (L.), winter moth (Lepidoptera: Geometridae). *In*: Biological control programmes against insects and weeds in Canada, 1959–68. *Commonwealth Institute of Biological Control Technical Communication* 4,167–175.

Gillespie, D.R.; Finlayson, T. (1981) Final-instar larvae of native hymenopterous and dipterous parasites of *Operophtera* spp. (Lepidoptera: Geometridae) in British Columbia. *Canadian Entomologist* 113,45–55.

Gillespie, D.R.; Finlayson, T.; Tonks, N.V.; Ross, D.A. (1978) Occurrence of the winter moth, *Operophtera brumata* (Lepidoptera: Geometridae), on southern Vancouver Island, British Columbia. *Canadian Entomologist* 110,223–224.

Hawboldt, L.S.; Cuming, F.G. (1950) Cankerworms and European winter moth in Nova Scotia. *Canadian Department of Agriculture, Forest Biology Division Bi-monthly Progress Report* 6(1),1–2.

Sterner, T.E.; Davidson, A.G. (1981) Forest insect and disease conditions in Canada 1980. *Environment Canada Publication,* Catalogue No. Fo. 21-1/1980, 43 pp.

Tonks, N.V.; Everson, P.R.; Theaker, T.L. (1980) Efficacy of insecticides against geometrid larvae, *Operophtera* spp., on southern Vancouver Island, British Columbia. *Journal of the Entomological Society of British Columbia* 75,6–9.

Orgyia leucostigma (J.E. Smith), Whitemarked Tussock Moth (Lepidoptera: Lymantriidae)

D.G. EMBREE, D.E. ELGEE and G.F. ESTABROOKS

Pest Status

The whitemarked tussock moth, *Orgyia leucostigma* (J.E. Smith), is a noxious insect in eastern Canada, particularly in Nova Scotia. It is primarily a forest pest, capable of killing entire stands of mature balsam fir, *Abies balsamea* (L.) Mill., in a year and of defoliating extensive areas of hardwoods. It is particularly devastating to balsam fir Christmas trees which it deforms, defoliates, or kills. The whitemarked tussock moth is also a pest in urban areas where it feeds on a wide variety of hardwood shade trees, crawls over houses and lawns, and spins unsightly cocoons on trees and shrubs. Some people are allergic to the larval setae. Moreover, *O. leucostigma* is a pest of blueberries and is the target of an extensive spray programme to protect the commercial crop.

Egg masses are laid on empty female cocoons and covered with froth. On hatching, larvae feed gregariously on the froth. Once this is consumed, larvae spin downwards on silken threads and are carried great distances by the wind. On occasion, during severe outbreaks, swarms of airborne larvae appear as a grey mist and the insects can cover the surface of small lakes and ponds when they land.

The larvae feed on a wide variety of hardwoods and conifers such as tamarack, *Larix laricina* (Du Roi) K. Koch, but their primary host is balsam fir. Larvae feed in the shade on the underside of leaves and twigs. This feeding habit causes a unique form of damage to balsam fir Christmas trees because larvae also feed on the thin bark of developing leaders and upper lateral branches causing them to warp downwards. The result is a pronounced crook in the stems of surviving trees, visible for years afterwards. In heavy infestations, trees are festooned with silk.

Pupation usually occurs on the underside of twigs and branches or on the stems of host trees, but it can occur almost anywhere, the yellowish white cocoons often being spun on fences, houses, or logs. Females remain on the surface of their cocoons when they emerge, mate with winged males, and lay from 100 to over 500 eggs.

Background

Outbreaks are controlled ultimately by a nuclear polyhedrosis virus (NPV). These outbreaks of *O. leucostigma*, at least in Nova Scotia, appear to originate in balsam fir stands at high elevations and occur at intervals of 6–10 years between outbreaks. The infestation spreads by wind-borne larvae and eventually large areas of balsam fir and hardwoods are defoliated. Such outbreaks become widespread and last for 5–6 years, then collapse because of the gradual spread of the virus disease. The initial infestation collapses in about 3 years, subsequent infestations in 2–3 years, and in the final stages of the outbreak infestations appear and collapse in the same year.

Over 25 species of parasitoids attack the whitemarked tussock moth but the only control they have over populations is to kill survivors following the collapse of outbreaks caused by NPV. As a result, whitemarked tussock moth populations are reduced suddenly to extremely low levels, which delays the build-up of subsequent outbreaks.

No attempt has been made to control this insect through the introduction of additional parasitoid species. The emphasis has been on establishing control through the manipulation of the virus (Cunningham 1972, Elgee 1975). Virus suspensions can be used as insecticides and have been tested successfully on small trees.

Field Trials

A major attempt at biological control was made in 1975 to initiate an epizootic in a large population of the whitemarked tussock moth in the hope of shortening the length of an outbreak; it has been reported by Kurstak (in press). During the previous year the suspected start of a general outbreak of the whitemarked tussock moth was detected in an 18-ha stand of balsam fir near Castlereagh, Colchester County, Nova Scotia. A virus suspension was prepared from third-instar larvae reared from eggs collected at the site and infected with the Nova Scotia strain of NPV: the virus had been saved and stored from the last outbreak in 1968.

On the evening of 9 July 1975, 204 l of an NPV suspension containing 10.8×10^6 polyhedral inclusion bodies (PIB) per millilitre mixed with 204 l of a virus spray adjuvant manufactured by Sandoz Inc. and a fluorescent dye were sprayed from 30 m on 18 ha of mature balsam fir with a Cessna Ag-truck aircraft equipped with a boom and nozzle sprayer. Whitemarked tussock moth larvae were mainly first instars with some early second instars.

There were no true controls (untreated areas) but larvae collected before the spray application and reared were free from disease. Twenty-four sample trees were established in a line bisecting the spray block. However, one edge of the block was not sprayed so that trees 20–24 were outside the block. No disease was detected in larvae collected on tree 24 and only light mortality (33%) occurred on trees 20–23. Mortality of larvae collected 1 day after spraying from trees 1–19, and reared on foliage from the sprayed area, averaged 84%, whereas that of survivors collected 2 weeks later was 64%. Field populations declined by 95%, as compared to 67% on the unsprayed trees, and defoliation was light. From the air, the outline of the sprayed area was clearly discernible in a severe outbreak that developed around the initial Castlereagh infestation. No appreciable population reappeared in the sprayed area. The outbreak in the surrounding areas intensified in 1976, covering a radius of 16 km, but collapsed from the virus disease in all but the periphery during the same year.

Early in 1976, virus suspensions were dispersed at two locations along the periphery of the 1975 outbreak using an explosive device. Canisters containing 1.9 l of suspension were fired into the air using a home-made mortar and exploded over the forest canopy. Virus particles were dispersed over an area of 1000–2000 m^2 depending on wind velocity. However, by this time the virus was already present in the population. Eventually the outbreak spread throughout the province, collapsing everywhere by 1978.

Evaluation of Control Attempts

The experiment showed that the virus can be used as an effective biological insecticide, but whether it can be used to initiate a widespread epizootic is uncertain. The rapid spread of the *O. leucostigma* infestation in 1975 resulted from dispersion of newly hatched larvae, most of which occurred before spraying took place. Moreover, low populations of larvae existed in the area immediately surrounding the initial 18 ha of detected outbreak. The outbreak in 1976 was widespread and extremely intense; even though it collapsed just 2 years after it appeared, it is unlikely that the source of the epizootic was the 18-ha spray block. Had the initial outbreak been detected in 1973 and sprayed in 1974 a clearer understanding of the role of the virus in whitemarked tussock moth epidemiology might have been gained.

The pattern of development of whitemarked tussock moth outbreaks presents an opportunity for an elegant approach to biological control; but there are major difficulties. The most promising strategy is to artificially accelerate the normal virus epizootic that has occurred consistently in every past outbreak of this insect. The obvious advantage of such a strategy is that the virus is used as an inoculum rather than as a blanket insecticide. The resulting epizootic will be one that inevitably would have occurred later, a consideration that might ease the registration approval of this NPV by Canadian authorities.

Recommendations

Historically each whitemarked tussock moth outbreak in Nova Scotia appears to have begun in one or two small locations (epicentres). Virus material from the 1974–79 outbreak has been stored and the techniques of introducing the virus into a developing outbreak have been shown to be workable. However, two major problems must be addressed: first, early detection of epicentres requires constant monitoring of likely sites for a period of up to 10 years; and second, whitemarked tussock moth hatching occurs over a period of at least 2 weeks, so although larvae feed on the froth covering of egg masses for 2 or 3 days, many larvae are likely to disperse before the virus is applied during peak egg hatching.

A second strategy might be to employ sex pheromones that could be used to disrupt mating in epicentres. Success would depend on the almost immediate detection of these epicentres before they become large enough to develop a reservoir of dispersing larvae.

Literature Cited

Cunningham, J.C. (1972) Preliminary studies of nuclear polyhedrosis viruses infecting the whitemarked tussock moth, *Orgyia leucostigma Canadian Forestry Service Insect Pathology Research Institute, Sault Ste. Marie, Ontario, Information Report.*

Elgee, D.E. (1975) Persistence of a virus of the whitemarked tussock moth on balsam fir foliage. *Canadian Forestry Service Bi-monthly Research Notes* 31(5),33-34.

Kurstak, E. (in press) Microbial and viral pesticides. New York; Marcel Dekker.

Orgyia pseudotsugata (McDunnough), Douglas-fir Tussock Moth (Lepidoptera: Lymantriidae)

J.C. CUNNINGHAM and R.F. SHEPHERD

Pest Status

The Douglas-fir tussock moth, *Orgyia pseudotsugata* (McDunn.), occurs in the semi-arid interior of British Columbia and in parts of Washington, Idaho, California, Nevada, Colorado, Arizona, and New Mexico in the United States. In Canada, Douglas fir, *Pseudotsuga menziesii* (Mirb.) Franco, is the preferred host, but Engelmann spruce, *Picea engelmannii* Parry, is also attacked and later instar larvae will feed on ponderosa pine, *Pinus ponderosa* Laws., when forced to abandon totally defoliated Douglas fir. Small larvae eat the underside of new needles. The later instar larvae may eat entire older needles or sever them near the base and leave them entwined in silk webbing. Heavily infested trees have a reddish-brown appearance. Trees may die after 1 year of defoliation, but mortality occurs more commonly after 2 or more years of severe defoliation (Johnson & Ross 1967). Severely damaged trees are also susceptible to attack by the Douglas-fir beetle, *Dendroctonus pseudotsugae* Hopkins.

Since 1916, when the first records were kept (Sugden 1957), six outbreaks of Douglas-fir tussock moth occurred in British Columbia. The latest outbreak began in 1971 and collapsed in 1976; a further outbreak was predicted for 1981 (Sterner & Davidson 1981).

Background

Characteristically, high populations of Douglas-fir tussock moth appear in relatively small patches of forest of 1–5 ha. The infestation normally extends from these areas in subsequent years, but more striking is the appearance of new outbreaks often many kilometres from the original sites. After an average of 5 years (3–6) outbreaks collapse, the factors responsible being nuclear polyhedrosis virus (NPV), egg parasitoids, and starvation. However, following several years of severe defoliation, the trees frequently die before this population collapse occurs. Hence attempts to regulate the population with biological or chemical agents should be made early in the outbreak cycle.

Field Trials

Nuclear polyhedrosis virus

Two types of nuclear polyhedrosis virus (NPV) have been found in populations of Douglas-fir tussock moth. In one type the rod-shaped virus particles are embedded singly in polyhedral inclusion bodies (PIB) and in the other they are embedded in bundles (Hughes & Addison 1970). These are referred to as single-embedded NPV (SNPV) and multiple-embedded NPV (MNPV). An SNPV isolated from the whitemarked tussock moth, *O. leucostigma* (J.E. Smith), is also pathogenic for Douglas-fir tussock moth larvae.

Small-scale ground-spray trials in 1962 with field-collected NPV (possibly a mixture of SNPV and MNPV) gave encouraging results (Morris 1963). The first aerial spray trial in British Columbia was conducted in 1974 using Douglas-fir tussock moth MNPV propagated in whitemarked tussock moth larvae. A 10-ha plot 35 km west of Kamloops was sprayed using a fixed-wing aircraft with boom and nozzle equipment. Two applications were made on second-instar larvae using a dosage of 250 x 10^9 PIB/ha and an emission rate of 18.8 l/ha. The aqueous suspension contained 3% IMC 90-001 (International Minerals Corp.) UV protectant and 0.001% Chevron® sticker (Chevron Chemical Co.). Twenty-three days after spraying, the population reduction due to treatment (Abbott 1925) was 96.8% (Tripp & Cunningham unpublished).

In 1975, the United States Forest Service, the Canadian Forestry Service, and the British Columbia Forest Service collaborated in conducting and evaluating aerial spray trials with NPV near Kamloops, British Columbia. A total of 1 071 ha were treated with three applications repeated three times on 8-ha plots and a further four non-repeated applications on larger blocks, which ranged in size from 80 to 507 ha. The same dosage, 250 x 10^9 PIB/ha, was used in all treatments and three different virus stocks, four tank mixes, and two emission rates were evaluated (Table 92) (Stelzer *et al.* 1977, Shepherd 1980). The three virus stocks were Douglas-fir tussock moth MNPV propagated in its homologous host, the same virus propagated in whitemarked tussock moth larvae, and whitemarked tussock moth SNPV propagated in its homologous host.

Table 92 Application of viruses on Douglas-fir tussock moth, *Orgyia pseudotsugata* (McDunn.), second- and third-instar larvae in 1975 using a fixed-wing aircraft fitted with boom and nozzle spray equipment

Treat-ment no.	Virus*	Host for virus production	Area treated (ha)	Tank mix	Emitted volume (l/ha)	Droplets/cm^2 on Kromekote cards
1	MNPV	*Orgyia pseudotsugata*	8 x 3**	25% molasses 6% Shade® (Sandoz Inc.)	18.8	38
2	MNPV	*O. leucostigma*	8 x 3	50% Sandoz® adjuvant V	18.8	48
3	SNPV	*O. leucostigma*	8 x 3	50% Sandoz® adjuvant V	18.8	80
4	MNPV	*O. pseudotsugata*	80	50% Sandoz® adjuvant V	9.4	19
5	MNPV	*O. leucostigma*	80	50% Sandoz® adjuvant V	9.4	12
6	MNPV	*O. pseudotsugata*	507	25% molasses 6% Shade®	9.4	27
7	SNPV	*O. leucostigma*	320	25% molasses	9.4	32

* MNPV, multiple-embedded nuclear polyhedrosis virus.
SNPV, single-embedded nuclear polyhedrosis virus.
** 8 ha treated three times.

A fixed-wing aircraft fitted with a boom and nozzle spray system was used to apply the viruses when larvae were in the second and third instars. Because of high larval populations, most of the current year's needles had already been destroyed. Results are shown in Table 93 and, with the exception of the SNPV treatment on the large block, excellent foliage protection and population control were achieved. Levels of virus infection were determined by rearing larvae in the laboratory until pupation or death occurred. Dead larvae were examined microscopically for the presence of NPV. The reduced efficacy of the one SNPV treatment was attributed to high winds during spray application (9.6 km/h gusting to 19 km/h), which resulted in poor spray deposit on the leeward side of the trees. No Douglas-fir tussock moth cocoons or egg masses were found in any of the virus-treated areas, but large numbers were found in untreated check plots and adjacent unsprayed areas. High levels of naturally occurring NPV were found in the untreated areas, but this epizootic occurred too late in the season to prevent

Table 93 Impact of NPV sprayed on Douglas-fir tussock moth, *Orgyia pseudotsugata* (McDunn.), in 1975 (For details see Table 92)

Treatment no.	Pre-spray population density larvae/6450 cm² foliage	Percentage population reduction due to treatment* Days post-spray 28	35	Percentage larvae infected with NPV Days post-spray 21	28	35	Percentage of older foliage saved**
1	618	72.6	95.6	74.8		92.3	82.6
2	274	80.2	88.8	67.3		78.1	87.6
3	392	75.1	78.8	62.0		80.8	84.3
4	330	59.5		94.0	90.8		81.3
5	338	60.8		80.5	88.8		92.1
6	331	68.7	84.4		94.6		93.4
7	288	0	0		65.7		0
Check	493			11.7		65.7	

* Calculated by Abbott's formula.

** Percentage foliage saved $= \dfrac{(\text{Expected defoliation - observed defoliation}) \times 100}{\text{Expected defoliation}}$

severe damage to the trees and was not intense enough to regulate the insect population the following year.

In 1976, the efficacy of SNPV propagated in 1975 in field populations of Douglas-fir tussock moth was compared with the same virus propagated in laboratory-reared whitemarked tussock moth. The two treatments were applied on 2-ha plots near Kamloops using a fixed-wing aircraft with boom and nozzle equipment. The aqueous tank mix contained 25% molasses, the dosage of each virus stock was 125 x 10^9 PIB/ha, and the emitted volume was 18.8 l/ha. Larvae were in the first and second instars at the time of application and, 38 days after spraying, the population reduction due to treatment (Abbott's formula) was 81.8% for the field produced NPV and 89.7% for the laboratory produced NPV (Ilnytzky *et al.* 1977).

***Bacillus thuringiensis* Berliner**

In 1975, aerial spray trials were also conducted with commercial preparations of *Bacillus thuringiensis* Berliner *(B.t.)*, as Thuricide® HPC (Sandoz Inc.) and Dipel® SC (Abbot Laboratories). A single application was made on 12 488 ha and a double application on 5 245 ha. A fixed-wing aircraft fitted with boom and nozzle spray equipment was used. The treatments were on second-instar larvae and are listed in Table 94 (Shepherd 1980).

Table 94 Application of *Bacillus thuringiensis* Berliner on Douglas-fir tussock moth, *Orgyia pseudotsugata* (McDunn.), second-instar larvae using a fixed-wing aircraft fitted with boom and nozzle equipment

Treatment no.	Status	Material	No. of applications	Volume of spray per application (l/ha)	Dosage I.U. x 10^9/ha each application	Anti-evaporant added	Percentage population reduction due to treatment
1	Experimental	Dipel® SC	1	4.74	25.5	25% molasses	48
2	Experimental	Dipel® SC	1	4.74	17.0	25% molasses	58
3	Operational	Dipel® SC	1	3.56	17.0	30% sorbitol	12
4	Operational	Dipel® SC	2	3.56	17.0	30% sorbitol	32
5	Experimental	Thuricide® HPC	1	9.47	17.0	None	57
6	Operational	Thuricide® HPC	1	4.74	17.0	None	34
7	Operational	Thuricide® HPC	2	4.36	17.0	None	56

Some of the treatments were designated operational and others experimental.

Twenty-one days after spraying, larval mortality due to the *B.t.* treatments averaged 34% with a high of 57% with Thuricide® and a low of 13% with Dipel®. A double application of both materials and a higher volume of Thuricide® gave about 20% more control than single applications or the lower volume. Increasing the dosage of Dipel® gave no significant change, but using molasses in the tank mix instead of sorbitol gave a significant increase in the effectiveness of Dipel® and brought it to the same level as Thuricide®.

Neither single nor double applications of Dipel® provided adequate foliage protection — 59% defoliation was recorded on untreated check plots and 50–58% on Dipel®-treated plots. Better foliage protection was recorded with Thuricide® with only 18–25% defoliation on treated trees.

The various operational and experimental treatments with *B.t.* gave a wide range of results. However, the best population reductions due to treatment were less than 60%. This did not provide adequate foliage protection and did not prevent the Douglas-fir tussock moth population increasing to high densities in the next generation.

Evaluation of Control Attempts

Results with *B.t.* on Douglas-fir tussock moth in 1975 were considered unsatisfactory and applications of *B.t.* on other defoliating lepidopterous forest pests during the last decade in British Columbia have generally proved disappointing. Recent improvements in application technology of *B.t.* have resulted in improved control of spruce budworm, *Choristoneura fumiferana* (Clem.), and similar improvements may eventually be possible with Douglas-fir tussock moth. On the other hand, results with NPV are already most encouraging and it appears that this biological control agent can now provide a useful tool for the regulation of Douglas-fir tussock moth populations.

Recommendations

Douglas-fir tussock moth MNPV was registered by the Environmental Protection Agency in the United States in 1975 under the name TM Biocontrol-1. It is proposed to apply for Canadian registration in 1982. At present this virus is not commercially produced and this poses two major problems that need solving: a source of supply; and production at a reasonable cost. Small amounts, sufficient to treat about 400 ha annually, are produced at the Forest Pest Management Institute, Sault Ste. Marie, Ontario, and larger amounts are produced for use in the United States by staff of the U.S. Forest Service, Corvallis, Oregon. A figure of $40 US per hectare was quoted for production at Corvallis in late 1980 (M.E. Martignoni personal communication) and this figure is probably even higher for material from Sault Ste. Marie. This cost figure is based on a dosage of 250×10^9 PIB/ha. This dosage can probably be reduced to 125×10^9 PIB/ha (Ilnytzky *et al.* 1977), and possibly even lower.

The mountainous terrain and consistently low relative humidity encountered in the interior of British Columbia pose problems for aerial application of pest control agents and particularly for application of aqueous spray formulations. Use of an oil-based formulation should be investigated, for this may well enhance the deposit and facilitate application of lower dosages.

Ideally, application of a virus initiates an epizootic in the pest insect population, regulating the pest either in the year of application if applied on early-instar larvae, or in the subsequent year. Some NPVs have the potential to initiate epizootics and others have not; the status of Douglas-fir tussock moth NPV has not been established and the virus has been applied in the same manner as a chemical pesticide. Although adequate control has been achieved, the material is too scarce and too costly to be applied in this way. Alternative methods should be investigated, such as seeding the NPV into the insect population. This could be tested by spraying widely spaced swaths to give partial coverage of the infested forest. Spread of the disease would then be relied on to regulate the Douglas-fir tussock moth population.

Among the few viruses currently being developed as biological control agents of forest insect pests in Canada, Douglas-fir tussock moth NPVs have proved to be effective pest management tools. Further testing and an attempt to register MNPV in Canada for operational use are strongly advocated. Improved production methods, lower dosage rates, and different strategies for its use may reduce the cost of treatment and make MNPV more attractive to forest managers.

Further efficacy trials with *B.t.* may be warranted, as better strains and improved application technology become available. This work, however, should have lower priority than the virus work described above.

Literature Cited

Abbott, W.S. (1925) A method of computing the effectiveness of an insecticide. *Journal of Economic Entomology* 18,265–267.

Hughes, K.M.; Addison, R.B. (1970) Two nuclear polyhedrosis viruses of the Douglas-fir tussock moth. *Journal of Invertebrate Pathology* 16,196–294.

Ilnytzky, S.; McPhee, J.R.; Cunningham, J.C. (1977) Comparison of field-propagated nuclear polyhedrosis virus from Douglas-fir tussock moth with laboratory-produced virus. *Canadian Department of Fisheries and Forestry Bi-monthly Research Notes* 33(1),5–6.

Johnson, P.C.; Ross, D.A. (1967) Douglas-fir tussock moth, *Hemerocampa (Orgyia) pseudotsugata* McDunnough. *In*: Davidson, A.G.; Prentice, R.M. (Comps. and Eds.) Important forest insects and diseases of mutual concern to Canada, the United States and Mexico. *Ottawa; Canadian Department of Forestry and Rural Development,* pp. 105–107.

Morris, O.N. (1963) The natural and artificial control of the Douglas-fir tussock moth, *Orgyia pseudotsugata* McDunnough, by a nuclear polyhedrosis virus. *Journal of Insect Pathology* 5,401–414.

Shepherd, R.F. (Ed.) (1980) Operational field trials against Douglas-fir tussock moth with chemical and biological insecticides. *Canadian Forestry Service, Victoria, British Columbia, Information Report* BC-X-201, 19 pp.

Stelzer, M.; Neisess, J.; Cunningham, J.C.; McPhee, J.R. (1977) Field evaluation of baculovirus stocks against Douglas-fir tussock moth in British Columbia. *Journal of Economic Entomology* 70,243–246.

Sterner, T.E.; Davidson, A.G. (comp.) (1981) Forest insect and disease conditions in Canada 1980. *Canadian Forestry Service Forest Insect and Disease Survey,* 43 pp.

Sugden, B.A. (1957) A brief history of the Douglas-fir tussock moth, *Hemerocampa pseudotsugata* McD., in British Columbia. *Proceedings of the Entomological Society of British Columbia* 54,37–39.

Pristiphora erichsonii (Hartig), Larch Sawfly (Hymenoptera: Tenthredinidae)

W.G.H. IVES and J.A. MULDREW

Pest Status

The larch sawfly, *Pristiphora erichsonii* (Htg.), is a major pest of larches, *Larix* spp., in North America. It is the principal defoliator of tamarack, *L. laricina* (Du Roi) K. Koch, and also attacks western larch, *L. occidentalis* Nutt., and alpine larch, *L. lyallii* Parl. Plantings of the various European and Asian species and their hybrids are also vulnerable (Turnock & Muldrew 1971*a*). Outbreaks have been recorded in Canada since late in the nineteenth century. Early outbreaks have been discussed by McGugan & Coppel (1962), and Turnock & Muldrew (1971*a*) summarized depredations for the decade 1959–68. Conditions since then have been outlined in the annual reports of the Forest Insect and Disease Survey (Anon. 1970–79). The following account of infestation history from 1969 to 1980 is based primarily on these reports. Areas in which infestations of the larch sawfly have caused moderate or severe defoliation during this period are shown in Fig. 18.

In Newfoundland, tamarack stands in the western and central parts of the province were moderately or severely defoliated in 1975, but most infestations collapsed in 1976. A resurgence occurred along the west coast in 1979. In Labrador, an outbreak developed in 1975, and by 1978 some tree mortality had been reported. In the maritimes, remnants of an earlier outbreak persisted into the period, particularly in southern New Brunswick and central Nova Scotia. In eastern Prince Edward Island, infestations increased in 1970 and populations did not collapse until 1976. By this time over 30% of the trees were dead, and many more had dead tops. In Nova Scotia, populations started increasing in 1970 and by 1975 tamarack stands in most of the province were affected. Populations in the eastern half of the province collapsed in 1976, but the decline was more gradual in the western half and some infestations still persisted in 1980. In New Brunswick, larch sawfly damage during the period 1969–80 was confined to a number of relatively small pockets. In Québec, several extensive areas with medium or high populations were reported between 1974 and 1977. In most of Ontario, outbreaks from the previous decade persisted into the early part of the present period, but most populations had collapsed by 1972. However, an upsurge in populations was soon noted in southern Ontario, where severe defoliation of larch was reported from 1974 to 1978, particularly in plantations of European larch. In Manitoba, old infestations also persisted into the early part of the period, particularly in the southeastern part of the province. A moderate to severe infestation occurred south of The Pas between 1972 and 1975. Only pockets of tamarack defoliation were noted in Saskatchewan between 1969 and 1980. Small pockets of defoliation were also reported in Alberta and in the Northwest Territories during the latter part of the period. A few infestations in British Columbia persisted in the early part of the period. The larch sawfly was extremely scarce on western larch until 1976, when a population build-up occurred near Sparwood. This infestation collapsed in 1980.

Background

The origin of larch sawfly in North America remains a matter of debate. Early entomologists (Fletcher 1885, 1906, Fyles 1892, 1906, Hewitt 1912) considered the insect to be a recent introduction, and a number of workers still share this belief (Turnock 1972, Turnock & Muldrew 1973). The first larch sawfly outbreaks were noted in 1880 on the

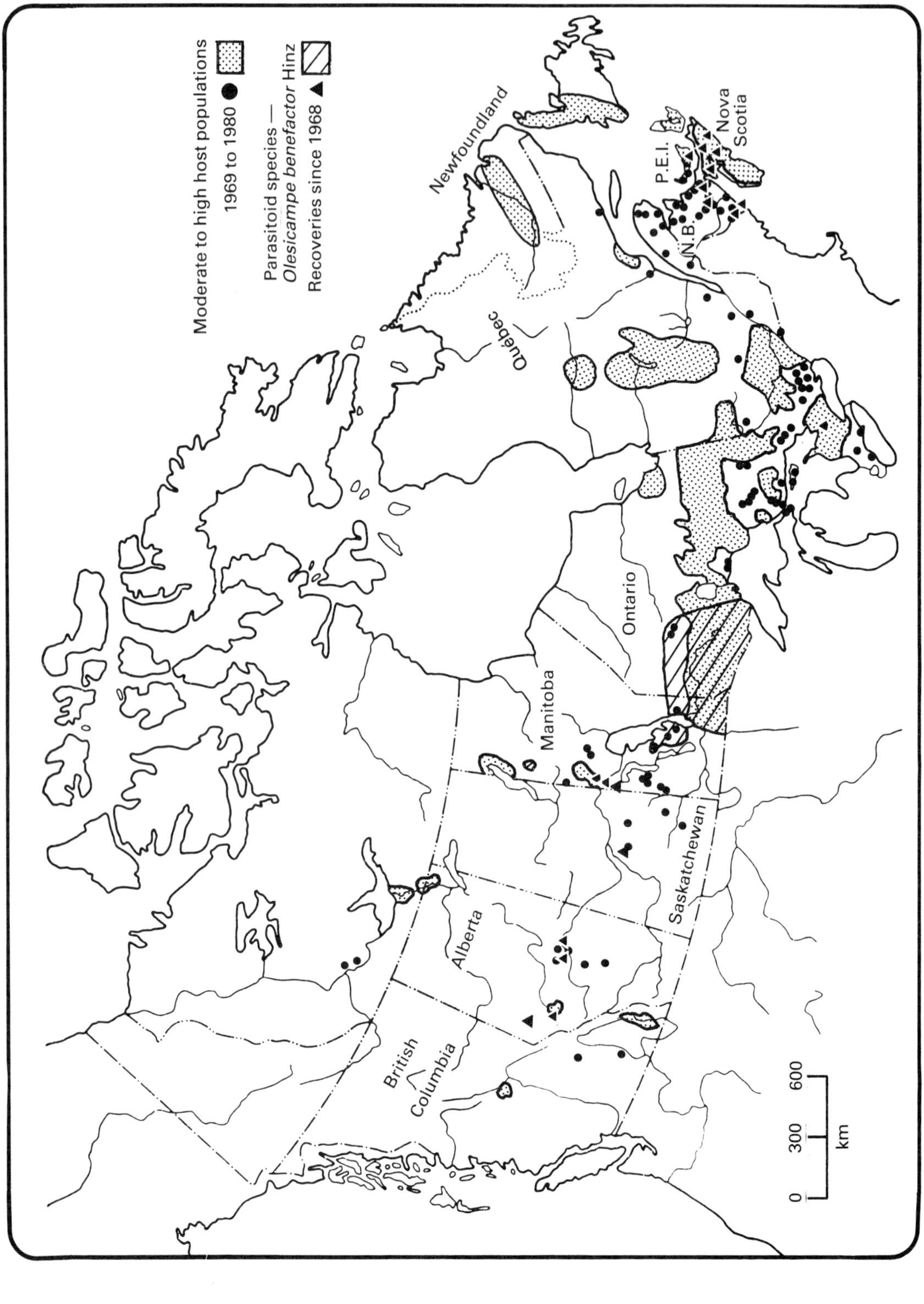

Fig. 18. *Pristiphora erichsonii* (Htg.) - defoliation areas 1969-1980 and recoveries of the parasitoid *Olesicampe benefactor* Hinz since 1968

eastern seaboard of the United States, and the insect appeared to spread westward (Coppel & Leius 1955, Nairn *et al.* 1962). Reports of these early outbreaks indicated that their effects were particularly devastating. However, some entomologists have doubted the validity of the assumption of recent introduction, and feel that the insect has been here for a long time. Graham (1956) examined growth rings of an old tamarack from northern Michigan and believed that he was able to trace evidence of sawfly attack as far back as 1734. Graham (1930) and Graham & Knight (1965) also reported growth reductions in tamarack before 1880, which they attributed to larch sawfly defoliation; however, the periods of supposed defoliation do not agree completely. Similarly, Daviault (1948, 1974) examined some old tamarack from Québec and noted severe growth suppression, which he attributed to larch sawfly attack, about 1829 to 1845. However, Nairn *et al.* (1962) pointed out that it is practically impossible to separate the effects of larch sawfly defoliation from growth suppression caused by unfavourable environmental conditions. Although some of their growth records, particularly those for Isle Pierre, British Columbia, show periods of growth suppression prior to the known occurrence of larch sawfly in the area, they did not claim that this suppression was due to larch sawfly attack. Lejeune & Martin (1948) also examined discs cut from trees up to 159 years old. They found evidence of outbreaks in Manitoba from 1908 to 1915 and from 1941 to the time of examination (1948) but apparently there was no evidence of earlier attacks.

Wong (1974) believed that the larch sawfly reached North America during the Miocene Period via the Bering land bridge. He examined large numbers of larch sawfly adults and identified, on the basis of morphological characteristics, five distinct strains. Two strains occur only in North America, one is confined to Eurasia, and two occur in both North America and Eurasia, presumably as a result of an unintentional introduction in about 1910.

The early attacks by the larch sawfly (Fyles 1892, Graham 1956) clearly demonstrate the severe impact that the insect can have on the successful production of larch. As a consequence, long-term population studies were undertaken in southeastern Manitoba to evaluate the effects of various factors on survival of the different life stages. These studies showed that none of the "native" agents behaved in a density-dependent manner and the populations fluctuated unpredictably (Ives 1976). The "native" factors are as follows. Egg survival is affected by adverse temperatures and several invertebrate predators, but both types of mortality are relatively unimportant. Feeding larvae are attacked by various vertebrate and invertebrate predators, larvae may be dislodged by storms, and starvation may cause mortality during severe outbreaks. Some disease organisms, including fungi, bacteria, microsporidia, and rickettsiae attack feeding larvae, but their effects are usually negligible. The only "native" parasitoid of any consequence in these studies was the tachinid *Bessa harveyi* (Tns.), which attacks late-instar larvae. However, the rates of attack were not related to population trends. Long-term sampling of parasitoids of larch sawfly in Manitoba and Saskatchewan by the Forest Insect and Disease Survey also failed to show any relationship between parasitism and population trends (Ives personal communication). Although parasitism was quite high, it was identical (44%) for both increasing and decreasing larch sawfly populations. Elsewhere in Canada, *Eclytus ornatus* Hlmgr. (which may be a recent introduction from Europe) is a larval parasitoid of some importance in Newfoundland, and the pteromalid *Tritneptis klugii* (Ratz.) is occasionally an important cocoon parasitoid in drier sites in British Columbia (Turnock & Muldrew 1971*a*). Fully fed larvae dropping to spin cocoons are susceptible to predation and heat prostration, and some may drown in pools of surface water. Survival in the cocoon phase is affected by a number of factors. Those of minor importance include invertebrate predation and fungal diseases. The principal sources of mortality in this phase, apart from introduced parasitoids discussed later, are high water tables and predation by small mammals. The combined effects of these were key factors in determining population trends (Ives 1976).

These factors clearly demonstrated the paucity of regulating factors in Canadian larch sawfly populations. Hewitt (1910) had been studying the larch sawfly in England before his appointment as Dominion Entomologist. He had found the ichneumonid *Mesoleius tenthredinis* Morl. to be an effective parasitoid of the larch sawfly and, as it was absent from Canada, he made arrangements for its importation and release between 1910 and 1913. The releases in Manitoba were successful and parasitism by *M. tenthredinis* gradually increased to nearly 90% by 1927 (Graham 1931) and seemed to be a factor in terminating the outbreak. However, when the larch sawfly again reached outbreak proportions, about 1940, it was noted that *M. tenthredinis* was no longer an effective parasitoid due to encapsulation of the eggs by host blood cells which prevented hatching (Muldrew 1953). This encapsulation reaction gradually spread across continental North America, and at present only larch sawfly populations in Newfoundland and cordilleran British Columbia remain highly susceptible to *M. tenthredinis* (Bronskill 1960, Carroll 1964). The lack of quarantine precautions at the time of the initial release was probably the cause of this change in susceptibility. Instead of releasing adult parasitoids, the release in Manitoba was made by placing host cocoons in a bog near Aweme to let the parasitoids emerge naturally (Hewitt 1917). Because not all the cocoons were parasitized, sawfly adults were released as well. This seemed harmless enough, as an outbreak was in progress at the time. However, subsequent research (Maw 1960) indicated that some of these adults probably possessed the encapsulation ability. Their progeny gradually spread across Canada, and have now become the predominant strain in most areas (Wong 1974).

The decrease in effectiveness of *M. tenthredinis* prompted a renewed interest in biological control. Redistribution of *B. harveyi* and *T. klugii* was undertaken (Turnock & Muldrew 1971*a*) but the results were disappointing. A decision was therefore made in 1957 to renew the search for suitable candidates in Europe and Japan. A number of species were considered, but only two releases were successful; a Bavarian strain of *M. tenthredinis* and another ichneumonid, *Olesicampe benefactor* Hinz. In addition, the masked shrew, *Sorex cinereus cinereus* Kerr, was successfully introduced into Newfoundland from New Brunswick.

During the collection of possible parasitoids in Europe, a number of *M. tenthredinis* were obtained. Laboratory experiments showed that those from Bavaria were only slightly encapsulated by their hosts, as were the progeny of back-crosses of this strain with *M. tenthredinis* originating from the earlier introductions. Two releases were made in Manitoba in 1963 near Hodgson and in 1964 near Rennie (Turnock & Muldrew 1971*a*). Increases in parasitism (the two "races" are morphologically indistinguishable) clearly indicated that the Rennie release was successful. *M. tenthredinis* from the early release were more abundant in the Hodgson area and trends following the 1963 release of the Bavarian *M. tenthredinis* were inconclusive.

A number of successful releases of *O. benefactor* were made between 1961 and 1968 (Turnock & Muldrew 1971*a*). Parasitoids collected in Europe were released in Manitoba near Pine Falls in 1961 and near Riverton in 1962 and 1963, and in Saskatchewan near Crutwell in 1964. Relocations of parasitoids collected near Riverton and Pine Falls were made as follows: Manitoba in 1967 and 1968; Saskatchewan in 1965; and the maritimes in 1967 and 1968.

A hyperparasitoid, *Mesochorus dimidiatus* Hlmgr., has been responsible for limiting the effectiveness of *O. benefactor* in Europe, and extreme care was exercised in the initial releases to ensure that none of the hyperparasitoids were inadvertently released. However, *M. dimidiatus* was recovered from the Pine Falls release area in 1966. Apparently *M. dimidiatus* was a rare but widely distributed holarctic species in Canada before the release of *O. benefactor*, and has now been recorded in most areas where this parasitoid has been released. Rates of hyperparasitism by *M. dimidiatus* usually increase rapidly and there is reason to be concerned about the future effectiveness of *O. benefactor*.

The masked shrew, *S. cinereus cinereus*, was successfully introduced into Newfoundland in 1958, when 12 females and 10 males were released near St. George's. Most of the individuals in the initial release survived the first winter and their progeny have prospered and spread and now probably occupy most of the suitable habitats throughout the island.

Releases and Recoveries

Biological control studies of the larch sawfly during the period 1969 to 1980 have consisted of the redistribution of *O. benefactor* and *Mesoleius tenthredinis,* plus follow-up studies of these parasitoids and of the masked shrew in Newfoundland.

Olesicampe benefactor Hinz. (Hymenoptera: Ichneumonidae)

Most of the redistributions of parasitoids involved *O. benefactor* (Table 95). The population increase and dispersal of *O. benefactor* are discussed in detail by Muldrew (in press). He has incorporated additional data that were not analysed by Turnock & Muldrew (1971*a*), so that his values for maximum dispersal in 1966 and 1967 are greater. He reports that the rate of dispersal was about 1 km after 4 years, 3 km after 5 years, and 5 km after 6 years. Follow-up sampling until 1974 showed that the rate of dispersal increased dramatically between 1969 and 1971 (Fig. 19). In 1969, the maximum recorded distance was 87 km (at the Rennie life-table plot). Sampling in 1970 was inadequate for assessing dispersal but *O. benefactor* was recovered in northern Minnesota 293 km south of the release point (Kulman *et al.* 1974, Thompson & Kulman 1976). Extensive sampling by Muldrew in 1971 detected *O. benefactor* at Ignace, Ontario, 357 km from the release point. The maximum detected dispersal distance by 1972 was 476 km. However, larch sawfly populations began declining throughout the area in 1972, while hyperparasitism by *Mesochorus dimidiatus* increased. Apparently these factors influence the rate of spread, as the maximum recorded distance increased by only 64 km, to 540 km, by 1974.

The dotted lines in Fig. 19 show the approximate maximum rates of parasitism at different distances from the release point. Until 1969, the points are clustered around the lines, but the scatter increases for 1971, 1972, and 1974. Part of this increased variability was attributable to the relatively slow spread of *O. benefactor* into southern Manitoba. Muldrew (in press) surmised that the direction of the prevailing winds and abundance of tamarack may have been the main reasons for the slow dispersal into this area. However by 1974 the rates of parasitism in southern Manitoba had increased considerably, and there was already some indication of a relative decline in rates of parasitism elsewhere, possibly due to hyperparasitism by *M. dimidiatus*. Although not shown in Fig. 19 the rates of *M. dimidiatus* parasitism increased dramatically from 1969 to 1974. In 1969, about 80% of *O. benefactor* near the release point were parasitized by *M. dimidiatus*. In 1971, high rates of hyperparasitization had extended to nearly 200 km from the release point. By 1972, the high levels of *M. dimidiatus* attack had extended to over 300 km. Muldrew (personal communication) indicated *M. dimidiatus* probably transferred to *O. benefactor* at a number of locations. The apparently rapid rate of dispersal between 1969 and 1971, and the variable levels of hyperparasitism from 1971 to 1974 both agree with this hypothesis. There was little additional increase of *M. dimidiatus* parasitism in 1974, but the apparent decrease in *O. benefactor* parasitism is probably attributable to the hyperparasitoid.

Information on increases in parasitism and dispersal from other release points is rather limited. A release of 694 males and 737 females was made near The Pas, Manitoba

Table 95 Redistribution releases and recoveries of *Olesicampe benefactor* Hinz and *Mesoleius tenthredinis* Morl. against *Pristiphora erichsonii* (Htg.)

Species and province	Year	Origin	Location of release Lat. °N	Location of release Long. °W	Numbers released Males	Numbers released Females	Numbers released Total	Year of recovery
Olesicampe benefactor Hinz								
Prince Edward Island	1971	Nova Scotia	46°01′	62°44′	157	117	274	1974
	1972	Nova Scotia	46°29′	63°58′	198	189	387	
Nova Scotia	1973	Nova Scotia	45°14′	62°49′	393	457	850	
	1975	Nova Scotia	45°43′	63°09′		63		1976
	1975	Nova Scotia	45°16′	63°18′		20		1976
	1976	Nova Scotia	45°21′	61°54′		45		1977
	1976	Nova Scotia	45°19′	62°15′		69		1977
Ontario	1976	Minnesota	79°18′	44°04′	182	213	395	1977
	1978	Manitoba	75°11′	45°12′	30	32	62	
Manitoba	1969	Manitoba	49°48′	97°08′	60	53	113	1974
	1970	Man. and Sask.	49°19′	96°00′	421	443	864	1970
	1971	Manitoba	50°03′	96°13′	245	262	507	Already present
	1972	Manitoba	49°19′	96°00′	75	74	149	
Alberta	1972	Manitoba	54°43′	110°04′			1283*	
	1972	Manitoba	54°27′	113°58′			1139*	1973
	1972	Manitoba	55°01′	119°06′			469*	1973
	1973	Alberta	54°30′	112°57′	118	122	240	1974
	1975	Manitoba	53°32′	117°04′	51	86	137	1975
	1981	Alberta	59°55′	111°43′			36	
Northwest Territories	1972	Manitoba	60°36′	116°06′			856*	
	1981	Alberta	60°01′	112°07′			332	1981
	1981	Alberta	60°01′	110°58′			30	
British Columbia	1980	Alberta	49°42′	114°55′			170	
	1980	Alberta	49°29′	115°04′			145	
	1980	Alberta	49°38′	114°56′			150	
Mesoleius tenthredinis Morley								
Manitoba	1970	Man. and Sask.	4919′	96°00′	63	85	148	
	1971	Manitoba	49°19′	96°00′	98	170	268	
	1972	Manitoba	49°19′	96°00′	9	42	51	

* Estimates only – "small" cocoons were placed in the field, adult parasitoid emergence holes were counted and proportions of *O. benefactor* and *Mesochorus dimidiatus* were based on reared samples.

in 1968. *O. benefactor* had dispersed 18 km by 1973 and 57 km by 1975, 7 years after the release. Another release, consisting of 51 males and 86 females, was made near Obed Lake, Alberta, in 1975. *O. benefactor* parasitism near the release area reached 88% by 1979 and 93% by 1980. Seventeen per cent of the larch sawfly larvae in a bog 15 km east of the release site were parasitized by *O. benefactor* in 1980. None had been found in this bog in 1979.

Dispersal from The Pas and Obed Lake releases, 18 km and 15 km in 5 years, is considerably greater than the 3 km in 5 years recorded for the initial release at Pine Falls. This may simply be due to a different set of environmental conditions, but it could indicate that *O. benefactor* has become better adapted to the Canadian environment.

***Mesoleius tenthredinis* Morley (Hymenoptera: Ichneumonidae)**

Redistributions are shown in Table 95.

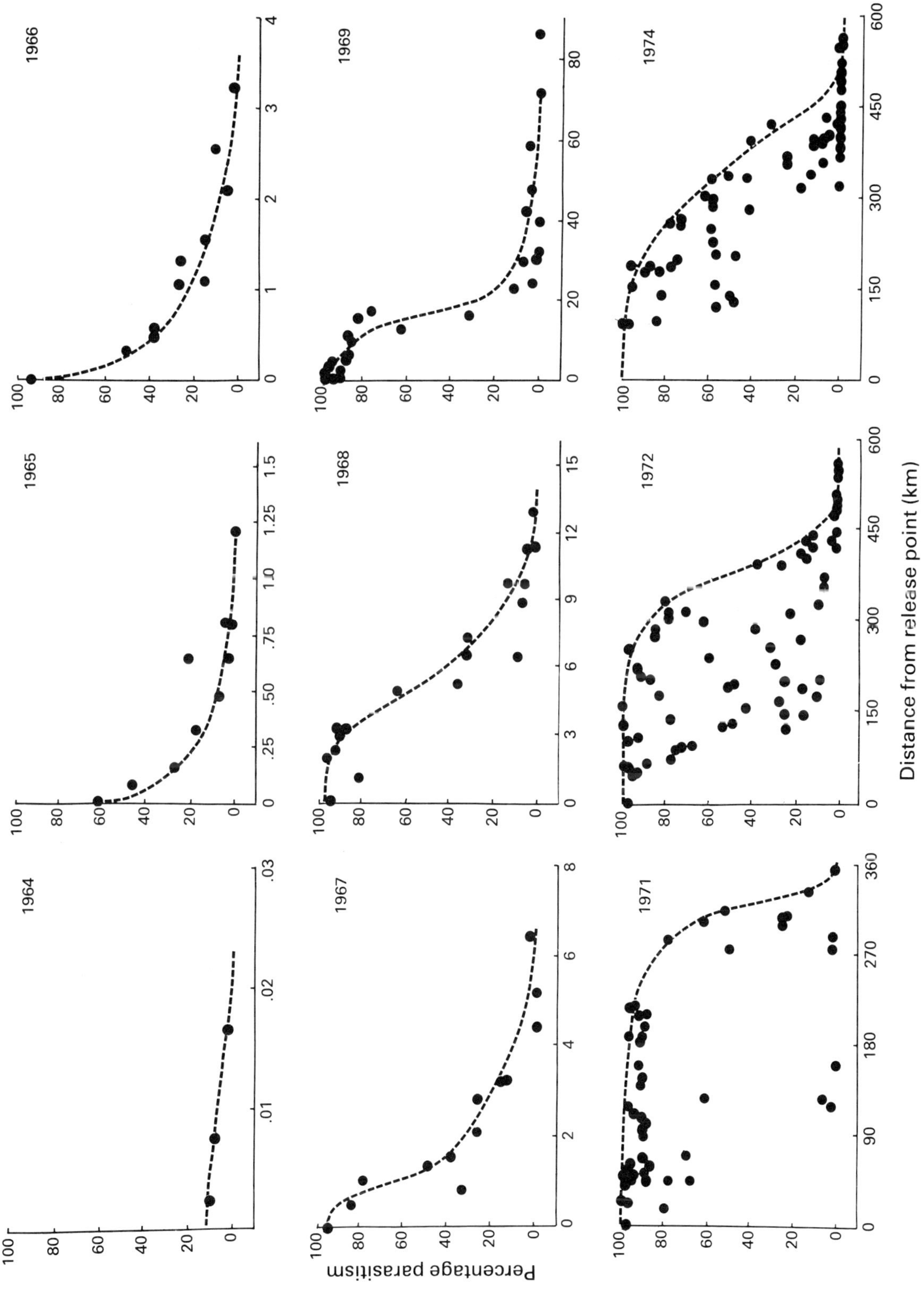

Fig. 19. Percentage parasitism and dispersal distance from Pine Falls release point of *Olesicampe benefactor*

***Sorex cinereus cinereus* Kerr (Insectivora: Soricidae)**

The masked shrew, introduced into Newfoundland in 1958, has continued to spread. Survey reports for 1969 and 1970 (Anon. 1970–79) mentioned the concern of local residents as high initial numbers were noted in each area: by 1971 the shrew had spread over 95% of the island. Populations on the island, excluding the first 2 or 3 years' data, averaged about eight shrews per hectare, which seems to be unusually high. However, the 18-year average populations of the masked shrew in the Rennie life-table plot (Ives 1981) was 5.2/ha. If one includes populations of the closely related arctic shrew, *Sorex arcticus* Kerr, which averaged 1.8/ha, the total of 7.0/ha is comparable. Populations of the masked shrew in Newfoundland therefore seem to have stabilized, and can be expected to fluctuate in the future in response to environmental conditions.

Evaluation of Control Attempts

O. benefactor and larch sawfly populations have probably not yet reached an equilibrium, even at the original release point near Pine Falls. As shown in Fig. 19, the level of *O. benefactor* parasitism at or near the release point reached 90% in 1967. It remained at or near this level until 1972, but larch sawfly populations collapsed in 1973, presumably as a result of the continued high rates of parasitism (Ives 1976). No larvae or cocoons were obtained near the plot in 1973 or 1974, although intensive sampling was conducted (Muldrew in press). Detailed studies were discontinued after 1974, and further trends are based on limited sampling. In 1977, a collection of 72 fourth- and fifth-instar larvae was made near the original release site at Pine Falls. None was parasitized by *O. benefactor*. In 1978, collections of larvae from the same area yielded 1730 cocoons. Parasitism by *O. benefactor* was 1.5%, hyperparasitism by *Mesochorus dimidiatus* was 86%. A total of 237 third- to fifth-instar larvae collected 8 km north of the release point in 1978 contained no *O. benefactor,* but 19% of 424 fourth- and fifth-instar larvae from an area 8 km south of the release point were parasitized. Ninety-three per cent of these *O. benefactor* larvae were parasitized by *M. dimidiatus*. In 1980, a total of 535 cocoons was obtained from the same location; 15% of these contained *O. benefactor*, of which 75% were parasitized by *M. dimidiatus*. No *O. benefactor* were present in 150 cocoons obtained by rearing late-instar larvae collected in 1980 near the Seddon's Corner life-table plot, about 70 km south of the Pine Falls release point.

Although the above data are rather fragmented, there is some evidence that *O. benefactor* is in danger of local extinction over a fairly wide area. Should this happen, it is possible that the larch sawfly may increase in abundance if environmental conditions are favourable. No population sampling of the larch sawfly was done in the Pine Falls release area in 1980, but observations by Ives indicated that this population increase may already be taking place. It remains to be seen whether or not *O. benefactor* will be able to prevent an outbreak in spite of the adverse influence of *M. dimidiatus. O. benefactor* is host specific, but *M. dimidiatus* is not (Pschorn-Walcher & Zinnert 1971). Because of this, one would expect *M. dimidiatus* to be able to persist on alternate hosts, thus obtaining a "head start" on any re-invading *O. benefactor.* Under these circumstances, *M. dimidiatus* would probably be able to control *O. benefactor*, thus permitting larch sawfly populations to reach outbreak proportions. However, a crude model developed by Ives (1976) indicated that the two parasitoids should reach an equilibrium with each other and with the larch sawfly. According to the model, the larch sawfly populations would fluctuate in response to density-independent environmental factors, but should not reach outbreak proportions in the presence of *O. benefactor,* even when *M. dimidiatus* is present. It is too early to tell which hypothesis is correct.

The impact of the masked shrew on larch sawfly populations in Newfoundland has not been assessed in detail (Warren 1971). The fact that larch sawfly infestations reached outbreak proportions late in the decade 1971–1980 clearly indicates that the shrews were not able to control the insect. However, as Warren (1971) indicated, "the shrew has been a valuable predator of larch sawfly, and, despite some evidence of reduced parasitism, there has been an important net gain in cocoons destroyed". Ives (1976) showed that mortality during the cocoon and adult stages was largely responsible for determining population trends. Small mammal predation is often a major component of this mortality, thus agreeing with Warren's assessment of the effect of the masked shrew in Newfoundland.

Although there is some indication that larch sawfly populations in and around the Pine Falls release site may now be on the increase, there has been a period of several years when this insect was extremely difficult to find. It was probably not as rare as in Europe, but it was much less common than it has been in recent memory. Two recent papers illustrate just how rare the larch sawfly can be in alpine areas of Europe. Lovis (1975) collected only six larch sawfly (out of a total of 2816 larch insects) during a 3-year study of insects attacking native European larch in Switzerland. Auer (1971) studied the population dynamics of the larch bud moth, *Zeiraphera diniana* Gn., in the Upper Engadine Valley in Switzerland between 1949 and 1968. He presented population curves for eight larch insects for the period 1952 to 1968. The larch sawfly was apparently too rare to be included.

The infestation patterns of the larch sawfly have been discussed by Turnock (1972). He did not consider the effects of *O. benefactor*, but with slight changes his discussion is still valid. He recognized three types of life systems: 1) the stable latent type with a diverse environment and a rich parasitoid complex, characteristic of infestations on larch in alpine Europe; 2) the stable permanent type with a paucity of specific parasitoids, characteristic of many North American outbreaks before 1920; and 3) the temporary type with an intermediate parasitoid complex which includes one (or two) very effective larval parasitoid(s), characteristic of European plantings, infestations on western larch in British Columbia and on tamarack in the rest of Canada following the first successful introduction of an ichneumonid parasitoid between 1910 and 1913 (and in southeastern Manitoba following the successful introduction of *O. benefactor* there in 1961). As this discussion of life systems illustrates, entomologists have been able to achieve at least partial biological control of the larch sawfly in North America. To continue this control, it may be necessary to seek ways in which biotic agents in the environment can be manipulated.

Turnock & Muldrew (1971*b*) and Turnock *et al.* (1976) recognized four ways in which insect parasites or predators might be manipulated in order to help achieve biological control: 1) colonization releases; 2) inoculative releases; 3) inundative releases; and 4) enhancement of biotic agents by environmental manipulation. In the case of the larch sawfly, the last three approaches offer the most promise. Because of the presence of *M. dimidiatus*, any releases of *O. benefactor* would need to be relatively large if they were to be effective. Releases intermediate in size between inoculative and inundative, say 2 000–3 000 mated females, should be able to control infestations in plantations in 2 or 3 years, if released when defoliation first becomes noticeable. However, this approach would necessitate a permanent facility capable of producing parasitoids on demand. Reliance on field populations would be too uncertain, especially as *M. dimidiatus* will almost certainly reduce the numbers of *O. benefactor* adults obtainable. The Bavarian strain of *Mesoleius tenthredinis* could probably be managed in a similar manner. Both species might even be able to sustain themselves in the field, and thus afford more or less permanent protection. In Manitoba *O. benefactor* appeared to suppress *M. tenthredinis* by sheer numbers, but this seems unlikely to happen again.

Enhancement of environmental conditions might be a particularly effective method for increasing the effectiveness of avian and small mammalian predators in plantations. Buckner (1971) discussed this situation as it related to forest insects in general. Birds might be encouraged to nest in plantations by providing suitable nesting sites. Bruns

(1960) reported that this approach appeared to be effective in controlling pine insects in Germany. In the case of small mammals, research would have to be conducted to ensure that voles do not attack larches. Neither of us has observed evidence of this, but some tree species are severely damaged by voles. If high vole populations are not a threat to the trees it should be possible to enhance the environment by providing adequate cover, either in the form of slash or between-row vegetation, as several workers have shown that mouse and shrew populations are highest in areas with adequate cover (Morris 1955, Beer 1961, Coulianos & Johnels 1962). Any such manipulation would have to be done carefully, to avoid creating an environment favouring snowshoe hares. However, it should be possible to improve the environment of the plantations as habitats for voles and shrews but not hares. This would increase the amount of small mammal predation, which should reduce the frequency of larch sawfly outbreaks.

In low-lying situations it may be possible to flood the plantations at critical times in the larch sawfly's life cycle, as suggested by Buckner (1971). Flooding the plantations for short periods in the fall, soon after cocoon formation, or in the spring when the insects are in the pupal stage would have a marked effect on larch sawfly survival (Lejeune *et al.* 1955) but would not seriously damage the tamarack trees if carefully timed. It would also adversely affect small mammal populations, but this could be minimized by careful timing. If flooding could be done just before parasitoids are released, it might increase their effectiveness considerably, as the initial rate of parasitism would be higher.

Finally, should serious infestations develop that may endanger the trees in plantations, it may still be possible to control the larch sawfly without unduly affecting the other biotic agents. In the case of *M. tenthredinis*, spraying with a larval insecticide (preferably one that does not kill adult Hymenoptera) when most of the larch sawfly larvae are in the first to third instars would reduce the sawfly populations but should not have an undue adverse effect on the parasitoids, as they attack mainly fourth- and fifth-instar larvae. Because of the prolonged period of larch sawfly emergence it would not be feasible to kill all the larvae, but it should be possible to reduce populations appreciably and thus enhance the effectiveness of *M. tenthredinis*. The use of chemicals in the presence of *O. benefactor* would be more difficult. Probably the most productive approach would be to reduce sawfly populations only to the extent that defoliation would not exceed 70%. This would save the trees, but should not have an unduly adverse effect on the parasitoids.

Recommendations

The Bavarian strain of *M. tenthredinis* was released near Rennie, Hodgson, and St. Labre. It is known to have become established at Rennie, but how far it spread before being overwhelmed by *O. benefactor* is not known. The extremely low density of larch sawfly in the vicinity of the Rennie plot for several years after its establishment may mean that its numbers were markedly reduced, if not eliminated. No follow-up studies were conducted in the St. Labre area, so it is not known if the Bavarian strain (from the Rennie plot) became established or not.

It is therefore recommended that the Bavarian strain of *M. tenthredinis* be released in the Pine Falls area as soon as larch sawfly populations in the Bavarian plantations are high enough to make collection feasible. The results of the release at the Rennie plot looked extremely encouraging and a second attempt to establish this strain seems warranted. In Europe, *M. tenthredinis* is the most important insect parasite of the larch sawfly (Pschorn-Walcher & Zinnert 1971) and there seems to be a reasonable chance of the Bavarian strain assuming this position in Canada, now that *Mesochorus dimidiatus* seems destined to limit the effectiveness of *O. benefactor*.

O. benefactor has been widely redistributed in various parts of Canada, from the maritimes to British Columbia. There does not seem to be much point in continuing with any further releases, except possibly into more or less isolated plantations. However, if *O. benefactor* is to be used to suppress larch sawfly in plantations, as suggested in the previous section, it is recommended that a small permanent facility be established that has the capability of producing the parasitoid when needed. Rearing of larch sawfly under laboratory conditions is not unduly difficult if specialized techniques are used (Heron & Drouin 1979). A similar recommendation applies to the rearing of *Mesoleius tenthredinis*.

Literature Cited

Anonymous. (1970–1979) Annual reports of the Forest Insect and Disease Survey, 1969–1976. Ottawa; Canadian Forestry Service.

Auer, C. (1971) Some analyses of the quantitative structure in populations and dynamics of the larch bud moth 1949–1968. In: Patil, G.P.; Pielou, E.C.; Waters, W.E. (*Eds.*) Statistical ecology. 2. Sampling and modelling biological populations and population dynamics. University Park, Pennsylvania; Pennsylvania State University Press, pp. 151–173.

Beer, J.R. (1961) Winter home ranges of the red-backed mouse and white-footed mouse. *Journal of Mammalogy* 42,174–180.

Bronskill, J.F. (1960) The capsule and its relation to the embryogenesis of the ichneumonid parasitoid *Mesoleius tenthredinis* Morl. in the larch sawfly, *Pristiphora erichsonii* (Htg.) (Hymenoptera: Tenthredinidae). *Canadian Journal of Zoology* 38,769–775.

Bruns, H. (1960) The economic importance of birds in forests. *Bird Study* 7,193–208.

Buckner, C.H. (1971) Vertebrate predators. In: Toward integrated control. *US Department of Agriculture, Forest Service Research Paper* NE-194. Upper Darby, Pennsylvania; Northeastern Forest Experiment Station, pp. 21–31.

Carroll, W.J. (1964) The larch sawfly in Newfoundland. *Canadian Department of Forestry, Forest Entomology and Pathology Branch Bi-monthly Progress Report* 20(5),1.

Coppell, H.C.; Leius, K. (1955) History of the larch sawfly, with notes on origin and biology. *Canadian Entomologist* 87,103–111.

Coulianos, C.C.; Johnels, A.G. (1962) Note on the subnivean environment of small mammals. *Arkiv für Zoologie* 15,363–370.

Daviault, L. (1948) Note pour servir à l'histoire de la tenthrède européenne (sic) du mélèze (*Pristiphora ericksonii* (sic) (Htg.)). *Annals de l'ACFAS* 14,85.

Daviault, L. (1974) Notes sur le développement de l'entomologie forestière au Québec. *Annals of the Entomological Society of Québec* 79,45–61.

Fletcher, J. (1885) The larch sawfly — *Nematus erichsonii* (Hartig). *Annual Report of the Entomological Society of Ontario* 15,72–77.

Fletcher, J. (1906) Report of the Entomologist and Botanist, 1905. *In*: Appendix to the report of the Minister of Agriculture on experimental farms. Ottawa; Canadian Department of Agriculture, pp. 159–204.

Fyles, T.W. (1892) *Nematus erichsonii*: a retrospect. *Annual Report of the Entomological Society Ontario* 22,28–30.

Fyles, T.W. (1906) The tamarack growth in the eastern township of the Province of Québec. *Canadian Forestry Journal* 2,132–134.

Graham, A.R. (1931) The present status of the larch sawfly (*Lygaeonematus erichsonii* Hartig) in Canada, with special reference to its specific parasite, *Mesoleius tenthredinis* Morley. *Canadian Entomologist* 63,99–102.

Graham, S.A. (1930) The larch sawfly and forestry. *Proceedings of the 4th International Congress of Entomology, Ithaca, August 1928.* 1¹, 401–407.

Graham, S.A. (1956) The larch sawfly in the Lake States. *Forest Science* 2,132–160.

Graham, S.A.; Knight, F.B. (1965) Larch sawfly. In: Principles of forest entomology. New York; McGraw-Hill, pp. 256–257.

Heron, R.J.; Drouin, J.A. (1979) Methods of collecting, rearing and handling the larch sawfly for experimental studies. *Canadian Department of Fisheries and Forestry, Forest Research Laboratory, Winnipeg, Information Report* MS-X-15.

Hewitt, C.G. (1910) Insects destructive to Canadian forests. Reprinted from the First Annual Report of the Committee of Conservation, 12 pp.

Hewitt, C.G. (1912) The large larch sawfly (*Nematus erichsonii*) with an account of its parasites, other natural enemies and means of control. *Dominion of Canada Department of Agriculture Experimental Farms, Division of Entomology Bulletin* No. 10., Second series, Ottawa.

Hewitt, C.G. (1917) The introduction into Canada of the ichneumon fly, *Mesoleius tenthredinis*, a parasitic enemy of the larch sawfly, *Nematus erichsonii. Agricultural Gazette* 4,355–357.

Ives, W.G.H. (1976) The dynamics of larch sawfly (Hymenoptera: Tenthredinidae) populations in southeastern Manitoba. *Canadian Entomologist* 108,701–730.

Ives, W.G.H. (1981) Environmental factors affecting 21 forest insect defoliators in Manitoba and Saskatchewan, 1945–1969. *Canadian Forestry Service, Northern Forest Research Centre, Edmonton, Information Report* NOR-X-233.

Kulman, H.M.; Thompson, L.C.; Witter, J.A. (1974) Introduction of parasitoids of the larch sawfly in Minnesota. *Great Lakes Entomologist* 7,23–25.

Lejeune, R.R.; Martin, J.B. (1948) Studies on annual rings of tamarack. *Dominion of Canada Department of Agriculture Science Service Division of Entomology Forest Insect Investigation Bi-monthly Progress Report* 4(6),2.

Lejeune, R.R.; Fell, W.H.; Burbridge, D.P. (1955) The effect of flooding on development and survival of the larch sawfly *Pristiphora erichsonii* (Tenthredinidae). *Ecology* 36,63–70.

Lovis, C. (1975) Contribution à l'étude des tenthrèdes du mélèze (Hymenoptera: Symphyta) en relation avec l'évolution dynamique des populations de *Zeiraphera diniana* Guenée (Lepidoptera: Tortricidae) en Haute-Engadine. *Mitteilungen der Schweizerischen entomologische Gesellschaft* 48,181-192.

Maw, M.G. (1960) Notes on the larch sawfly, *Pristiphora erichsonii* (Htg.) (Hymenoptera: Tenthredinidae), in Great Britain. *Entomologist's Gazette* 11,43–49.

McGugan, B.M.; Coppel, H.C. (1962) Biological control of forest insects — 1910–1958. *In*: A review of the biological control attempts against insects and weeds in Canada. *Commonwealth Institute of Biological Control Technical Communication* 2,35–127.

Morris, R.F. (1955) Population studies on some small forest mammals in eastern Canada. *Journal of Mammalogy* 36,21–35.

Muldrew, J.A. (1953) The natural immunity of the larch sawfly (*Pristiphora erichsonii* (Htg.)) to the introduced parasite *Mesoleius tenthredinis* Morley, in Manitoba and Saskatchewan. *Canadian Journal of Zoology* 31,313–332.

Muldrew, J.A. (In press) Dispersal of *Olesicampe benefactor* Hinz, an introduced parasite of the larch sawfly. *Canadian Forestry Service, Northern Forest Research Centre, Edmonton, Information Report.*

Nairn, L.D.; Reeks, W.A.; Webb, F.E.; Hildahl, V. (1962) History of larch sawfly outbreaks and their effects on tamarack stands in Manitoba and Saskatchewan. *Canadian Entomologist* 94,242–255.

Pschorn-Walcher, H.; Zinnert, K.D. (1971) Investigations on the ecology and natural control of the larch sawfly (*Pristiphora erichsonii* Htg., Hym.: Tenthredinidae) in central Europe. Part II: Natural enemies: their biology and ecology, and their role as mortality factors in *P. erichsonii. Commonwealth Institute of Biological Control Technical Bulletin* 14,1–50.

Thompson, L.C.; Kulman, H.M. (1976) Parasite complex of the larch sawfly in Minnesota. *Environmental Entomology* 5,1121–1127.

Turnock, W.J. (1972) Geographical and historical variability in population patterns and life systems of the larch sawfly (Hymenoptera: Tenthredinidae). *Canadian Entomologist* 104,1883–1900.

Turnock, W.J.; Muldrew, J.A. (1971*a*) *Pristiphora erichsonii* (Hartig), larch sawfly (Hymenoptera: Tenthredinidae). *In*: Biological control programmes against insects and weeds in Canada, 1959–1968. *Commonwealth Institute of Biological Control Technical Communication* 4,175–194.

Turnock, W.J.; Muldrew, J.A. (1971*b*) Parasites. In: Toward integrated control. *US Department of Agriculture Forest Service Research Paper* NE-194. Northeastern Forest Experiment Station, Upper Darby, Pennsylvania, pp. 59–87.

Turnock, W.J.; Muldrew, J.A. (1973) Characteristics of *Bessa harveyi* (Diptera: Tachinidae) suggesting the historic introduction of the larch sawfly to North America. *Manitoba Entomologist* 6,49–53.

Turnock, W.J.; Taylor, K.L.; Schroder, D.; Dahlsten, D.L. (1976) Biological control of pests of coniferous forests. In: Huffaker, C.B.; Messenger, P.S. (*Eds.*). Theory and practice of biological control. New York; Academic Press, pp. 289–311.

Warren, G.L. (19u1) Introduction of the masked shrew to improve control of forest insects in Newfoundland. *Proceedings of the Tall Timbers Conference on Ecological Animal Control by Habitat Management* 2,185–202.

Wong, H.R. (1974) The identification and origin of the strains of the larch sawfly, *Pristiphora erichsonii* (Hymenoptera: Tenthredinidae), in North America. *Canadian Entomologist* 106,1121–1131.

Pristiphora geniculata (Htg.), Mountain-Ash Sawfly (Hymenoptera: Tenthredinidae)

F.W. QUEDNAU

Pest Status

The mountain-ash sawfly, *Pristiphora geniculata* (Hartig), is a major defoliator of mountain-ash, mainly *Sorbus americana* Marsh, and S. *aucuparia* L. These tree species have little industrial importance but are planted widely as shade and ornamental trees. The mountain-ash sawfly was accidentally introduced into the United States in 1926 (Schaffner 1936) and was first recorded in Canada in southern Québec in 1934 (Petch 1935). Its present Canadian range is from Newfoundland to southwestern Ontario (Fig. 20). There has never been a serious outbreak. The sawfly causes noticeable and occasionally complete defoliation of mountain-ash, but this seldom causes mortality of the tree. To owners of such trees and to municipal park authorities, the insect is a considerable nuisance and often requires localized control by chemical methods. An account of the biology of mountain-ash sawfly in eastern Canada was given by Forbes & Daviault (1964). *P. geniculata* usually has one generation a year, but there may be a partial second generation depending on the climate.

Background

The decision to embark on a biological control programme in Québec against *P. geniculata* was determined by the policy to reduce chemical control in urban forestry and by the success of biological control in Manitoba against the closely related larch sawfly, *P. erichsonii* (Htg.), on tamarack, *Larix laricina* (Du Roi) K. Koch (Muldrew 1967, 1973). Data furnished by the CIBC European Station suggested that the ichneumonid wasp *Olesicampe geniculatae* Quednau & Lim, would be a primary candidate for introduction into Canada. Another parasitoid species under consideration was the ichneumonid wasp *Rhorus* sp. No. 3.

During the first years of attempts to colonize these parasitoid species in Québec, infestation levels of mountain-ash sawfly were very low. Furthermore, in 1972 *O. geniculatae* was not found in sufficient numbers in Europe to warrant a shipment to Canada. However, in 1973, heavy infestations of *P. geniculata* occurred in several localities near Québec City. Because of the scarcity in Europe of *P. geniculata*, laboratory mass-rearings of the sawfly were carried out in Québec in order to obtain host material that could be sent to Europe to increase the host populations there. The rearing techniques were essentially the same as those described for the larch sawfly by Heron & Drouin (1969). In 1974 the Laurentian Forest Research Centre sent 6 000 cocoons of Canadian *P. geniculata* to CIBC in Switzerland and another 30 000 cocoons in 1976. This insect material was reared to the adult stage in Europe and the sawfly females were placed in plastic bags tied over branches of mountain-ash trees in the Waldviertel area in Austria, where *O. geniculatae* and other parasitoid species of the mountain-ash sawfly were known to occur in significant numbers during the previous years. As a result, in 1977 large numbers of *O. geniculatae* and *Rhorus* sp. No. 3 were received in Québec.

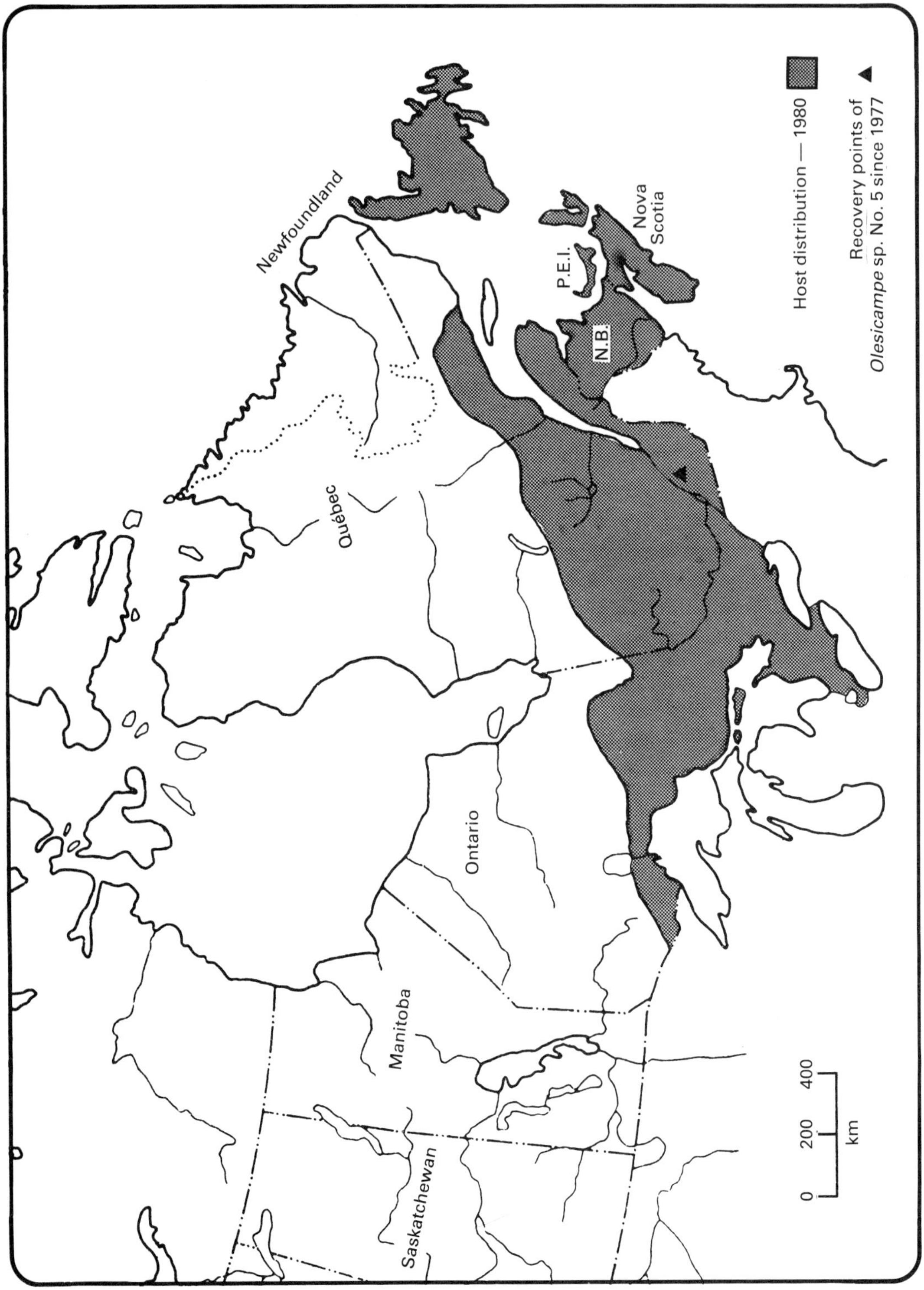

Fig. 20. Distribution of *Pristiphora geniculata* (Htg.) in 1980 and recoveries of *Olesicampe geniculatae* Quednau & Lim. since 1977

Releases and Recoveries

The utilization of the parasitoid material is summarized in Table 96.

Table 96 Parasitoids utilized for releases against mountain-ash sawfly, *Pristiphora geniculata* (Htg.), at Beaumont, Québec (lat. 46°50′N, long. 71°02′W)

Species	Year	Origin	No. and sex of parasitoids received	
			Male	Female
Rhorus sp. No. 3	1973	Austria	5	5
	1974	Switzerland and Austria	29	32
	1975	Austria	12	16
	1976	Austria	67	83
	1977	Austria	35	35
Olesicampe geniculatae	1974	Switzerland and Austria	22	16
	1975	Austria	2	3
	1976	Austria	92	117
	1977	Austria	315	551

Rhorus sp. No. 3 (Hymenoptera: Ichneumonidae)

This is a solitary endoparasitoid, probably specific to *P. geniculata*. The egg is laid through the ocellus into the head capsule of the host larva. This wasp is closely related to *R. lapponicus* Roman, a parasitoid of the larch sawfly, the biology of which was described by Pschorn-Walcher & Zinnert (1971). *Rhorus* sp. No. 3 was received in Québec from 1973 to 1977. Cage releases were made at Beaumont nurseries. A male was recovered in the field in 1978, but the parasitoid apparently did not become permanently established. It is difficult to mate this insect in captivity.

Olesicampe geniculatae Quednau & Lim (Hymenoptera: Ichneumonidae)

This is also an internal solitary parasitoid that prefers the first- and second-instar larvae of the mountain-ash sawfly for attack. It is specific to this host. In the early colonization work in Québec only small numbers of the parasitoid were available and shipments from Europe were often not well synchronized with host populations in Québec. No recoveries were obtained from these first releases. Major implantations of this parasitoid were made in 1977 at Beaumont nurseries. The female parasitoids were mated in the laboratory and released in batches in large tents of woven plastic screening built over mountain-ash trees, in which host densities were artificially increased by adding young colonies of *P. geniculata* larvae collected from other areas. Small bottles of water were fastened to the branches of the trees in the tents. The host material was exchanged and new parasitoids added each day. The parasitized *P. geniculata* larvae were brought to the laboratory, given plenty of mountain-ash foliage that was kept fresh in water bottles, and reared to about fourth instar. This method excluded natural predators. The sawfly larvae were then returned to the field and placed in wire baskets (100 per basket) containing some foliage. The baskets were tied to the branches of mountain-ash trees in the area chosen for implantation. Parasitized larvae eventually fell to the ground and formed cocoons.

Successful establishment of this parasitoid depends on massive initial implantations to give the parasitoid a good chance to multiply in the field. A female *O. geniculatae* can lay up to 400 eggs and live up to 3 weeks in the laboratory at 22°C and 75% relative humidity (F.W. Quednau unpublished). The biology of this species is essentially the same as that reported for *O. benefactor* Hinz by Pschorn-Walcher & Zinnert (1971). *O. geniculatae* may have a partial second generation each year in the field.

Evaluation of Control Attempts

Whereas the colonization of *Rhorus* sp. No. 3 was unsuccessful, *O. geniculatae* became firmly established at Beaumont nurseries, and within 4 years parasitized a large proportion of hosts. A brief history of this event follows.

At the beginning of massive implantations in 1977, population densities of *P. geniculata* were medium to heavy, and 1 year later were light to medium. In the fall of 1978 it was found from dissections of sawfly cocoons that the percentage of parasitism at the release centre had dropped to 25–30%, compared to 60–75% in 1977, as a result of the dilution of the parasitoid population in the area. It was estimated that within 1 year *O. geniculatae* spread about 500 m from the original release centre. In 1979 infestations by the mountain-ash sawfly were only light at Beaumont. Of 30 soil samples of 0.1 m^2 each, taken near the original release point of the parasitoid, only 10 contained living cocoons (range 1–5) and parasitism was about 73%, much higher than in 1978. Other sampling plots about 500 m from the release point yielded 30–35% parasitism in 1979. In 1980, in the same locality, dissections of mountain-ash sawfly larvae that were still in colonies (about third instar) were made. The infestation level was from 700 to 1 250 young colonies of the sawfly per hectare. Percentage parasitism obtained from these dissections was 88.7–93.5% in all the plots containing mountain-ash in the nursery (up to 500 m radius from the release centre). No second generation of *P. geniculata* was observed in 1980. From an evaluation of host cocoons collected in the fall of 1980 it was found that the number of *P. geniculata* cocoons had further diminished as compared with 1979 and few were unparasitized.

Preliminary indications for 1981 are that *O. geniculatae* has spread about 30 km from the original release site and could be recovered at Ste-Foy, Charlesbourg, La Durantaye, St-Charles de Bellechasse, St-Michel de Bellechasse, St-Etienne de Lauzon, and Chutes Mt. Ste-Anne where parasitism was 81, 24, 50, 30, 82, 70, and 4% respectively. A quantitative evaluation of the impact of the parasitoid on its host population has not been carried out so far. However, preliminary measures of pest numbers in consecutive years in the original release site at Beaumont indicate a drastic decline of the host populaton (from 500 colonies/ha in 1980 to an estimated 30 colonies/ha in 1981), which must be attributed primarily to the action of the parasitoid. Comparative population counts of *P. geniculata* at St-Édouard de Frampton, where the environmental structure is very similar to that at Beaumont, but is situated outside the action radius of *O. geniculatae,* indicate very high levels of infestation in 1981.

The cost of chemical insecticide treatment against *P. geniculata* would be at least $200/ha. For this reason most owners of property with mountain-ash do not use chemical treatments. The mass production of *O. geniculatae* would need extensive manual labour and costs cannot be calculated accurately at present. However, as the parasitoid spreads on its own, its implantation could be highly beneficial.

During the evaluation work in summer 1980, a large number of the larvae of *O. geniculatae* found in the hosts were dead or appeared sick. At this stage it was decided to rear adults from *P. geniculata* cocoons collected from Beaumont in order to determine the percentage parasitism by *O. geniculatae* and to find out whether a hyperparasitoid was present. Emergence of the insects in 1981 revealed up to 90% total parasitism, but about 60% of the parasitoids obtained were *Mesochorus globulator* Thnb. This latter species is a hyperparasitoid with a wide distribution in Canada. It attacks various other species of *Olesicampe* associated with nematine sawflies.

In summary, no detailed account of damage control at the Beaumont nurseries can be given. However, *O. geniculatae* became firmly established and the percentage parasitism reached high levels in subsequent years, followed by a drastic decline in the host population that can only be attributed to the biological control agent. Thus *O. geniculatae* must be rated as a successful control agent.

Recommendations

Rearings of *P. geniculata* cocoons from the Beaumont area should be made in consecutive years to obtain exact data on the incidence of *M. globulator* at different population levels in various areas where *O. geniculatae* became established. From these data it will be seen whether *M. globulator* seriously interferes with the primary parasitoid by repressing it to insignificant numbers or whether it acts in a beneficial way by helping to preserve its host supply. Preliminary hypotheses by Muldrew (1973) on the interaction between *O. benefactor* Hinz and *M. globulator* indicate that the hyperparasitoid may perform a useful role in preventing the local extermination of the host and consequently of itself. Muldrew points out, however, that the danger of such local extinctions may result in re-invasions of an area by the sawfly unimpaired by the primary parasitoid, and outbreaks may develop before *O. geniculatae* can return.

For the near future, relocations of *O. geniculatae* should be made in Newfoundland where a problem with mountain-ash sawfly also exists. At the same time the situation at Beaumont should be watched closely and the possible dispersal of *O. geniculatae* into other areas, such as the north shore of the St. Lawrence River, should be monitored.

A challenging approach to assessing the biological control mechanism would be periodic sampling of host material on a limited scale for at least two consecutive years to obtain life tables that reflect the key factors of mortality in *P. geniculata* populations in which the parasitoid became endemic. However, cocoon examinations would require considerable manual labour because mountain-ash sawfly material is difficult to obtain from soil samples.

Literature Cited

Forbes, R.S.; Daviault, L. (1964) The biology of the mountain-ash sawfly, *Pristiphora geniculata* (Htg.) (Hymenoptera: Tenthredinidae), in Eastern Canada. *Canadian Entomologist* 96,1117–1133.

Heron, R.J.; Drouin, J.A. (1969) Methods of collecting, rearing and handling the larch sawfly for experimental studies. *Canadian Department of Fisheries and Forestry, Forest Research Laboratory, Winnipeg, Information Report* MS-X-15, pp. 1–31.

Muldrew, J.A. (1967) Biology and initial dispersal of *Olesicampe (Holocremnus)* sp. nr. *nematorum* (Hymenoptera: Ichneumonidae) a parasitoid of the larch sawfly recently established in Manitoba. *Canadian Entomologist* 99,312–321.

Muldrew, J.A. (1973) Larch sawfly. A biological control success story. *Canadian Forestry Service Northern Forest Research Centre, Edmonton, Information Report* 3(2),1–2.

Petch, C.E. (1935) Forest and shade tree insects. *Canadian Agricultural Insect Pest Review* 13,147.

Pschorn-Walcher, H.; Zinnert, K.D. (1971) Investigations on the ecology and natural control of the larch sawfly (*Pristiphora erichsonii* Htg. Hym. Tenthredinidae) in Central Europe. Part II: Natural enemies: their biology and ecology, and their role as mortality factors in *P. erichsonii*. *Commonwealth Institute of Biological Control Technical Bulletin* 14,1–50.

Schaffner, J.V. Jr. (1936) European sawfly *Pristiphora geniculata* attacks mountain ash in United States. *Journal of Economic Entomology* 29,469.

Rhyacionia buoliana (Schiff.), European Pine Shoot Moth (Lepidoptera: Tortricidae)

P.D. SYME

Pest Status

The European pine shoot moth, *Rhyacionia buoliana* (Schiff.), has been known as a destructive pest of hard pine, *Pinus* spp., plantations in North America since it was identified in 1914 (Busk 1914). Its history and biology have been the subjects of extensive investigations prior to 1969 and are well summarized by McGugan & Coppel (1962), Pointing & Green (1962), Miller (1967), and Syme (1971). The status of *R. buoliana* has been under continual surveillance and has been recorded in the Annual Reports of the Forest Insect and Disease Survey. The known distribution is shown in Fig. 21 for eastern Canada and British Columbia.

The European pine shoot moth has one generation per year and adults appear in the field over a 4-week period that begins between the first and third weeks in June. Eggs are laid singly or in small groups on the twigs or needle sheaths of the new growth. Incubation requires 2 weeks. First- and second-instar larvae mine needle bases on the new growth. Later they migrate to the new buds where they feed before overwintering as half-grown larvae within the bud. Feeding is resumed in the spring and is extended to additional buds before pupation takes place in May.

R. buoliana is primarily a pest of pine plantations. In North America, red pine, *Pinus resinosa* Ait., is the most seriously damaged. Scots, Austrian, ponderosa, and ornamental Mugho pines, *P. sylvestris* L., *P. nigra* Arnold, *P. ponderosa* Laws., and *P. mugo* Turra, respectively, are moderately susceptible; whereas pitch, Virginia, jack, and eastern white pines, *P. rigida* Mill., *P. virginiana*, *P. banksiana* Lamb., and *P. strobus* L., respectively, are relatively resistant. The needle-mining larvae cause a browning of the foliage, but this apparently has little effect on the vigour of the host. Feeding on the terminal bud causes serious injury to the main stem form and repeated destruction of the entire terminal cluster during the summer months initially causes profusion of shoot growth (witch's broom) and eventually a spike top when the leader dies. Various degrees of crookedness result from the lateral branches or buds assuming dominance over terminal shoots. Attack usually becomes less intensive as trees increase in height.

In Canada, the European pine shoot moth is an important pest of pine plantations in southern Ontario, but it also exists in southern Québec, primarily on ornamentals (Béique 1960). However, the Forest Insect and Disease Survey has not collected the moth in Québec since 1968. Presumably the distribution within that province is unchanged but the prevalence is markedly decreased since 1968. In British Columbia the insect is well established, mostly on ornamental stock, in the Vancouver–Victoria area and to a lesser degree in the Okanagan Valley north to Kamloops (C.E. Brown 1981 personal communication). The European pine shoot moth was first reported in the maritimes at Bear River, Nova Scotia, in 1925 (Reeks *et al.* 1951), and has spread to cover all Nova Scotia, Prince Edward Island, and most of the southern half of New Brunswick. It has become a serious pest of Scots and red pine plantations in Prince Edward Island and Nova Scotia, where increased plantings have intensified the problem. However, there has been little spread in New Brunswick in the last decade and abundance has remained at a low level (L.P. Magasi 1981 personal communications). Since 1968 the European

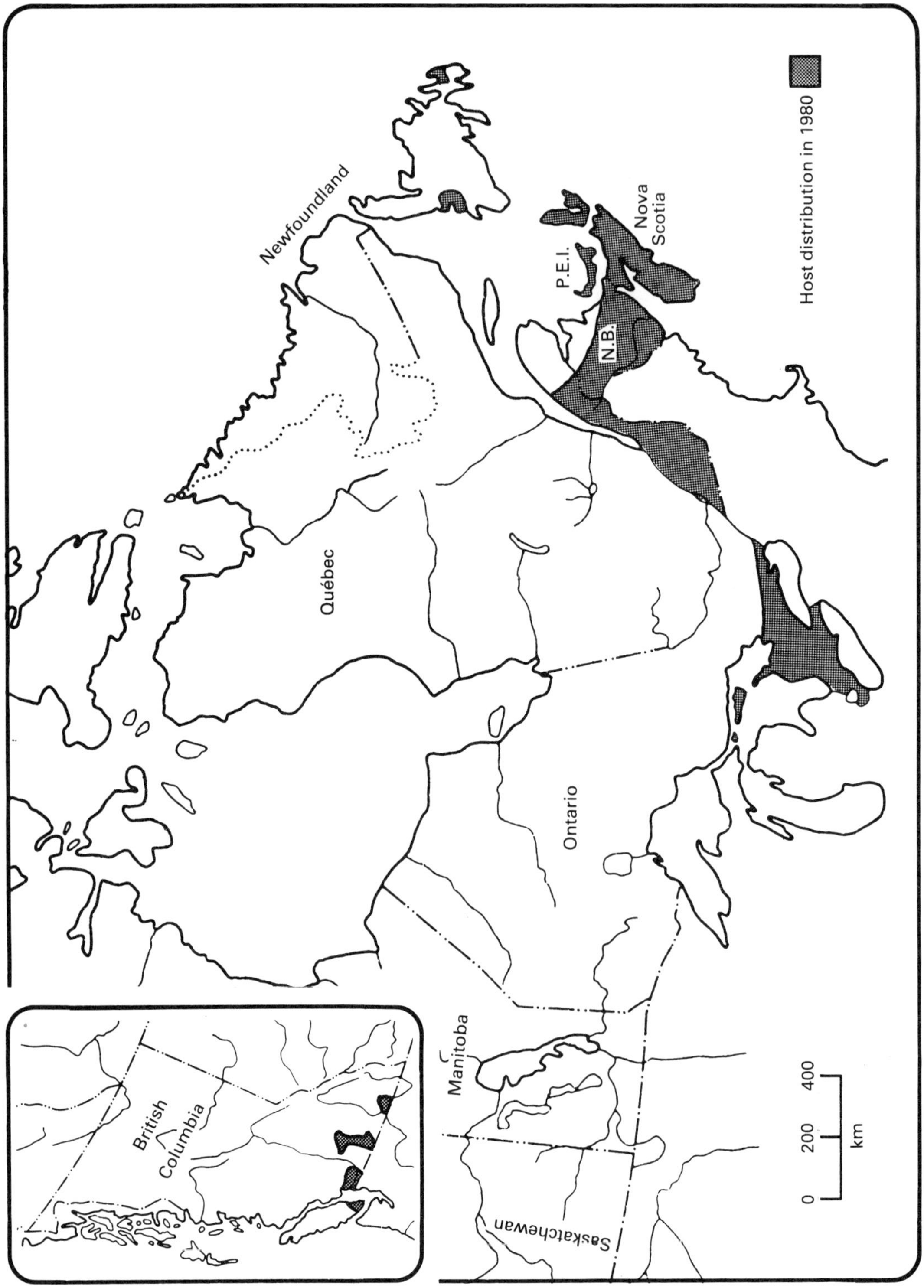

Fig. 21. Distribution of *Rhyacionia buoliana* (Schiff.) in 1980

pine shoot moth has remained an unimportant pest of ornamental pines in the cities of St. John's and Corner Brook in Newfoundland (A.G. Raske 1981 personal communication).

The progressive distribution of the shoot moth in North America since 1968 suggests that the shipping of infested nursery stock has contributed strongly to its dispersal, resulting in a rapidly spreading but spotty distribution. This was particularly evident in the Maritime Provinces. The Forest Insect and Disease Survey has recorded the spread in that area as well as in British Columbia. Declining populations were noted in Québec. In Ontario, the nothern limit of continuous distribution has held, as predicted (Green 1962), at about the −29°C mean minimum temperature isotherm. The populations on Manitoulin and Cockburn islands in the North Channel of Lake Huron have collapsed because the host trees have outgrown susceptibility, and populations declined generally throughout the province during the 1970s. Changes in the economics and marketing of red pine, at least in Ontario, have further reduced the importance of the European pine shoot moth; deformed stems may be acceptable under some circumstances.

In the rest of Canada, the situation remains static. In New Brunswick it is not known why the insect has not penetrated to the northern part of the province, although a lack of susceptible hosts could be an important factor.

Background

Although considerable work was done on various biological aspects of the European pine shoot moth before 1968 (Syme 1971), the research emphasis and the parasitoid monitoring capability changed after 1969. For example, studies on *R. buoliana* in Ontario were virtually confined to the effects of wild carrot, *Daucus carota* L., and other flowering plants on the effectiveness of *Orgilus obscurator* (Nees), the most effective introduced parasitoid of the pest. Meanwhile, the earlier literature on the biology, behaviour, history, distribution, ecology, damage, and control were well reviewed (Pointing & Green 1962, Pointing & Miller 1967, Miller 1967, Syme 1971). Since 1973 virtually no new research has been done in Canada on either the European pine shoot moth or its parasitoids.

Releases and Recoveries

The following brief descriptions of species of parasitoid include recoveries or significant range extensions of species released before 1968, and significant developments and findings on their possible effectiveness determined since 1968. The origin and numbers of released parasitoids are summarized in Tables 97 and 98. With the exceptions of *O. obscurator* (Nees), *Lypha dubia* (Fall.), and *Parasierola nigrifemur* (Ashm.), little can be reported.

Table 97 Open releases and recoveries in Ontario of parasitoids against *Rhyacionia buoliana* (Schiff.)

Species	Year of release	Origin	No. Released	Year of recovery
Agathis binominata Mues.	1969	Ontario	52	
	1970	Ontario	33	1973
Lypha dubia (Fall.)	1969	Germany	496	1970
Orgilus obscurator (Nees)	1969	Germany	218	
	1970	Germany	45	
	1971	Germany	225	
	1972	Germany	82	1978
Parasierola nigrifemur (Ashm.)	1974	Argentina	695	

Table 98 Laboratory and field cage studies in Ontario or parasitoids against *Rhyacionia buoliana* (Schiff.)

Species	Year	Origin	Number
Orgilus obscurator (Nees)	1969	Germany	778
	1970	Germany	449
	1971	Germany	500
	1972	Germany	181
	1973	Germany	6
Parasierola nigrifemur (Ashm.)	1973	Argentina	56

Eulimneria rufifemur (Thom.) (Hymenoptera: Ichneumonidae)

E. rufifemur was released before 1959 at Toronto, Niagra Falls, and Elmira (McGugan & Coppel 1962) but has not been released since. Although it had spread significantly west of the release point by 1968 (Syme 1971), collapse of host insect populations made the parasitoid difficult to find. It was taken only once (1972) at Elmira and twice (1969, 1971) near Drayton, 16 km NW of Elmira, always in small numbers. Because this parasitoid interacts with *O. obscurator,* to the latter's detriment, and in view of the relatively greater importance of *O. obscurator, E. rufifemur* is considered detrimental to the control effectiveness of the two parasitoids.

Lypha dubia (Fall.) (Diptera: Tachinidae)

This species was described in 1810 from Sweden (Stone *et al.* 1965) and is recorded by Thompson (1951) from many species of Lepidoptera in Europe. It is a primary larval parasitoid that deposits living larvae on twigs or foliage near host larvae. The wide range of host habitats suggests that a complex of closely related species may be involved. Small numbers were released in Nova Scotia against the winter moth, *Operophtera brumata* (L.), but none was recovered (Embree 1971).

L. dubia attacks late-instar larvae of the European pine shoot moth, and after rapid development within the host the parasitoid larva drops to the ground to pupate. It spends most of the summer and all winter in the ground as a fully developed adult, which emerges at about the time the *R. buoliana* begins feeding in the spring. There is a pre-oviposition period of about 4 weeks.

Studies in Europe showed that there was no serious interference by this species with *Orgilus obscurator,* the most effective introduced parasitoid in Ontario, and that *L. dubia* is one of the more effective parasitoids in northern Germany (Adlung & Sailer 1963, D. Schroeder, CIBC, Delémont, Switzerland, personal communication).

The parasitoid had also been released against *R. buoliana* in the United States in the years 1933–38 and again in 1962 and 1963, but has not been recorded (Dowden 1962). Although 90% of 1 200 puparia set out at Elmira in 1968 successfully emerged (Syme 1971), only a few were recovered and only enough in 1970 to say that at least one pair of flies had successfully emerged and mated! No parasitoids have been recaptured from a release at Vance Tract, Wellington County, Ontario. Thus *L. dubia*, despite its effectiveness in northern Germany, does not seem to be effective against the European pine shoot moth in North America.

Orgilus obscurator (Nees) (Hymenoptera: Braconidae)

O. obscurator is an internal solitary parasitoid that attacks first-instar larvae. Its life history has been described by Juillet (1960). Early introduction of the parasitoid and movement of nursery stock infested with *R. buoliana* resulted in a wide distribution of *O. obscurator* in Ontario (McGugan & Coppel 1962). In 1968 it was present virtually

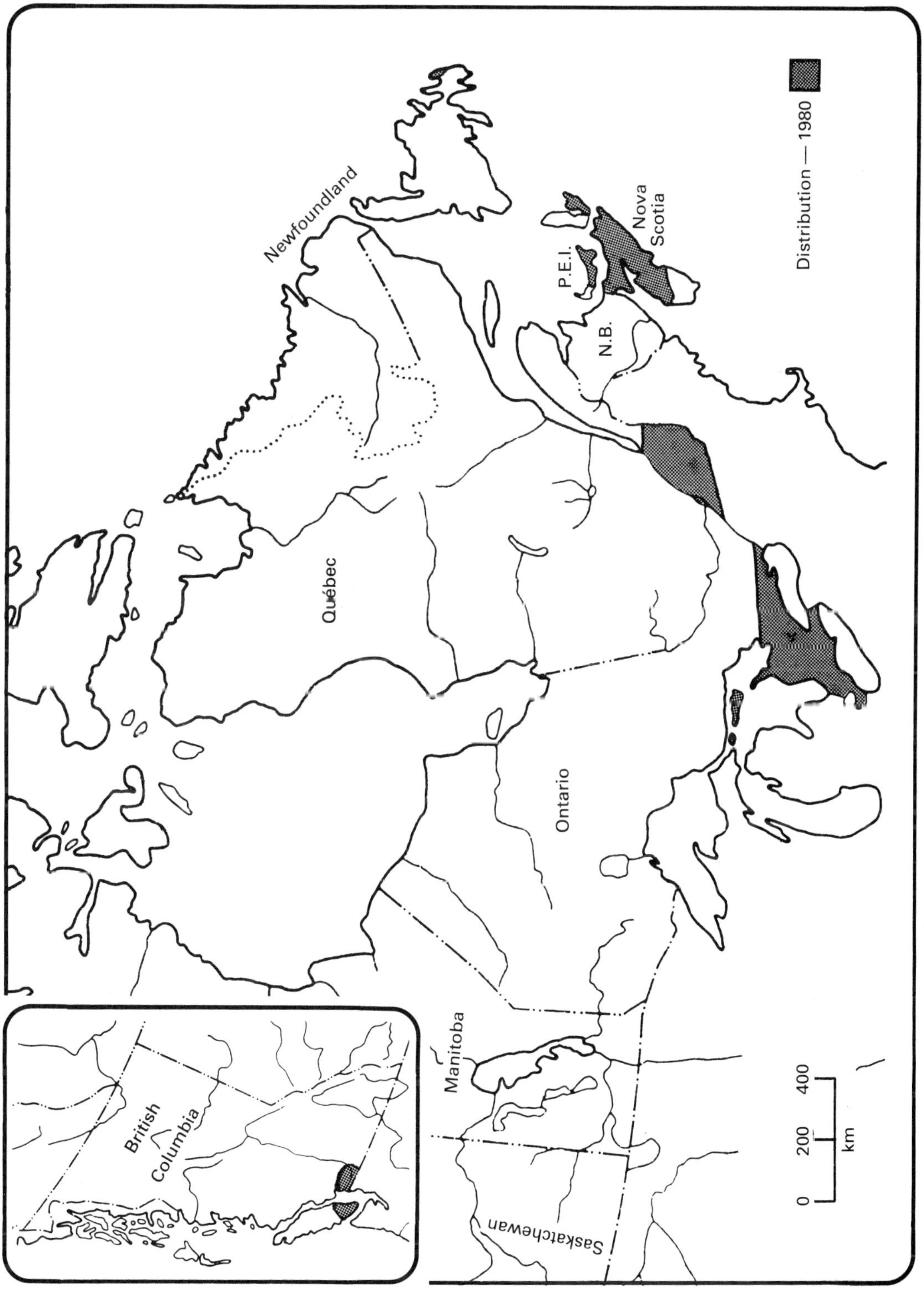

Fig. 22. Distribution of *Orgilus obscurator* (Nees) in 1980

throughout the range of European pine shoot moth in Ontario and collections since that date have confirmed this (Fig. 22). It was also well established throughout Québec, in the maritimes, and in Michigan, where it had not been released (Syme 1971).

Since 1968, this parasitoid has become widespread throughout Nova Scotia and Prince Edward Island and has been found in the southeastern corner of New Brunswick (L.P. Magasi 1981 personal communication). In Newfoundland it has been recovered twice, both times at St. John's (A.G. Raske 1981 personal communication). No releases have ever been made of this parasitoid in either the maritimes or Newfoundland. Thus, the hypothesis expressed by McGugan & Coppel (1962) and Syme (1971), that the parasitoid and its host are easily dispersed by planting infested nursery stock, is clearly reinforced by the events of later years. It is also known that *O. obscurator* disperses well on its own, for at Cockburn Island in Lake Huron, where it had been released in 1966 (Syme 1971), it was recovered 5 years later (1971) at Sand Bay, 12 km from the release point. Here it parasitized 39% of the European pine shoot moth, indicating it had been present for some years. In 1972 no *R. buoliana* could be found at Sand Bay.

In British Columbia, where again no intentional releases have ever been made, *O. obscurator* occurs in the general area of Vancouver–Victoria (C.E. Brown 1981 personal communication) (Fig. 22).

In Ontario, where further studies have been made of the beneficial effects of wild carrot and other flowering plants on the longevity and increased fecundity of *O. obscurator* (Syme 1977), several situations were studied since 1968 that supported this technique of enhancing parasitoid attack. At one small plantation in Brantford Township, Brant County, where *D. carota* was plentiful, *O. obscurator* was released during the summer of 1972. The population was not sampled in 1973 or 1974, but by 1975 parasitism had risen to 60% and no European pine shoot moth population remained in 1976. In three other plantations replete with *D. carota*, parasitism rose to over 50% in 3 to 4 years with the concurrent demise of the European pine shoot moth population.

It seems apparent that the presence of a suitable source of food in the form of *D. carota* can, under otherwise favourable conditions, increase the effectiveness of *O. obscurator* to a level virtually able to eliminate the European pine shoot moth population. On the basis of these results, workers in Nova Scotia are attempting to establish wild carrot in pine plantations, but the results are not yet fully known (L.P. Magasi 1981 personal communication).

***Temelucha interruptor* (Grav.) (Hymenoptera: Ichneumonidae)**

This primary internal larval parasitoid with cleptoparasitic habits (Syme 1971) has not been released since 1961. It persists at Elmira, Ontario, where it competes to the detriment of *O. obscurator*. *T. interruptor* was virtually always in a host with *O. obscurator*, resulting in the death of the latter. Thus, in a field experiment designed to show the effects of *D. carota*, on *O. obscurator* at Elmira, from 1966 to 1978, *T. interruptor* parasitized up to 50% of the hosts already attacked by *O. obscurator*, preventing the latter from attaining levels of parasitism over 40% necessary to cause the collapse of host populations. Thus it confounded the field trial that otherwise showed promise of demonstrating the beneficial effects of *D. carota* on *O. obscurator*.

***Parasierola nigrifemur* (Ashm.) (Hymenoptera: Bethylidae)**

This external, multivoltine gregarious parasitoid had been found to be very effective against European pine shoot moth in Argentina (Brewer & Varas 1971). With background supplied by E.M. deBrewer, Department of Entomology, Cordoba National University, laboratory tests in Canada (Sault Ste. Marie) demonstrated that *P. nigrifemur* showed no tendency to select for hosts parasitized by *O. obscurator*. Thus it has no clepto-parasitic tendencies (Syme 1974). Consequently, on 21 August 1974, 695 adults (unsexed) were released at a site 4.5 km north of Elsinore, Amabel Township, Bruce County,

Ontario. The trees averaged 1.4 m in height and there was an abundance of flowering plants including *D. carota*. In that same year, in October, two adult *P. nigrifemur* were found during the annual fall sample, but they were not associated with any apparent hosts. No parasitized hosts were found during the next 2 years. We can only conclude that the attempted introduction was a failure.

Agathis binominata (Mues.) (Hymenoptera: Braconidae)

Although this native parasitoid was released at Vance Tract, Wellington County, Ontario in 1969 and 1970, only a single larva was recovered in 1973 and none in 1975, 76 or 77. Reasons for its failure to establish are speculative.

Evaluation of Control Attempts

The difficulties of evaluating the impact of biological control agents on the European pine shoot moth were discussed by Syme (1971) and can be summarized by stating that the host–pest relationship and pest damage change drastically with time. In the absence of any further trials of exotic parasitoids or further behavioural studies, the well-established European species *O. obscurator* seems to offer the greatest promise for effective localized biological control in Canada. Several cases have been documented in which, in the presence of *D. carota* flowers, *O. obscurator* has brought about complete collapse of the *R. buoliana* infestation within a few years.

As *O. obscurator* is a univoltine internal parasitoid, chemical control applied to the host larval stage in the spring will remove a proportionate part of the parasitoid population. Chemical control of the early-instar needle-mining stage will have a negative impact on the parasitoid adult population, which would be actively searching for hosts. This would depress the parasitoid population disproportionately. Thus if chemical control is required within an integrated pest management scheme, the spring wandering stage of the European pine shoot moth would be the time to spray and cause the least potential damage to *O. obscurator*. Similarly, a systemic insecticide applied in the spring would have the least effect on *O. obscurator*, but any insecticide will depress the parasitoid population to some extent.

It should be pointed out here, however, that wild carrot is listed in the secondary noxious class of the Canada Seed Act of 1961 and is listed in The Weed Control Acts of Manitoba, Nova Scotia, Ontario, and Québec. It is detrimental to cattle when they ingest large quantities, and taints the milk, but Dale (1974) has pointed out that it is not a weed of cultivated fields and prime agricultural land and that there are no reports of cattle grazing on it from choice. Further, it does not compete in a well-established turf and is not found where ploughing is an annual event.

Wild carrot may also affect commercial production of carrot either by transmitting pests or harbouring diseases common to both cultivated and wild varieties, or by causing the production of poor seed of commercial varieties when hybridization occurs (Frankton 1955). However, in Canada carrot seed is grown commercially only in British Columbia (Dale 1974).

On the positive side Dale (1974) indicates that *D. carota* has a relatively high nutritive value and should be regarded with greater tolerance when found in pastures among plants of low nutritive value. Thus foresters concerned with the establishment of red pine might well question the status of wild carrot as a noxious weed.

Recommendations

The parasitoid *O. obscurator* appears to be ubiquitous and the methods for manipulating *D. carota* are published (Syme 1976), so it remains for the system to be practised and refined. The method may in some instances be labour-intensive. However, if a plantation is established where *D. carota* already occurs, the plantation manager may preserve the wild carrot to favour *O. obscurator* at virtually no cost.

This unique method whereby the probability of control can be enhanced by manipulating the environment into which a parasitoid is introduced, thus favouring its survival and effectiveness, represents a new working principle that requires emphasis in future research as well as in methodology associated with biological control attempts.

Literature Cited

Adlung, K.G.; Sailer, R.I. (1963) Forest entomological studies in German afforestation areas, with special regard to pine shoot moth parasitoids of *Rhyacionia buoliana* (Schiff.). *Allgemeine Forst- und Jagdzeitung* 134,229–238.

Béique, R. (1960) The importance of the European pine shoot moth *Rhyacionia buoliana* (Schiff.) in Quebec City and vicinity. *Canadian Entomologist* 92,858-862.

Brewer, M.; Varas, D. (1971) Cria masiva de *Parasierola nigrifemur* (Ash.), (Hym., Bethylidae) — primeras liberuciones en Calamuchita, Cordoba, Argentina. *Revista Peruana de Entomologia* 14,352–361.

Busk, A. (1914) A destructive pine moth introduced from Europe. *Journal of Economic Entomology* 7,340–341.

Dale, H.M. (1974) The biology of Canadian weeds. 5. *Daucus carota. Canadian Journal of Plant Science* 54,673–685.

Dowden, P.B. (1962) Parasites and predators of forest insects liberated in the United States through 1960. *US Department of Agriculture Handbook* 226, 70 pp.

Embree, D.G. (1971) *Operophtera brumata* (L.), winter moth (Lepidoptera: Geometridae). *In*: Biological control programmes against insects and weeds in Canada, 1959–1968. *Commonwealth Institute of Biological Control Technical Communication* 4,167–175.

Frankton, C.(1955) Weeds of Canada. Ottawa, Ontario; Queen's Printer, 196 pp.

Green, G.W. (1962) Low winter temperatures and the European pine shoot moth, *Rhyacionia buoliana* (Schiff.) in Ontario. *Canadian Entomologist* 94,314–36.

Juillet, J.A. (1960) Immature stages, life histories, and behaviour of two hymenopterous parasites of the European pine shoot moth, *Rhyacionia buoliana* (Schiff.) (Lepidoptera: Olethreutidae). *Canadian Entomologist* 92,342–346.

McGugan, B.M.; Coppel, H.C. (1962) Biological control of forest insects, 1910–1958. *In*: A review of the biological control attempts against insects and weeds in Canada. *Commonwealth Institute of Biological Control Technical Communication* 2,35–127.

Miller, W.E. (1967) The European pine shoot moth — ecology and control in the Lake States. *Forest Science Monograph* 14, 72 pp.

Pointing, P.J.; Green, G.W. (1962) A review of the history and biology of the European pine shoot moth, *Rhyacionia buoliana* (Schiff.) (Lepidoptera: Olethreutidae) in Ontario. *Proceedings of the Entomological Society of Ontario* 92,58–69.

Pointing, P.J.; Miller, W.E. (1967) European pine shoot moth. In: Important forest insects and disease of mutual concern to Canada, the United States and Mexico. *Canadian Department of Forestry and Rural Development Publication* 1180,162–166.

Reeks, W.A.; Forbes, R.S.; Cuming, F.G. (1951) *Canadian Department of Forestry and Rural Development Annual Report, Forest Insect and Disease Survey 1950,* pp. 5–20.

Stone, A.; Sabrosky, C.W.; Wirth, W.W.; Foote, R.H.; Coulson, J.R. (1965) A catalogue of the Diptera of America, north of Mexico. *US Department of Agriculture Handbook* 276, 1696 pp.

Syme, P.D. (1971) *Rhyacionia buoliana* (Schiff.), European pine shoot moth (Lepidoptera: Olethreutidae). In: Biological control programmes against insects and weeds in Canada, 1959–1968. *Commonwealth Institute of Biological Control Technical Communication* 4,194–205.

Syme, P.D. (1974) Interaction between three parasitoids of the European pine shoot moth. *Canadian Forestry Service Bi-monthly Research Notes* 30(2),9-10.

Syme, P.D. (1976) Red pine and the European pine shoot moth in Ontario. *Canadian Forestry Service Information Report* 0-X-244, 17 pp.

Syme, P.D. (1977) Observations on the longevity and fecundity of *Orgilus obscurator* (Hymenoptera: Braconidae) and the effects of certain foods on longevity. *Canadian Entomologist* 109,995–1000.

Thompson, W.R. (1951) A catalogue of the parasites and predators of insect pests. Sect. 2, Pt. 1. Hosts of the Coleoptera and Diptera. *Commonwealth Institute of Biological Control,* 147 pp.

Addresses of Contributors

Burgess, L., Research Station, Agriculture Canada, 107 Science Crescent, Saskatoon, Saskatchewan S7N 0X2.

Cheng, H.H., Research Station, Agriculture Canada, P.O. Box 186, Delhi, Ontario N4B 2W9.

Costello, R.A., British Columbia Ministry of Agriculture, 17720–57th Avenue, Surrey, British Columbia V3S 4P9.

Craig, C.H., Research Station, Agriculture Canada, 107 Science Crescent, Saskatoon, Saskatchewan S7N 0X2.

Cunningham, J.C., Forest Pest Management Institute, Canadian Forestry Service, Environment Canada, Box 490, Sault Ste. Marie, Ontario P6A 5M7.

Elgee, D.E., Maritimes Forest Research Centre, Canadian Forestry Service, Environment Canada, Box 4000, Fredericton, New Brunswick E3B 5P7.

Ellis, C.R., Department of Environmental Biology, University of Guelph, Guelph, Ontario N1G 2W1.

Embree, D.G., Maritimes Forest Research Centre, Canadian Forestry Service, Environment Canada, Box 4000, Fredericton, New Brunswick E3B 5P7.

Ennis, T.J., Forest Pest Management Institute, Canadian Forestry Service, Environment Canada, Box 490, Sault Ste. Marie, Ontario P6A 5M7.

Estabrooks, G.F., Maritimes Forest Research Centre, Canadian Forestry Service, Environment Canada, Box 4000, Fredericton, New Brunswick E3B 5P7.

Ewen, A.B., Research Station, Agriculture Canada, 107 Science Crescent, Saskatoon, Saskatchewan S7N 0X2.

Finnamore, D.B., New Brunswick Department of Agriculture and Rural Development, Box 6000, Fredericton, New Brunswick E3B 5H1.

Finnegan, R.J., Laurentian Forest Research Centre, Canadian Forestry Service, Environment Canada, Box 3800, Ste-Foy, Québec G1V 4C7.

Frazer, B.D., Research Station, Agriculture Canada, 6660 N.W. Marine Drive, Vancouver, British Columbia V6T 1X2.

George, J.A., Department of Zoology, University of Western Ontario, London, Ontario N6A 5B7.

Gerber, G.H., Research Station, Agriculture Canada, 195 Dafoe Road, Winnipeg, Manitoba R3T 2M9.

Green, G.W., Forest Pest Management Institute, Canadian Forestry Service, Environment Canada, Box 490, Sault Ste. Marie, Ontario P6A 5M7.

Griffiths, K.J., Great Lakes Forest Research Centre, Canadian Forestry Service, Environmental Canada, Box 490, Sault Ste. Marie, Ontario P6A 5M7.

de Groot, P., Forest Pest Management Institute, Canadian Forestry Service, Environment Canada, Box 490, Sault Ste. Marie, Ontario P6A 5M7.

Guibord, M.O'c., Service de recherches en défense des cultures, Ministère de l'agriculture du Québec, 2700 rue Einstein, Ste-Foy, Québec G1P 3W8.

Guppy, J.C., Research Station, Agriculture Canada, Ottawa, Ontario K1A OC6.

Harcourt, D.G., Research Station, Agriculture Canada, Ottawa, Ontario K1A OC6.

Harris, J.W.E., Pacific Forest Research Centre, Canadian Forestry Service, Environment Canada, 506 West Burnside Road, Victoria, British Columbia V8Z 1M5

Harris, P., Research Station, Agriculture Canada, Box 440, Regina, Saskatchewan S4P 3A2.

Howse, G.M., Great Lakes Forest Research Centre, Canadian Forestry Service, Environment Canada, Box 490, Sault Ste. Marie, Ontario P6A 5M7.

Hulme, M.A., Pacific Forest Research Centre, Canadian Forestry Service, Environment Canada, 506 West Burnside Road, Victoria, British Columbia V8Z 1M5

Ives, W.G.H., Northern Forest Research Centre, Canadian Forestry Service, Environment Canada, 5320–122nd Street, Edmonton, Alberta T6H 3S5.

Jaques, R.P., Research Station, Agriculture Canada, Box 370, Harrow, Ontario N0R 1G0.

Kelleher, J.S., Research Program Service, Agriculture Canada, Ottawa, Ontario K1A 0C6.

Loan, C.C., Research Station, Agriculture Canada, Ottawa, Ontario K1A 0C6.

Laing, J.E., Department of Environmental Biology, University of Guelph, Guelph, Ontario N1G 2W1.

Magasi, L.P., Maritimes Forest Research Centre, Canadian Forestry Service, Environment Canada, Box 4000, Fredericton, New Brunswick E3B 5P7.

Maw, M.G., Research Station, Agriculture Canada, Box 440, Regina, Saskatchewan S4P 3A2.

McClanahan, R.J., Research Station, Agriculture Canada, Box 370, Harrow, Ontario N0R 1G0.

McEwen, F.L., Department of Environmental Biology, University of Guelph, Guelph, Ontario N1G 2W1.

McNeil, J.N., Département de Biologie, Université Laval, Québec, Québec G1K 7P4.

Miller, C.A., Maritimes Forest Research Centre, Canadian Forestry Service, Environment Canada, Box 4000, Fredericton, New Brunswick E3B 5P7.

Morris, O.N., Research Station, Agriculture Canada, 195 Dafoe Road, Winnipeg, Manitoba R3T 2M9.

Morris, R.F., Research Station, Agriculture Canada, Box 7098, St. John's West, Newfoundland A1E 3Y3.

Mukerji, M.K., Research Station, Agriculture Canada, 107 Science Crescent, Saskatoon, Saskatchewan S7N 0X2.

Muldrew, J.A., Northern Forest Research Centre, Canadian Forestry Service, Environment Canada, 5320–122nd Street, Edmonton, Alberta T6H 3S5.

Myers, J.H., Animal Resource Ecology Institute, University of British Columbia, Vancouver, British Columbia V6T 1W5.

Otvos, I.S., Pacific Forest Research Centre, Canadian Forestry Service, Environment Canada, 506 West Burnside Road, Victoria, British Columbia V8Z 1M5.

Pendrel, B., Maritimes Forest Research Centre, Canadian Forestry Service, Environment Canada, Box 4000, Fredericton, New Brunswick E3B 5P7.

Peschken, D.P., Research Station, Agriculture Canada, Box 440, Regina, Saskatchewan S4P 3A2.

Quednau, F.W., Laurentian Forest Research Centre, Canadian Forestry Service, Environment Canada, Box 3800, Ste-Foy, Québec G1V 4C7

Raske, A.G., Newfoundland Forest Research Centre, Canadian Forestry Service, Environment Canada, Box 6028, St. John's, Newfoundland A1C 5X8.

Schooley, H.O., Petawawa National Forestry Institute, Canadian Forestry Service, Environment Canada, Chalk River, Ontario K0J 1J0.

Schroeder, D., European Station, Commonwealth Institute of Biological Control, 1 Chemin des Grillons, CH-2800 Delémont, Switzerland.

Shemanchuk, J.A., Research Station, Agriculture Canada, Box 3000, Lethbridge, Alberta T1J 4B1.

Shepherd, R.F., Pacific Forest Research Centre, Canadian Forestry Service, Environment Canada, 506 West Burnside Road, Victoria, British Columbia V8Z 1M5.

Smirnoff, W.A., Laurentian Forest Research Centre, Canadian Forestry Service, Environment Canada, Box 3800, Ste-Foy, Québec G1V 4C7.

Syme, P.D., Great Lakes Forest Research Centre, Canadian Forestry Service, Environment Canada, Box 490, Sault Ste. Marie, Ontario P6A 5M7.

Thompson, L.S., Research Station, Agriculture Canada, Box 1210, Charlottetown, Prince Edward Island C1A 7M8.

Turnock, W.J., Research Station, Agriculture Canada, 195 Dafoe Road, Winnipeg, Manitoba R3T 2M9.

Tyrrell, D., Forest Pest Management Institute, Canadian Forestry Service, Environment Canada, Box 490, Sault Ste. Marie, Ontario P6A 5M7.

Van Sickle, G.A., Pacific Forest Research Centre, Canadian Forestry Service, Environment Canada, 506 West Burnside Road, Victoria, British Columbia V8Z 1M5.

Varty, I.W., Maritimes Forest Research Centre, Canadian Forestry Service, Environment Canada, Box 4000, Fredericton, New Brunswick E3B 5P7.

Watson, A.K., Macdonald Campus, Faculty of Agriculture, McGill University, Ste-Anne de Bellevue, Québec H9K 1C0.

Whisler, H.C., Department of Botany, University of Washington, Seattle, Washington, USA 98195.

Wilkinson, A.T.S, Research Station, Agriculture Canada, 6660 N.W. Marine Drive, Vancouver, British Columbia V6T 1X2.

Wilson, G.G., Forest Pest Management Institute, Canadian Forestry Service, Environment Canada, Box 490, Sault Ste. Marie, Ontario P6A 5M7.

Wylie, H.G., Research Station, Agriculture Canada, 195 Dafoe Road, Winnipeg, Manitoba R3T 2M9.

Zebold, S.L. Department of Botany, University of Washington, Seattle, Washington, USA 98195.